Advances in Hydrodynamics of Water Pump Station System

Advances in Hydrodynamics of Water Pump Station System

Editors

Changliang Ye
Yuan Zheng
Kan Kan
Ran Tao
Huixiang Chen

MDPI

Basel • Beijing • Wuhan • Barcelona • Belgrade • Novi Sad • Cluj • Manchester

Editors

Changliang Ye
School of Electrical and Power Engineering
Hohai Univeristy
Nanjing
China

Yuan Zheng
School of Electrical and Power Engineering
Hohai University
Nanjing
China

Kan Kan
School of Electrical and Power Engineering
Hohai University
Nanjing
China

Ran Tao
College of Water Resources and Civil Engineering
China Agricultural University
Beijing
China

Huixiang Chen
College of Agricultural Science and Engineering
Hohai University
Nanjing
China

Editorial Office
MDPI AG
Grosspeteranlage 5
4052 Basel, Switzerland

This is a reprint of articles from the Special Issue published online in the open access journal *Water* (ISSN 2073-4441) (available at: https://www.mdpi.com/journal/water/special_issues/91ZX9BUO4D).

For citation purposes, cite each article independently as indicated on the article page online and as indicated below:

Lastname, A.A.; Lastname, B.B. Article Title. *Journal Name* **Year**, *Volume Number*, Page Range.

ISBN 978-3-7258-1650-7 (Hbk)
ISBN 978-3-7258-1649-1 (PDF)
doi.org/10.3390/books978-3-7258-1649-1

Contents

About the Editors

Changliang Ye

Ye Changliang, male, born in Lu'an, Anhui Province, holds a doctoral degree in engineering and is a lecturer at Hohai University. In June 2015, he received a Bachelor of Engineering degree from China Agricultural University (Thermal Energy and Power Engineering). In June 2021, he received a doctorate degree in engineering (Hydraulic Engineering) from China Agricultural University. From October 2018 to June 2020, he was jointly cultivated and exchanged at Eindhoven University of Technology in the Netherlands and mainly engaged in research on the internal flow mechanism, flow control, and optimization design of fluid machinery. He led a National Natural Science Foundation Youth Project and an Open Fund Project for Key Laboratories of the Ministry of Education and participated in multiple general and key projects of the National Natural Science Foundation of China. In recent years, he has published more than 20 academic papers in journals and conferences, and he has authorized more than 10 invention patents and software copyrights.

Yuan Zheng

Professor and Doctoral Supervisor at Hohai University, Deputy Director of the University's Innovation Research Institute, and a high-level talent in Jiangsu Province's "333 Project" and "Top Six Talent Peaks". He is the head of a national quality course and a recipient of the Baogang Education Award.

Kan Kan

Dr. Kan is an Associate Professor in the Department of Power Engineering of the School of Electrical and Power Engineering, Hohai University. Main interests focus on design and optimization of hydraulic machinery; high fidelity CFD simulations; fluid Mechanics and fluid-structure interactions among hydraulic machinery; renewable energy like wind and current turbines.

Ran Tao

Prof. Tao is an esteemed scholar with a Ph.D. from China Agricultural University and the Swiss Federal Institute of Technology in Lausanne (EPFL), and he also holds a postdoctoral position at Tsinghua University and a visiting scholar role at the University of Tokyo. His academic voyage has been truly global, and his research has been dedicated to advancing the field of hydraulic machinery.

From 2014 to 2018, he concentrated on experimental and numerical studies of cavitation within hydraulic machinery, devising methods to design pump-turbine blades that resist cavitation erosion, aiming to make an intelligent design to achieve cavitation-free operation.

During his time from 2018 to 2020 at Tsinghua and Tokyo University, he integrated computational models of two-phase flow with supercomputing capabilities, performing high-precision large eddy simulations (LESs) to address the balance between computational accuracy and resource management, particularly focusing on the twin-vortex structure and cavitation issues in pump-turbine cascades.

Currently, at China Agricultural University, he is enhancing computer-aided techniques in flow research by incorporating CFD, modal decomposition, machine learning, and intelligent optimization. His work is pivotal for the efficient exploitation of clean energy sources like hydropower and tidal energy, aligning with the critical goals of "carbon peak and carbon neutrality".

Huixiang Chen

Chen Huixiang, female, associate professor, master's supervisor, fourth level "Great Yu Scholar" at Hohai University. In 2012, she obtained her Bachelor's degree in Thermal Energy and Power Engineering (Hydrodynamics) from Hohai University. From 2012 to 2014, she studied for her Master's degree in Fluid Machinery and Engineering at Hohai University, and later pursued her Master's and Doctoral degrees. From 2016 to 2018, she jointly trained at Michigan State University for 18 months. In 2019, she obtained a Ph.D. in Water Resources and Hydropower Engineering from Hohai University. In July 2019, she entered the School of Agricultural Science and Engineering at Hohai University to engage in teaching and research work.

She is mainly engaged in research on the hydraulics of pumping stations (hydropower stations) and safety control and transition processes of water pumps and pumping station systems (hydraulic units). She has hosted nine projects, including the National Natural Science Foundation Youth Program, Jiangsu Provincial Natural Science Foundation Youth Program, China Postdoctoral Foundation General Program, Jiangsu Provincial Postdoctoral Research Funding Program, Nanjing Overseas Chinese Science and Technology Innovation Program, Nantong Municipal Science and Technology Program, and Central University Basic Undergraduate Research Business Fee Program. As the main participant, she has participated in multiple projects, including the National Natural Science Foundation Key Program, National Natural Science Foundation General Program, and National Natural Science Foundation Youth Program. she has published over 20 academic papers, including 16 SCI papers as the first author/corresponding author, 5 EI papers, and 3 core papers. She has also authorized 15 national invention patents.

Preface

This reprint, titled "Advances in the Hydrodynamics of Water Pump Station Systems", represents a comprehensive exploration of the intricate dynamics within water pump station systems, a critical infrastructure for water resource management and distribution. The subject of our study is the hydrodynamic performance and stability of these systems, which are essential for ensuring efficient water conveyance and energy conservation.

Scope and Aims: The scope of this work encompasses the latest advancements in the field, with a particular focus on operational stability enhancement and the integration of energy conversion technologies. Our aim is to synthesize the findings from a series of original research papers that have been meticulously peer reviewed and accepted for publication, highlighting the contributions of various scholars in this domain.

Purpose: The purpose of this scientific endeavor is twofold: firstly, to disseminate knowledge that can inform the design, optimization, and management of water pump stations; secondly, to contribute to the body of literature that supports the sustainable development of water resources and the infrastructure that serves them.

Reasons and Motivations: The motivation for undertaking this work stems from the recognition of the pivotal role that water pump stations play in modern society. As demands on water resources grow with economic development and population increases, the need for efficient and reliable water pump stations becomes ever more critical. We are driven by the desire to address the challenges posed by climate change, energy crises, and the quest for sustainable solutions in water management.

Audience: This reprint is addressed to a diverse audience, including academic researchers, engineers, policymakers, and students in the fields of hydraulic engineering, environmental science, and water resource management. We also extend our reach to professionals and institutions involved in the planning, construction, and maintenance of water infrastructure.

Involved Authors: The authors of this reprint are a collective of experts from various institutions, each bringing their unique insights and expertise to the table. The lead authors, Changliang Ye, Yuan Zheng, and Huixiang Chen, are joined by Kan Kan, the corresponding author, and Ran Tao, whose contributions have been instrumental in shaping the content and direction of this work.

Acknowledgments: We extend our sincere gratitude to the National Natural Science Foundation of China for their financial support, grant numbers 52209109, No. 52379086, and No. 52271275, which have enabled this research. Our thanks go to the peer reviewers and editorial staff for their meticulous reviews and valuable feedback. We also acknowledge the contributions of our colleagues and collaborators who have provided support, expertise, and critical discussions throughout the research process.

Changliang Ye, Yuan Zheng, Kan Kan, Ran Tao, and Huixiang Chen
Editors

water

Editorial

Advances in Hydrodynamics of Water Pump Station System

Changliang Ye [1,2], Yuan Zheng [1,2], Kan Kan [1,2,*], Ran Tao [3] and Huixiang Chen [2,4]

1 School of Electrical and Power Engineering, Hohai University, Nanjing 211100, China; yechangliang@hhu.edu.cn (C.Y.); zhengyuan@hhu.edu.cn (Y.Z.)
2 College of Water Conservancy & Hydropower Engineering, Hohai University, Nanjing 210098, China; chenhuixiang@hhu.edu.cn
3 College of Water Resources and Civil Engineering, China Agricultural University, Beijing 100107, China; randytao@cau.edu.cn
4 College of Agricultural Science and Engineering, Hohai University, Nanjing 211100, China
* Correspondence: kankan@hhu.edu.cn

Citation: Ye, C.; Zheng, Y.; Kan, K.; Tao, R.; Chen, H. Advances in Hydrodynamics of Water Pump Station System. *Water* **2024**, *16*, 1430. https://doi.org/10.3390/w16101430

Received: 24 April 2024
Accepted: 6 May 2024
Published: 17 May 2024

1. Introduction

As an indispensable part of water conservancy engineering construction, the importance of pumping stations is reflected in several aspects. First of all, pumping stations undertake the important tasks of regional flood control, flood cleanup, irrigation, water transfer, and water supply [1,2]. They can meet the needs of farmland irrigation, urban water supply, etc., and also help to preserve the ecological environment and improve the utilization rate of water resources. In particular, in modern society, the escalating demand for water resources is a consequence of the burgeoning economy and the enhancement of living standards. Consequently, pumping stations play a critical role in ensuring water resource availability, fostering agricultural progress, and sustaining the regular operations of entire cities. Secondly, the construction and management of pumping stations are also of great significance. The backwardness of pumping station technology, the decline in unit efficiency, the increase in energy consumption, and other problems not only affect the normal operation and efficiency of pumping stations but also restrict the progress of water conservation [3]. Therefore, the implementation of energy-saving measures for water conservation and the upgrading of pumping stations is an important method for improving the operational efficiency of pumping stations and reducing energy consumption. In addition, the construction and management of pumping stations also involve the maintenance and renewal of electromechanical equipment. Much of the electromechanical equipment that has run for long periods of time has already failed, but due to a shortage of funds and other reasons, this equipment is still in use, which not only affects the functioning of pumping stations but also increases the operational risks involved [4,5]. Consequently, it is critical to enhance pumping stations' construction and administration and ensure the timely replacement and maintenance of station equipment in order to guarantee secure, stable, and effective operations. Finally, from the point of view of economic and social benefits, the construction and management of pumping stations also represent an important means of achieving the efficient utilization of water resources and facilitating the sustainable growth of water conservation. By improving the quality of pumping station construction and management levels, water conservancy projects can effectively utilize their advantages and contribute to the sustainable growth of the economy and society [6].

The aim of this Special Issue is to focus on the latest advances in pumping station system research, with a particular focus on operational stability enhancement and the potential applications of energy conversion technologies. Since the December 2022 call-for-papers announcement, a total of 17 original papers have been accepted for publication following a thorough peer-review process (Manuscripts 1–17); the contents thereof cover strategies for the upgrade of pumping stations, the exploration of the water pumping station phenomenon, the fine-grained study of pumping efficacy and stability, the safe and stable operation of pumps as turbines, the vibration characterization of pump turbine units, and hydraulic transients in pipeline network systems. To enhance the comprehension of this particular edition, we have succinctly outlined the key points of the published articles below.

2. Overview of This Special Issue

In the field of pumping station research, scholars have endeavored to enhance system performance and efficiency through a variety of research methods and technological approaches. These studies focus not only on the optimization of individual components but also on the coordinated operation and long-term stability of the entire system.

In the study of pumping station buildings, in order to optimize the flow distribution in the pumping station forebay, Contribution 1 conducted research using numerical simulation methods to examine the impact of rectification facilities on the water flow in the pumping station forebay. Two rectification methods were suggested: one includes a solitary bottom sill and another involves a combination of a bottom sill and a diversion pier. These methods were specifically created to enhance the flow conditions in the forebay; minimize undesired flow structures, such as backflows and eddies; and ensure a more consistent flow distribution in each part. The results show that the bottom sill controls the central beam of water coming out of the water distribution pipe while the installation of the diversion pier allows for the flow to be rectified twice, thus facilitating the dispersion of the incoming water and making the flow distribution more uniform in each part.

In the study of pumping station water pipelines, Contribution 2 is the first systematic study of the water-filling process of water–gas two-phase flow in right-angle elbow pressure pipelines, especially under different water-filling velocity conditions. The integration of numerical simulations with physical model tests offers a novel approach to comprehending and forecasting water–gas two-phase flow. A staged water-filling method is proposed according to the different stages of the process to efficiently decrease the pressure peak and shorten the water-filling time. The comparative analysis of experimental and numerical simulation results provides a theoretical and practical basis for controlling the water–gas two-phase flow in actual engineering applications. Contribution 3 showed detailed numerical simulations of gas–liquid two-phase flow in a high-elevation water pipe carried out using the CLSVOF model and the RNG k-ε turbulence model, thereby providing a new perspective for understanding and predicting the flow phenomena in the water pipe system. The paper analyzes the effects of bubble size and distribution, flow velocity, and air content on the evolution of flow patterns and pressure pulsations. The analysis reveals the effects of random bubble distribution on flow patterns and pressure pulsations in high-pressure water piping systems. The results help predict the rupture risk and water hammer protection in stagnant gas pipeline systems, thereby providing theoretical support and computational parameters for engineering safety.

In the pump unit study, Contribution 4 provides a detailed analysis of pumps' internal flow characteristics under different start-up combination scenarios, thereby uncovering the impacts of varied start-up conditions on the internal hydrodynamic behavior of the centrifugal pump. Combined with entropy generation and a vortex analysis, the paper provides a new perspective for evaluating and optimizing the internal flow and energy losses witnessed in pumping stations. The paper examines the performance of pumping stations under various start-up combinations and suggests an optimal start-up combination scheme. This scheme is crucial for enhancing the energy efficiency and stability of pumping stations. The impeller blades of water pumps can usually be simplified to be studied as hydrofoils; Contribution 5 proposes a bionic pumping device based on two oscillating hydrofoils, which aims to improve the fluidity of water bodies in the river network area. The hydraulic performance in both in-phase and out-of-phase oscillation modes was analyzed via numerical simulations and experimental validation. The finite volume method and overlapping mesh technique were employed for this purpose. The findings of this study indicate that the out-of-phase oscillation mode exhibits a superior hydraulic performance in comparison to the in-phase oscillation mode, particularly at a constant frequency. Furthermore, the pump efficiency of out-of-phase oscillation is notably higher than that of in-phase oscillation. Contribution 6 improves the hump characteristic at low flow rates through adjustments to the shapes of the impeller inlet edge and the blade. It also optimizes the matching relationship between the guide vanes and the impeller to improve pressure pulsation characteristics and optimize the width of the leafless zone. At the same time, the performance curves of the centrifugal pump at different speeds were studied, and the maximum input shaft power curves under various working conditions were determined. Based on this, an analysis was performed to identify the safe and efficient operating range of the pump. The occurrence of pump stalling is a frequent issue that arises during the real functioning of centrifugal pumps. Therefore, Contribution 7 used the scale-adaptive simulation (SAS) model and found the existence of double-hump characteristics in the head curve. Through the comparison of the flow field characteristics under various flow conditions, it was determined that the peak region near the optimal operating condition is a result of hydraulic losses, whereas the peak region further away from the optimal operating condition is caused by the combined influence of Euler head and hydraulic losses. The clock effect mechanism of the guiding vanes is examined by simplifying the model, thus revealing that the disturbance caused by the wake of the upstream vane on the downstream vane's boundary layer influences the transition; this in turn affects the friction stress of the vanes and leads to a change in the pressure amplitude. The results of this study are important for understanding and controlling the performance of centrifugal pumps under stall conditions, especially in the secure and dependable functioning of large pumping stations. In terms of vibration characteristics, Contribution 8 details a comprehensive study of the modal characteristics of the pump impeller, examining its behavior in both air and water environments. Dynamic stress simulations demonstrate that the cyclic pattern of the highest stress point is mostly influenced by the number of blades and the rotational speed of the impeller during a single rotation cycle. Using this approach, the behavior of the pump under different operating conditions can be better understood, thus improving its performance and life. Centrifugal pumps are more suited for situations that require high pressure and have low flow rates and high heads. On the other hand, axial-flow pumps are more appropriate in situations with large flow rates and low-to-medium heads. Contribution 9 evaluates the structural strength and durability of the rotor system in a large-scale vertical axial pump when operating in both pump mode and PAT (pump-as-turbine) mode, which provides strong support for the optimal design and safe operation of centrifugal pumps. An analysis was conducted to examine the deformation and distribution of equivalent stress in the blades under various flow conditions. The results revealed that the greatest deformation and concentration of stress primarily occurred near the root of the blades. The fatigue life of the blades was evaluated, and it was found that the cycle times of the blades exceeded 106 cycles under all operating conditions, which indicates that the blades have a service life and cycle times that fall within the acceptable limit for safety. Comparisons of blade safety factors in pump and PAT modes indicated that the blade root in PAT mode necessitated a careful evaluation of its material strength and safety stability. Contribution 10 proposed a method for predicting the Q–H curve of PAT at different speeds, which was validated via numerical simulations and new analytical expressions. The results show that these methods provide satisfactory results when the operating point is in proximity to the BEP (best efficiency point).

On some occasions, turbines serve as the power source of pumping stations, using the kinetic energy of the water to drive the pumps to operate in service of water extraction and transfer [7]. Contribution 11 details a systematic analysis of the flow-induced noise of the Francis turbine under different operating conditions. The analysis was carried out through the application of a combination of CFD simulations and sound power-level assessment methods. Through a comparison of experimental and simulated values, three distinct loading circumstances situated in the permissible, limited, and forbidden zones were chosen for the investigation. This selection facilitated a more comprehensive understanding of the creation and distribution of flow-induced noise. The relationship between the flow-induced noise and the internal flow characteristics of the turbine (e.g., vortex, guide vane opening, etc.) is revealed, which provides a scientific basis for the low-noise design of the turbine and the diagnosis of noise problems. By analyzing the flow characteristics under different operating conditions, a design method to reduce the flow-induced noise is proposed, which is of great significance for the design and optimization of hydraulic turbines. In addition, Contribution 12 compares the pressure pulsation characteristics of Francis turbines with varying draft tube arrangement directions. The VMD approach is used here for the first time to analyze pressure pulsations in the Francis turbines, which helps clearly extract the signal characteristics and obtain more explicit results. Based on a real engineering case, the findings offer a hydrodynamic benchmark for the design of the deflector casing of a Francis turbine. Meanwhile, in order to have a more comprehensive understanding of the performance characteristics of the turbine, Contribution 13 conducted a cavitation characterization study of the axial-flow paddle turbine for both the prototype and the model in the surrounded vortex rope. It is suggested that the critical cavitation coefficients between the prototype and the model remain similar under varied cavitation coefficients. This similarity serves as a significant reference for the design and operation of the prototype. The article also examines the alterations in the cavitation properties of the runner chamber of both the model and the prototype when the model critical cavitation coefficient is reduced below the prototype critical cavitation coefficient, which is of practical significance for the understanding and prevention of cavitation damages in the operation of hydraulic turbines.

With renewable energy receiving global emphasis and investment, the current base of intermittent renewable energy sources like wind and solar will continue to increase. The power generation of these energy sources is unstable and requires energy storage technology to balance power supply and demand. Pumped storage power plants, as one of the key technologies and economically feasible energy storage methods, have a significant impact on the transition of energy structure [8,9]. In an in-depth study of pumped storage unit performance optimization, Contribution 14 proposes the use of fractional-order PID (FOPID) controllers to improve the control performance of pumped storage units. The FOPID controller differs from standard PID controllers in its incorporation of two additional adjustable parameters, namely differential order and integral order, thus providing higher control freedom and better robustness; this provides a new solution to the control problem of pumped storage units and helps improve the stability and regulation performance of the system. In order to have a more comprehensive understanding of the pumped storage unit's performance in actual operation, Contribution 15 details a field study of the pressure fluctuation and vibration stability characteristics of the prototype pump turbine under multiple operating conditions. In turbine mode, the pressure variations exhibit distinct characteristics; in pump mode, the pressure fluctuations at different measurement locations exhibit changes in the low-frequency region. The vibration acceleration characteristics are affected by the interaction between the rotor and stator and the frequencies of the structural modes. These findings are crucial for guaranteeing the stability and dependability of the unit. Contribution 16 describes an analysis of the energy loss mechanism in a pump turbine equipped with diverter vanes at various characteristic heads. The study employed the entropy generation theory in conjunction with the flow field distribution to quantitatively evaluate the energy loss and internal flow state changes of the pump turbine at various characteristic heads. The investigation aimed to uncover the underlying causes and distinctive features of energy loss. Through the application of the entropy generation theory, a new approach is provided to analyze and optimize the hydraulic design of pump turbines, which helps improve their efficiency and operational performance. Contribution 17, in order to explore in depth the internal flow characteristics of a single-pier, front-loaded pump turbine in a draft tube at partial load, investigates the interactions between vortex ropes and return zones and their effects on the pressure fluctuation characteristics within the tailrace. A preliminary explanation of the phenomenon of fluid kinematic progression based on vortex theory is presented, and a theoretical analysis of the distortion of vortex tubes under the influence of geometric potential is presented. This study not only helps us understand the operating characteristics of the pump turbine under partial load but also provides an important reference for optimizing the design of the unit and improving the operating stability.

3. Conclusions

This Special Issue not only presents the latest research results in the field of pumping stations but also provides a wealth of practical experience and insights. The guest editors believe that the following concerns in the field of pumping stations need continued attention:

(1) For the lack of applicable hydraulic pump models for water conservancy pumping stations, continued research and the development of high-performance hydraulic models with excellent cavitation performance and little pressure pulsation are necessary in order to effectively support the efficient operation of pumping stations.
(2) In light of the numerous space vortices and insufficient submergence depth of the pump suction flapper in the front and intake pools of water conservancy pumping stations, a flow control system must be implemented in order to enhance the pumping stations' intake conditions.
(3) Improving the control performance of water conservancy pumping stations by enhancing the performance of gate valves should receive additional prioritization.

We believe that through future research, pumping station technology will continue to make new breakthroughs, thereby providing a strong guarantee for urban and rural water supply security and sustainable development.

Author Contributions: Conceptualization, C.Y.; validation, K.K.; resources, Y.Z.; writing—original draft preparation, C.Y.; writing—review and editing, R.T.; supervision, H.C. All authors have read and agreed to the published version of the manuscript.

Funding: This research was funded by the National Natural Science Foundation of China, grant number 52209109, No. 52379086, No. 52271275.

Data Availability Statement: Not applicable.

Conflicts of Interest: The authors declare no conflicts of interest.

List of Contributions:

1. Zheng, X.; Zhang, P.; Zhang, W.; Yu, Y.; Zhao, Y. Numerical Study on the Influence of Combined Rectification Facilities on the Flow in the Forebay of Pumping Station. *Water* **2023**, *15*, 3847.
2. Hu, J.; Wang, Q.; Zhang, Y.; Meng, Z.; Zhang, J.; Fan, J. Numerical and Experimental Study on the Process of Filling Water in Pressurised Water Pipeline. *Water* **2023**, *15*, 2508.
3. Wang, Q.; Hu, J.; Song, M.; Shen, H.; Zhou, Y.; Li, D.; Xie, F. Study on the Transient Flow Characteristics of a Hump Water Pipeline Based on the Random Distribution of Bubbles. *Water* **2023**, *15*, 3831.
4. Zheng, X.; Wang, W.; Zhang, P.; Pu, Y.; Zhao, Y. Internal Flow Characteristics of Centrifugal Pumps under Different Startup Combination Schemes. *Water* **2024**, *16*, 1087.
5. Hua, E.; Qiu, L.; Xie, R.; Su, Z.; Zhu, W. Comparative Analysis of the Hydrodynamic Performance of Dual Flapping Foils with In-Phase and Out-of-Phase Oscillations. *Water* **2023**, *15*, 3275.
6. Zheng, Y.; Meng, L.; Zhang, G.; Xue, P.; Wang, X.; Zhang, C.; Tian, Y. Study on Impeller Optimization and Operation Method of Variable Speed Centrifugal Pump with Large Flow and Wide Head Variation. *Water* **2024**, *16*, 812.
7. Ye, C.; An, D.; Huang, W.; Heng, Y.; Zheng, Y. Investigation on stall characteristics of centrifugal pump with guide vanes. *Water* **2023**, *15*, 21.
8. Li, S.; Tu, Y.; Ye, C.; Yan, H.; Dai, J.; Dang, M.; Yang, C.; Zheng, Y.; Li, Y. Analysis of Stress Characteristics of a Vertical Centrifugal Pump Based on Fluid-Structure Interaction. *Water* **2023**, *15*, 426

9. Li, H.; Cai, Z.; Zheng, Y.; Feng, J.; Xu, H.; Chen, H.; Binama, M.; Kan, K. Investigation of Structural Strength and Fatigue Life of Rotor System of a Vertical Axial-Flow Pump under Full Operating Conditions. *Water* **2023**, *15*, 3041.
10. Plua, F.A.; Sánchez-Romero, F.J.; Hidalgo, V.; López-Jiménez, P.A.; Pérez-Sánchez, M. Variable Speed Control in PATs: Theoretical, Experimental and Numerical Modelling. *Water* **2023**, *15*, 1928.
11. Zhang, T.; He, G.; Guang, W.; Lu, J.; Song, X.; Zhu, D.; Wang, Z. Investigation of the Internal Flow in a Francis Turbine for Comparing the Flow Noise of Different Operation Conditions. *Water* **2023**, *15*, 3461.
12. Zhang, T.; Hu, Z.; Liu, X.; Lu, J.; Song, X.; Zhu, D.; Wang, Z. Comparison of Pressure Pulsation Characteristics of Francis Turbine with Different Draft Tube Arrangement Direction. *Water* **2023**, *15*, 4028.
13. Zhang, Y.; Luo, W.; Chen, T.; Zhou, L.; Wang, Z. Study on the Cavitation Characteristics of Shroud Clearance in Prototype and Model of a Kaplan Turbine. *Water* **2023**, *15*, 3960.
14. Zhou, X.; Zheng, Y.; Xu, B.; Liu, W.; Zou, Y.; Chen, J. Fractional-PID and Its Parameter Optimization for Pumped Storage Units Considering the Complicated Conduit System. *Water* **2023**, *15*, 3851.
15. Hu, H.; Xia, M.; Song, X.; Zhao, W.; Wang, W.; Wang, Z. A Field Investigation of Stability Characteristics of Pressure Fluctuation and Vibration in Prototype Pump Turbine at Multiple Working Points. *Water* **2023**, *15*, 3378.
16. Gui, Z.; Xu, Z.; Li, D.; Zhang, F.; Zhao, Y.; Xu, L.; Zheng, Y.; Kan, K. Analysis of the Energy Loss Mechanism of Pump-Turbines with Splitter Blades under Different Characteristic Heads. *Water* **2023**, *15*, 2776.
17. Hu, H.; Xia, M.; Song, X.; Wang, Z.; Qiao, M. Numerical Investigation of Inner Flow Characteristics of a Prototype Pump Turbine with a Single Pier in Draft Tube at Part Load Conditions. *Water* **2024**, *16*, 13.

References

1. Tse, H. Challenges for Pumping Station Design in Water Industries: An Overview of Impacts from Climate Change and Energy Crisis. *Water Res.* **2024**, *253*, 121250. [CrossRef] [PubMed]
2. Kow, P.Y.; Liou, J.Y.; Yang, M.T.; Lee, M.H.; Chang, L.C.; Chang, F.J. Advancing climate-resilient flood mitigation: Utilising transformer-LSTM for water level forecasting at pumping stations. *Sci. Total Environ.* **2024**, *927*, 172246. [CrossRef] [PubMed]
3. Bettaieb, N.; Taieb, E.H. Assessment of failure modes caused by water hammer and investigation of convenient control measures. *J. Pipeline Syst. Eng. Pract.* **2020**, *11*, 04020006. [CrossRef]
4. Elgammi, M.; Hamad, A.A. A feasibility study of operating a low static pressure head micro pelton turbine based on water hammer phenomenon. *Renew. Energy* **2022**, *195*, 1–16. [CrossRef]
5. Ye, C.; Tang, Y.; An, D.; Wang, F.; Zheng, Y.; van Esch, B.P.M. Investigation on stall characteristics of marine centrifugal pump considering transition effect. *Ocean Eng.* **2023**, *280*, 114823. [CrossRef]
6. Zi, D.; Wang, F.; Wang, C.; Huang, C.; Shen, L. Investigation on the air-core vortex in a vertical hydraulic intake system. *Renew. Energy* **2021**, *177*, 1333–1345. [CrossRef]
7. Anagnostopoulos, J.S.; Papantonis, D.E. Pumping station design for a pumped-storage wind-hydro power plant. *Energy Convers. Manag.* **2007**, *48*, 3009–3017. [CrossRef]
8. Jurasz, J.; Piasecki, A.; Hunt, J.; Zheng, W.; Ma, T.; Kies, A. Building integrated pumped-storage potential on a city scale: An analysis based on geographic information systems. *Energy* **2022**, *242*, 122966. [CrossRef]
9. Lopes, R.A.; Junker, R.G.; Martins, J.; Murta-Pina, J.; Reynders, G.; Madsen, H. Characterisation and use of energy flexibility in water pumping and storage systems. *Appl. Energy* **2020**, *277*, 115587. [CrossRef]

water

MDPI

Article

Numerical Study on the Influence of Combined Rectification Facilities on the Flow in the Forebay of Pumping Station

Xiaobo Zheng [1], Pengli Zhang [2], Wenjing Zhang [1], Yue Yu [1] and Yaping Zhao [1,*]

1 Institute of Water Resources and Hydropower, Xi'an University of Technology, Xi'an 710048, China; zxb@xaut.edu.cn (X.Z.); 18309299001@163.com (W.Z.); 18871105987@163.com (Y.Y.)
2 Hanjiang to Weihe River Valley Water Diversion Project Construction Co., Ltd., Xi'an 710024, China; xazpl@163.com
* Correspondence: zyp0168@163.com

Abstract: The flow pattern of the forebay of the pumping station has a considerable effect on the operating efficiency and stability of the pump unit. A good forebay flow pattern can enable the pump unit to improve efficiency and operating conditions. This study takes a large pumping station as the research object and considers two rectification schemes, namely, a single bottom sill and a "bottom sill + diversion pier". Without rectification facilities under different start-up schemes, the forebay flow pattern after the addition of rectification facilities is calculated, and the influence of single and combined rectification facilities is analyzed. Results show large-scale undesirable flow structures such as backflow and vortex in the forebay of the original design that without rectification facilities and uneven flow distribution occurs in the operating unit. The addition of a bottom sill in the forebay can control the central water beam from the water diversion pipe. The flow is divided to spread to both sides of the forebay and can be rectified twice after installing the diversion piers. The combined rectifier facility of "bottom sill + diversion pier" is beneficial to disperse incoming flow and make the flow distribution of each unit more uniform. The backflow and vortex inside the forepond are basically eliminated, and the flow state of the forepond is significantly improved.

Keywords: forebay; numerical simulation; rectification facilities; bottom sill; diversion pier

Citation: Zheng, X.; Zhang, P.; Zhang, W.; Yu, Y.; Zhao, Y. Numerical Study on the Influence of Combined Rectification Facilities on the Flow in the Forebay of Pumping Station. *Water* **2023**, *15*, 3847. https://doi.org/10.3390/w15213847

Academic Editor: Ahmad Shakibaeinia

Received: 6 October 2023
Revised: 29 October 2023
Accepted: 1 November 2023
Published: 3 November 2023

1. Introduction

The forebay of the pumping station is set up as a connection between the diversion channel and the water inlet pool to enable a smooth and even flow and thus provide good conditions for the pump unit. However, after the water flows into the forebay, adverse phenomena such as backflow and vortex may occur due to the influence of various factors, resulting in not only sedimentation in the forebay but also flow disorder in the water inlet channel of the pump unit. Thus, the operating efficiency of the pump decreases and severely affects the safe and stable operation of the unit [1–3].

For the study of the flow pattern of the forepool of the pumping station, in the past, the direct experiment mainly relied on the physical model, and the velocity measurement at the distribution point was mainly adopted by the current meter or particle image velocimetry (PIV) and laser Doppler anemometry (LDA), which show flow fields at character level [4,5]. With the development of computer technology, numerical simulation using computational fluid dynamics (CFD) has become a common research method, which can improve the efficiency of analysis and better study the flow field. Kim et al. [6] studied the flow distribution in the intake channel of the pump based on experiments and CFD technology to find out the cause of the eddy current in detail. At present, a large number of studies have been carried out on the flow state of the forebay. Harding et al. [7] established a mathematical model of the movement of the forebay in the velocity field and analyzed the internal flow field of the forebay by using an acoustic Doppler current profiler to quantify the errors caused by spatial changes in velocity. Amin et al. [8] believe that the efficiency of the pumping station depends to a large extent on the structural design of the inlet pool, not only on the performance of the selected pump. Numerical and experimental studies were conducted on a rectangular inlet pool to predict the swirl angle and the formation of free-surface vortices, and the swirl angle and average tangential velocity estimated by CFD simulation were consistent with the experimental results. In the hydraulic model of seawater intake at the Aliveri power plant in Greece, Dimas and Vouros [9] studied the influence of cross flow in the front pool on the eddy current angle in the suction pipe of the pump and found that when the average cross flow velocity dropped below the critical value, the eddy current angle only depended on the shape of the front pool. Based on the above research, many scholars have carried out a great deal of analysis and research on the flow pattern. Zhang et al. [10] carried out experimental research on a lateral inflow pumping station at different water levels and found large-scale backflow areas in the forebay at the surface or the layers. Ying et al. [11] used a two-phase flow scheme to study the flow regime of a forward-influent forebay and found that the water flow separates at the side wall and large-scale backflows occur on both sides. For these flow characteristics of the forebay, domestic and foreign scholars have carried out research on improving the flow pattern of the forebay and carried out a lot of rectification measures for the forebay. Rtimi et al. [12] and Karami et al. [13] optimized the base splatter; Li et al. [14] and Luo et al. [15] added columns to the front pool and designed the columns; Ahmed et al. [16] and Zhou et al. [17] set different diversion pier layout methods; Xu et al. [18] used pressurized water plates for rectification of the front pool. Mi et al. [19] set a bottom sill in the forebay to eliminate the influence of eddy current and circulation generated by diffusion flow on pump performance and sediment deposition, and the research results can provide a theoretical basis for improving the flow pattern of the forebay and avoiding sediment deposition. Zhou et al. [20] analyzed the adverse flow state in the forebay of a certain lateral pumping station, such as large-scale backflow area and severe deflection of water inlet angle, and used various rectification measures to carry out numerical simulations; their results show that diversion piers can lead to a reasonable flow velocity area, the bottom sill can change the flow structure, and diversion walls can weaken the interference of backflow on the mainstream.

In addition, Nasr et al. [21] used various rectification measures to change the flow pattern of the front pool of the pumping station and found that when the parabola wall and part of the rectifier pier were well set, the flow pattern of the inlet pool was better and the uniformity of velocity distribution was improved. Yang et al. [22] studied and compared the rectifier flow pattern in the forepond, the uniformity of velocity distribution in the measured section, and the reduction rate of the eddy current area, and found that the combination scheme of the rectifier wall and the diversion wall had a good effect.

Existing literature has shown that adding bottom sills and diversion piers to the forebay can effectively improve the flow state, but it mainly focuses on the influence of a single rectification facility. Few studies examine the application of the combined rectification facility of the bottom sill and diversion piers. Therefore, the present study takes a large-scale pumping station as the research object and considers two rectification facilities: A single bottom sill and a "bottom sill + diversion pier". A three-dimensional (3D) numerical analysis is carried out on the flow with different numbers of units opened to examine the improvement effect of the bottom sill and the combination of the bottom sill and diversion pier on the flow. The findings can provide a reference for the design and renovation of rectification facilities for the same type of pumping stations.

2. Research Object

This study examines a large pumping station with five units, which are symmetrically arranged (4 working and 1 standby). The design flow rate of a single machine is 3.75 m^3/s. The forebay of the pumping station is 38.38 m long and is diverted by two steel pipes with a diameter of 2.6 m. The slope of the pool bottom is 0.049, the divergence angle of the forebay is 39.02°, and the inlet pool is 15 m long and 39.2 m wide. The center distance between two adjacent pumps is 8.5 m, and the design water level is 19.5 m. The diameter of the trumpet tube is $D = 2.6$ m, the suspended height is 0.4 D, and the rear wall distance is 0.08 D. From top to bottom, these are units 1–5. The 3D model of the pumping station is shown in Figure 1.

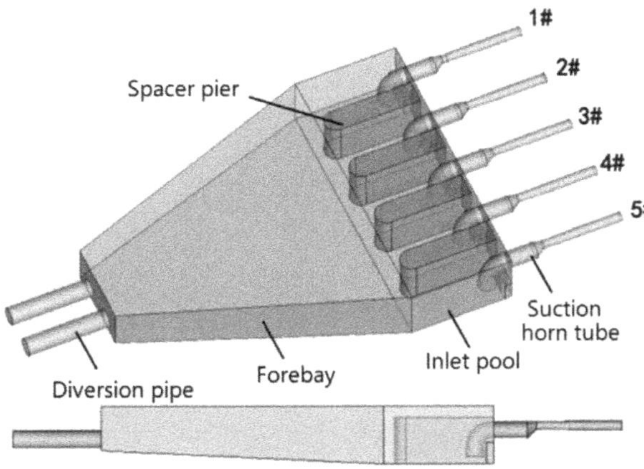

Figure 1. 3D model of pumping station.

3. Numerical Simulation

3.1. Turbulence Model

According to the hydraulic characteristics of the pumping station, the water movement in the forebay and the inlet pond of the pumping station is a turbulent flow with a large Reynolds number and a wide range of backflow and shedding. The RNG k-ε turbulence model can manage flows with swirling, high strain rates, and large streamline bending, and thus this study adopts the RNG k-ε turbulence model [23].

In this study, the CFX software is used to numerically simulate the flow in the forebay and inlet pond of the pumping station. The convection item of the turbulence model adopts a high-order precision format, and the convergence precision of each monitoring parameter is 10^{-4}.

3.2. Calculation Region and Grid

Taking the water diversion pipe, forebay, water inlet pool, and water-absorbing trumpet for the calculation, the hexahedral structured grid is used to divide the different components into meshes, as shown in Figure 2.

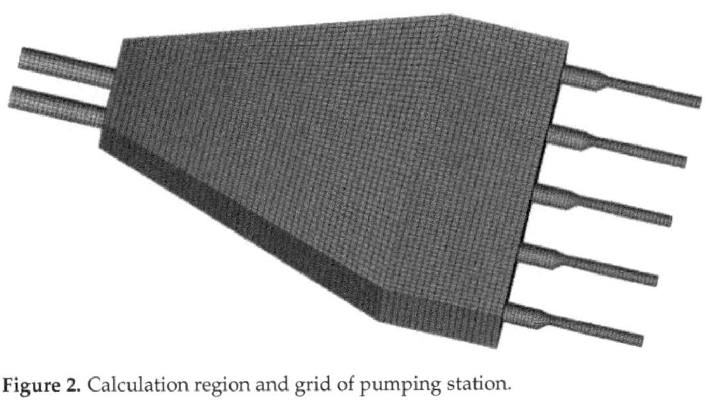

Figure 2. Calculation region and grid of pumping station.

Accurate numerical simulation results are obtained by verifying the calculation domain for grid independence. A total of six sets of grid schemes with different densities are generated, with the number of grids ranging from 1.76–3.05 million. The calculation results are shown in Figure 3.

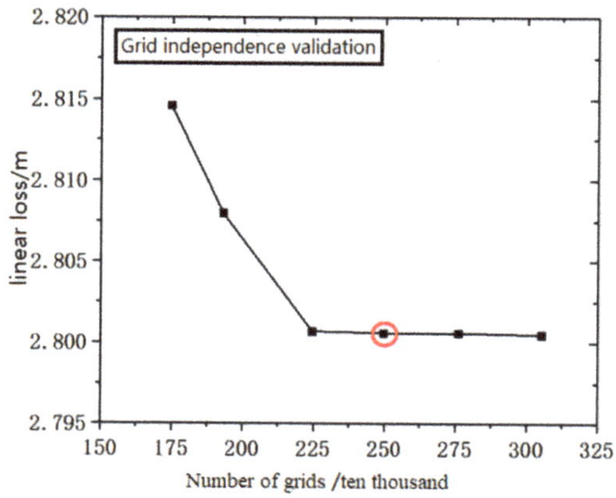

Figure 3. Grid independence validation.

The figure shows that when the number of grids increases from 1.76 to 2.24 million, the hydraulic loss in the calculation domain changes significantly, but when the increase is from 2.24 to 3.05 million, the hydraulic loss hardly changes, which meets the grid independence verification requirements. Therefore, 2.49 million meshes are selected for numerical simulation. The number of meshes for each flow-through component is shown in Table 1.

Table 1. Grid condition of each component.

Water Diversion Pipe + Forebay	Inlet Pool	Suction Trumpet	Total
Number of grids (10,000)	165.14	34.48	249.55

3.3. Boundary Condition

Inlet and outlet conditions: This study defines the inlet boundary as the inlet of the water diversion pipe, adopts the mass flow inlet boundary condition, and sets the design flow rate of a single pump as 3.75 m^3/s. The outlet is set at the outlet of the water-absorbing horn tube, and the static pressure outlet is set to 0 atm.

Wall conditions: All solid walls are smooth and non-slip (including side walls, bottom of the front pool, bottom of the water inlet pool, water diversion pipes, and water-absorbing trumpet pipes).

The surfaces of the forebay and the inlet pool are free liquid surfaces, treated symmetrically, assuming the use of the steel cover. The shear stress and heat exchange generated by air on the water surface are ignored.

4. Simulation Analysis of Forebay Flow in Original Scheme

The numerical calculation and analysis of the flow field of the research object is carried out by taking the operating conditions of Nos. 1, 2, 4, and 5 units in operation and No. 3 unit on standby as the calculation mode. Figure 4 shows the overall flow diagram of the pumping station, and after the water flow enters from the diversion pipe, the main flow is concentrated in the middle of the forebay. The flow velocity on both sides is relatively small. A large-scale backflow, distributed on both sides, and a vortex near the side wall are observed.

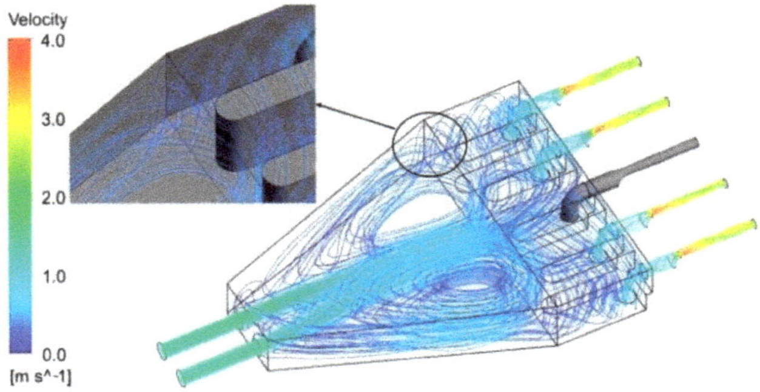

Figure 4. Flow pattern of the pumping station.

For further analysis of the internal flow of the inflow field of the pumping station, three sections were cut along the height direction of the forebay (the heights are: bottom layer $z = 13.5$ m, middle layer $z = 16.5$ m, surface layer $z = 19.5$ m). The flow velocity vector diagrams of the three sections are shown in Figure 5. As the position moves down, the range of unfavorable flow patterns such as backflow and vortex expands and even goes deep into the pier of the inlet pool.

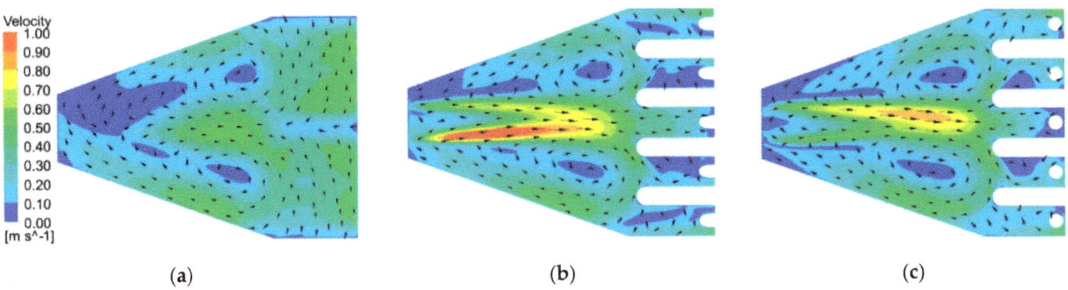

Figure 5. Streamlines of horizontal sections of forebay and pump sump: (**a**) Surface layer; (**b**) middle layer; (**c**) bottom layer.

Figure 6 shows the vorticity diagram of the bottom surface of the water inlet pool of the original scheme, and Figure 7 shows the vortex structure diagram under the water suction horn tube of each operating unit. Vortices of different scales form near and below the water-absorbing horns of each operating unit and at the head of the pier. The bottom vortex is clearly shown in Figure 7. The pier is formed by the impact of water flow, and the attached bottom vortices under the water-absorbing horns of units 1 and 5 are relatively strong.

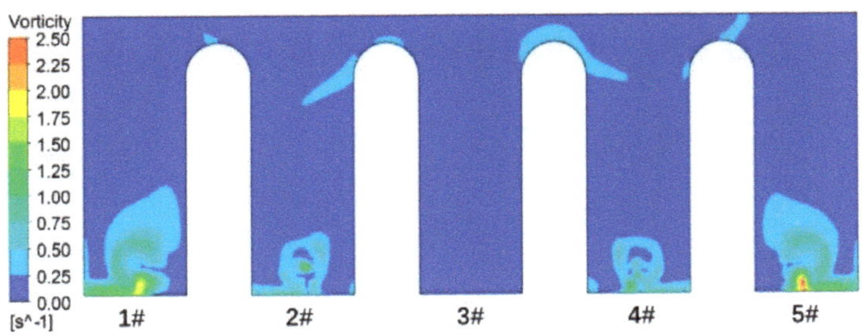

Figure 6. Vorticity distribution of pump sump.

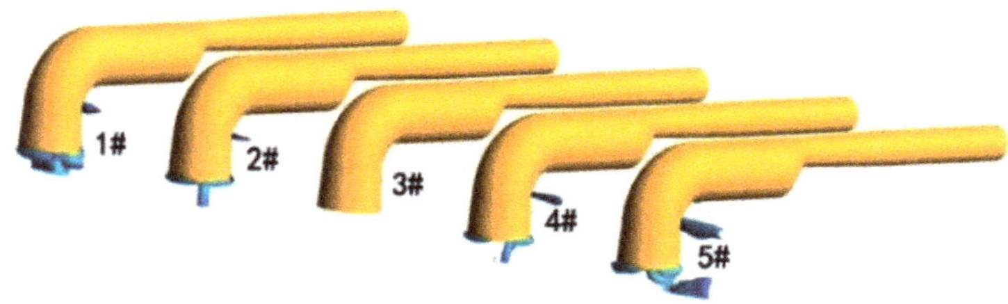

Figure 7. Vorticity of the original scheme operation unit.

From the above analysis, the reasons for the adverse flow conditions such as backflow and vortex in the forebay can be attributed to the following:

(1) The diffusion angle of the forebay is too large, close to the critical value of the pumping station design specification [24] (the forebay diffusion angle of the forward water pumping station is <40°).

(2) The forebay of the pumping station uses pressurized steel pipes to divert water, which causes high-speed water jets to form and impact the pier, thereby forming reverse water flow. This occurrence causes large-scale backflow, vortexing, and other undesirable flow patterns in the forebay.

5. Analysis of Influence of Rectification Facilities on Forebay Flow

5.1. Rectification Facilities and Operating Conditions

The principle of bottom sill rectification is to use appropriate engineering measures to cause the façade behind the sill to swirl, disrupt the plane backflow, and then use the diversion function of the diversion pier to homogenize the incoming flow. Thus, bad flow patterns such as backflow and vortexing are effectively eliminated. Based on the original design, this study proposes two rectification schemes, as shown in Table 2 and Figure 8.

Table 2. Rectification scheme.

Program Number	Rectification Measures	Scheme Description
Scheme 1	Bottom sill	Located at 10 m in front of the forebay, across the bottom, width × height: 1 × 1.5 m
Scheme 2	Bottom sill + diversion pier	Data The diversion pier is located 7 m behind the bottom sill, and the distance between the heads of the two diversion piers is 8.7 m. The length × width × height: 10 × 1 × 5 m, and the included angle with the center line of the forebay is 18°

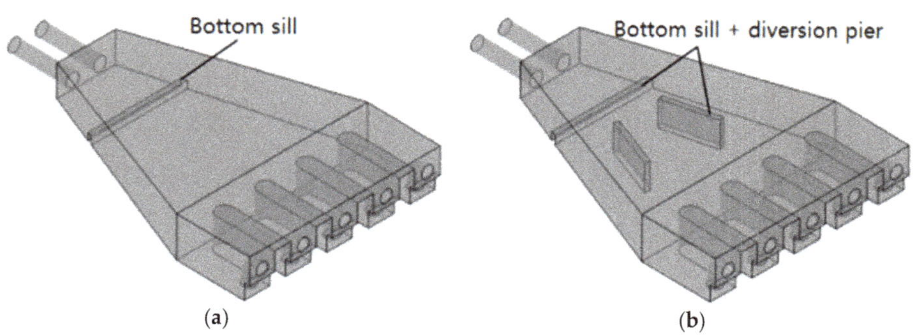

Figure 8. Structure diagram of rectification measures: (a) Scheme 1; (b) Scheme 2.

Given that the research object is the parallel operation of multiple units, this study determines the influence of different numbers of units on the flow state of the forebay of the pumping station. The parallel operations of 3, 4, and 5 units are examined, and the operating conditions and start-up combinations are shown in Table 3.

Table 3. Operating conditions.

Operating Conditions	Rectification Measures	Flow m^3/s	Boot Group Number
1	Original design scheme	11.25	1#, 3#, 5#
2		15	1#, 2#, 4#, 5#
3		18.75	1#, 2#, 3#, 4#, 5#
4	Scheme 1	11.25	1#, 3#, 5#
5		15	1#, 2#, 4#, 5#
6		18.75	1#, 2#, 3#, 4#, 5#
7	Scheme 2	11.25	1#, 3#, 5#
8		15	1#, 2#, 4#, 5#
9		18.75	1#, 2#, 3#, 4#, 5#

5.2. Analysis of the Influence of Rectification Facilities on Forebay Flow

5.2.1. Comparative Analysis of Forebay Flow under Different Operating Conditions

Figure 9 is a vector diagram of the flow velocity in the forebay and inflow tank of the pumping station under various operating conditions. The velocity vector of each point in the flow field can be seen, so as to understand the velocity distribution of all parts in the flow field. Figure 10 is a vorticity diagram of the flow velocity in the forebay and inflow tank of the pumping station under various operating conditions, and the vorticity distribution in each region can be obtained.

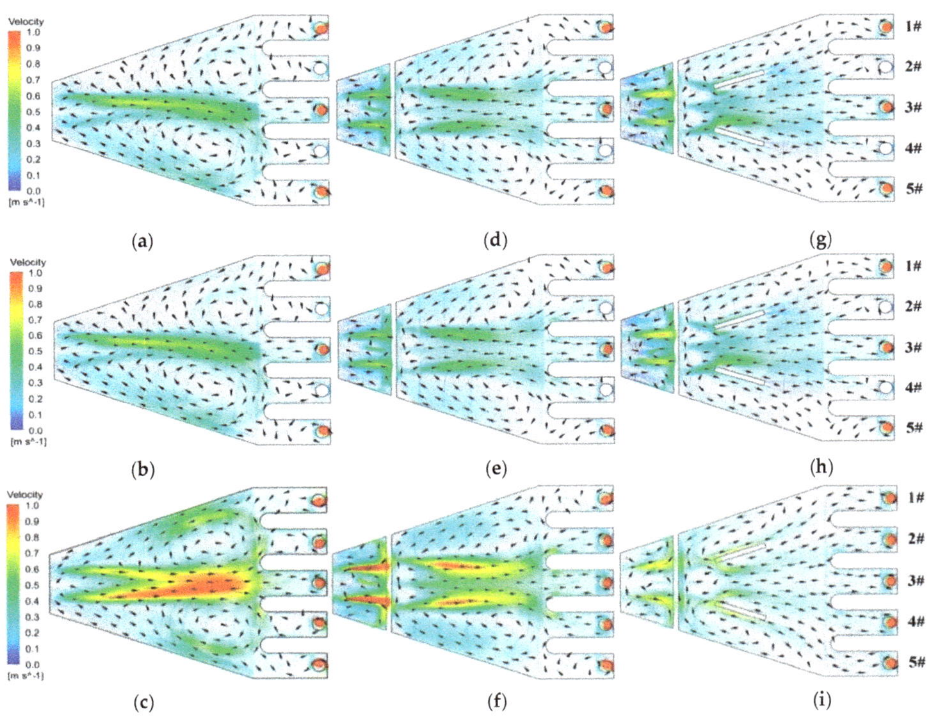

Figure 9. Velocity vector diagram of horizontal section (z = 13.5 m) of forebay and pump sump under different operating conditions: (**a**) Condition 1; (**b**) condition 2; (**c**) condition 3; (**d**) condition 4; (**e**) condition 5; (**f**) condition 6; (**g**) condition 7; (**h**) condition 8; (**i**) condition 9.

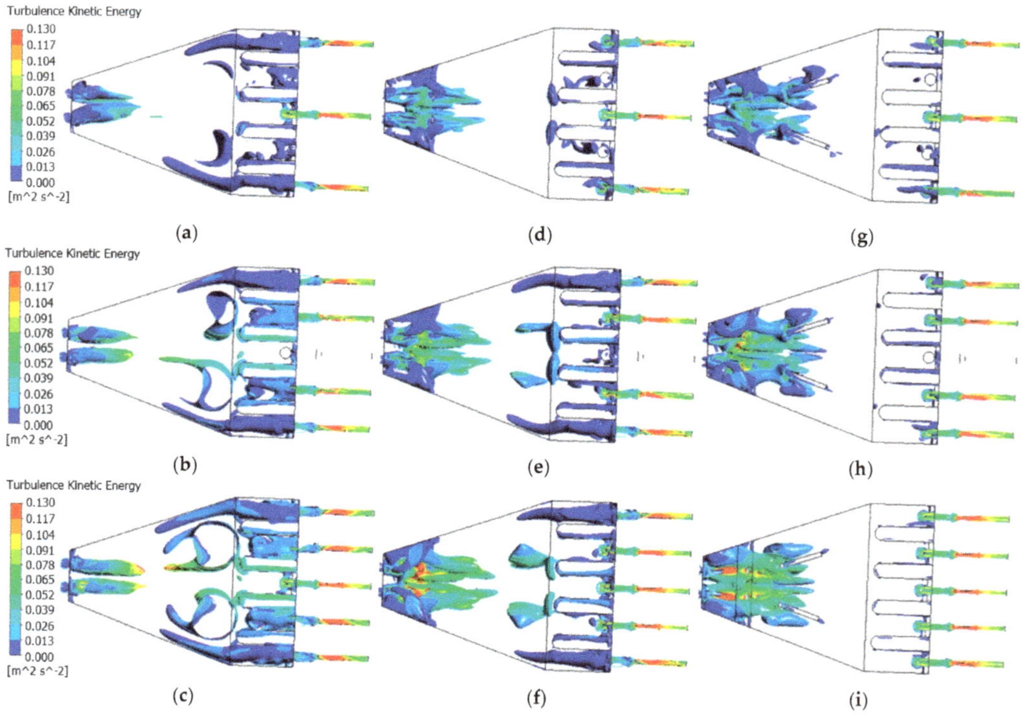

Figure 10. Vorticity diagram of horizontal section of front pool and inlet pool under different operating conditions: (**a**) Condition 1; (**b**) condition 2; (**c**) condition 3; (**d**) condition 4; (**e**) condition 5; (**f**) condition 6; (**g**) condition 7; (**h**) condition 8; (**i**) condition 9.

According to the analysis of the original design scheme of the forebay (i.e., working conditions 1–3 in Figure 9), when the pumping station starts 3, 4, and 5 units at the same time, a large-scale backflow occurs and is symmetrical on both sides of the forebay, and the scope of the spread goes deep to the pier of the inlet pool. Except for the No. 3 unit, the water inlet channels of the other operating units have a lateral side flow.

In view of the problems of the original scheme, the bottom sill is added to the forebay, and the water flow with large kinetic energy is intercepted by the bottom sill so that the water flow in the forepond is evenly distributed. Figure 9 Working Cases 4–6 show the vector diagram of flow velocity in the forebay section of Scheme 1. A bottom sill is added to the forebay, and the water flow enters through the water diversion pipe. The bottom sill diverts the central water beam from the water diversion pipe and diffuses it to both sides of the forward pool, which improves the diffusion of the water flow on the plane and disrupts the backflow. The improvement effect is mainly reflected in the water inlet channels of Nos. 2 and 4, which have improved horizontal side flows. The lateral side flow remains in the inlet channel of Nos. 1 and 3 units.

By comparing the original scheme with Scheme 1, it can be found that after adding the bottom sill to the original scheme, the flow line in the middle area of the forebay has tended to be stable, and the flow line in the inlet channel has also been improved, but there are still transverse flow measurements on both sides, so the diversion pier continues to be added to divert the incoming flow through the bottom sill and carry out secondary rectification. Figure 9 Working cases 7–9 show the cross-sectional flow velocity vector diagram of Scheme 2. Based on Scheme 1, diversion piers are added and carry out a secondary rectification of the water flow, further enhancing the diffusion of turbulent kinetic energy. The large-scale backflow areas on both sides of the forebay basically disappear, and the flow of the water before entering the inlet pool is relatively stable. On the bottom section, the flow state of the water inlet pool is smooth, and the suction conditions of the water pump considerably improve.

Figure 10 is the comparison diagram of vorticity between the front pool and the inlet pool of the pumping station under different operating conditions. It can be seen from the vortex intensity distribution diagram of working conditions 1, 2, and 3 that when 3, 4, and 5 units are opened at the same time in the pumping station, large-scale vortices are generated at the inlet pool and the inlet channel of the unit. After adding the base sill rectification, it can be seen from the vortex intensity distribution diagram of working conditions 4, 5, and 6 that the scale of the vortices on both sides of the front pool is reduced, and the vortices at the inlet pool and the unit inlet channel are greatly improved. The vortex distribution diagram for working conditions 7, 8, and 9 is obtained after the diversion pier is added. In the diagram, the large-scale vortices at the inlet pond and the inlet channel of the unit basically disappear, and the flow pattern at the inlet of the unit is greatly improved. It can be seen from Figures 9 and 10 that, compared with the original scheme, the scale and quantity of adverse flow patterns such as backflow and vortex in the forebay in each operating condition of the two rectification schemes have been reduced, and the flow patterns have also been significantly improved. Therefore, the study shows that the combined rectification method with base sill and diversion pier has the best rectification effect in the scheme.

Figure 11 shows the vorticity diagram of the bottom surface of the inlet pool after the addition of rectification facilities when four units are turned on. By comparison with Figure 6, after the addition of rectification facilities, the range of vortex attached to the bottom of the operating unit decreases, but the single bottom sill increases the local (Nos. 2, 4 units) water flow vortices; and the addition of "bottom sill + diversion pier" in the forebay effectively improves the inflow conditions of the pump unit.

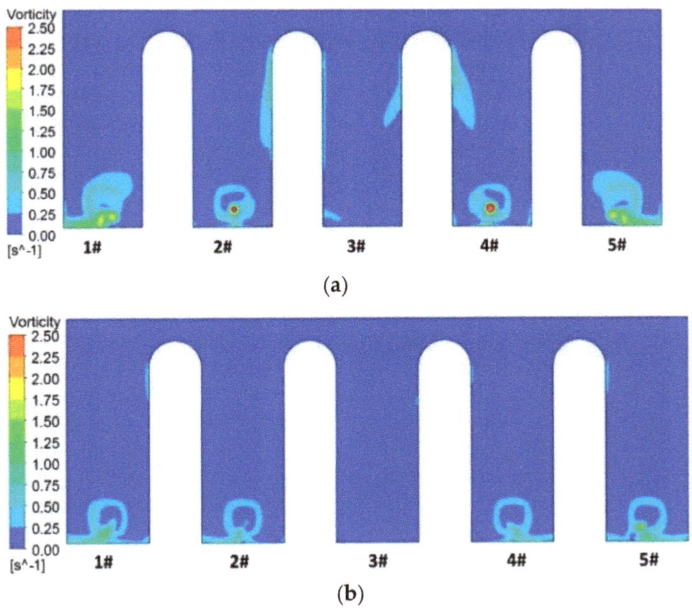

Figure 11. Vorticity distribution of pump sump: (**a**) Bottom sill; (**b**) bottom sill + diversion pier.

5.2.2. Comparative Analysis of Unit Flow Distribution

This study further analyzes the flow distribution among the operating units after the addition of rectification facilities. The flow distribution coefficient λ is introduced, defined as the ratio of the flow rate Q_i of the suction horn tube corresponding to the operating unit to the design flow rate of a single pump $Q = 3.75\ m^3/s$ (i represents the number of operating unit) and expressed as follows:

$$\lambda = \frac{Q_i}{Q} \tag{1}$$

The flow distribution coefficients of units under various operating conditions before and after the addition of rectification facilities are shown in Figure 12. As the λ approaches 1, the flow distribution uniformity of the unit increases.

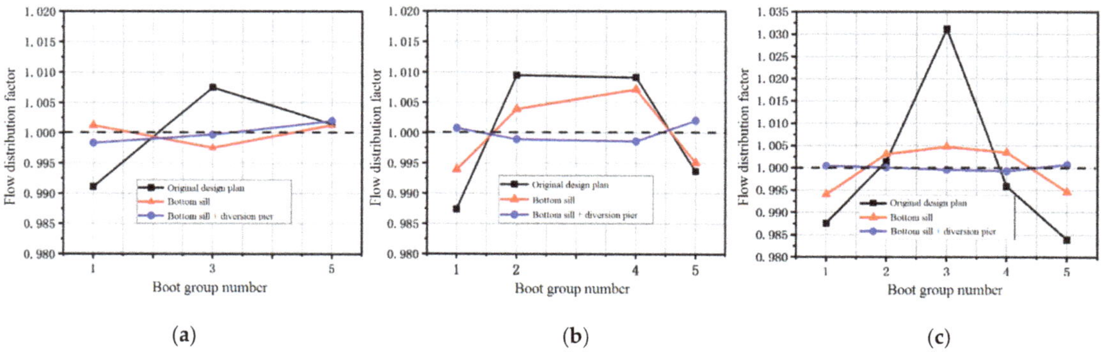

Figure 12. Flow distribution coefficient of operating units before and after rectification: (**a**) 3 units running in parallel; (**b**) 4 units running in parallel; (**c**) 5 units running in parallel.

The flow distribution coefficient curve shows that the original scheme has an uneven flow distribution, and the working flow of the unit operating in the middle is greater than that of units operating on both sides. In Scheme 1, a single bottom sill is set in the forebay, and the uneven flow distribution of each operating unit improves, but the effect is not apparent when four units are turned on at the same time. In Scheme 2, after the addition of the combined rectification facility of "bottom sill + diversion pier", the uneven flow distribution of each operating unit almost disappears.

5.2.3. Comparative Analysis of Flow Velocity Distribution Uniformity and Average Drift Angle

The flow velocity distribution uniformity and average drift angle at the inlet section of the water pump suction trumpet are important indicators to test the effect of the forebay renovation of the pumping station. This study introduces the velocity distribution uniformity V_a and the cross-sectional average drift angle θ [25] and uses them as indicators to evaluate the actual effect of the rectification device.

The closer the cross-section flow velocity uniformity V_a is to 100% and the average drift angle θ is closer to 0°, the more uniform is the axial flow velocity distribution of the water pump impeller inlet section. Thus, the inflow conditions of the water flow introduced into the pump unit improve, and the safe and stable pump operation and the operating efficiency of the pump device are also enhanced. The expression is as follows:

$$V_a = \left[1 - \frac{1}{\overline{u}_a}\sqrt{\frac{\sum(u_{ai} - \overline{u}_a)^2}{m}}\right] \times 100\% \tag{2}$$

$$\theta = \frac{\sum u_{ai}\left[\arctan\left(\frac{u_{ti}}{u_{ai}}\right)\right]}{\sum u_{ai}} \tag{3}$$

In the above formula, V_a is the uniformity of flow velocity distribution; θ is the average drift angle; m is the number of units; $\overline{u}_a$ is the average axial velocity of the horn tube inlet; u_{ai} is the axial velocity of each unit in section i; u_{ti} is the lateral velocity of the ith calculation unit, and $u_{ti} = \sqrt{u_{wi}{}^2 + u_{ri}{}^2}$ (u_{wi} and u_{ri} are the tangential and radial velocities of the ith calculation unit, respectively).

Table 4 shows the flow velocity distribution uniformity and average deflection angle at the inlet section of the suction horn tube under various operating conditions before and after the addition of rectification facilities. Compared with Table 3 with the original scheme, after adding the rectification facilities, the uniformity of cross-sectional flow velocity distribution and the average drift angle have improved. This result shows that both optimization schemes can improve the unfavorable flow state in the forebay of the pumping station. However, the improvement of the second scheme is better achieved by adding a "bottom sill + diversion pier" combined rectification in the forebay, and the effect increases with the increase in the number of start-up units. With the addition of "bottom sill + diversion pier" combined rectification facilities, when Nos. 3, 4, and 5 units are turned on at the same time, the flow velocity distribution uniformity increases by 3.8%, 5.51%, and 7.46%, respectively.

Table 4. The uniformity and average drift angle of the inlet flow velocity distribution of the suction horn in various operating conditions.

		Original Scheme	Bottom Sill	"Bottom Sill + Diversion Pier"
Three units run in parallel	Velocity distribution uniformity V_a	73.58%	76.20%	77.38%
	Mean drift angle θ	18.21°	16.89°	15.42°
Four units run in parallel	Velocity distribution uniformity V_a	73.37%	74.85%	78.88%
	Mean drift angle θ	22.87°	19.88°	14.43°
Five units run in parallel	Velocity distribution uniformity V_a	70.52%	75.62%	77.98%
	Mean drift angle θ	21.09°	20.33°	15.30°

6. Conclusions

Based on numerical simulation technology, this study analyzes the water flow state of the forebay of the original design scheme of the pumping station. According to the simulation results of the forebay, we propose a combined rectification scheme of two rectification schemes that influence the flow of the forebay of the pumping station under the condition of parallel operation of different sets of pumping stations.

1. In the original scheme, the forebay of the research object has large-scale backflow, vortex, and other adverse flow structures when the units are turned on and running under the design working conditions. When multiple units run in parallel, uneven flow distribution occurs and the pumps have poor water inlet conditions, and this situation becomes more and more serious as the position moves down.
2. The bottom sill is set in the forebay and causes the following benefits: Improves the centering of the mainstream, facilitates the diffusion of water flow on the plane, reduces the range of the forebay recirculation zone, and improves the uniformity of flow velocity distribution at the entrance of the suction horn pipe and the flow distribution uniformity of each operating unit. However, there are still backflow and vortices on both sides of the front pool, which need to be further improved.
3. The combined facility of "bottom sill + diversion pier" is added to the forebay. The bottom sill diverts the central water beam from the water diversion pipe to spread to both sides of the forebay. The diversion pier then performs secondary rectification on the diverted forebay flow, which effectively improves the centering of the mainstream.

At present, the rectification effect of the combined rectifier facility on the forebay of the pump station is only the model stage, which proves that the technology is feasible, and more experimental verification should be conducted in the future. At the same time different factors, such as incoming flow speed and flow rate, will also have an impact Finally, the size and location of the rectifier facility need to be specifically designed.

Author Contributions: Y.Z. and Y.Y. carried out the numerical simulations and analyzed the data P.Z. and W.Z. conducted the experiment and wrote the first draft of the manuscript; W.Z. and X.Z conceived and supervised the study and edited the manuscript; X.Z. and Y.Z. guided the experiment and the manuscript. All authors have read and agreed to the published version of the manuscript.

Funding: This work was supported by the Joint Foundation of Shaanxi [Grant No. 2019JLP-25].

Data Availability Statement: The data presented in this study are available on request from the corresponding author.

Conflicts of Interest: Author Pengli Zhang was employed by the company Hanjiang to Weihe River Valley Water Diversion Project Construction Co., Ltd. The authors declare that this study received funding from the company Hanjiang to Weihe River Valley Water Diversion Project Construction Co., Ltd. The funder had the following involvement with the study: research object, technical scheme, formal analysis, data curation. The other authors declare no conflicts of interest.

References

1. Ezz-Aldeen, M.; Al-Ansari, N.; Abd-Albaki, Y. Sediment control strategies for sustainable water intake. *Dams Reserv.* **2021**, *31*, 21–30. [CrossRef]
2. Samsudin, M.L.; Munisamy, K.M.; Thangaraju, S.K. Review of Operational Challenges and Changing Conditions associated with Offshore Submerged Vertical Pumping Station. In Proceedings of the 7th International Conference on Cooling & Heating Technologies (ICCHT 2014), Selangor, Malaysia, 4–6 November 2014; Volume 88.
3. Moreno-Rodenas, A.M.; Duinmeijer, A.; Clemens, F.H.L.R. Deep-learning based monitoring of FOG layer dynamics in wastewater pumping stations. *Water Res.* **2021**, *202*, 117482. [CrossRef]
4. Hiromasa, S.; Hara, S.; Inaoka, K. The velocity and thermal fields of two parallel plane jets using simultaneous PIV and two-color LIF measurements. *Int. J. Heat Fluid Flow* **2023**, *103*, 109192. [CrossRef]
5. Chesnakas, C.J.; Dancey, C.L. Three-component LDA measurements in an axial-flow compressor. *J. Propuls. Power* **1990**, *6*, 474–481. [CrossRef]
6. Kim, C.G.; Choi, Y.D.; Choi, J.W. A study on the effectiveness of an anti vortex device in the sump model by experiment and CFD. *IOP Conf. Ser. Earth Environ. Sci.* **2012**, *15*, 072004. [CrossRef]
7. Harding, S.F.; Richmond, M.C.; Romero-Gomez, P.; Serkowski, J.A. Effects of non-homogeneous flow on ADCP data processing in a hydroturbine forebay. *Flow Meas. Instrum.* **2016**, *52*, 1–9. [CrossRef]
8. Amin, A.; Kim, B.H.; Kim, C.G.; Lee, Y.H. Numerical Analysis of Vortices Behavior in a Pump Sump. *IOP Conf. Ser. Earth Environ. Sci.* **2019**, *240*, 032020. [CrossRef]
9. Dimas, D.A.A.; Vouros, A.P. Effect of Cross-Flow Velocity at Forebay on Swirl in Pump Suction Pipe: Hydraulic Model of Seawater Intake at Aliveri Power Plant in Greece. *J. Hydraul. Eng.* **2012**, *138*, 812–816. [CrossRef]
10. Zhang, X.; Zhou, J.; Liang, J.; Zhang, H. Experimental research on flow pattern of forebay of side-inlet pumping station. *S. N. Water Transf. Water Sci. Technol.* **2016**, *14*, 101–105.
11. Ying, J.; Yu, X.; He, W.; Zhang, J. The flow pattern and combined rectification scheme of the forebay of pumping station based on fluid volume model. *J. Drain. Irrig. Mach. Engin* **2020**, *38*, 476–480+493.
12. Rtimi, R.; Sottolichio, A.; Tassi, P. The Rance tidal power station: Toward a better understanding of sediment dynamics in response to power generation. *Renew. Energy* **2022**, *201*, 323–343. [CrossRef]
13. Karami, H.; Farzin, S.; Sadrabadi, M.T. Simulation of flow pattern at rectangular lateral intake with different dike and submerged vane scenarios. *Water Sci. Eng.* **2017**, *10*, 246–255. [CrossRef]
14. Li, Y.; Gu, J.; Guo, C.; Zhou, C. Flow Patterns and Rectification Measures in the Forebay of Pumping Station. *IOP Conf. Ser. Mater. Sci. Eng.* **2020**, *794*, 012058. [CrossRef]
15. Luo, C.; He, Y.; Shang, Y.; Cong, X.; Ding, C.; Cheng, L.; Lei, S. Flow Characteristics and Anti-Vortex in a Pump Station with Laterally Asymmetric Inflow. *Processes* **2022**, *10*, 2398. [CrossRef]
16. Moussa, A.M. Solving the problem of sedimentation at water intake of Rowd El-Farag pump station using 2D model. *Ain Shams Eng. J.* **2010**, *1*, 103–114. [CrossRef]
17. Zhou, J.; Zhao, M.; Wang, C.; Gao, Z. Optimal Design of Diversion Piers of Lateral Intake Pumping Station Based on Orthogonal Test. *Shock Vib.* **2021**, *2021*, 6616456. [CrossRef]
18. Xu, C.; Wang, R.; Liu, H.; Zhang, R.; Wang, M.; Wang, Y. Flow pattern and anti-silt measures of straight-edge forebay in large pump stations. *Int. J. Heat Technol.* **2018**, *36*, 1130–1139. [CrossRef]
19. Mi, Z.; Zhou, D.; Mao, Y. Three-dimensional CFD simulation of inlet structure flow in pumping station based on Eulerian solid-liquid two-phase flow model. *J. Drain. Irrig. Mach. Eng.* **2015**, *33*, 494–498.

20. Zhou, J.; Zhong, Z.; Liang, J.; Shi, X. Three-dimensional Numerical Simulation of Side-intake Forebay of Pumping Station. *J. Irrig. Drain.* **2015**, *34*, 52–55+80.
21. Nasr, A.; Yang, F.; Zhang, Y.; Wang, T.; Hassan, M. Analysis of the Flow Pattern and Flow Rectification Measures of the Side-Intake Forebay in a Multi-Unit Pumping Station. *Water* **2021**, *13*, 2025. [CrossRef]
22. Yang, F.; Zhang, Y.; Liu, C.; Wang, T.; Jiang, D.; Jin, Y. Numerical and Experimental Investigations of Flow Pattern and Anti-Vortex Measures of Forebay in a Multi-Unit Pumping Station. *Water* **2021**, *13*, 935. [CrossRef]
23. Yanase, S.; Yamasaki, R.; Kouchi, T.; Hosoda, S.; Nagata, Y.; Shunji, H.; Kawabe, T.; Takami, T. Numerical study of air-entraining and submerged vortices in a pump sump. In Proceedings of the 29th IAHR Symposium on Hydraulic Machinery and Systems, Kyoto, Japan, 16–21 September 2018.
24. *GB50265-2010*; Design Code for Pumping Station. Standards Press of China: Beijing, China, 2010.
25. Lu, W.; Yao, T.; Xia, H.; Xu, B. Numerical simulation of diversion piers in forebay under asymmetric operation of pumping station units. *J. Yangzhou Univ. Nat. Sci. Ed.* **2018**, *21*, 62–67.

water

MDPI

Article

Numerical and Experimental Study on the Process of Filling Water in Pressurized Water Pipeline

Jianyong Hu [1,2], Qingbo Wang [1,3], Yuzhou Zhang [1,2,*], Zhenzhu Meng [2,4], Jinxin Zhang [2,4] and Jiarui Fan [5]

1 School of Geomatics and Municipal Engineering, Zhejiang University of Water Resources and Electric Power, Hangzhou 310018, China; hujy@zjweu.edu.cn (J.H.); wangqb@zjweu.edu.cn (Q.W.)
2 Engineering Research Center of Digital Twin Basin of Zhejiang Province, Hangzhou 310018, China; mengzhzh@zjweu.edu.cn (Z.M.); zhangjx@zjweu.edu.cn (J.Z.)
3 School of Electric Power, North China University of Water Resources and Hydropower, Zhengzhou 450045, China
4 School of Water Conservancy and Environment Engineering, Zhejiang University of Water Resources and Electric Power, Hangzhou 310018, China
5 Northwest Electric Power Design Institute of China Power Engineering Consulting Group, Xi'an 710000, China; fanjr@zjweu.edu.cn
* Correspondence: zhangyzh@zjweu.edu.cn

Abstract: As an important working condition in water conveyance projects, the water filling process of pipelines is a complex hydraulic transition process involving water–air two-phase flow with sharp pressure changes that can easily cause pipeline damage. In light of the complex water–air two-phase flow during pipeline water filling, this study explores the water filling process of right-angle elbow pressure pipelines using CFD numerical simulations and physical model experiments, analyzing changes in water phase volume fraction, water-gas two-phase flow patterns, and hydraulic parameters in the pipeline under low flow rate conditions of 0.6 m/s and high flow rate conditions of 1.5 m/s. Results show that under low flow rate conditions, there is more local trapped gas at the top of the pipeline, causing negative pressure at local high points in the pipeline and forming a vacuum. Under high velocity conditions, water-gas two-phase flow changes more frequently in the pipeline, with a large number of bubbles collapsing at the top, resulting in large fluctuations in pipeline pressure. Finally, through physical experiments, the main flow patterns during water filling in right-angle elbows are verified and analyzed. These results have certain reference significance for formulating safe and efficient water filling velocity schemes for pressurized pipelines.

Keywords: right-angle elbow; water filling process; water–air two-phase flow; numerical simulation; physical model experiment

Citation: Hu, J.; Wang, Q.; Zhang, Y.; Meng, Z.; Zhang, J.; Fan, J. Numerical and Experimental Study on the Process of Filling Water in Pressurized Water Pipeline. *Water* **2023**, *15*, 2508. https://doi.org/10.3390/w15142508

Academic Editors: Ran Tao, Changliang Ye, Kan Kan, Huixiang Chen and Yuan Zheng

Received: 4 May 2023
Revised: 27 June 2023
Accepted: 5 July 2023
Published: 9 July 2023

1. Introduction

Water filling of pressurized water pipelines is an important part of long-distance water transmission projects. It occurs when water pipelines are first put into operation, when empty pipes are filled with water, when water supply is restored after maintenance, and when water supply is restored after troubleshooting. The pipelines reach stable operation through water filling [1–3]. The water filling process is a complex and dynamic hydraulic process that involves the interaction of water, air, and pressure. The pressure in the pipeline fluctuates greatly, resulting in a complex and changeable water-gas two-phase flow pattern. This directly affects the safe operation of the water supply system [4]. Many water supply projects to be built have the characteristics of long pipelines, large flows, and great elevation fluctuations [5]. In order to ensure the safe and efficient operation of water supply projects, many scholars have conducted numerical simulations and experimental research on the water filling process of pressurized water pipelines.

In a numerical simulation of the water filling process, both one-dimensional Method of Characteristics (MOC) models and three-dimensional Computational Fluid Dynamics

(CFD) models are commonly used to calculate unsteady flow. Within the one-dimensional calculation model, the rigid water column model and weak water column model are frequently employed [6], and many researchers have improved and optimized these models. Liu et al. [7,8] proposed a complete rigid model that incorporates local head loss, valve opening, and pipeline elevation changes, taking into account pipeline characteristics and establishing a mathematical model with variable characteristics suitable for long pipelines. Zhou et al. [9] introduced two variables, air release rate and water column length change, to perform numerical simulations and experimental verification of rapid water filling in horizontal pipelines. The theoretical value of pressure oscillation agrees well with experimental data when there is no hole or a small hole. Jin et al. [10] treated the water column as elastic in the elastic water column model, incorporating the propagation process of pressure waves in the water flow, resulting in more accurate numerical simulations of velocity and pressure in the pipeline. Zhou et al. [11] added water elasticity and gas compressibility characteristics to derive and establish a mathematical model of the pipeline water filling process containing stagnant air mass, which can accurately simulate the transient pressure of air mass in the pipeline. Wang [12] used the interface tracing method to verify that the elastic water column model is more applicable than the rigid water column model, with more practical simulation results for the pipeline water filling process. Compared to one-dimensional models, three-dimensional models can intuitively display the flow state of gas and water phases. Currently, the Volume of Fluid (VOF) model is the most commonly used multiphase flow model in pipeline water filling processes [13]. Feng et al. [14] used a VOF model to simulate water-gas two-phase flow in a pressurized pipeline of a pumping station, with simulation results showing an obvious transition phenomenon between flow patterns during the water filling process. Zhou et al. [15] used a VOF method to simulate the water filling process of a pipeline containing an air mass without deflation, with results showing that bottom flow first caused a sudden change in pressure, confirming the limitations of one-dimensional models and the advanced nature of VOF models. Bai et al. [16] used a VOF model to perform three-dimensional numerical simulations of concave pipelines, with result analysis showing that gas-liquid velocity differences in pipeline sections caused water-gas mixing. Warda et al. [17] used three-dimensional numerical calculations based on a VOF model to visually study phenomena such as column separation and reconnection in pipelines.

In terms of physical model experiments of the water filling process, Ciro et al. conducted a water filling experiment using a pipeline with a 30° inclination and proposed a standard for predicting maximum pressure fluctuations during rapid water filling [18]. Balacco et al. studied the dynamic characteristics of air valves during water filling under different boundary conditions for various pipeline layouts with pipeline inclination angles of 11°, 22°, and 30°, respectively [19]. Vasconcelos and Wright found that intercepted air directly impacts the shape of the leading edge of the water filling column after conducting water filling tests on pipes with different diameters, and that hydraulic jumps may occur during water filling [20]. Hou et al. performed an experimental study on two-phase pressurization moving characteristics during the rapid water filling process of large pipelines and found that the front edge of the flowing water phase did not completely fill the pipeline cross-section, with water flow being stratified and mixed [21]. Patrick et al. [22] conducted an experimental study on the influence of entrained air on pressure wave velocity in pipelines during rapid water filling using transparent closed PVC pipes with three different diameters and three different flow rates. Chen et al. [23] established a pipeline water filling simulation test system with an adjustable pipeline inclination range (0~30°) to address the air plug problem in undulating pipelines, with results showing that pipelines exhibit different flow patterns and hydraulic characteristics under different pipeline inclination angles. Guo et al. [24] performed a water-filling model test on a pipe section of approximately 5 km, with results showing that under existing vent design conditions, the influence of trapped bubbles on water delivery capacity could be ignored.

In summary, although some progress has been made in research on the water filling process of pressurized water pipelines, related research is not sufficient. In terms of numerical simulation, research on one-dimensional models is relatively mature, while research on three-dimensional numerical simulations mainly targets specific working conditions, with less research on unsteady flow characteristics of pipeline water filling processes [25]. In physical model tests, most tests are conducted under conditions of small diameter and low inclination, but there is no in-depth study on water filling tests of large diameter and high inclination. In terms of water filling velocity, it is generally believed that the water filling velocity should be between 0.3~0.6 m/s in engineering [26]. Currently, only the water filling velocity is about 0.3 m/s [27], but this water filling velocity is too small, greatly slowing down project completion progress. In this study, we provide insights into the water filling process of right-angle elbow pressurized pipelines using numerical simulations and physical model experiments. The characteristics of water–air two-phase transient flow are analyzed under two working conditions, with inlet water velocities of 0.6 m/s and 1.5 m/s, respectively.

2. Numerical Simulation of the Water Filling Process of Pressurized Pipeline

2.1. Model Development

As shown in Figure 1, we develop a geometric model of a pipeline which consists of a horizontal section at the left side of the pipeline, L1 = 2.2 m, a vertical upward section, L2 = 0.2 m, an upper horizontal section, L3 = 1.5 m, a vertical downward section, L4 = 0.2 m, and a horizontal section at the right side of the pipeline, L5 = 2.2 m, in which the left end of the pipeline is the water inlet, the upper horizontal section is an air inlet, the right end of the pipe is the outlet, and the rest are solid walls. The diameter of the pipeline is d = 0.2 m. As shown in Figure 2, our monitoring points P1, P2, P3, and P4 are selected to analyze the time evolution of the pressure in the pipeline.

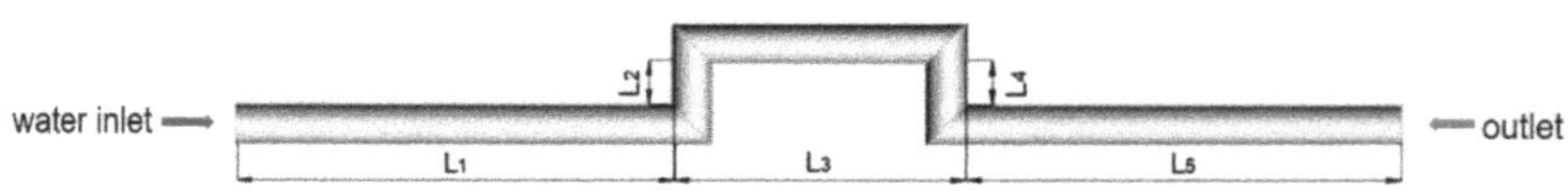

Figure 1. Schematic diagram of the numerical model.

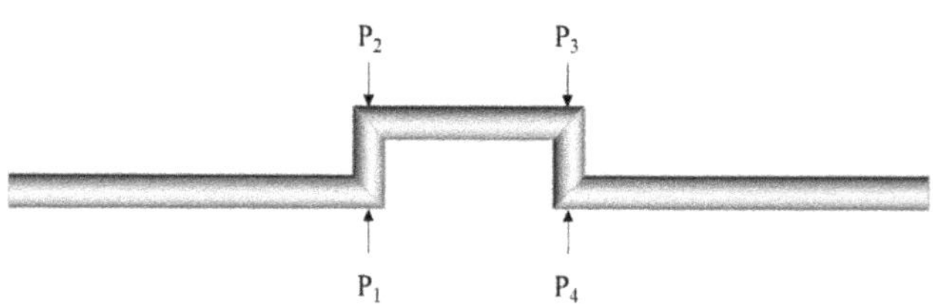

Figure 2. Layout of the monitoring points P1, P2, P3, and P4.

In this study, ICEM software is used to divide the grids. Since the model in the present study is a simple and regular three-dimensional cylindrical pipeline, the grid division is conducted using the O-shaped subdivision in ICEM and the grid's right-angle turning area is densified. The meshes of the numerical model is shown in Figure 3. In order to ensure the high accuracy of numerical simulation, the grid independence is verified by the pressure of P1 monitoring point under the steady condition of water inlet velocity of 0.6 m/s. The verification results of grid independence are shown in Figure 4. When the number of grids reaches 130,000, the data accuracy tends to be stable and the final number of grids is 137,802. The minimum value of grid orthogonal quality is 0.55.

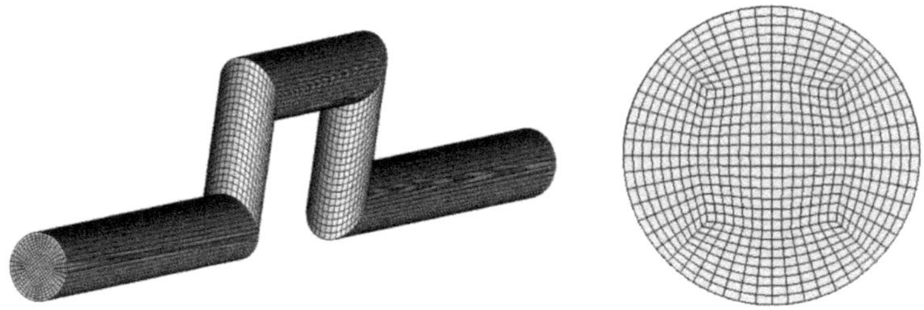

Figure 3. The meshes of the numerical model.

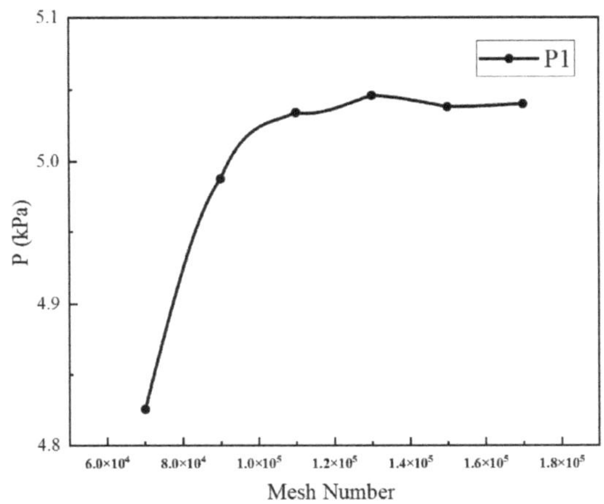

Figure 4. Grid independence verification.

For the water filling process, the inlet of the pipeline is the velocity inlet, the inlet velocity is constant and independent with time, and the velocity value is set according to the needs of different working conditions. The outlet of the pipeline is connected to the atmosphere, the outlet boundary condition is set as the pressure outlet, and the pressure is the atmospheric pressure $p = 1.01 \times 10^5$ pa. In the process of pipeline filling water, the inlet of air is not considered; we thus assume that the boundary of the air inlet is same as the side wall. The standard wall function method is adopted to calculate the side wall of the pipe and no slip boundary conditions are adopted. The pipe system is insulated and has no heat exchange with the fluid. At the beginning of water filling process, the water volume fraction in the calculation area is set to 0, the relative humidity of air is 50%. The volume fraction at the boundary of the inlet is set to 1, indicating that the fluid at the inlet is the physical parameters of the water and air phases under the temperature of 20 °C. The standard atmospheric pressure is shown in Table 1.

Table 1. Physical parameters of water and air.

Fluid	Density (kg/m^3)	Dynamic Viscosity (pa·s)	Surface Tension (N/m)
Water	998.2	1.003×10^{-3}	0.072
Air	1.2	1.79×10^{-5}	0.072

2.2. Mathmatical Details of the Numerical Simulation

For the water–air two-phase transient flow in a pressurized pipeline, the proportions of air in the pipeline affect the flow characteristics. At present, the water–air two-phase flow in pipelines mainly include horizontal pipe flow and vertical pipe flow.

The flow patterns in horizontal pipes are generally asymmetric due to the influence of gravity of fluids with different densities. This leads to a vertical stratification, which means that the water tends to occupy the lower part of the pipe and force the air to flow upward. Specifically, it can be divided into seven types: bubble flow, plug flow, stratified smooth flow, stratified undulate flow, slug flow, annular flow, and spray flow. Compared with the two-phase flow pattern in a horizontal pipeline, the flow pattern of the two-phase flow in a vertical pipe is more symmetrical. It can be divided into six types: bubble flow, plug flow, slug flow, agitation flow, annular flow, and spray flow.

We use the standard k–ε turbulence model to establish a mathematical model for solving the water filling problem of the water–air two-phase flow in a pressurized pipeline. The control equations are as follows:

1. Continuity equation

$$\frac{\partial \rho}{\partial t} + \nabla(\rho \mathrm{u}) = 0 \tag{1}$$

Since the water–air two-phase fluid exist simultaneously in each grid, the continuity equation can be expanded as:

$$\frac{\partial}{\partial t}(\alpha_g \rho_g) + \nabla\cdot(\alpha_g \rho_g u_g) = 0 \tag{2}$$

$$\frac{\partial}{\partial t}(\alpha_l \rho_l) + \nabla\cdot(\alpha_l \rho_l u_l) = 0 \tag{3}$$

$$\alpha_g + \alpha_l = 1 \tag{4}$$

2. Momentum equation

$$\frac{\partial(\rho u)}{\partial t} + \nabla(\rho \mathrm{uu}) = -\nabla \mathrm{p} + \nabla\left[\mu\left(\nabla \mathrm{u} + \nabla u^T\right)\right] + \rho g + F \tag{5}$$

3. Energy equation

$$\frac{\partial(\alpha_p T)}{\partial t} + \nabla\cdot(\alpha_p U T) = \nabla\cdot(\lambda \nabla T) + S_T \tag{6}$$

where α_l,α_g are the volume fractions of the liquid (water) and gas (air) phases; u_l, u_g are the velocities of the liquid phase (water) and gas phase (air), m/s; ρ_l, ρ_g are the densities of the liquid phase (water) and the gas phase (air), kg/m^3; t is time, s; μ is the dynamic viscosity coefficient; g is the gravitational acceleration, m/s^2; F is the volumetric force, N; p is the pressure, Pa; T is the temperature, °C; S_T is the source term.

By solving the continuity equation, the position of the interface of two phases and the proportions of water and gas in each grid can be obtained. The hydraulic parameters including the pressure and velocity of water and gas can be obtained by solving the momentum equation and energy equation.

4. Turbulence equation

Turbulent kinetic energy k equation:

$$\frac{\partial(\rho k)}{\partial t} + \frac{\partial(\rho k u_i)}{\partial x_i} = \frac{\partial}{\partial x_j}\left(\left(\mu + \frac{\mu_t}{\sigma_k}\right)\frac{\partial k}{\partial x_j}\right) + G_k + G_b - \alpha\varepsilon - Y_m + S_k \tag{7}$$

where σ_k is the turbulent Ludwig Prandtl number corresponding to turbulent kinetic energy k, which is taken by default as $\sigma_k = 1.0$; μ is the dynamic viscosity coefficient; G_k represents the generation term of turbulent kinetic energy k caused by the average velocity gradient; G_b represents the generation term of turbulent kinetic energy k caused by buoyancy; Y_m represents the effect of compressible turbulent pulsation expansion on the total dissipation rate; g is the gravitational acceleration, m/s^2; S_k is the source term.

The finite volume method, which has been commonly used to solve the governing equations of two-phase flow, is selected to simulate the water filling process of the pressurized pipeline. The transient hydraulic calculation is carried out based on the pressure transient solver. In order to ensure the accuracy of the calculation, the three-dimensional and double precision calculation mode is selected. The most widely used standard k-1 ε turbulence model is selected as the turbulence model. For the pressure velocity coupling method, we adopt the pressure implicit with splitting of operators (PISO) algorithm, which is proficient at solving transient problems. The gradient term is given based on the element volume least square method. The volume force weighting format is selected as the pressure interpolation format. The geo reconstruct format is selected for volume fraction interpolation, which can accurately track the calculation format of the interface of the water–air two-phase flow. The second-order upwind scheme is selected for the interpolation schemes of the continuity equation, momentum equation, turbulence equation, and energy equation. By verifying the grid-independent solution, we finally select 130,000 grids.

2.3. Results of the Numerical Simulation

Based on the three-dimensional numerical model that has been proposed in the above section, we select two working conditions with the inlet water flow velocity set to 0.6 m/s (named as case 1) and 1.5 m/s (named as case 2), respectively. The characteristics of the water–air two-phase transient flow in the pipeline are observed under these two working conditions.

We first determine the water–air two-phase transient flow state in the pipeline at varying time. Figures 5 and 6 illustrate the time variation of the water–air interaction under the working condition of case 1 and case 2, respectively. The red parts represent the water flow, the blue parts denote the air bag, and the green parts represent large air bubbles.

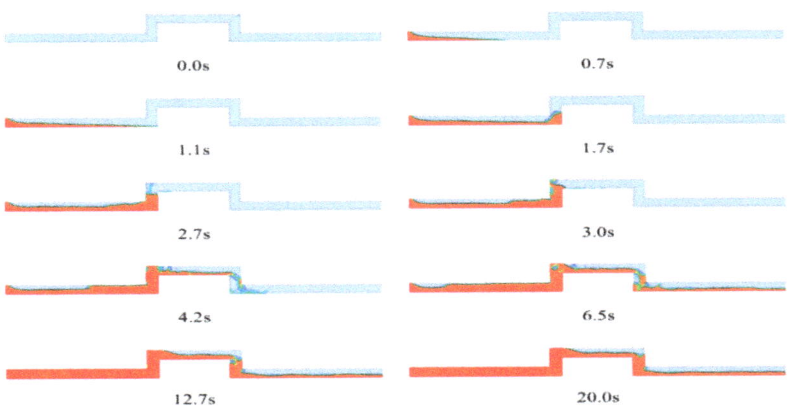

Figure 5. Time variation of the water–air interaction under the working condition of case 1 obtained from numerical simulation.

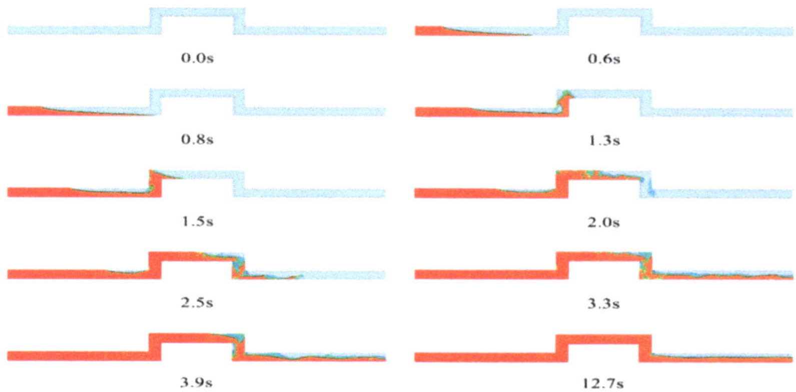

Figure 6. Time variation of the water–air interaction under the working condition of case 2 obtained from numerical simulation.

By comparing the water filling process under two different working conditions, it can be seen that the time required for water filling process decreases with the increase of water filling velocity. From the perspective of gas stagnation in the pipeline, with the increase of flow rate, the air retained in the pipeline decreases. When the speed reaches 1.5 m/s, except for the horizontal pipe section L5, all other sections of the pipelines are filled with water.

Water filling is a process in which the volume fraction of the water phase changes with time. Figure 7 shows the time evolution of the volume fraction of the water phase under case 1 and case 2. Equations (8) and (9) are fitting functions for the curve of water phase volume fraction with time.

$$y = \begin{cases} 0.089t & 0 \le t < 5.7 \\ 6 \times 10^{-5}t^3 + 0.0034t^2 + 0.0661t + 0.2398 & 5.7 \le t < 18.6 \\ 0.69 & t \ge 18.6 \end{cases} \quad (8)$$

$$y = \begin{cases} 0.149t & 0 \le t < 4.1 \\ 6 \times 10^{-5}t^3 + 0.0028t^2 + 0.0491t + 0.4656 & 4.1 \le t < 16.7 \\ 0.69 & t \ge 16.7 \end{cases} \quad (9)$$

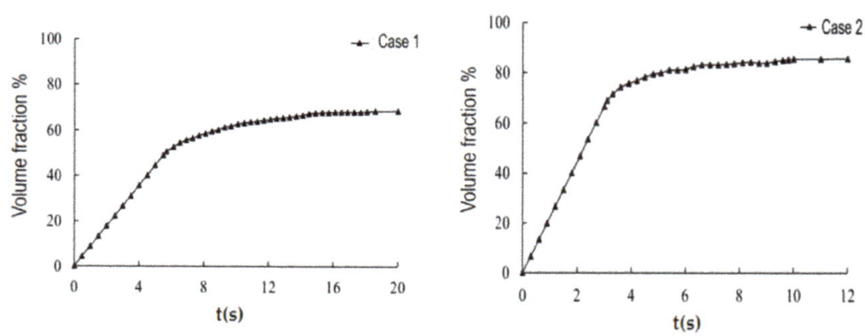

Figure 7. Time variation of the volume fraction of the water phase for case 1 and case 2.

In order to facilitate the observation of the flow pattern changes of water and gas at different times during the water filling process, the central sectional views of the pipeline with the same flow direction are shown in Figure 8, where the red parts denote the water phase and the blue parts represent the air phase. It can be seen from Figure 7 that the flow patterns in the pipeline water filling process under case 1 include stratified flow, plug flow, bubble flow, slug flow, and wavy flow. Due to the slow water filling flow rate, the mutual conversion of different flow patterns is also slow. The flow patterns in the process of pipeline water filling under case 2 include stratified flow, plug flow, bubble flow, slug flow, and wavy flow. In this case, the conversion of different flow patterns is very frequent.

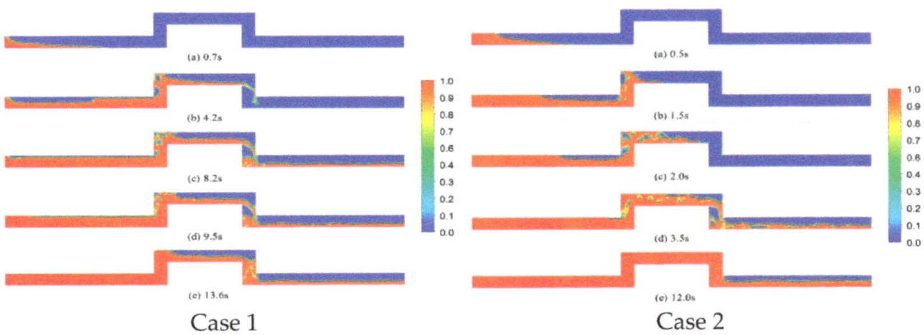

Figure 8. Water–air two-phase flow pattern of the pipeline.

According to the flow patterns of the water–air two-phase flow observed in the pipeline, we select the flow state of case 1 as an example and analyze the velocity field during the water filling process. Figure 9 is a velocity vector diagram of the two-dimensional plane flow field along the flow direction of the pipeline, and Figure 10 is a partial image of the flow field. The difference of the velocity vector arrow size in the velocity vector diagram shows the difference of its velocity. The larger the arrow, the greater the flow velocity. The dense arrows and the twisted streamline direction indicate that the area is chaotic. It is noted that the velocity field and streamline distribution of the stratified flow and wavy flow are uniform and close to linearity. As the sizes of bubbles in the bubble flow are relatively small, the motion of the air bubbles has a very small effect on the streamline and flow direction of the flow pattern. For plug flow, the oscillation of large air bubbles carried in the aqueous phase makes the streamline of the flow pattern wavy and distorted, and the velocity direction changes significantly. In the slug flow, as the air bag with a large volume fraction is retained in the flow pattern and the water phase overflow area is greatly reduced, the distribution of the streamline of the water phase is relatively dense. The gas phase in the air bag moves, driven by the water phase, and its streamline generates vortices. In addition, an obvious velocity gradient is observed at the intersection of water and air on the flow field vector diagram of the pipeline.

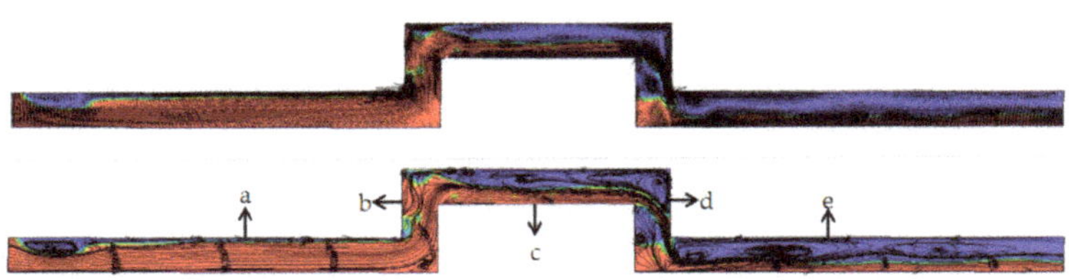

Figure 9. Pipeline velocity vector diagram (**upper**) and streamline diagram (**lower**).

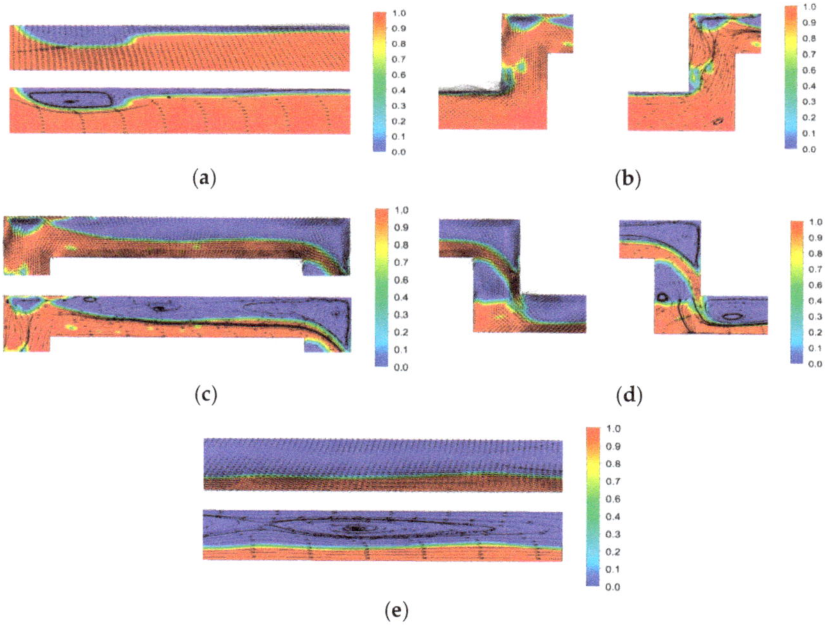

Figure 10. Local velocity vector diagram and streamline diagram. (**a**) a horizontal section at the left side, (**b**) a vertical upward section, (**c**) an upper horizontal section, (**d**) a vertical downward section, (**e**) a horizontal section at the right side.

Figure 11 illustrates the time evolution of the relative pressure at the indicated four monitoring points. At the beginning of water filling process, the air that remains in horizontal pipe (i.e., section L1 in Figure 1) affects the variation of pressure in the pipeline. The greater the amount of air, the more intense the pressure variation in the vertical upward pipe (i.e., section L2), and the greater the fluctuation range. In addition, the water filling velocity is an important factor affecting the pressure in the pipeline. By comparing the pressure under these two working conditions, it can be seen that a small water filling velocity causes negative pressure at the high points of the pipeline. The greater the water filling velocity, the higher the maximum pressure in the pipeline, and the greater the stable pressure when the water flow is relatively stable. However, the maximum and minimum pressure in the pipeline under the water filling flow rate conditions are within the pressure bearing range of the pipeline. Therefore, for the water transmission pipeline like the right-angle elbow model, in order to shorten the water filling cycle and reduce the retention of air at local high points after water filling, the water filling velocity can be increased under the condition of ensuring the safety of the pipeline.

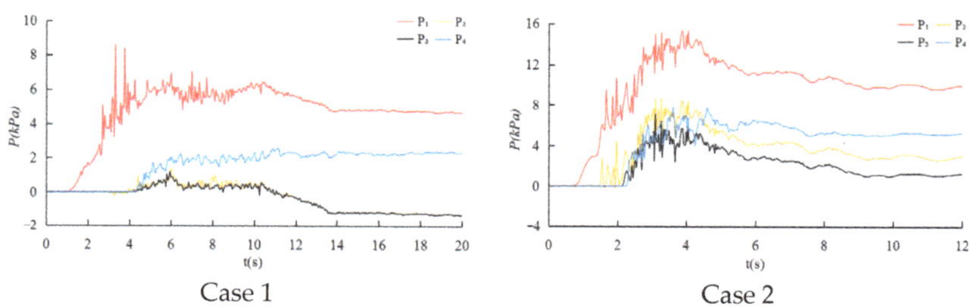

Figure 11. Time variation of the pressure at different monitoring points for case 1 and case 2.

3. Physical Model Experiments

3.1. Experimental Method

The water–air two-phase transient flow test system mainly includes a water supply module, air supply module, and data acquisition module. The experimental facilities are shown in Figure 12. The facilities are established to study complex two-phase flow in the pressurized pipeline in water supply systems. The water supply module is established to control the water filling velocity in the pipeline, where the water pump used in experiments is a horizontal pipeline centrifugal pump with variable frequency. For the air supply module, to provide different flow streams and ensure the control accuracy of the gas flow, three parallel pipelines are designed to connect gas flow meters with different ranges. One gas pipeline can be selected, the other two pipelines are closed, and the gas flow can be finely adjusted through the cooperation of the valve and flow meter. Pipelines with multi-inclination and multiple diameters are established. The water and air are mixed and transported into different types of two-phase flow by connecting the water supply and gas supply modules through multi-line pipelines.

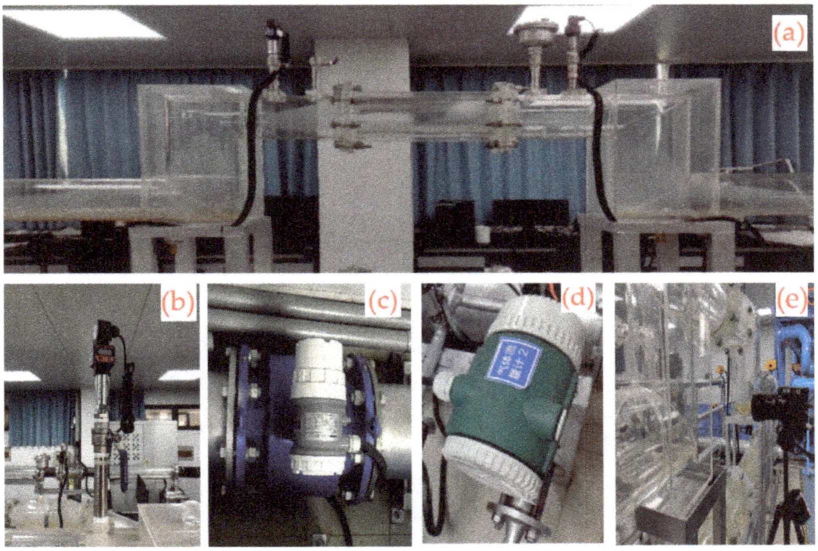

Figure 12. Experimental facilities: (**a**) pipeline, (**b**) the reassure measurement system, (**c**) the water flow measurement system, (**d**) the air flow measurement system, (**e**) high speed camera.

The data acquisition module consists of three parts: the reassure measurement system, flow measurement system, and image acquisition system. The reassure measurement system is shown in Figure 12b. Multiple sets of pressure measurement components are used to monitor the pressure in the pipeline in real time. Among them, the intelligent pressure transmitter with the model of hr3202 is used, with a range of 0~1.6 Mpa, accuracy of 0.5% FS, output of a 4~20 mA current signal, a power supply of 12–24 VCD, and a pressure sensor customized by Shaanxi Hengrui Measurement and Control System Company. For the water flow measurement system, the MGG / KL electromagnetic flowmeter is adopted. Its main parameters are accuracy ±0.3% R, velocity measurement range 0.1~15 m/s, velocity resolution 0.5 mm/s, which is customized by Jiangsu Chuanghui Braking Instrument Company. Its layout is shown in Figure 12c. For the air flow measurement system, the CH-LU20F vortex flowmeter is adopted. Its measuring range is 8~800 L/min, the accuracy is 1.5%, the output model is 4–20 mA, and the power supply is 24 V, which is customized by Jiangsu Chuanghui Braking Instrument Company. Its layout is shown in Figure 12d. For the image acquisition system, the high-speed high-definition camera FASTEC IL5 is used, and the electronic shutter speed of high-speed high-definition camera is 3 microseconds to 41.667 milliseconds, which can track the operation state of the gas–liquid two-phase flow test through image acquisition at a high acquisition frequency. The image acquisition is shown in Figure 12e.

3.2. Experimental Results

The experiments mimic the water filling process of a pressurized pipeline with a right-angle elbow with a diameter of DN200 and a water filling flow rate of 0.6 m/s and 1.5 m/s. When the water filling velocity at the inlet of the pipeline is 0.6 m/s, the water–gas two-phase flow pattern image of each part in the two-phase flow observation section is as shown in Figure 13.

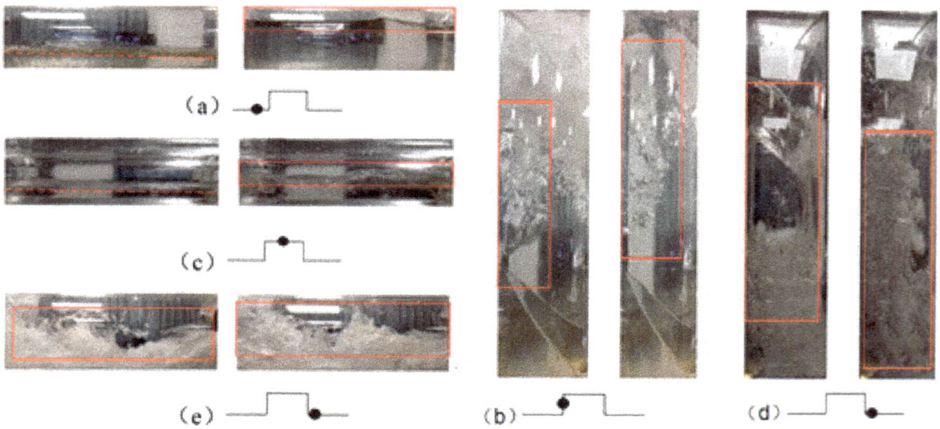

Figure 13. Two-phase flow pattern distribution of each pipe section (flow rate 0.6 m/s).

When the water filling velocity at the inlet of the pipeline is 1.5 m/s, the water–gas two-phase flow pattern image of each part in the two-phase flow observation section is as shown in Figure 14.

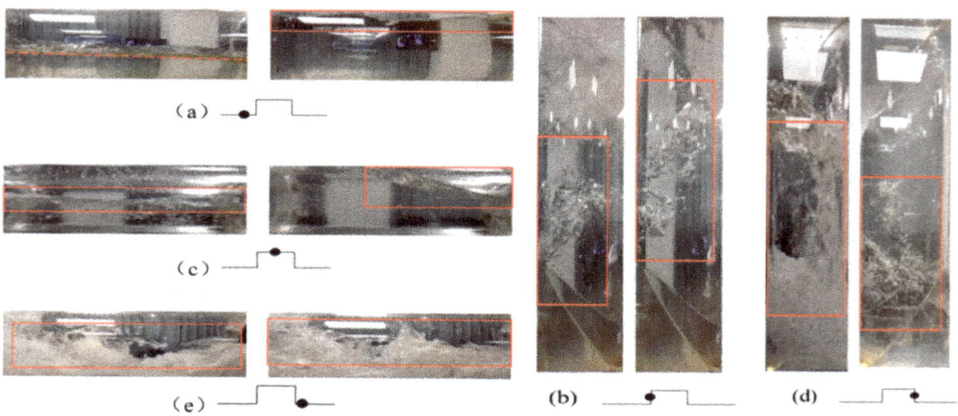

Figure 14. Two-phase flow pattern distribution of each pipe section (flow rate 1.5 m/s).

Figure 15 indicates the time variation of the pressure at the four indicated monitoring points with the water filling velocity at the inlet of the pipeline is 0.6 m/s and 1.5 m/s, respectively. It can be seen that the largest pressure fluctuation in the two working conditions appears at the monitoring point P1. When the flow rate is 0.6 m/s, the pressure fluctuation at P1 is significantly greater than that under the flow rate of 1.5 m/s. The overall pressure at P1 is significantly greater than the former. With a flow rate of 0.6 m/s, the time evolution curves of the pressure at monitoring points P2 and P3 fluctuate and rise at the very beginning, and then slowly decline. The varying tendencies of pressure are basically the same with different flow rates, and finally the pressure is basically stable in the state of negative pressure. Under the water filling condition with the flow rate of 1.5 m/s, the pressure curves at monitoring points P2 and P3 vary obviously, and there is no negative pressure.

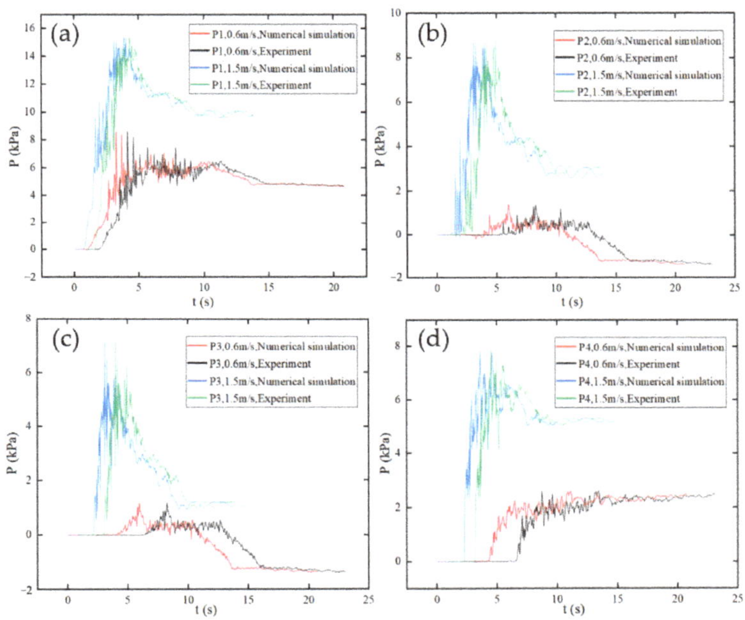

Figure 15. Time evolution of the pressure at the four indicated monitoring points: (**a**) P1 monitoring point, (**b**) P2 monitoring point, (**c**) P3 monitoring point, (**d**) P4 monitoring point.

3.3. Comparison of Experimental Results and Numerical Simulation Results

As shown in Table 2, for the two working conditions with flow rates of 0.6 m/s and 1.5 m/s, the water–air two-phase flow patterns obtained from physical model experiments are compared with numerical simulation results. By comparing the two-phase flow patterns, we observed stratified flow, wavy flow, plug flow, bubble flow, and slug flow both in numerical simulation and experiments, the appearance time and structure distribution of each flow pattern are very consistent. For the pressure change of each monitoring point during water filling, it can be seen that the experimental results obviously lag behind the numerical simulation, which is caused by ignoring the friction resistance of the pipeline in the numerical simulation process. The overall pressure change law is very similar, and the pressure value of the experimental results is slightly higher than that of the numerical simulation. Additionally, with the increase in the water filling velocity, the interaction between the water and gas is more intense and the flow pattern changes more frequently.

Table 2. Comparison of numerical and experimental results under different working conditions.

Condition	Inlet Velocity (m/s)	Calculated Pressure (kPa)	Measured Pressure (kPa)	Error (%)
Water filling	Velocity 0.6 m/s	8.60	8.48	1.39
Water filling	Velocity 1.5 m/s	15.37	14.86	3.32

4. Conclusions

This study provides insights into the water filling process of right-angle elbow pressurized pipelines using numerical simulations and physical model experiments. The characteristics of water–air two-phase transient flow are analyzed under two working conditions with inlet water velocities of 0.6 m/s and 1.5 m/s, respectively. The main conclusions and prospects are as follows:

(1) Under low flow rate conditions, there is more local trapped gas at the top of the pipeline, causing negative pressure at local high points in the pipeline. Under high velocity conditions, there is no gas stagnation at local high points in the pipeline, with a large number of bubbles collapsing at the top of the pipeline, causing large fluctuations in pipeline pressure. Therefore, in practical engineering, air valves should be installed at local high points in water pipelines to not only discharge trapped gas but also allow gas to enter when there is negative pressure in the pipeline.

(2) The main flow patterns during the water filling process of right-angle elbow pipelines include stratified flow, slug flow, bubble flow, plug flow, and wavy flow. The larger the water filling velocity, the more frequent the conversion of water–air two-phase flow patterns in the pipeline. Pipeline pressure changes violently due to slug flow.

(3) The pipeline water filling process can be divided into two stages. In the first stage, flow patterns and pressure in the pipeline change dramatically, while in the second stage, flow patterns and pressure in the pipeline gradually stabilize. Based on this principle, a phased water filling method can be adopted in practical engineering to effectively reduce pressure peaks and shorten water filling time.

Author Contributions: Conceptualization, J.H. and Y.Z.; methodology, J.H., Q.W. and J.Z.; software, J.H., Q.W. and Z.M.; validation, Z.M.; formal analysis, J.H.; investigation, J.H.; resources, J.H.; data curation, J.Z.; writing—original draft, J.H. and Y.Z.; writing—review and editing, J.H., Q.W. and Z.M. visualization, J.F.; supervision, Q.W.; project administration, J.H.; funding acquisition, J.H. All authors have read and agreed to the published version of the manuscript.

Funding: This work is supported by the Key Joint Funds of the Zhejiang Provincial Natural Science Foundation of China (LZJWZ22E090004).

Data Availability Statement: The numerical simulation data and experimental data used to support the findings of this study are available from the corresponding author upon request.

Acknowledgments: We are very grateful to anonymous reviewers for their constructive comments and suggestions.

Conflicts of Interest: The authors declare no conflict of interest.

References

1. Guo, Y.X.; Wu, H.T.; Yang, K.; Guo, X.L.; Wang, T. Analysis of Dynamic Characteristics of Gas-liquid Flow During Water-filling Process in Pipeline. *S-N Water Transf. Water Sci. Tech.* **2013**, *11*, 65–69.
2. Li, L.; Zhu, D.Z.; Huang, B. Analysis of Pressure Transient Following Rapid Filling of a Vented Horizontal Pipe. *Water* **2018**, *10*, 1698. [CrossRef]
3. Zhang, J.; Zhu, X.Q.; Qu, X.H.; Ma, S.B. Arrangement of air-valve for water hammer protection in long-distance pipelines. *J. Hydraul. Eng.* **2011**, *39*, 1025–1033. [CrossRef]
4. Liu, J.; Xu, D.; Zhang, S.; Bai, M. Experimental and Numerical Study on Water Filling and Air Expelling Process in a Pipe with Multiple Air Valves under Water Slow Filling Condition. *Water* **2019**, *11*, 2511. [CrossRef]

5. Shi, L.; Zhang, J.; Ni, W.X.; Chen, X.Y.; Li, M. Water Hammer Protection of Long-Distance Water Supply Project with Special Terrain Conditions. *J. Hydraulic. Eng.* **2019**, *38*, 81–88. [CrossRef]
6. Wang, F.J.; Wang, L. Research progress of transient flow in water filling process of large pipeline water conveyance system. *J. Hydraulic. Eng.* **2017**, *36*, 12. [CrossRef]
7. Liu, D.Y.; Suo, L.S. Rigid mathematical model of air mass trapped in pipeline impacted by water flow. *Adv. Water Sci.* **2004**, *15*, 6. [CrossRef]
8. Liu, D.Y.; Suo, L.S. Rigid mathematical model of water flow impacting air mass in long pipeline with variable characteristics. *Res. Pro. Hydrodyn.* **2005**, *20*, 6. [CrossRef]
9. Zhou, L.; Liu, D.Y.; Karney, B.; Zhang, Q.F. Influence of Entrapped Air Pockets on Hydraulic Transients in Water Pipelines. *J. Hydraul. Eng.* **2011**, *137*, 1686–1692. [CrossRef]
10. Jin, Z.; Jiang, N.C.; Wang, X.H. *Pump Stop Water Hammer and Its Protection*, 2nd ed.; China Building Industry Press: Beijing, China, 2004; ISBN 978-71-1206-274-4.
11. Zhou, L.; Liu, J.; Huang, K.; Liu, D.Y. Numerical simulation of transient flow with trapped air mass during start-up and filling of water pipeline. *Adv. Sci. Tech. Water Resour.* **2021**, *41*, 1–7.
12. Wang, L. Numerical Study on Transient Flow of Rapid Water Filling in Pipeline. Ph.D. Thesis, China Agricultural University, Beijing, China, 2017.
13. Liu, B. *Fluent 19.0 Fluid Simulation from Entry to Mastery*; Tsinghua University Press: Beijing, China, 2019; ISBN 978-73-0252-575-2.
14. Feng, L.; Yao, Q.Y. Numerical simulation of gas-liquid two-phase flow in pressure pipeline of pumping station based on VOF model. *China Rural Water Hydropower* **2012**, *12*, 124–126.
15. Zhou, L.; Liu, D.Y.; Chuan, Q. Simulation of Flow Transients in a Water Filling Pipe Containing Entrapped Air Pocket with VOF Model. *Eng. Appl. Compu. Flu. Mech.* **2014**, *5*, 127–140. [CrossRef]
16. Bai, R.Y.; Bao, J.W.; Song, L. Transient numerical simulation analysis of gas-liquid two-phase flow in municipal water supply pipeline under water filling and exhaust conditions. *J. Inner. Mong. Uni. Tech.* **2017**, *36*, 9. [CrossRef]
17. Warda, H.A.; Wahba, E.M.; Salah El-Din, M. Computational Fluid Dynamics (CFD) simulation of liquid column separation in pipe transients. *Ale. Eng. J.* **2020**, *59*, 3451–3462. [CrossRef]
18. Apollonio, G.; Balacco, G.; Fontana, N.; Giugni, M.; Marini, G.; Piccinni, A.F. Hydraulic Transients Caused by Air Expulsion During Rapid Filling of Undulating Pipelines. *Water* **2016**, *8*, 25. [CrossRef]
19. Balacco, G.; Apollonio, C.; Piccinni, A.F. Experimental analysis of air valve behaviour during hydraulic transients. *J. Appl. Water Engi. Res.* **2015**, *3*, 1–3. [CrossRef]
20. Vasconcelos, J.; Wright, S. Investigation of rapid filling of poorly ventilated stormwater storage tunnels. *J. Hydra. Res.* **2009**, *47*, 547–558. [CrossRef]
21. Hou, Q.; Tijsseling, A.S.; Laanearu, J.; Annus, I.; Koppel, T.; Bergant, A.; Vučković, J.; Anderson, A.; Jos, M.C. Experimental investigation on rapid filling of a large–scale pipeline. *J. Eind. Uni. Tech.* **2013**, *1335*, 1–27. [CrossRef]
22. Patrick, A.C.; Vasconcelos, J.G. Air Entrainment Effects on the Pressure Wave Celerities Following Rapid Filling Pipe Flows. In Proceedings of the World Environmental and Water Resources Congress, Austin, TX, USA, 17—21 May 2015; pp. 1638–1647. [CrossRef]
23. Chen, Y.Y.; Chui, Y.Y.; Yu, D.; Gong, J. Loop system for simulation test of water filling and exhaust of pipeline containing gas plug. *Res. Explor. Labor* **2019**, *38*, 42–46. [CrossRef]
24. Guo, Y.X.; Yang, K.L.; Guo, X.L.; Fu, H. Influence of retained bubbles on water delivery capacity during water filling of large pipeline water delivery system. *J. Hydra. Eng.* **2013**, *44*, 262–267. [CrossRef]
25. Du, Y.Z.; Chen, Y.C.; Li, W.Q.; Liu, D.C.; Song, T.N. Introduction of Water Passing Test of Mopanshan Water Transfer Project in Harbin. *Water Supply Drain.* **2009**, *14*, 12–14. [CrossRef]
26. American Water Works Association. *Concrete Pressure Pipe: M9—1995*; American Water Works Association: New York, NY, USA, 1995.
27. Wang, L.; Wang, F.J.; Huang, J. Numerical investigation of filling transients in small-scale pipelines with submerged outlet. *J. Hydrodyn.* **2019**, *31*, 1–7. [CrossRef]

water

MDPI

Article

Study on the Transient Flow Characteristics of a Hump Water Pipeline Based on the Random Distribution of Bubbles

Qingbo Wang [1,2], Jianyong Hu [1,3,*], Mingming Song [1,3], Hui Shen [1,4], Yu Zhou [5], Dongfeng Li [5] and Feng Xie [5]

1 School of Geomatics and Municipal Engineering, Zhejiang University of Water Resources and Electric Power, Hangzhou 310018, China; wqb5113@163.com (Q.W.); songmm@zjweu.edu.cn (M.S.); huishen0807@163.com (H.S.)
2 School of Electric Power, North China University of Water Resources and Hydropower, Zhengzhou 450045, China
3 Engineering Research Center of Digital Twin Basin of Zhejiang Province, Hangzhou 310018, China
4 School of Water Resources and Hydropower, Hebei University of Engineering, Handan 056038, China
5 School of Water Conservancy & Environment Engineering, Zhejiang University of Water Resources and Electric Power, Hangzhou 310018, China; zhouy@zjweu.edu.cn (Y.Z.); lidf@zjweu.edu.cn (D.L.); xief@zjweu.edu.cn (F.X.)
* Correspondence: hujy@zjweu.edu.cn

Abstract: Aiming at establishing the transient flow characteristics of gas–liquid two-phase flow in high-undulation water pipelines, based on the bubble distribution law measured using physical tests, the bubble distribution law function was input into the hump-pipe fluid domain model, and CFD numerical simulation was carried out for different flow rates and different air contents. The CLSVOF two-phase flow model and the RNG k-ε turbulence model were used to analyze the flow pattern evolution and pressure pulsation propagation in the process of gas–liquid two-phase flow through a hump pipe. The results show that the bubble size has a lognormal distribution, the equivalent diameter is between 3 mm and 10 mm, and the evolution of the flow pattern in the hump pipe is complex and violent. In the horizontal pipe section, there are three main flow patterns: bubble flow, wavy flow and segment plug flow. In the vertical pipe, there are two main flow patterns, slug flow and churning flow, and the flow pattern is affected by the flow rate and the air content rate. When air bubbles or air pockets in the pipeline flow through a certain area, this leads to a steep increase and decrease in the pressure pulsation amplitude in the region, and the pressure fluctuation is extremely frequent. Compared with the water flow rate, the air content is the main factor affecting the relative pressure pulsation amplitude under the condition of a 0.15-air content operating mode, which is generally approximately two to six times that of the 0-air content operating mode. The results of the research should facilitate the prediction of stagnant gas pipeline system bursts and water hammer protection, providing a theoretical basis and calculation parameters.

Keywords: high-undulation water pipeline; gas–liquid two-phase flow; random distribution of bubbles; flow pattern evolution; pressure pulsation

Citation: Wang, Q.; Hu, J.; Song, M.; Shen, H.; Zhou, Y.; Li, D.; Xie, F. Study on the Transient Flow Characteristics of a Hump Water Pipeline Based on the Random Distribution of Bubbles. *Water* **2023**, *15*, 3831. https://doi.org/10.3390/w15213831

Academic Editors: Ran Tao, Changliang Ye, Kan Kan, Huixiang Chen and Yuan Zheng

Received: 17 September 2023
Revised: 27 October 2023
Accepted: 31 October 2023
Published: 2 November 2023

1. Introduction

Large-scale water-resource allocation projects often have the characteristics of large fluctuations and many operating conditions, their operating safety requirements are extremely high, and stagnant gas poses a threat to the safe operation of the project, representing one of the major potential risks [1]. In pressurized water pipelines, pumps and other mechanical equipment, bubbles can reach 5–10% of the volume of water, and undulating sections of pipelines usually have a higher air content, which accumulates to form stagnant air pockets [2,3]. Stagnant air pockets not only reduce the effective water cross-section, but also reduce the system's water transfer capacity, while at the end of the pipeline, if a valve closes quickly when stagnant gas is present, the pipe may burst, leading to water supply interruption and other serious accidents [4].

The motion characteristics of bubbles with water flow are key to studying the mechanism of stagnant gas in complex water-resource pressurized pipeline transportation systems [5]. In order to reduce the risk of pipe bursts in pipeline stagnation, many scholars have conducted a lot of research based on the bubble distribution model. In the bubble distribution model, using the gas–liquid two-phase transient flow in pressurized pipelines from the 1960s, there are two main computational models: the bubble uniform distribution model and the bubble discrete model [6]. Kranenburg et al. [7] proposed a bubble uniform distribution model in which bubbles are uniformly distributed throughout the entire region and the pressure pipeline liquid flow is divided into three regions. In the transient flow calculation process, the water-hammer wave velocity should be calculated according to the change in pressure and cavitation rate. Brown et al. [8] proposed a discrete model which considers that the bubbles are centrally distributed in each calculation section of the pipeline, the expansion law of each bubble with the change in pressure conforms to the perfect gas equation of state, there is no gas in the pipeline liquid between the two bubble cross-sections and the water-hammer wave velocity is constant. However, in reality, the distribution of bubbles in the pipeline is based on random distribution, so the use of discrete and uniform distribution models based on centralized distribution for calculation is bound to cause large errors in the calculation results [9,10]. Yang et al. [11,12] concluded that air is non-uniformly distributed in the pipeline, the position of bubbles is a variable that changes with time, only a portion of the pipeline has bubbles at any instant, and the pipeline with bubbles has a larger cavitation rate. In the physical test of bubble distribution, Tokuhiro [13] and others obtained the required flow field parameters using a CCD camera and workstation processing under laser irradiation and measured the trajectory and morphology change rule of bubbles. SRA et al. [14] photographed and processed the stagnant gas phenomenon in the pipeline and established the generation law and movement state of stagnant gas in the pressurized pipeline. Wan et al. [15] designed a physical test model with several consecutive bends and concluded through a large number of tests that an appropriate reduction in the inclination angle of the pipeline and an increase in the flow rate can effectively discharge the stagnant gas mass in the pipeline. Jansson et al. [16] studied a pipeline water hammer with column separation in a flow range through physical tests. The results showed that there was no obvious separation interface; the boundary was composed of dispersed small bubbles mixed with a larger steam structure, in which the bubbles seemed to become smaller after each rupture. Urbanowicz et al. [17] modified the discrete bubble cavity model (DBCM); the comparisons between computed and measured results showed that the influence of delayed strain is far greater than unsteady friction on pressure wave damping. Based on physical tests, He et al. [18] discussed the related changes in metal surface morphology through weightlessness and the cavitation erosion rate; the results showed that a reasonable surface morphology can inhibit cavitation erosion. Veisi et al. [19] designed a metering system for measuring liquid voidage and verified the accuracy of the system through the physical test of water–air two-phase annular flow. By analyzing the probability density function of dynamic pressure and the results of fast Fourier transform analysis, Khan et al. [20] proposed a new method to identify the flow pattern in horizontal pipes.

Combinations of CFD numerical simulation and physical tests are often used to study water–gas two-phase flow patterns [21]. According to the appearance of the shape of the fluid and the distribution characteristics of the gas phase, the horizontal pipeline gas–liquid two-phase flow pattern can be divided into six kinds: bubble flow, plug flow, laminar flow, wavy flow, slug flow and annular flow. The vertical pipeline gas–liquid two-phase flow pattern can be divided into five kinds: bubble flow, slug flow, churning flow, annular flow and thin-beam annular flow. A flow category schematic diagram is shown in Figure 1 [22,23].

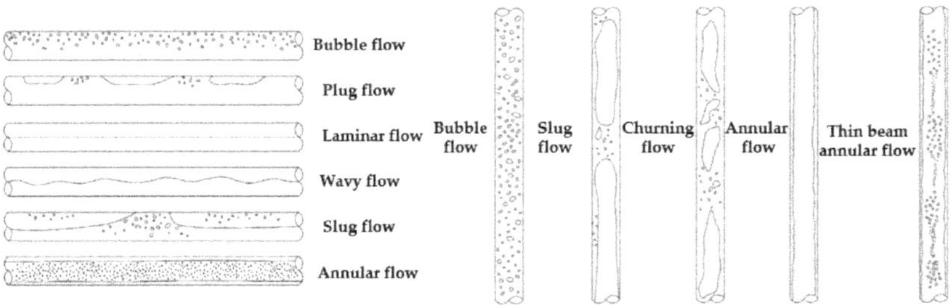

Figure 1. Water–air flow pattern categories from [22,23].

The water–gas two-phase flow pattern in the pipeline is mainly affected by the water–gas-phase flow rate, fluid physical parameters, pipe diameter and inclination and other factors, and for undulating pipelines, common stagnant parts mainly include the raised point in the pipeline, the horizontal pipe section and the downward inclined pipe section [24,25]. Based on the experimental test and numerical analysis, considering the basic dynamics of air–water interaction in unventilated pipes, Ramos, H.M. et al. [26] predicted air exchange through any installed valve and their roles in changing the system behavior during unsteady flow. Feng et al. [27] used the VOF two-phase flow model to conduct transient simulations of gas–liquid flow in a pipeline, explored the changes between flow patterns, and found that the water-phase integral number of the pipeline undergoes three periods of linear growth, irregular growth and no further growth due to the water-phase integral number of the pipeline. Wang et al. [28] performed numerical simulations and physical tests on a 90-degree gas–liquid elbow based on the VOF model and explored the effects of the flow pattern evolution on the pressure distribution, velocity distribution and air content of the cross-section. Compared with the VOF model alone, Yu et al. [29] used the VOSET method to simulate the characteristics of the gas–liquid flow pattern in a vertically ascending circular pipe, and the results show that the VOSET method not only maintains the conservation of the mass of the gas–liquid flow, but also discretizes the surface tension more efficiently, and the simulation results are more accurate and effective. Bourlioux et al. [30] proposed a CLSVOF (Coupled Level Set and Volume of Fluid) method, which solved the problems of mass non-conservation in the transport process of the Level Set method and discontinuity at the phase interface of the VOF model. Yang [31] used the CLSVOF method to study the effects of different flow rates, inclination angles, pipe diameters and other comprehensive factors on the critical flow rate of bubble initiation in gas–liquid two-phase flow, and the empirical formulas for the calculation of the critical flow rate of bubble initiation in three kinds of pipes, namely, horizontal, inclined and vertical pipes, were obtained after computational analysis and were completely matched with the physical experiments. Shang et al. [32] used a CLSVOF model to calculate the critical flow rate for gas–liquid two-phase flow in a vertical pipe. A numerical simulation of gas–liquid two-phase flow was carried out, and more accurate data than the VOF model were obtained by comparing them with physical tests. Tang et al. [33] studied the cavitation flow in the pipeline based on the computational fluid dynamics (CFD) method of Fluent 2020R2 software and successfully captured the formation, development and collapse process of cavitation bubbles.

In summary, the distribution of bubbles in pressurized water pipelines, the bubble morphology and pipeline layout, the water flow rate and other dynamic and static multiparameter coupling relationships reflecting the complexity of the pipeline system containing gas flow lead to bubble aggregation, airbag retention and air mass generation. The lack of fine experimental observations of the risk of stagnant gas in different sections of pipes demonstrates that the dynamic identification of these risks is difficult.

The current research on water–liquid two-phase flow in pipelines is mainly carried out under the assumption that bubbles are discretely and uniformly distributed, while in fact bubbles are randomly distributed in the pipeline, and such assumptions will inevitably lead to large errors in the calculation results. In addition, the gas–liquid flow through a highly undulating pipe is a complex transient process, and so it is necessary to consider a larger range of water–gas two-phase volume fractions so as to explore the flow pattern evolution and pressure pulsation propagation law in the pipe.

Therefore, a vertically undulating hump pipe is taken as the research object in this paper; firstly, the bubble random distribution model of the pressurized pipeline is established through physical tests, and then, based on the CLSVOF model and the RNG k-ε turbulence model, the typical flow rate and air content as physical variables are used for physical tests and numerical simulations so as to reveal the effect of bubble random distribution on the flow pattern and the pressure pulsation in the long pressurized pipeline system. The research results can provide theoretical support for the risk analysis of stagnant gas bursting in pressurized pipeline systems, improve the theory of the two-phase transient flow of water and gas in pressurized pipelines, and provide computational parameters for the safe and stable operation of the project.

2. Physical Tests of Gas–Liquid Two-Phase Pipe Flow

2.1. Test Platform

Figure 2 shows the system layout of the water–air flow test platform. The test system is mainly composed of a water supply system, a gas supply system and a data acquisition system.

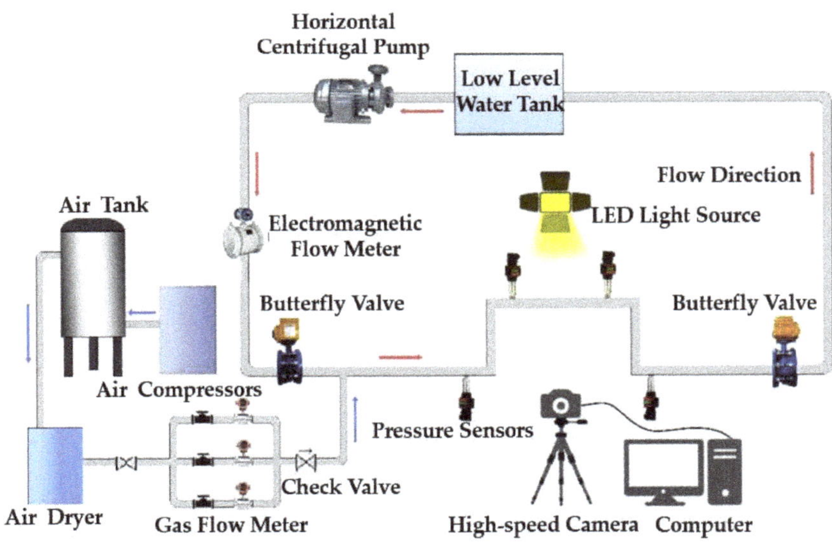

Figure 2. Layout of the water–air transient flow test system.

The water supply module mainly consisted of a low-level water tank, a horizontal centrifugal pump, pressurized pipelines and an electric control valve. The test selected the pressurized flow water supply mode by adjusting the centrifugal pump frequency to realize different flow rate operating modes. The characteristic parameters of the pump are shown in Table 1, and physical pictures of the relevant devices of the water supply module are shown in Figure 3.

Table 1. Characteristic parameters of the horizontal centrifugal pump.

Model Number	Design Flow Rate (m^3/h)	Design Head (m)	Rated Speed (r/min)	Motor Power (kW)	Cavitation Allowance (m)
QZHW200-250IA Pipeline centrifugal pump	358	14	1450	22	4

Figure 3. Physical pictures of the water supply module device: (**a**) QZHW200-250IA centrifugal pump; (**b**) motorized butterfly valve; (**c**) motorized regulating valve; (**d**) low-level water tank.

As shown in Figure 4, the gas supply module mainly consisted of an air storage tank, an air compressor and a vortex gas flow meter. The volume of the air storage tank was 600 L, and the maximum pressure value was 1 MPa. The maximum pressure of the air compressor was 8~10 bar, the maximum exhaust volume was 1000 L/min and air cooling was used. In order to improve the input gas flow accuracy control requirements, three groups of parallel pipelines were set up, and each group of pipelines were set up with a valve and a vortex gas flow meter, which could be used to control the gas flow rate according to the selection of different gas delivery pipelines and the setting of valve openings to control the size of the gas mass in the input pipeline.

Figure 4. Physical picture of the gas supply module unit: (**a**) gas storage tank; (**b**) air compressor; (**c**) vortex gas flow meter.

As shown in Figure 5, the data acquisition module was mainly composed of a DN100 hump pipe, an electromagnetic flowmeter, a pressure sensor, a high-speed camera image acquisition system and a complementary light meter. The DN100 hump-pipe section consisted of transparent plexiglass. The electromagnetic flowmeter used an MGG/KL type electromagnetic sensor to monitor the volume flow of water in the pipeline (with a flow rate measurement range of 0~15 m/s, a basic error of ±3% R (R represents rounding error) and a flow rate accuracy of 0.5%); in order to prevent the air mass in the pipeline from interfering with the accurate reading of the electromagnetic flowmeter, the electromagnetic flowmeter was located 10 m in front of the gas injection point. The model of pressure sensor is HR3202, the sensing range is −0.1~1.0 MPa, the accuracy is 0.5% FS (full scale) and the output current signal is 4~20 mA; this type of pressure sensor can continuously and accurately measure the pressure of gas, liquid and steam and has the advantages of high measurement accuracy and good working stability. The high-speed camera model FASTEC IL5, with a built-in 12-bit CMOS sensor and with Full SXGA 1024P shooting, could reach 668FPS, with a high-speed camera shutter speed of 3 μs to 41.654 ms, allowing image acquisition at high acquisition frequencies. In this study, the combination of a Pixel-p45c supplementary light lamp and an ordinary LED lamp was used to provide the required brightness for high-speed cameras. The color temperature of the Pixel-p45c supplementary light lamp could be adjusted between 3000 K and 5400 K, providing high color rendering and stability.

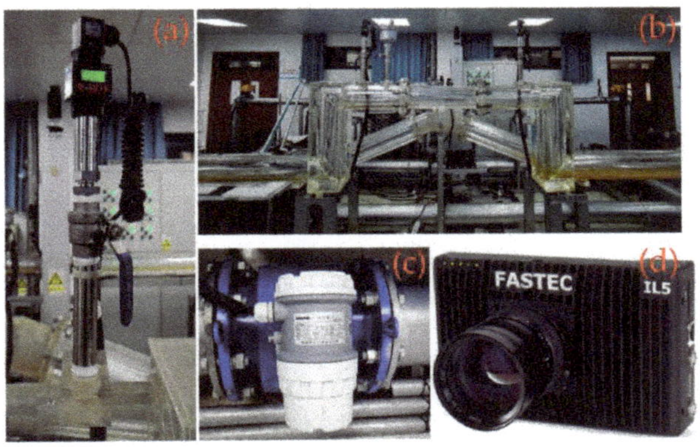

Figure 5. Physical picture of the air supply module device: (**a**) pressure sensor; (**b**) DN100 hump pipe; (**c**) electromagnetic flow meter; (**d**) high-speed camera.

2.2. Test Program

The test mainly aimed to study the random distribution of bubbles in the pump pressurized aqueduct and its effects on the two-phase transient flow characteristics of water and gas for different air contents and different flow rates within a DN100 right-angle elbow test section for two-phase water and gas transient flow research. First of all, the image acquisition and data acquisition systems were used to record the results for the random distribution of bubbles in the test section, providing the experimental basis for the model for the numerical simulation of the random distribution of bubbles. Then, we identified and classified the changes in the two-phase flow pattern and analyzed the effects of the random distribution of bubbles in the pressurized pipe on the water transfer process according to the measured pressure signal.

In engineering applications, the water filling rate is generally between 0.6 and 3.0 m/s, while the pressurized aqueduct system air content should be 5–10%, and undulating sections of pipes usually have a higher air content [2,34]. Therefore, three inlet flow rates of 1.0 m/s, 1.5 m/s and 2.0 m/s were selected for the test program, and four air content operating modes were chosen, namely, air-phase volume fractions of 0 (no air), 0.05, 0.10 and 0.15, and the specific operating modes of the test are shown in Table 2.

Table 2. Calculation operating mode table.

Operating Mode	Inlet Flow Rate (m/s)	Gas-Phase Volume Fraction	Outlet Absolute Pressure (Pa)
Case 1	1.0	0	101,325
Case 2	1.0	0.05	101,325
Case 3	1.0	0.10	101,325
Case 4	1.0	0.15	101,325
Case 5	1.5	0	101,325
Case 6	1.5	0.05	101,325
Case 7	1.5	0.10	101,325
Case 8	1.5	0.15	101,325
Case 9	2.0	0	101,325
Case 10	2.0	0.05	101,325
Case 11	2.0	0.10	101,325
Case 12	2.0	0.15	101,325

After the test of one working condition was finished, we slowly increased the frequency of the water pump to 15 Hz and flushed the residual bubbles of the previous working condition out of the test pipe section. We repeated the same working condition 3 times to ensure the correctness of the test and then carried out the test under other working conditions.

2.3. *Image Analysis of the Random Distribution of Bubbles*

As shown in Figure 6, in order to obtain the accurate random distribution of bubbles, the image acquisition system and data acquisition system were used to observe the morphology, size and distribution of bubbles with different air contents in the horizontal pipe section through the image processing methods of image gray scaling, image binarization and hole filling. It was concluded that the bubble size follows a lognormal distribution law; that the equivalent diameter is between 3 mm and 10 mm; that the higher the air content, the more obvious the phenomenon of small bubbles aggregating to generate large bubbles; that the distribution of bubbles is random; and that the relationship between the bubble size and the distribution of bubbles is random. The results for the distribution of bubbles are shown in Figure 7.

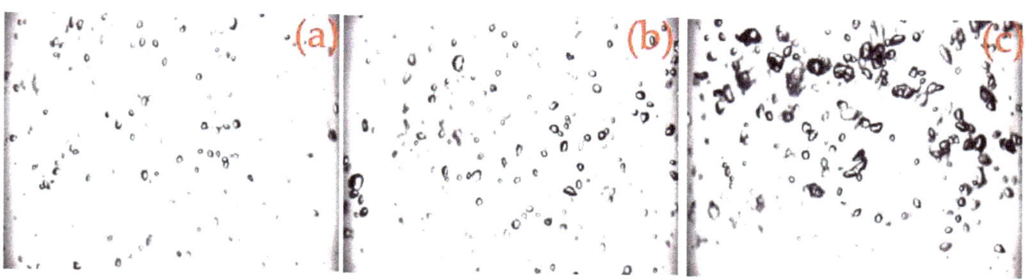

Figure 6. Distribution of bubbles with different air contents: (**a**) 0.05 air content; (**b**) 0.1 air content; (**c**) 0.15 air content.

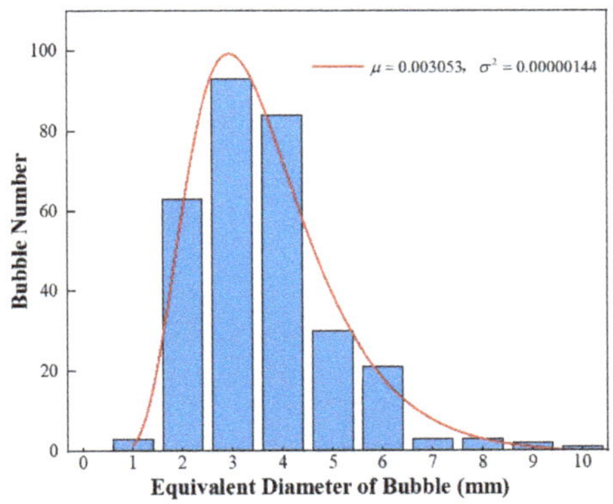

Figure 7. Statistics of random distribution of bubbles.

3. CFD Numerical Simulation

3.1. Hydraulic Model and Meshing

As shown in Figure 8, a three-dimensional fluid model was established, which was the same as the one used for the physical test. The test pipe section is a hump pipe with a pipe diameter of 0.1 m. The test pipe section is divided into five segments: the first one is the front horizontal section with a length of 1 m; the second one is the vertical rising section with a length of 0.3 m; the third one is the middle horizontal section with a length of 1.15 m; the fourth one is the vertical descending section with a length of 0.3 m; and the fifth one is the rear horizontal section with a length of 1 m. The pressure in the pipe is analyzed through the monitoring points at P1, P2, P3 and P4, with the change in pressure with time.

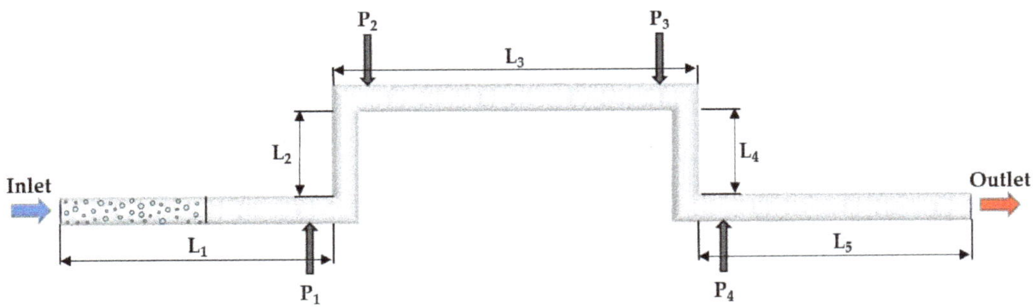

Figure 8. Three-dimensional fluid domain model of a hump pipe.

The software Fluent Meshing was used to mesh the model, an unstructured tetrahedral mesh was used for the whole flow channel, different local mesh scales were set to control the total number of meshes, the boundary layer was locally encrypted at the right-angled turn, the mesh independence was verified according to the steady-state pressure and head loss at point P1, and the final mesh delineation results are shown in Figure 9, with a total number of cells of 700,000 meshes. The minimum grid orthogonal quality was 0.45, and the grid independent verification results are shown in Figure 10.

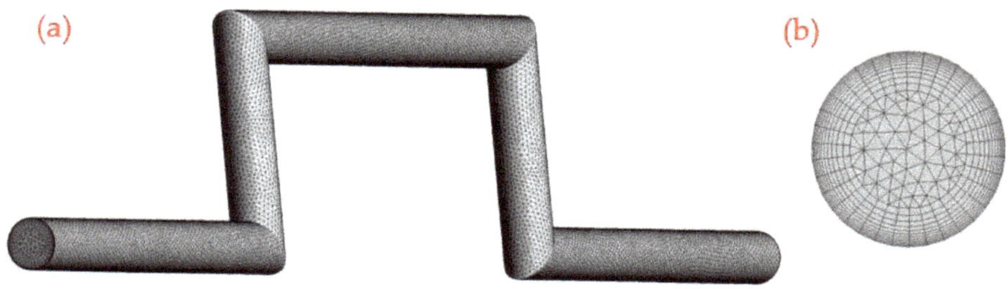

Figure 9. Mesh delineation results: (**a**) longitudinal section mesh delineation; (**b**) transverse section mesh delineation.

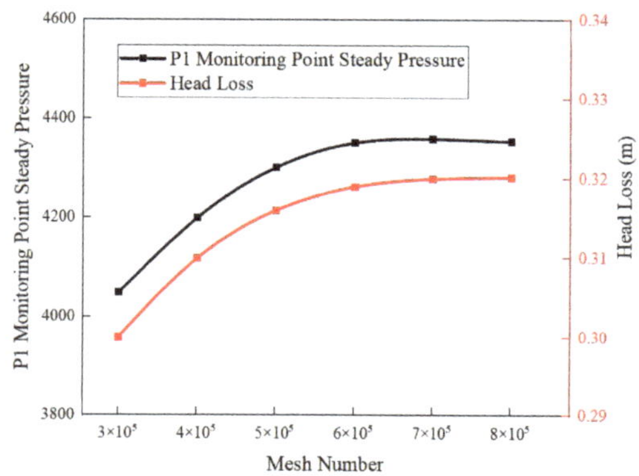

Figure 10. Grid-independent verification.

3.2. Numerical Simulation Methods

3.2.1. VOF Multiphase Flow Model

The VOF model [35] can simulate multiphase flow by tracking the volume fraction of each fluid, which is able to effectively track the trajectory of bubbles in the pipeline and changes in bubble morphology. This study focuses on the non-constant flow characteristics in the water transfer process and does not involve the law of conservation of energy, which is controlled by the following equations:

The continuity equation:

$$\frac{\partial(\alpha_q \rho_q)}{\partial t} + \nabla \cdot \left(\alpha_q \rho_q \vec{v}_q\right) = 0 \tag{1}$$

where α_q is the q fluid volume fraction of the phase; ρ_q is the fluid volume fraction of the first phase q fluid density of the phase; t represents time; and $\vec{v}_q$ is the velocity of the q phase.

In VOF, the volume fraction of each phase satisfies:

$$\sum_{q=1}^{n} \alpha_q = 1 \tag{2}$$

The momentum equation:

$$\frac{\partial\left(\rho \vec{v}\right)}{\partial t} + \nabla \cdot \left(\rho \vec{v}\right) = -\nabla p + \nabla \cdot \left[\mu\left(\nabla \vec{v} + \nabla \vec{v}^T\right)\right] + \rho \vec{g} + \vec{F} \tag{3}$$

where ρ is the volume-weighted average density; $\vec{v}$ is the mixing velocity; p is the pressure; μ is the volume-weighted average viscosity; $\vec{g}$ is the gravity acceleration; and $\vec{F}$ is the surface tension.

Among these,

$$\rho = \alpha_l \rho_l + \alpha_g \rho_g \tag{4}$$

$$\mu = \alpha_l \mu_l + \alpha_g \mu_g \tag{5}$$

where α_l and α_g denote the liquid-phase volume fraction and the gas-phase volume fraction, respectively; ρ_l and ρ_g denote the liquid-phase density and the gas-phase density, respectively; and μ_l and μ_g denote the liquid-phase viscosity and the gas-phase viscosity, respectively.

The continuous surface force model is used to simulate the surface tension of the gas–liquid interface, and the surface tension is added to Equation (3) as a source term [28].

$$\vec{F} = \sigma \frac{2\rho k_1 \left(\nabla \vec{s_1}\right)}{\rho_l + \rho_g} \tag{6}$$

$$k_1 = \nabla \cdot \vec{n_1} \tag{7}$$

where σ is the surface tension coefficient; k_1 is the surface curvature; s_1 is the surface normal vector at the interface; and n_1 is the unit normal vector.

3.2.2. RNG k-ε Turbulence Model

The flow pattern of the hump pipe is curved water with large curvature, which is fully developed turbulence with a high Reynolds number, so the RNG k-ε turbulence model with good simulation accuracy was chosen [36]. The turbulent kinetic energy (k) equation and turbulent dissipation rate (ε) equation are as follows:

$$\frac{\partial(\rho k)}{\partial t} + \frac{\partial(\rho k u_i)}{\partial x_i} = \frac{\partial}{\partial x_j}\left[\alpha_k \mu_{eff} \frac{\partial k}{\partial x_j}\right] + G_k + G_b - \rho\varepsilon - Y_M + S_k \tag{8}$$

$$\frac{\partial(\rho \varepsilon)}{\partial t} + \frac{\partial(\rho \varepsilon u_i)}{\partial x_i} = \frac{\partial}{\partial x_j}\left[\alpha_\varepsilon \mu_{eff} \frac{\partial \varepsilon}{\partial x_j}\right] + G_{1\varepsilon}\frac{\varepsilon}{k}(G_k + G_{3\varepsilon}G_b) + G_{2\varepsilon}\rho\frac{\varepsilon^2}{k} - R_\varepsilon + S_\varepsilon \tag{9}$$

$$\eta = S\frac{k}{\varepsilon} \tag{10}$$

$$S = \sqrt{2S_{ij} \cdot S_{ij}} \tag{11}$$

$$S_{ij} = \frac{1}{2}\left(\frac{\partial u_i}{\partial x_j} + \frac{\partial u_j}{\partial x_i}\right) \tag{12}$$

where α_k and α_ε correspond to the reciprocal of the effective turbulence Prandtl number for α and ε, respectively, with the default values of α_k = 1.39 and α_ε = 1.39; G_k is the production term for the turbulent kinetic energy (k) due to the mean velocity gradient; G_b is the production term for the turbulent kinetic energy (k) due to buoyancy effects; Y_M is the pulsating expansion term; $C_{1\varepsilon}$, $C_{2\varepsilon}$ and $C_{3\varepsilon}$ are empirical constants, taken as 1.42, 1.68 and 0.09, respectively (the default values in the Fluent calculation software were adopted); and S_k and S_ε are user-defined source terms (a source term represents an unstable term that cannot be included in the control equation). In general, a source term is not constant, and the default source term of Fluent software was adopted in this paper.

3.3. Boundary Conditions and Solution Settings

3.3.1. Boundary Conditions

The inlet boundary condition was set as the inlet velocity, at 1.0 m/s, 1.5 m/s and 2.0 m/s. The outlet pressure was set as a standard atmospheric pressure, and the air content in the inlet front end area was 0, 0.05, 0.10 and 0.15. The numerical simulation conditions were the same as the physical test conditions, and the specific calculation conditions are shown in Table 2. The fluid medium in the pipeline was water and air, and the specific physical properties of the two phases of water and gas are shown in Table 3.

Table 3. Parameters of physical properties of water and gas phases.

Medium	Density (kg/m^3)	Dynamic Viscosity (Pa·s)	Surface Tension (N/m)	Temp (°C)	Atmospheric Pressure (Pa)
Water	998.2	1.003×10^{-3}	0.072	20	101,325
Air	1.225	1.7894×10^{-5}			

The input of bubbles was based on the numerical simulation of constant flow by converting the equivalent diameters of the experimentally obtained bubbles and uniformly inputting circular bubbles to serve as initial conditions. In this paper, the radius of the bubble was programmed with a lognormal distribution by constraining the total volume of the bubble according to the mean and variance of the bubble size obtained from the experimental data, then by defining the distance between the location of the bubble and the center of the axis, and finally by varying the two parameters of the circular centroid and the radius, with some of the conditions shown in Equation (13).

$$\mathrm{p}(\theta)\begin{pmatrix} x(\theta) \\ y(\theta) \end{pmatrix} = r\begin{pmatrix} \cos\theta \\ \sin\theta \end{pmatrix} + \begin{pmatrix} x_0 \\ y_0 \end{pmatrix} \qquad \theta \in [\,0,\ 2\pi\,] \tag{13}$$

where (x_0, y_0) denotes the coordinates of the center of the bubble and r is the radius of the bubble.

3.3.2. Solution Setup

The numerical simulation was solved using Fluet 2020 R2 software, the coupled constant calculation of pressure and velocity was performed using the SIMPLEC algorithm, the non-constant calculation was performed using the PISO algorithm, the momentum and turbulent kinetic energy dissipation rate were discretized using the second-order windward format, the solid wall was set to a non-slip wall, and the accuracy of residual convergence was set to 10^{-5}. In order to facilitate the calculation of the curvature of the phase interface, the normal vector, the surface tension and other parameters, the CLSVOF coupled model was used. Firstly, the constant flow calculation without bubble input was carried out, and the random distribution of bubbles program was input into the front horizontal 0–0.5 m pipe section through the Journal file. Taking the air content rate of the 0.15 operating mode as an example, the distribution of bubbles after completing the input of bubbles is shown in Figure 11; the fixed time step (Δt) was set to 0.001 s, and the total computation time was 5 s.

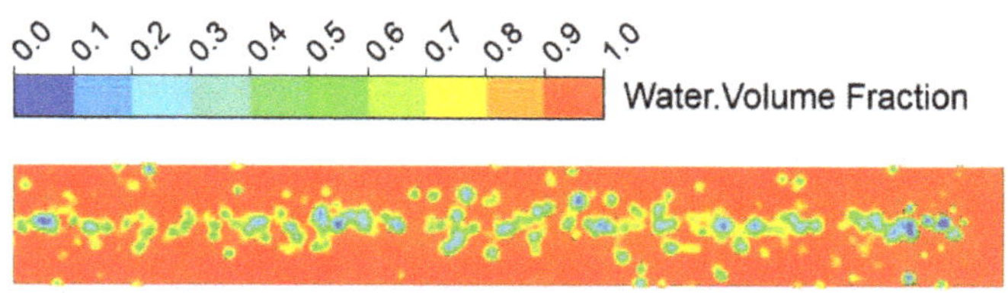

Figure 11. Random distribution of bubbles in the 0–0.5 m region.

4. Results and Analysis

4.1. Validation of the Numerical Simulation Results

In order to verify the accuracy of the numerical simulation calculations, it is necessary to compare and analyze the flow pattern changes in the pipe; in order to avoid redundancy in this article, the following is a comparison of the flow pattern changes in Case 12 (2.0 m/s, 0.15 air content), as shown in Figure 12. The flow pattern of the whole flow process changes into bubble flow–wavy flow–slug flow–churning flow–bubble flow–plug flow.

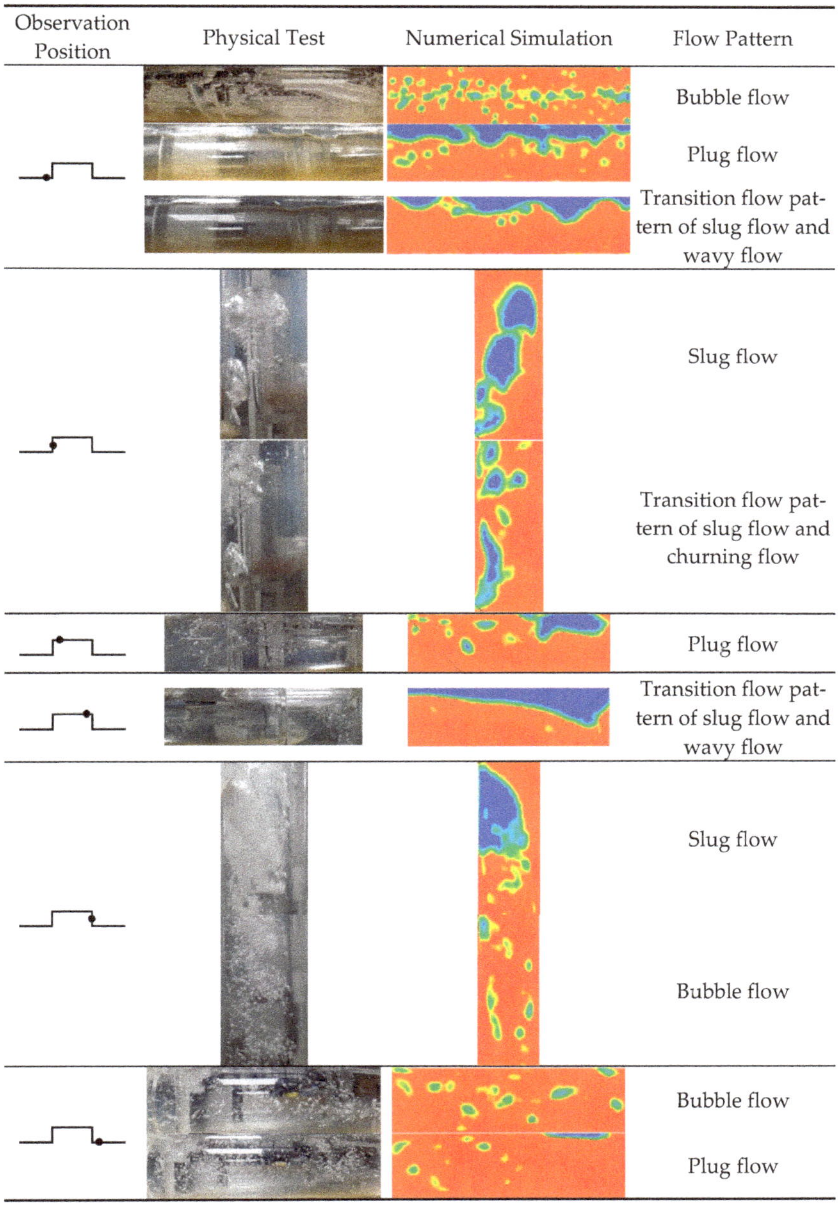

Figure 12. Comparison of experimental and numerical simulation flow patterns.

In order to further ensure the effectiveness of the CFD numerical simulation, the pressure of each monitoring point obtained via the numerical simulation of Case 12 is quantitatively compared with the experimental data. As shown in Figure 13, the simulation results are all within the error range, and the accuracy of the CFD numerical simulation is further confirmed by combining qualitative analysis with quantitative analysis.

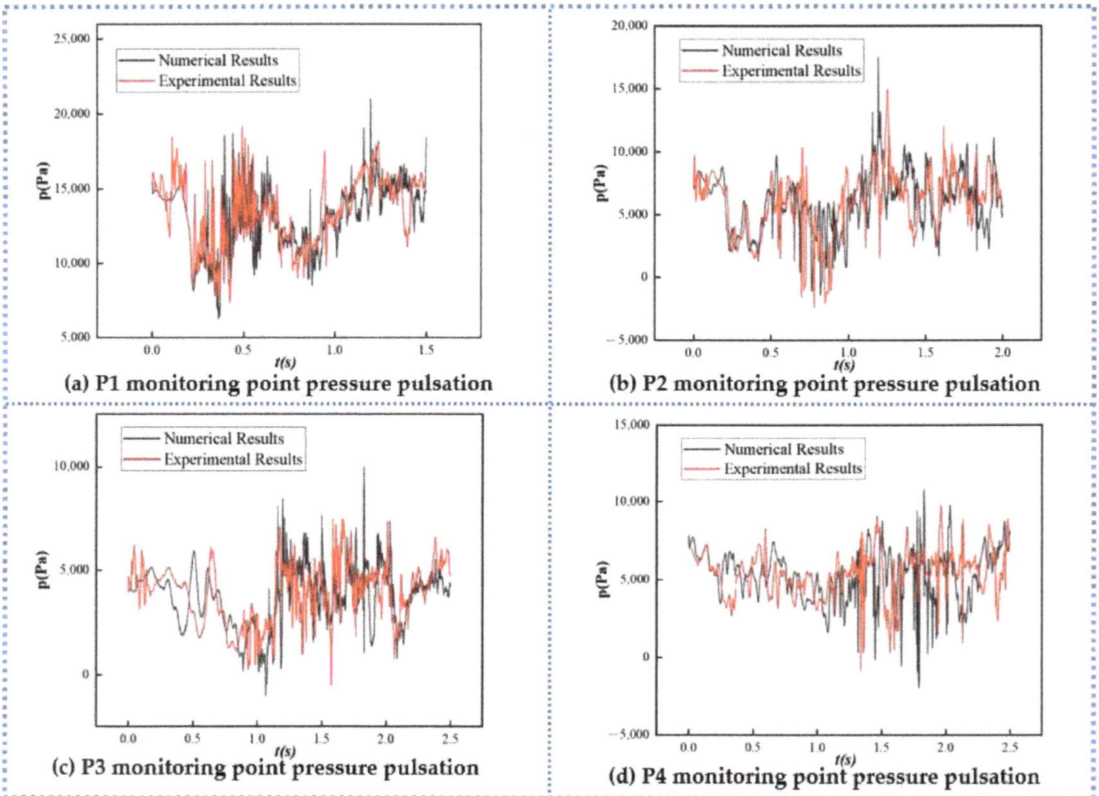

Figure 13. Comparison of experimental and numerical simulation pressure.

4.2. Flow Field Analysis

As shown in Figures 14–16, in order to observe the change in flow patterns in the pipe, the center cross-section of the pipe section is selected to observe the cloud map of the water- and gas-phase distribution, where red is defined as water and blue is defined as air.

Figure 14A shows the flow process of Case 2, where, under the influence of buoyancy, the small bubbles of bubble flow gradually merge into air sacs at the top of the tube and the whole horizontal section L_1 water and gas two-phase stratification is obvious, presenting a wave-like flow; the air sacs arrive at the vertical section L_2, splitting into a number of large air masses, and L_2 is presented as a bullet-like flow; after L_2 and L_3, the vertical corner, the large air mass is further split into small bubbles, and the balloon flow is gradually transformed into churning flow; the small bubbles in L_3 gather into air pockets again, forming an obvious plug flow; as the test pipe section is a vertical bend, the air pockets forming in the latter half of the vertical corner of L_3 struggle to be quickly washed away, and the stagnant air pockets are slowly washed into a smaller air mass, which rotates and accumulates at the left wall of the vertically descending section of L_4 to form a slug flow; and when passing through the vertical bend, it further divides and flows out of the test tube section in a bubble flow type. When the flow rate of each operating mode is certain, the gas-phase motion law is basically consistent with Case 2. Figure 14B shows the flow process of Case 3—due to the increase in air content, the volume of the air pocket formed through the accumulation in the horizontal section is larger, the gas output continuity is better and the stagnation phenomenon is more likely to occur in the vertical corner. The flow pattern of the whole flow process changes into bubble flow–slug flow–churning flow–plug flow–slug flow–bubble flow–plug flow. Figure 14C shows the flow process of Case 4—with a further increase in air content, the whole process changes into bubble flow–wavy flow–slug flow–plug flow–churning flow–bubble flow.

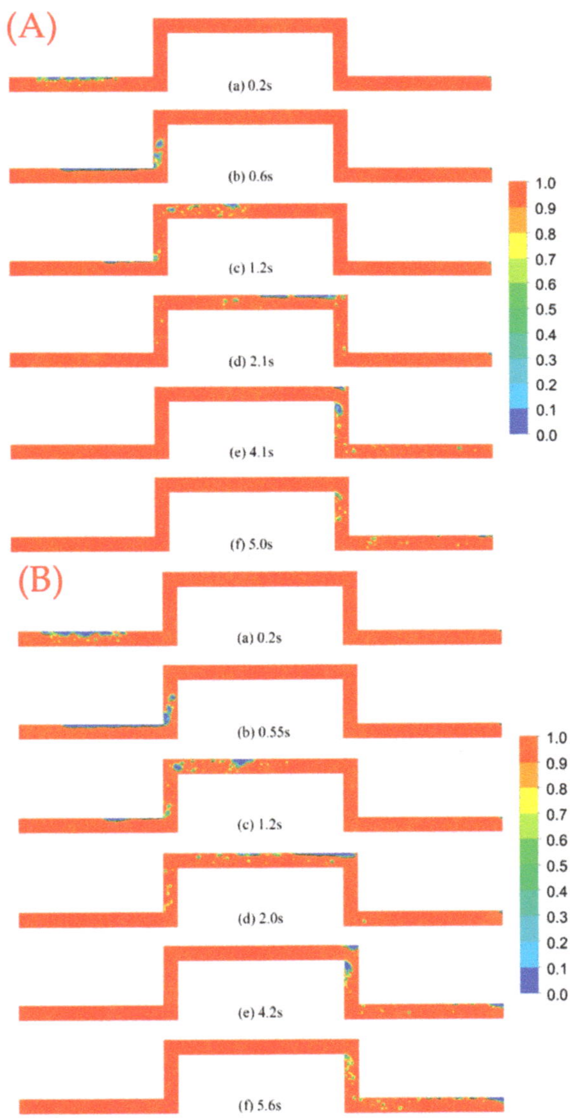

Figure 14. *Cont.*

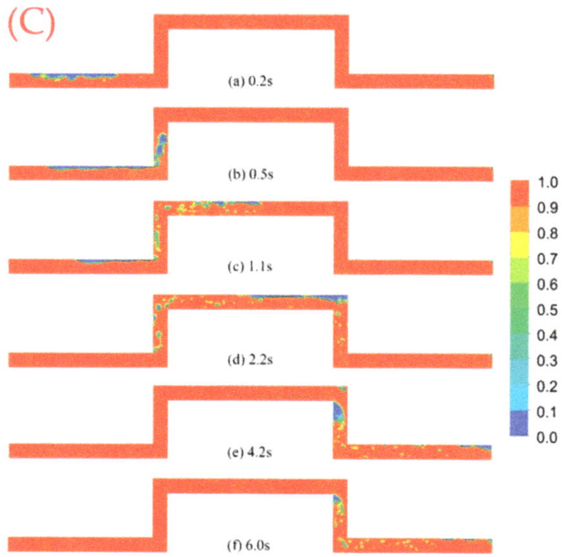

Figure 14. Water–air flow pattern of the pipe with flow rate 1.0 m/s: (**A**) Case 2, 0.05 air content; (**B**) Case 3, 0.10 air content; (**C**) Case 4, 0.15 air content.

Figure 15A shows the flow process of Case 6—with the acceleration of the water velocity, bubbles quickly gather in the horizontal section of the L_1 top of the pipe, leading to the formation of plug flow; due to the operating mode with less air content, the water velocity is faster, meaning that the output of the gas continuity is poorer, and the L_1 top of the pipe in the gas capsule demonstrates faster speed, leading to the formation of slug flow and churning flow of the transition flow type L_2; in the L rising process of water and gas mixed with each other, significant churning flow forms; around the vertical corner, churning flow breaks into small bubbles in L_3, and small bubbles, by means of buoyancy, accumulate at the top of the tube, leading to the emergence of a local plug flow; in L_4, after the vertical corner, the stagnation of the airbag in the water impacts differentiation in the form of bubbles that escape from the bag in the middle part of L_4 to form the slug flow; and broken bubbles make it to the top of the L_5 tube, bubbling out of the test flow. Figure 15B shows the flow process of Case 7—due to the increase in the air content, the whole flow process changes into bubble flow–plug flow–slug flow–bubble flow–slug flow–bubble flow–plug flow. Figure 15C shows the flow process of Case 8—with the further increase in air content, the whole process changes into bubble flow–wavy flow–slug flow–plug flow–churning flow–slug flow–bubble flow–plug flow.

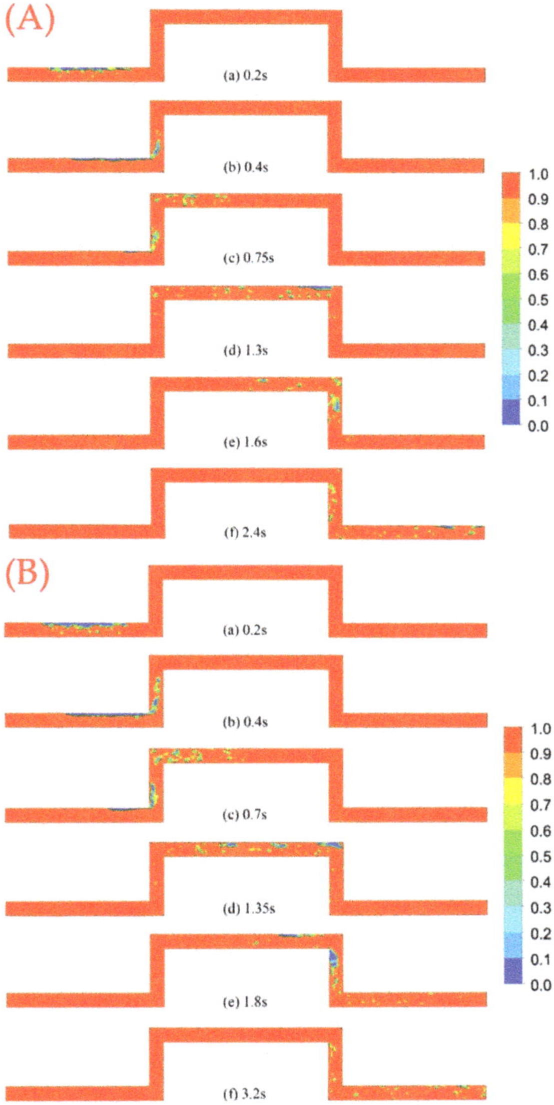

Figure 15. *Cont.*

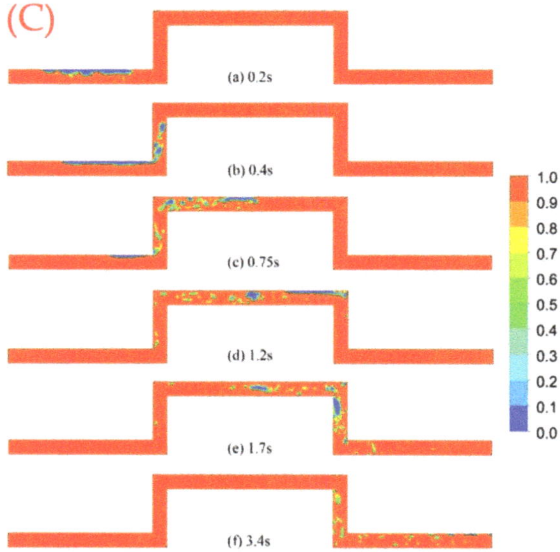

Figure 15. Water–air flow pattern of the pipe with flow rate 1.5 m/s: (**A**) Case 6, 0.05 air content; (**B**) Case 7, 0.10 air content; (**C**) Case 8, 0.15 air content.

Figure 16A shows the flow process of Case 10. In the initial state, the randomly distributed bubbles rise due to buoyancy with the increase in the distance to the gas bag, and bubbles accumulate at the top of the pipe to form plug flow; due to the flow rate being larger, the gas along L_2 moves, leading to the emergence of a slug flow pattern, and the slug flow in the L_2 rising process begins to break, splitting to form churning flow; small bubbles at the front of the L_3 rapid flow, leading to the emergence of bubble flow and churning flow transition; bubbles from L_3 flow through the vertical corner into L_4, and due to the water flow velocity being higher and the air content being low, only some of the bubbles accumulate to form a local churning flow; in L_5, after some of the bubbles in the top of the tube accumulate in the airbag, a local plug flow occurs out of the test tube section. Figure 16B shows the flow process of Case 11—with the increase in bubble density, the whole flow process changes into bubble flow–slug flow–plug flow–slug flow–churning flow–bubble flow–plug flow. Figure 16C shows the flow process of Case 12—with the further increase in the air content, the whole flow process changes into bubble flow–wavy flow–slug flow–churning flow–bubble flow–plug flow.

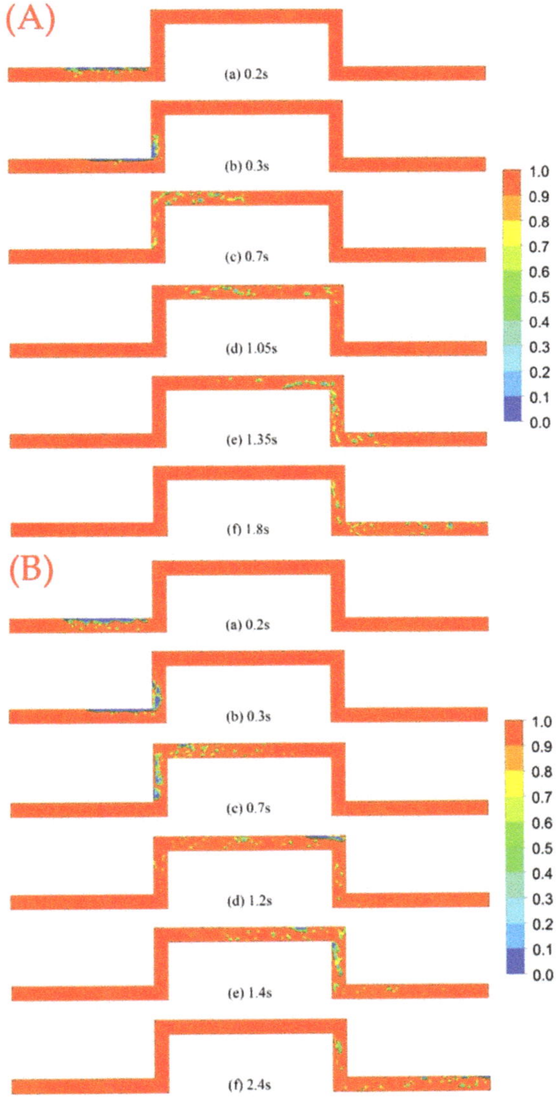

Figure 16. *Cont.*

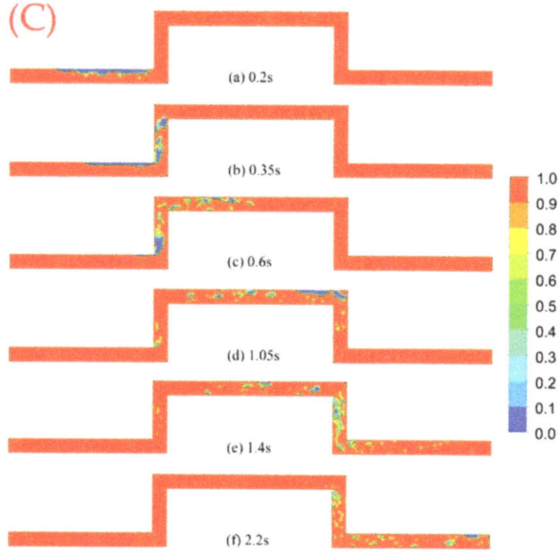

Figure 16. Water–air flow pattern of the pipe with flow rate 2.0 m/s: (**A**) Case 10, 0.05 air content; (**B**) Case 11, 0.10 air content; (**C**) Case 12, 0.15 air content.

4.3. Pressure Pulsation Analysis

Figure 17 shows the time-domain variation in pressure pulsation in the 1.0 m/s operating mode, and it can be observed that when the air content is low, the pressure curves at each monitoring point are generally smooth, which indicates that the crushing and fusion of bubbles in the pipeline enhance the turbulence of the disturbing effect, and the greater the number of bubbles, the more intense the process and the greater the fluctuation in the pressure amplitude. Comparing the pressure fluctuations at different moments, it can be seen that there exists a stage of extremely frequent changes in pressure pulsation at each monitoring point—for example, 0.5–2.0 s at point P1 and 1.5–3.5 s at point P2—and these characteristic stages are exactly the time ranges of the flow of bubbles or air sacs in the pipeline through the region, which indicates that in the process of collapsing, aggregating, separating and mingling with the aqueous phase of the bubbles, the gas–liquid phases exhibit disorder and uncertainty, causing the pressure pulsation to increase and decrease steeply.

Figure 18 shows the pressure pulsation time-domain changes at each monitoring point under the 1.5 m/s operating mode. Compared with the 1.0 m/s operating mode, with the increase in the inlet flow rate, the pressure amplitude generally increases, and the characteristic phase of high-frequency pressure changes is advanced, but the pressure waveforms inside the pipeline still have a high similarity. The frequency of pressure pulsation changes is more frequent in the 1.5 m/s operating mode, and the phase of high-frequency and high-amplitude pressure changes caused by bubbles passing through a certain monitoring point is earlier than that of the other monitoring points, but the characteristic phase is shortened obviously, which indicates that although bubbles are retained in certain areas for a shorter period of time, the processes of bubbles collapsing, aggregating, separating and so on are more violent, and the flow structure in the pipe is more turbulent.

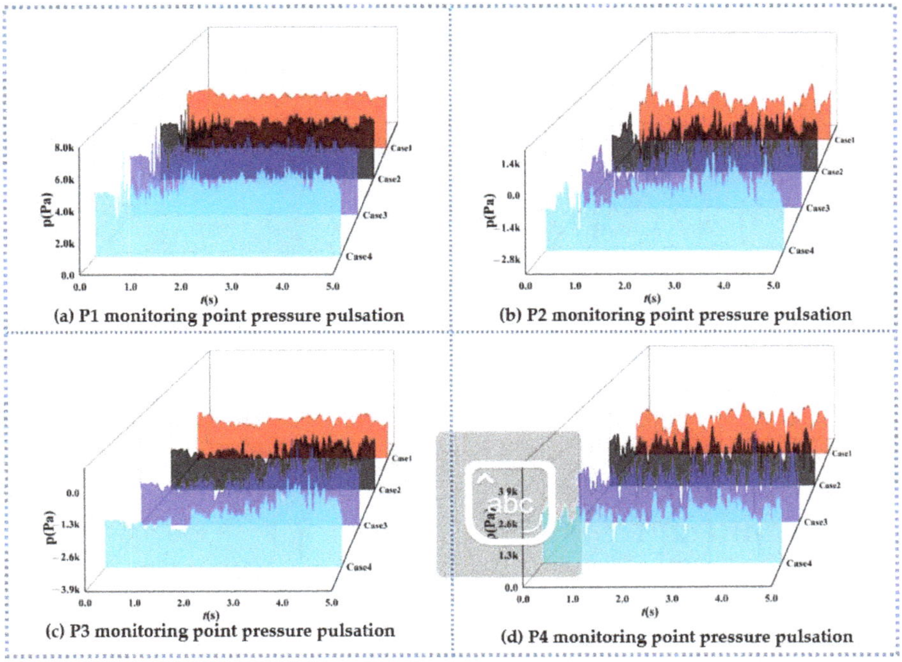

(a) P1 monitoring point pressure pulsation (b) P2 monitoring point pressure pulsation (c) P3 monitoring point pressure pulsation (d) P4 monitoring point pressure pulsation

Figure 17. Time-domain plot of pressure pulsation at each monitoring point (1.0 m/s).

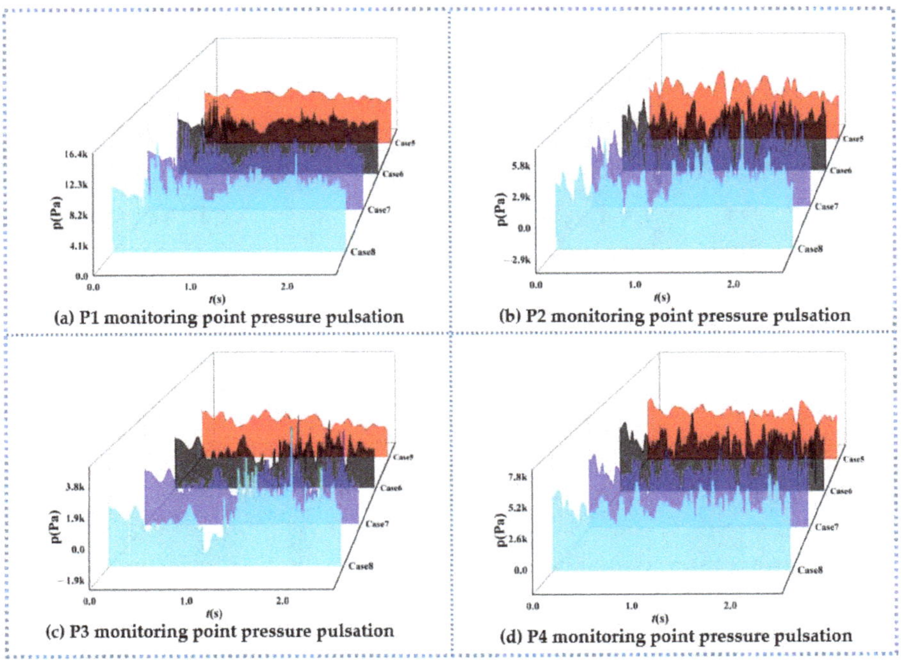

(a) P1 monitoring point pressure pulsation (b) P2 monitoring point pressure pulsation (c) P3 monitoring point pressure pulsation (d) P4 monitoring point pressure pulsation

Figure 18. Time-domain plot of pressure pulsation at each monitoring point (1.5 m/s).

Figure 19 shows the time-domain variation in pressure pulsation at each monitoring point under the 2.0 m/s operating mode. With the further increase in the flow rate, the pressure inside the pipe also increases further, and the negative pressure condition at the monitoring points P2 and P3 is obviously improved. As for other flow rates, the higher the air content and the higher the pressure change frequency, the more obvious the turbulence effect caused by bubble breakup and fusion. Unlike other flow rates, under the impact of faster water flow, the time that bubbles remain in a certain region is shorter, so the flow pattern changes caused by bubble collapse and aggregation are more rapid, the flow pattern is more turbulent, and the fluctuation in pressure pulsation amplitude is more obvious.

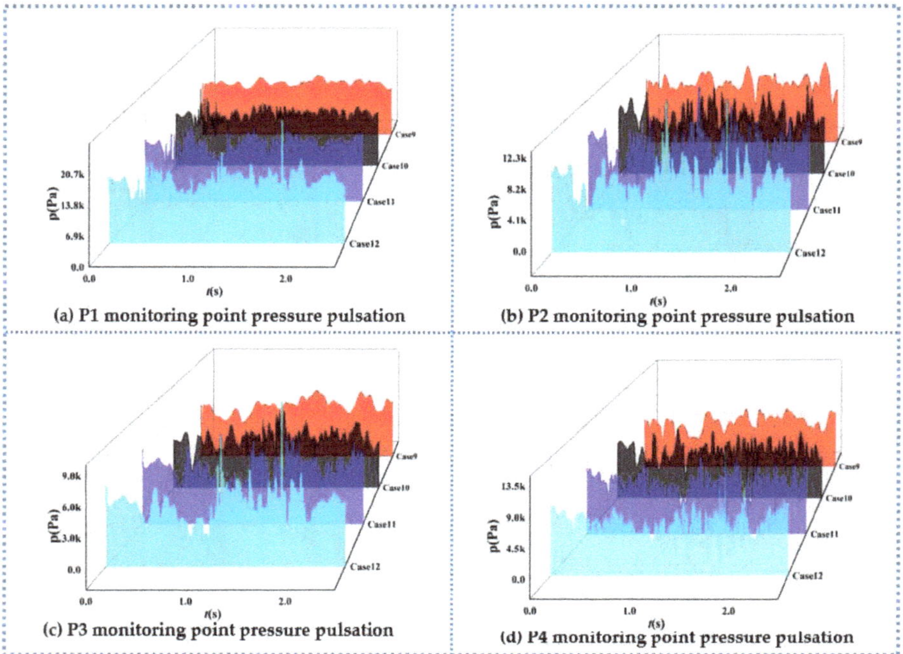

Figure 19. Time-domain plot of pressure pulsation at each monitoring point (2.0 m/s).

In order to further analyze the variation in the main characteristics of pressure pulsation, a dimensionless number, $\Delta H'$, is introduced to characterize the pressure pulsation properties [37].

$$\Delta H' = \frac{\Delta H}{H} = \frac{P_{imax} - P_{imin}}{\rho g H} \tag{14}$$

where $\Delta H'$ is the relative pressure pulsation amplitude; ΔH is the peak pressure pulsation, m; H is the hydraulic head, m; and P_{imax} and P_{imin} are the corresponding maximum and minimum values of pressure at point i, Pa.

Figure 20 shows the variation in the relative pressure pulsation amplitude at each monitoring point, and it can be observed that the relative pressure pulsation amplitude decreases with the decrease in air content. Under the condition of the 0.15-air content operating mode, the highest relative pressure amplitude is observed, which is generally two to six times that of the 0-air content operating mode and which indicates that the bubbles greatly enhance the turbulence perturbation in the process of movement. The higher the number of bubbles, the stronger the pressure pulsation caused. At the same time, the velocity of water flow has less influence on the relative pressure pulsation amplitude, which indicates that the air content is the main factor affecting the relative pressure pulsation amplitude compared with the velocity of water flow. For the monitoring points at different locations, the relative pressure pulsation amplitude is significantly higher than that at the other monitoring points because the air pockets generated by the convergence of air bubbles at the P1 monitoring point have a longer retention time and the flow pattern changes drastically.

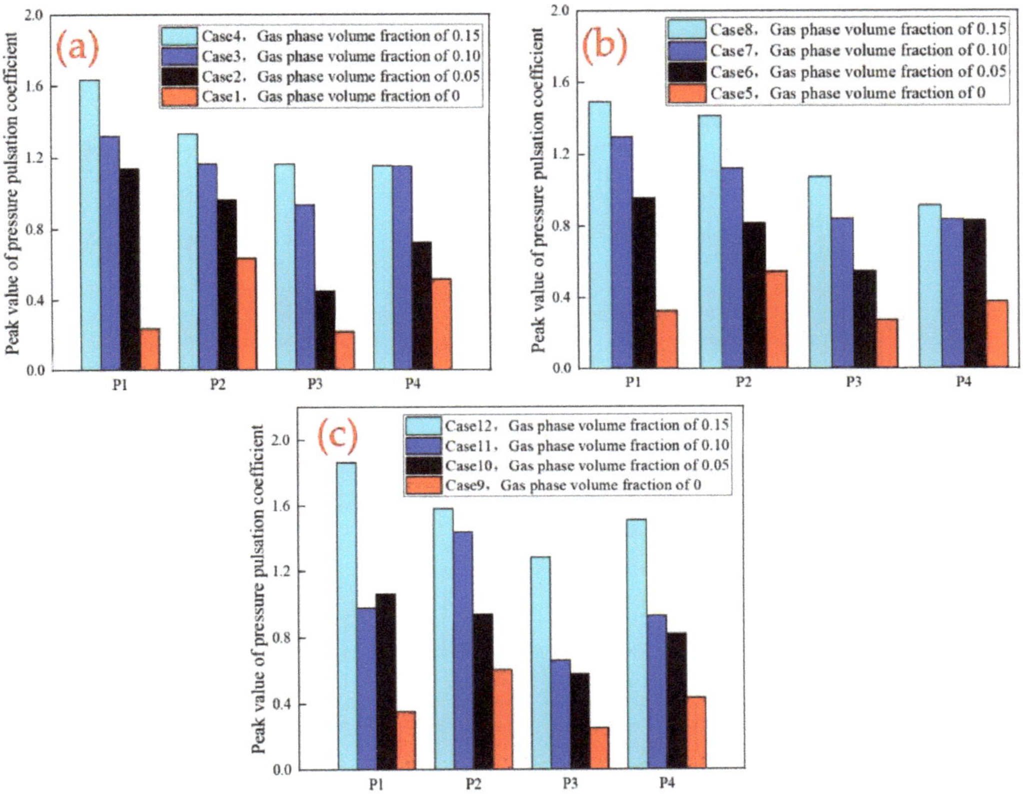

Figure 20. The relative pressure pulsation amplitude: (**a**) 1.0 m/s; (**b**) 1.5 m/s; (**c**) 2.0 m/s.

5. Conclusions

Based on the physical model test and the CLSVOF gas–liquid two-phase pipe flow numerical simulation, different flow rates and different air contents within the hump pipe flow transient process were studied in this paper, and the hump pipe pressure pulsation propagation law was compared and analyzed. The research results can provide a theoretical basis and calculation parameters for stagnant gas burst prediction and water hammer protection in aqueduct systems. The main conclusions are as follows.

(1) The random distribution of bubbles in the test tube section was analyzed, the random distribution of bubbles law was measured, the final bubble size was obtained as a lognormal distribution law, and the equivalent diameter was between 3 mm and 10 mm. The higher the air content, the more obvious the phenomenon of small bubbles aggregating to form large bubbles. However, the distribution position of bubbles is random, and the relationship between bubble size and bubble distribution position is also random.

(2) The flow pattern in the hump pipe is rich and varied, and there are five main types of flow patterns. After the 90-degree bend, the flow pattern is rapidly transformed,

experiencing a complex flow pattern transformation, such as bubble flow–wavy flow–slug flow–churning flow–slug flow–bubble flow. The higher the flow rate, the more intense the flow pattern evolution process, the more disordered the flow pattern and the higher the air content. In addition to the increase in the typical flow pattern, there are also many transitional flow patterns.

(3) When pipeline bubbles or airbags flow through a region, due to gas–liquid two-phase disorder, they will cause the region pressure pulsation amplitude to steeply increase or decrease, and pressure fluctuations are extremely frequent, in which the flow rate mainly affects the pressure pulsation amplitude, and the air content has an effect on the pressure pulsation frequency and pressure pulsation amplitude. The relative pressure pulsation amplitude decreases as the air content decreases, and the air content is the main factor affecting the relative pressure pulsation amplitude compared to the flow rate.

Author Contributions: Conceptualization, Q.W. and J.H.; methodology, Q.W. and M.S.; software, Q.W. and H.S.; validation, Q.W.; formal analysis, J.H.; investigation, Q.W.; resources, Q.W.; data curation, H.S.; writing—original draft, Y.Z. and F.X.; writing—review and editing, Q.W. and D.L.; visualization, J.H.; supervision, Q.W.; project administration, J.H.; funding acquisition, J.H. All authors have read and agreed to the published version of the manuscript.

Funding: This work is supported by the Key Joint Funds of the Zhejiang Provincial Natural Science Foundation of China (LZJWZ22E090004).

Data Availability Statement: The original contributions presented in the study are included in the article, further inquiries can be directed to the corresponding author.

Acknowledgments: We are very grateful to the anonymous reviewers for their constructive comments and suggestions.

Conflicts of Interest: The authors declare no conflict of interest.

References

1. Valve, V.M.; Corp, M. *Protecting Drinking Water Pipelines with Inflow Prevention*; American Society of Sanitary Engineering: Elmhurst, IL, USA, 2011.
2. Pozos, O.; Gonzalez, C.A.; Pedrozo, A.; Sanchez, A.; Rodal, E. Boundary between air entrainment and air transport downstream of a hydraulic jump in circular conduits. In Proceedings of the 5th IAHR International Symposium on Hydraulic Structures, Brisbane, Australia, 25–27 June 2014. [CrossRef]
3. Lu, K.M.; Zhou, L.; Cao, B.; Wang, H. Three-dimensional dynamic characteristics simulation of water flow impacting the trapped air mass in undulating pipeline. *J. Irr. Dra. Mac. Eng.* **2020**, *38*, 384–389+402.
4. Guo, Y.X.; Yang, K.L.; Guo, X.L.; Fu, H. Influence of retained bubbles on water delivery capacity during water filling of large pipeline water delivery system. *J. Hydraul. Eng.* **2013**, *44*, 6. [CrossRef]
5. Lu, K.M.; Zhou, L.; Liu, J. Three-dimensional numerical simulation of water impact on multi-stage stagnant air mass. *J. Irr. Dra. Mac. Eng.* **2021**, *39*, 6.
6. Zheng, Y.; Liu, D.Y.; Zhang, J.; Suo, L.S. Summary of research on gas-liquid two-phase transient flow in pressurized water pipeline system. *J. Hohai Univ.* **2002**, *30*, 5. [CrossRef]
7. Kranenburg, C. The Effects of free gas on cavitation in pipe tines. In *1st Inter Conference on Pressure Surges*; British Hydromechanics Research Assoc: Cranfield, UK, 1972; pp. 41–52.
8. Brown, R.J. Water column separation at to pumping pants. *J. Basic. Eng. Trans.* **1968**, *90*, 521–531. [CrossRef]
9. Simpson, A.R.; Bergant, A. Numerical comparison of pipe-column-separation models. *J. Hydraul. Eng.* **1994**, *120*, 361–377. [CrossRef]
10. Bergant, A.; Simpson, A.R. Pipeline column separation flow regimes. *J. Hydraul. Eng.* **1999**, *125*, 835–848. [CrossRef]
11. Yang, K.L.; Dong, X.L. Study on dynamic characteristics of bubbles in long-distance water conveyance pipeline of hydropower station. *J. Hydraul. Eng.* **1998**, *29*, 6–16. [CrossRef]
12. Yang, K.L. Research progress and frontier scientific problems of hydraulic control for long-distance water conveyance. *J. Hydraul. Eng.* **2016**, *47*, 424–435. [CrossRef]
13. Tokuhiro, A.; Maekawa, M.; Iizuka, K. Turbulent flow past a bubble and an ellipsoid using shadow-image and PIV techniques. *Int. J. Multiph. Flow* **1998**, *24*, 1383–1406. [CrossRef]
14. Rezapour, S.; Riasi, A. Experimental investigation of viscoelastic turbulent fluid hammer in helical tubes, considering column-separation. *Int. J. Press. Ves. Piping* **2021**, *194*, 104489. [CrossRef]

15. Wan, W.; Li, C.; Yu, Y. Investigation on critical equilibrium of trapped air pocket in water supply pipeline system. *J. Zhejiang Univ.* **2017**, *18*, 167–178. [CrossRef]
16. Jansson, M.; Andersson, M.; Karlsson, M. High-Speed Imaging of Water Hammer Cavitation in Oil–Hydraulic Pipe Flow. *Fluids* **2022**, *7*, 102. [CrossRef]
17. Urbanowicz, K.; Bergant, A.; Kodura, A.; Kubrak, M.; Malesińska, A.; Bury, P.; Stosiak, M. Modeling Transient Pipe Flow in Plastic Pipes with Modified Discrete Bubble Cavitation Model. *Energies* **2021**, *14*, 6756. [CrossRef]
18. He, J.; Liu, X.; Li, B.; Zhai, J.; Song, J. Cavitation Erosion Characteristics for Different Metal Surface and Influencing Factors in Water Flowing System. *Appl. Sci.* **2022**, *12*, 5840. [CrossRef]
19. Veisi, A.; Shahsavari, M.H.; Roshani, G.H.; Eftekhari-Zadeh, E.; Nazemi, E. Experimental Study of Void Fraction Measurement Using a Capacitance-Based Sensor and ANN in Two-Phase Annular Regimes for Different Fluids. *Axioms* **2023**, *12*, 66. [CrossRef]
20. Khan, U.; Pao, W.; Sallih, N. Numerical Gas–Liquid Two-Phase Flow Regime Identification in a Horizontal Pipe Using Dynamic Pressure Data. *Appl. Sci.* **2023**, *13*, 1225. [CrossRef]
21. Zhao, L.; Song, Z.C.; Wu, R.C.; Wang, J.P. Theoretical Analysis of Gas Release during Transient Process of Hydraulic Transients. *DEStech Trans. Soc. Sci. Educ. Hum. Sci.* **2017**. [CrossRef]
22. Li, X.Q.; Meng, L.P.; Li, P.; Cai, Z.H.; Liu, J.W.; Wang, F.J. Influence of trapped gas on operation stability of diversion system of hydropower station. *J. Hydroelectr. Eng.* **2017**, *43*, 4.
23. Zhou, L.; Liu, J.; Huang, K.; Liu, D.Y. Numerical simulation of transient flow with trapped air mass during start-up and filling of water pipeline Influence of trapped gas on operation stability of diversion system of hydropower station. *Adv. Sci. Tech. Water Resour.* **2021**, *41*, 1–7.
24. Apollonio, C.; Balacco, G.; Fontana, N.; Giugni, M.; Marini, G.; Piccinni, A.F. Hydraulic Transients Caused by Air Expulsion during Rapid Filling of Undulating Pipelines. *Water* **2016**, *8*, 25. [CrossRef]
25. Vasconcelos, J.G.; Wright, S.J. Rapid Flow Startup in Filled Horizontal Pipelines. *J. Hydra. Eng.* **2008**, *134*, 984–992. [CrossRef]
26. Ramos, H.M.; Fuertes-Miquel, V.S.; Tasca, E.; Coronado-Hernández, O.E.; Besharat, M.; Zhou, L.; Karney, B. Concerning Dynamic Effects in Pipe Systems with Two-Phase Flows: Pressure Surges, Cavitation, and Ventilation. *Water* **2022**, *14*, 2376. [CrossRef]
27. Feng, L.; Yao, Q.Y. Numerical simulation of gas-liquid two-phase flow in pressure pipeline of pumping station based on VOF model. *China Rural. Water Hydropower* **2012**, *12*, 124–126+130.
28. Wang, Z.W.; He, Y.P.; Li, M.Z.; Qiu, M.; Huang, C.; Liu, Y.D. Numerical simulation and flow pattern evolution of gas-liquid two-phase flow in 90 elbow based on computational fluid dynamics. *J. Shanghai Jiaotong Univ.* **2022**, *56*, 1159–1167. [CrossRef]
29. Yu, Q.Q.; Shi, H.H.; Dong, R.L.; Peng, S.S. Numerical simulation of flow pattern characteristics of gas-liquid two-phase flow in vertical upward circular tube. *J. Zhejiang Sci.-Tech. Univ.* **2022**, *47*, 397–404.
30. Bourlioux, A. A coupled level-set volume-of-fluid algorithm for tracking material interfaces. In Proceedings of the 6th International Symposium on Computational Fluid Dynamics, Lake Tahoe, CA, USA, 4–8 September 1995; p. 15.
31. Yang, C.Z. Study on Critical Velocity of Bubble Initiation in Long Water Pipeline Pressurized by Pump. Master's Dissertation, North China Univ. Water Res. & Electric Power, Zhengzhou, China, 2021.
32. Shang, Z.; Lou, J.; Li, H. Simulations of flow transitions in a vertical pipe using coupled level set and VOF method. *Int. J. Comp. Meth.* **2017**, *14*, 135–141. [CrossRef]
33. Tang, X.; Duan, X.; Gao, H.; Li, X.; Shi, X. CFD Investigations of Transient Cavitation Flows in Pipeline Based on Weakly-Compressible Model. *Water* **2020**, *12*, 448. [CrossRef]
34. Hu, J.; Wang, Q.; Zhang, Y. Numerical and Experimental Study on the Process of Filling Water in Pressurized Water Pipeline. *Water* **2023**, *15*, 2508. [CrossRef]
35. Ling, P.; Ting, Z.; Jian, L. Three-Dimensional Numerical Study of Dam-Break Flood Impacting Problem with VOF Method and Different Turbulence Closures. *Water Res. Manag.* **2023**, *37*, 1–21. [CrossRef]
36. Devade, K.D.; Pise, A.T.; Urade, A.R. Numerical Analysis of Flow Behavior in Vortex Tube for Different Gases. *Mech. Eng. Res.* **2017**, *7*, 18–39. [CrossRef]
37. Xu, L.C.; Peng, Y.J.; Tang, W.; Liu, D.M.; Liu, X.B. Flow Characteristics and Pressure Pulsation in the S Characteristic Area of Model Pump Turbine. *Chin. J. Hydro.* **2022**, *37*, 213–225. [CrossRef]

water

Article

Internal Flow Characteristics of Centrifugal Pumps under Different Startup Combination Schemes

Xiaobo Zheng [1], Wei Wang [1], Pengli Zhang [2], Yongjian Pu [1] and Yaping Zhao [1,*]

1 Institute of Water Resources and Hydropower, Xi'an University of Technology, Xi'an 710048, China; zxb@xaut.edu.cn (X.Z.); 2220420038@stu.xaut.edu.cn (W.W.); 2200420102@stu.xaut.edu.cn (Y.P.)
2 Hanjiang to Weihe River Valley Water Diversion Project Construction Co., Ltd., Xi'an 710024, China; xazpl@163.com
* Correspondence: zyp0168@xaut.edu.cn

Abstract: Pump station engineering is a water conservancy project used for long-distance water transfer, irrigation and drainage, and urban living and industrial water supply. Centrifugal pumps are one of the main pump types commonly used in pumping stations, and their operation is of considerable importance for the safety, stability, and efficient operation of pumping stations. This paper takes a large pumping station with seven centrifugal pump units as the research object and combines experimental research and numerical simulation. The axial flow velocity uniformity, average cross-sectional deviation angle, and hydraulic loss of the pump inlet section are evaluated, and the internal flow characteristics of the pump under different startup combination conditions are analyzed based on entropy generation and vorticity. This study also explores the operational performance of the pump station under different startup combination conditions, revealing the mutual influence mechanism between different startup combinations of pump stations and the internal and external characteristics of centrifugal pumps and introducing the optimal startup combination scheme for the pump station system. Research results indicate that the difference in energy loss of centrifugal pumps under different startup combinations is mainly manifested in the impeller and guide vane flow channels. For the two existing inlet flow channel structures in the pump station, the unit effectively operates when the inlet flow channel is tilted to the left. The optimal startup combination method of the pump station under different startup combinations is determined.

Keywords: pump station; centrifugal pump; internal flow characteristics; startup combinations

Citation: Zheng, X.; Wang, W.; Zhang, P.; Pu, Y.; Zhao, Y. Internal Flow Characteristics of Centrifugal Pumps under Different Startup Combination Schemes. *Water* **2024**, *16*, 1087. https://doi.org/10.3390/w16081087

Academic Editor: Helena M. Ramos

Received: 27 February 2024
Revised: 30 March 2024
Accepted: 4 April 2024
Published: 10 April 2024

1. Introduction

Considering the total water resources of China, despite the uneven spatial and temporal distribution, the "14th Five-Year Plan" proposes the following to solve the increasingly tense water resource phenomenon: "to be based on the basin as a whole and the spatial balance of water resources allocation". Pumping station projects aim to provide an effective means for achieving the aforementioned goal. Therefore, the safe, stable, and efficient operation of pumping stations is also a concern. Most of the large pumping stations are generally installed with the same hydraulic characteristics as the pump unit. Thus, in the same flow and head conditions, the pumping station of the optimal combination of power on and the optimal operating conditions can help realize the minimum total energy consumption of the pumping station or its maximum total efficiency [1–3].

For the pumping station system, different start combinations and sequences have a remarkable impact on the pump inflow, which, in turn, will induce changes in the internal flow state of the pump. At present, numerous experts and scholars have conducted relevant research on the start combination of multi-unit pumping stations. Mahdi et al. [4] optimized the operating cost and efficiency of the pumping station based on the combination of high-efficiency pumps and the selection of pump types and established an optimization model.

Chang Pengcheng et al. [5] simulated the lateral inlet forebay of a multi-unit pumping station and found that its original model had a wide range of vortices inside the inlet pool. Xu Cundong et al. [6] studied the changing law of the flow pattern of the forebay under different combinations of switching on. They found that the best flow pattern improvement of the forebay in the lateral pumping station unit lies in the symmetrical switching on of the unit. Wu Peifeng et al. [7] used numerical simulation to analyze the symmetrical distribution of unit intake flow pattern in Dazhai River lock station. Their results revealed that two centered pumping station units arranged on each side of the unit intake conditions are highly favorable. Manuel et al. [8] proposed an optimal pump scheduling algorithm based on the ant colony algorithm, with the number of pumping station startups and the startup time as the optimization variables. Zidan et al. [9] studied different startup combinations and the distribution of sediment concentrations in the intake system. They found that when the number of start units increases, the sediment concentration on the sidewalls on both sides of the intake system decreases; when the units close to the wall do not operate, the sidewall area will appear as a dead water or vortex zones, which increases the sediment concentration. Wang Weilin [10] studied the hydraulic characteristics of the two lake sections of the terrace pumping station through hydraulic simulation to address the optimal efficiency of the switching machine and unit flow combination of the pumping station. He Yong et al. [11] numerically simulated the flow field in the front pool of the pumping station under different startup combinations and proposed the optimal startup combination scheme by combining the distribution uniformity of the water flow and the undesirable flow conditions such as vortex and walling off. Li Yingchun [12] conducted numerical simulations of the flow pattern in the forebay under different start combinations and combined the flow uniformity with the distribution of undesirable flow patterns to obtain the optimal startup combination scheme under high flow conditions. Qi Dunzhe et al. [9] studied the operating performance of the pumping station under different numbers of start units and found that increasing the number of start units worsens the operating performance of the pumping station. Guohui Cong et al. [13] investigated the vortex structure in the intake pool of the pumping station for different water levels and pump combinations and discovered the location and intensity of the vortex in different conditions. Wang et al. [14] developed a mathematical model to solve the problem of large energy consumption of the start–stop scheme in the pumping station. Chen et al. [15] combined numerical simulation and experimental research to analyze the hydraulic characteristics of the pumping station intake pool with an eccentric outlet. They found poor flow patterns in the intake pool and optimized these patterns by adding a flow-guiding pier. M T Rahman et al. [16] combined numerical simulation and experiments to analyze the internal flow characteristics of centrifugal pumps and combined the simulation results with the experiments results to obtain the location and intensity of vortices under different operating conditions. Byskov et al. [17] studied the internal flow field of centrifugal pumps through the large vortex simulation and found that the impeller flowed best at the design condition. The flow distribution of fluid in each flow channel was mainly uniform, while that when the pump was under small flow conditions was homogeneous. The flow distributions in the impeller and fluid were both highly homogeneous when the pump was under small flow conditions. Numerous stall vortices appear in the impeller channel when the centrifugal pump is in the small flow condition, blocking the impeller channel. The simulation results are finally compared with the particle image velocimetry (PIV) test results, yielding a high similarity. Ding et al. [18] simulated the internal flow characteristics of the centrifugal pump by changing the outlet angle of the vanes and conducted an experimental validation. Their results showed that the hydraulic loss of the centrifugal pump increases with the outlet angle of the vanes with the rise of the flow rate.

The results of the above studies reveal that different startup combination conditions have an important impact on the operational performance of the pumping station. Therefore, for a specific pumping station system, the safe and stable operation of the pumping station is of considerable importance for analyzing the hydraulic performance of the pumping station under the conditions of different startup combinations and studying the optimal startup combination scheme. A large pumping station is taken in this paper as the research object, and the hydraulic performance of the pump unit under different startup combinations is studied by combining experimental and numerical research methods. The internal flow and energy characteristics of the centrifugal pump under different startup combinations are analyzed to determine the optimal startup combination scheme of the pumping station system. This scheme provides theoretical guidance for the safe and efficient operation of the pumping station.

2. Research Background and Experimental Model

2.1. Research Background

This paper takes a large pumping station system in Yangxian County, Shaanxi, as the research object. The pumping station has a total storage capacity of 221 million m^3 and a regulating storage capacity of 0.98 billion m^3. The powerhouse of the pumping station is at the dam toe, which has a total of seven vertical single-stage, single-suction centrifugal pumps with a water-lift design of 108.5 m and a single-pump flow rate of 12.6 m^3/s. Among these pumps, six are operational and one is on standby.

2.2. Model Pumping Station Test Stand

Considering the feasibility of conducting the test, owing to the large size of the pumping station system prototype, this paper is based on the similarity principle: that is, the Euler number E and the Strouhal number Sr are equal, and the ratio of the size of the prototype pump to that of the model pump is 1:20. Sr and E are, respectively, defined as follows [19]:

$$Sr = \frac{D_2^3 n}{Q}, \tag{1}$$

$$E = \frac{gH}{D_2^2 n^2}, \tag{2}$$

where D_2 indicates the impeller outlet diameter (m); n denotes the rotational speed, (r/min); Q is the flow (m^3/s); g is the acceleration of gravity (m/s^2); and H is the head of the pump (m).

According to similarity theory, the parameter conversion relationship between the prototype and model pumps is shown in the following equation:

$$\frac{Q_p}{Q_m} = \frac{D_{2p}^3}{D_{2m}^3}\frac{n_p}{n_m}, \tag{3}$$

$$\frac{H_p}{H_m} = \frac{D_{2p}^2}{D_{2m}^2}\frac{n_p^2}{n_m^2}, \tag{4}$$

where the subscripts p and m represent prototype and model pumps, respectively.

The model test device comprises the forebay, circulating water supply pipe, model pump, electromagnetic flowmeter, flow control valve, water pump, and outlet pond. The model is generally transparent plexiglass, and its specific parameters are as follows: impeller inlet diameter of 74.67 mm, impeller outlet diameter of 138.67 mm, and rated speed of 1500 r/min. Figures 1 and 2 show the model test device and the site layout.

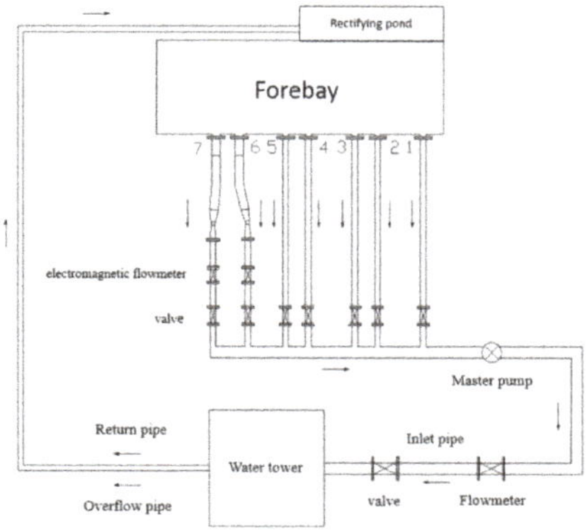

Figure 1. Schematic diagram of the overall model device. (1–7 is the pipe number).

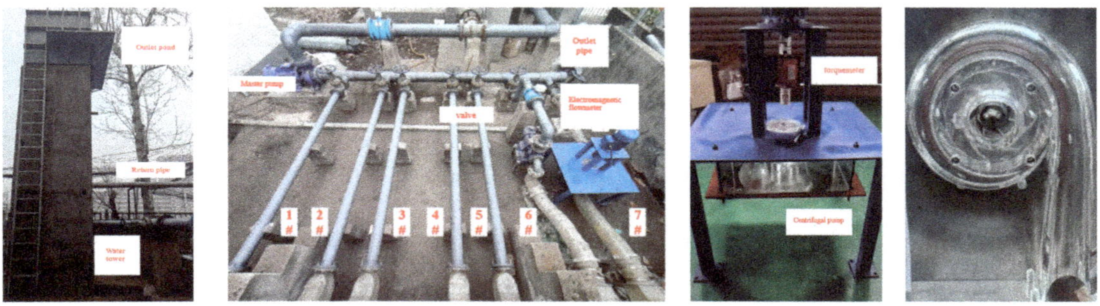

(**a**) (**b**) (**c**) (**d**)

Figure 2. Model site layout. (**a**) The layout of water towers, return pipes, and outlet pond. (**b**) The layout of pumps, outlet pipes, and electromagnetic flowmeter. (**c**) Torquemeter and centrifugal pump. (**d**) Centrifugal pump. (1#–7# is the pipe number).

3. Numerical Calculation Method

3.1. Computational Modelling and Meshing

Figure 3 shows the numerical calculation area in this study, which contains the pumping station intake sump, inflow runner, seven vertical single-stage single-suction centrifugal pumps (impellers, guide vanes, and spiral casing), and the discharge pipe.

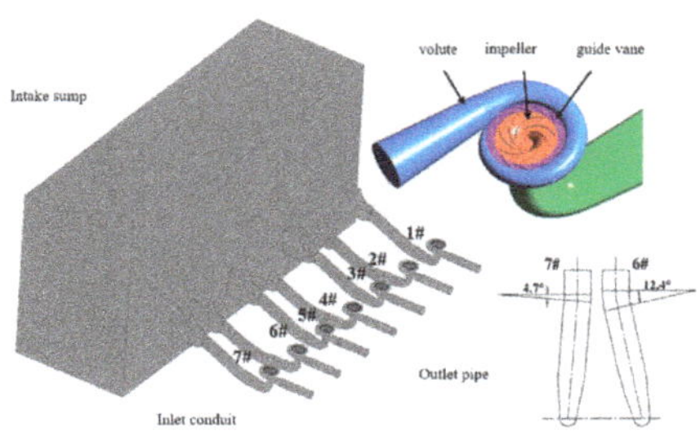

Figure 3. Geometric model of the model pumping station. (1#–7# is the pipe number).

The pumping station prototype single-pump design flow rate is 12.6 m^3/s, the design of the water lift is 108.5 m, the maximum water lift is 116.5 m, and the minimum water lift is 100.9 m. According to similar theoretical conversions of a single-pump design with a flow rate of 22.68 m^3/h, the design of the water lift is 4.34 m, the impeller inlet and outlet diameters are 74.67 and 138.67 mm, respectively, and the rated speed is 1500 r/min. The shaped inlet passage of the elbow draft tube has two forms: one is 4.7° to the left of the central axis (units 1, 3, 5, and 7), and the other is 12.4° to the right of the central axis (units 2, 4, and 6).

The hexahedral mesh is used in this study to mesh each flow pump component to select the appropriate number of meshes that can guarantee the calculation accuracy and avoid the slow calculation speed. Under the design condition, with the pump head as the judgment index, the centrifugal pump unit section (including the intake sump, inflow runner, impeller, guide vane, spiral casing, and outlet pipe section) for a single unit of the mesh-independent verification calculations is shown in Figure 4. The figure reveals that the change in pump head amplitude is small and stable when the number of grids is 3.2 million in terms of the calculation cycle and the calculation capability. Finally, the number of grids is 4.26 million to complete the calculation after the grid distribution of various parts of the overcurrent, as shown in Figure 5.

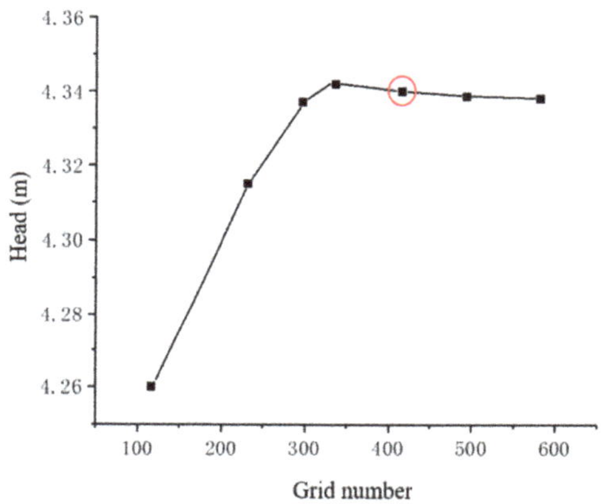

Figure 4. Grid-independent validation. (The red circle is the final number of meshes).

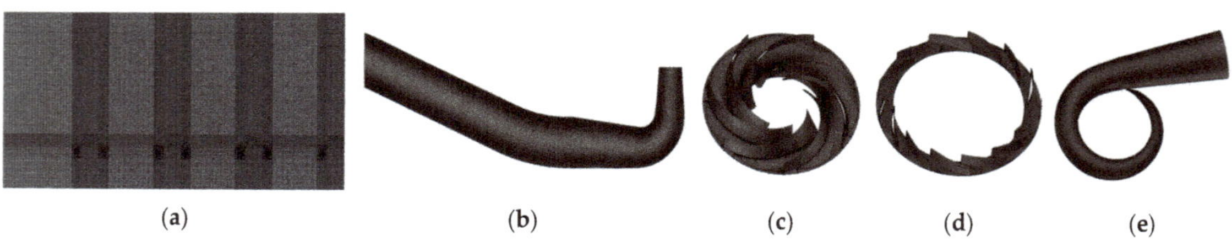

Figure 5. Grid schematic diagrams of each flow pump component. (**a**) Forebay grid; (**b**) inflow runner grid; (**c**) impeller grid; (**d**) guide vane grid; and (**e**) volute grid.

3.2. Numerical Calculation Model

With the pumping station centrifugal pump to water as the working medium, incompressible flow in the heat exchange calculation can be ignored, and only the continuity and momentum equations are considered [20].

(1) Continuity equation:

$$\frac{\partial \rho_m}{\partial t} + \frac{\partial (\rho_m u_i)}{\partial x_i} = 0, \quad (5)$$

where ρ is the fluid density (kg/m^3), t is the time (s), and u_i is the velocity tensor (m/s).

The research fluid in this paper is an incompressible fluid. Thus, the fluid density will not change with time, and the density remains constant. Equation (5) can then be simplified as follows:

$$div(\rho u) = 0, \tag{6}$$

(2) Momentum equation:

$$\frac{\partial(\rho u_i)}{\partial t} + \frac{\partial(\rho u_i u_j)}{\partial x_j} = -\frac{\partial p}{\partial x_i} + \frac{\partial}{\partial x_j}\left(\mu\frac{\partial u_i}{\partial x_j} - \rho\overline{u_i' u_j'}\right) + S_i, \tag{7}$$

where p is the pressure on the microbody (Pa), μ is dynamic viscosity ($Pa{\cdot}s$), $-\rho\overline{u_i' u_j'}$ is the Reynolds stress (N), and S_i is the generalized source term.

A Reynolds stress term in the Reynolds-averaged Navier–Stokes equation remains unknown in the calculation of numerical simulation; thus, the turbulence model must be introduced when inserting the equations [21]. The issue studied in this paper belongs to the actual large-scale engineering problems; therefore, the standard model can be used to meet the computational requirements [22].

Numerical simulation software ANSYS (https://www.ansys.com/) is used in this paper to solve the flow field, and the boundary conditions and physical parameters are set as follows: the selected fluid medium is 25 °C water, the mass flow inlet is set at the inlet of the forebay, the flow rate changes with the number of startup units, and the specified value is 6.28 kg/s under the design condition of a single unit. Based on several pumping station calculation examples, the boundary condition at the outlet of the pumping station is selected as the hydrostatic outlet and set to a standard atmospheric pressure. The rotational speed is 1500 r/min. The grid interface is connected by a general grid interface (GGI) grid, and the convergence accuracy is 10^{-4} with a maximum residual.

4. Internal Flow Characteristics of Centrifugal Pumps under Different Combinations of Start Conditions

A certain deflection angle in the water flow from the forebay into the pumping station intake sump will result in poor flow conditions, thereby inducing deterioration of the pump suction conditions. Therefore, in the study of startup combinations, the most economical and reasonable startup scheme is selected, the inlet water flow pattern is improved, and the stable operation of the pumping unit is then ensured. This paper selects the section flow uniformity and the average deflection angle as the evaluation indexes (good or poor) of water flow in the flow channel. The value of the axial flow uniformity of the section is close to 1, while the value of the average deflection angle is close to 0°, which indicates that the axial flow velocity distribution of the pump impeller inlet section is highly uniform: vertical water flow into the pump leads to superior inlet conditions of the internal flow of the water pump [23].

The flow velocity uniformity λ of the water pump section is

$$\lambda = \left[1 - \frac{1}{\overline{v}_a}\sqrt{\frac{\sum(v_{ai} - \overline{v}_a)^2}{m}}\right] \times 100\%, \tag{8}$$

where: λ—inlet section axial flow uniformity, %; $\overline{v}_a$—average axial velocity of inlet section, m/s; m—number of calculation units in the inlet section; v_{ai}—axial velocity of unit i of the inlet section, m/s.

The average deflection angle of pump section θ is computed as follows:

$$\theta = \frac{\sum v_{ai}\left[\arctan\left(\frac{v_{ti}}{v_{ai}}\right)\right]}{\sum v_{ai}}, \tag{9}$$

where: θ—mean cross-section deflection angle, °; ν_{ti}—axial velocity of unit *i* of the inlet section, m/s; $\nu_{ti} = \sqrt{\nu_{wi}^2 + \nu_{ri}^2}$ (ν_{wi}, ν_{ri} are the tangential and radial velocities of the *i* unit of the inlet section, respectively).

When more than one pumping unit simultaneously runs in the pumping station, the inlet conditions between different units vary, affecting the flow pattern of adjacent or separated pumping units. This paper aims to identify the economic and stable start conditions of multi-unit pumping stations. Numerical simulation and experimental analysis of all different start combinations of opening two, three, four, five, and six units are conducted, and the inlet cross-section of the pumps is selected as the characteristic cross-section. The most economical combination of operating conditions is determined by combining the evaluation indexes of various schemes, the flow velocity distribution of the characteristic cross-section, as well as the entropy production and vortex volume distribution. The velocity cloud diagram, entropy production diagram, and vorticity diagram were obtained through ANSYS. The velocity cloud diagram represented the velocity distribution of the inlet section of the pump, and the entropy production diagram and vorticity diagram represented the flow pattern of the inlet section of the pump.

4.1. Parallel Operation of Two Units

All the start combination conditions are simulated during the operation of the two units, and experimental tests on the hydraulic losses in the inlet section of the pumping station are simultaneously conducted. Figure 6 shows the distribution of flow uniformity and average deflection angle of the cross-section for different start combination scenarios. Figure 7 shows the distribution of hydraulic losses in the inlet section of the pumping station for different start combination scenarios and the total hydraulic losses.

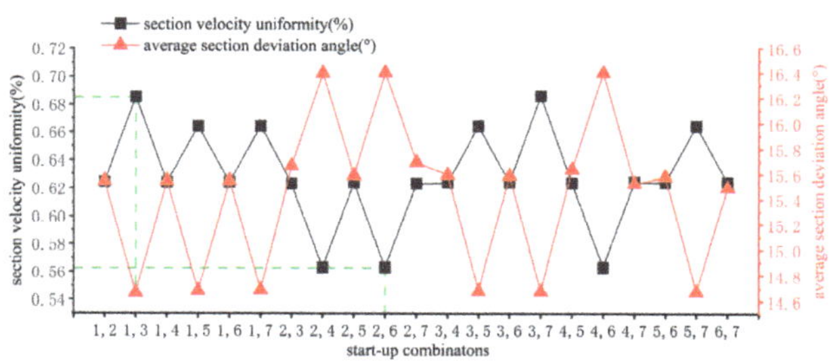

Figure 6. Uniformity of flow velocity and average section deviation angle distribution of cross-section for different start-up combination schemes. (The dotted line represents the coordinates of the point).

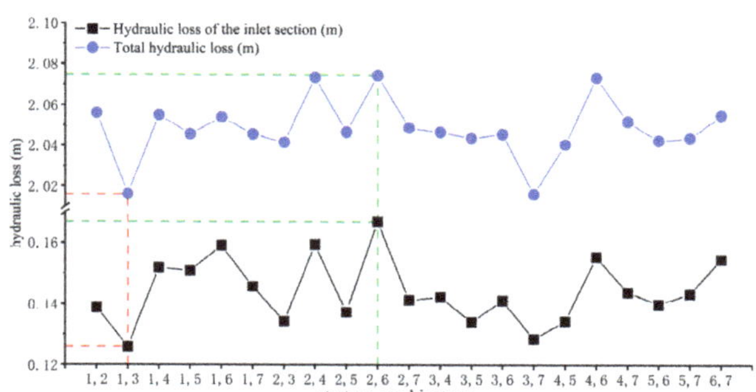

Figure 7. Hydraulic loss of the inlet section and total hydraulic loss of the pump station system for two units operating. (The dotted line represents the coordinates of the point).

The figures show that for two units running simultaneously, opening units 1 and 3 as well as 3 and 7 of the evaluation indexes are almost the same, demonstrating a flow rate uniformity of up to 68.55%. Opening units (2, 4), (2, 6), and (4, 6) of the evaluation index are no different. In these units, the flow rate uniformity is the lowest at 56.27%, the section flow rate uniformity of the maximum value and the minimum difference is within 12.28%, and the change in average cross-section deflection angle is minimal. The difference in maximum and minimum values is only 1.75°. The cross-section flow rate uniformity of the law of change and the pump unit hydraulic loss law are different from the law of change. When units 1 and 3 are switched on, the total hydraulic loss is the smallest (2.016 m); when units 2 and 6 are switched on, the total hydraulic loss is the largest (2.075 m). According to the test and simulation results, when the pumping station needs to switch on two units, the priority is to switch on the unit combination with the inlet flow path deviated to the left. Thus, the benefits of units 1 and 3 are optimal at this time. However, the benefits are worse when switching on the units with the right deviation of the inlet flow path. Meanwhile, the benefits of units 1 and 3 are optimal. The benefit is worse when the inlet flow path deviates to the right, and the worst benefit is achieved when the combination of units 2 and 6 is switched on.

Figure 8 shows the cross-section flow velocity distributions for the worst startup combination for units 2 and 6 and the best startup combination for units 1 and 3. Under the operating conditions of the two units, the following conditions are observed. When the inlet runner is switched on to the left of the unit, the corresponding flow velocity distribution area is average. When the inlet runner is switched on to the right of the unit, the flow velocity near the wall changes abruptly (high flow velocity is suddenly reduced to 0), and a large area of low-speed zone emerges at the wall, which is prone to cause water disturbance and induces an unstable flow, resulting in the emergence of vortex and increasing hydraulic losses. Differences in the flow state in the intake pipe will inevitably lead to variations in energy distribution within the pump.

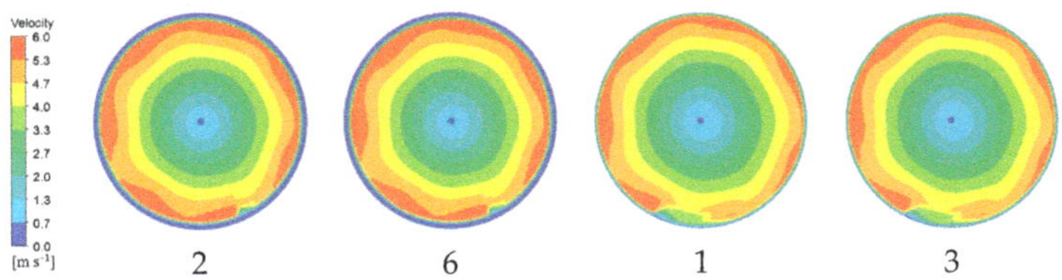

Figure 8. Velocity cloud diagram of characteristic section.

Figures 9 and 10, respectively, show the entropy production and vortex distribution in the middle section of the main overflow parts (impeller, guide vane, spiral casing) of the centrifugal pump with the optimal and worst combinations when two units are simultaneously switched on. The analysis shows that under the same working conditions, different combinations of opening the centrifugal pump affect its operating performance. In the impeller channel, the impeller energy loss is mainly concentrated in the impeller outlet due to the dynamic and static interferences between the impeller and guide vane, and local high entropy production area also occurs in individual impeller channels. In the guide vane and the spiral case, the energy loss is mainly concentrated in the guide vane flow channel due to the impingement of the high-speed flow of the fluid medium on the guide vane blades and the flow disorder due to the dynamic and static interferences between the guide vane and the impeller. The suction surface of the blade produces a local low-pressure area, thus forming a reverse pressure gradient; under the action of centrifugal force, the suction surface of the impeller flow separation phenomenon generates a vortex. Numerous vortices are generated at the inlet of the guide vane, which impedes the flow of water in the flow channel, resulting in increased energy loss. The entropy production value at this time is higher than that at the impeller, indicating that the energy loss at the guide vane is higher than the energy loss in the impeller. Compared with the worst start combination, the entropy production value of the optimal start combination is remarkably reduced, and the vortex volume is also remarkably smaller than the worst start combination. This finding indicates that when the optimal start combination of units 1 and 3 is switched on, the fluid flow is relatively stable, and the flow loss is small. Overall, when two units must be switched on, priority is given to switching on units 1 and 3.

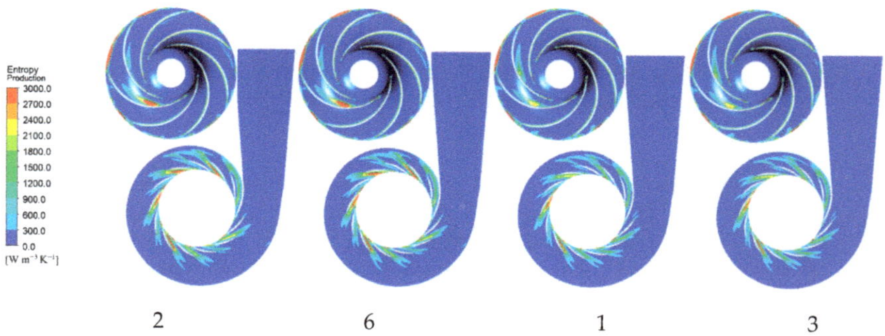

Figure 9. Distribution of entropy production of centrifugal pump's main overflow parts when two units are turned on.

Figure 10. Vorticity cloud diagram of main flow components of centrifugal pump when two units are started.

4.2. Three Units in Parallel Operation

Figures 11 and 12 show the distribution of section flow uniformity and average section deviation angle as well as the distribution of hydraulic losses in the inlet section and total hydraulic losses for different start combination scenarios when the three units are operated in parallel.

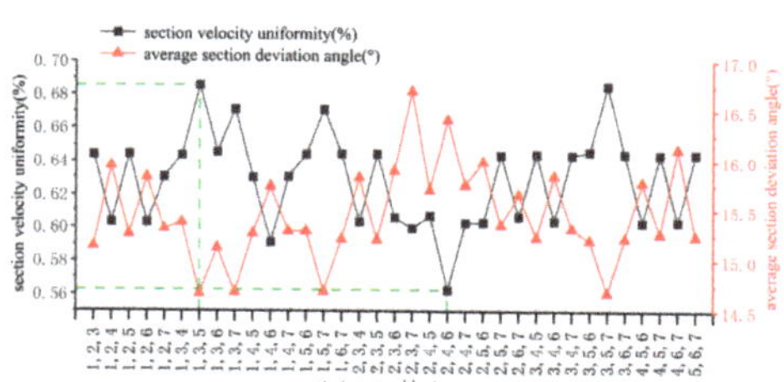

Figure 11. Uniformity of flow velocity and average section deviation angle distribution of cross-section for different start-up combination schemes. (The dotted line represents the coordinates of the point).

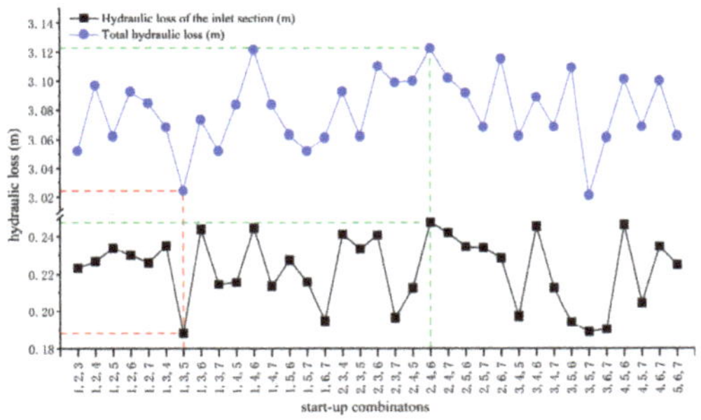

Figure 12. Hydraulic loss of the inlet section and total hydraulic loss of the pump station system for three units operating. (The dotted line represents the coordinates of the point).

According to the calculation results, for three units running simultaneously, opening units 1, 3, and 5 as well as 3, 5, and 7 of the evaluation indexes are almost the same, revealing the highest flow rate uniformity of up to 68.54%. Meanwhile, opening units 2, 4, and 6 yields the lowest flow rate uniformity of 56.27%. The maximum and maximum values of the section flow rate uniformity demonstrated a difference of 12.27%. The relative change in the average cross-section deflection angle is minimal. The difference in maximum and minimum values is only 2.03°. Cross-section flow rate uniformity of the law of change and the pumping unit hydraulic loss law is different from the law of change: opening units 3, 5, and 7 yielded the smallest total hydraulic loss of 3.021 m, while opening units 2, 4, and 6 produced the largest total hydraulic loss of 3.123 m. The test and simulation results of the trend are the same. When units 1, 3, and 5 are opened, the difference in maximum and minimum values is 12.27%. The hydraulic loss in the inlet section of the pumping station system is the smallest (0.1882 m) when units 1, 3, and 5 are switched on. In contrast, the hydraulic loss in the inlet section of the pumping station system is the largest (0.2473 m) when units 2, 4, and 6 are switched on.

Figure 13 shows the characteristic cross-sectional flow velocity distributions for the worst startup combination (opening units 2, 4, and 6) and the optimal startup combination (opening units 1, 3, and 5). The characteristic cross-section flow velocity plots reveal that, under the operating conditions of three units, the corresponding flow velocity distribution is better when the unit with the inlet runner to the left is switched on than when the unit with the inlet runner to the right is switched on. Compared with switching on two units simultaneously, the uniformity of flow velocity in the characteristic cross-section of switching on three units simultaneously is almost unchanged.

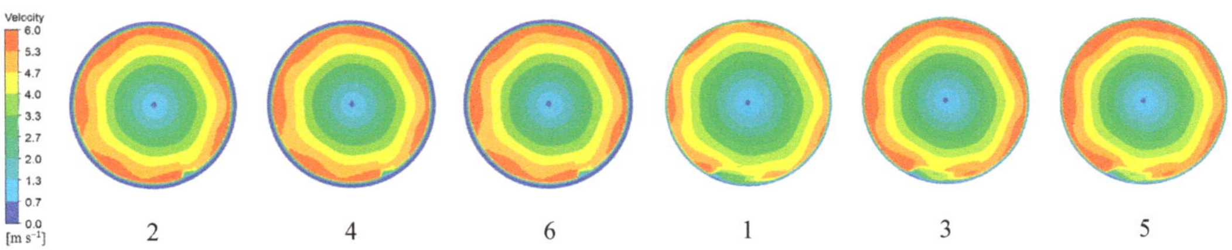

Figure 13. Velocity cloud diagram of characteristic section.

Figure 14 reveal the entropy production inside the centrifugal pump. Figure 15 shows the entropy production distribution of the main overflow parts (impeller, guide vane, volute) in the middle section of the centrifugal pump for the optimal and worst combinations when three units are switched on simultaneously. Compared with switching on two units simultaneously, the entropy production in the flow channel of the main overflow parts of the centrifugal pump is increased when the three units are switched on simultaneously.

Therefore, when the three units are simultaneously switched on, the energy loss inside the centrifugal pump is larger than when the two units are simultaneously switched on; the optimal start combination is smaller than that of the worst start combination compared with that of the optimal start combination. The optimal start combination has a smaller entropy production value than the worst start combination in the impeller channel and the channel near the nose of the spiral case. The maximum value of the vortex distribution of the centrifugal pump when three units are switched on is mainly concentrated in the impeller suction surface and guide vane inlet and flow channel. Meanwhile, the vortex of the worst combination significantly increases in the impeller channel near the wall of the vane suction surface, as well as in the guide vane flow channel and the spiral casing near the outlet of the guide vane. The vortex increase is particularly observed in the guide vane flow channel. The flow of the optimum combination is better than that of the worst combination in all flow channels. Therefore, when three units must be switched on, priority is given to switching on units 1, 3, and 5.

Figure 14. Entropy production distribution cloud diagram of main flow parts of the centrifugal pump when three units are operated.

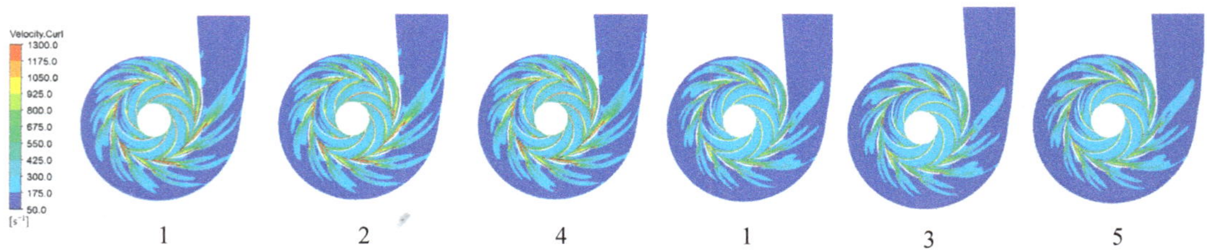

Figure 15. Vorticity cloud diagram of main flow components of the centrifugal pump when three units are started.

4.3. Parallel Operation of Four Units

All the start combination conditions are simulated in this paper during the operation of the four units. Figure 16 shows the calculation results of the four units. Meanwhile, Figure 17 reveals the distribution of hydraulic and total hydraulic losses in the inlet section of the pumping station during the operation of the four units.

According to the calculation results, for the case of four units running simultaneously, the flow rate uniformity is the highest when units 1, 3, 5, and 7 are opened, reaching 67.49%. The flow situation is similar when the following units are opened: 1, 2, 4, and 6; 2, 3, 4, and 6; 2, 4, 5, and 6; 2, 4, 6, and 7. When flow rate uniformity is the lowest at 59.32%, the difference in flow velocity section uniformity of the maximum and minimum values is 8.17%. Meanwhile, the average cross-section deflection angle of the relative change is minimal, and the difference in the maximum and minimum values is only 1.50°. The cross-section flow rate uniformity of the law of change and the pump unit hydraulic loss law is different from the law of change. When units 1, 3, 5, and 7 are switched on, the total hydraulic loss is the smallest at 4.06628 m; when units 1, 2, 4, and 6 are switched on, the total hydraulic loss is the largest at 4.1318 m. The test and numerical simulation results demonstrate the same trend: when units 1, 3, 5, and 7 are switched on, the hydraulic loss of the pumping station system inlet section is the smallest at 0.2535 m; when units 1, 2, 4, and 6 are switched on, the hydraulic loss of the pumping station system inlet section is the largest at 0.3479 m.

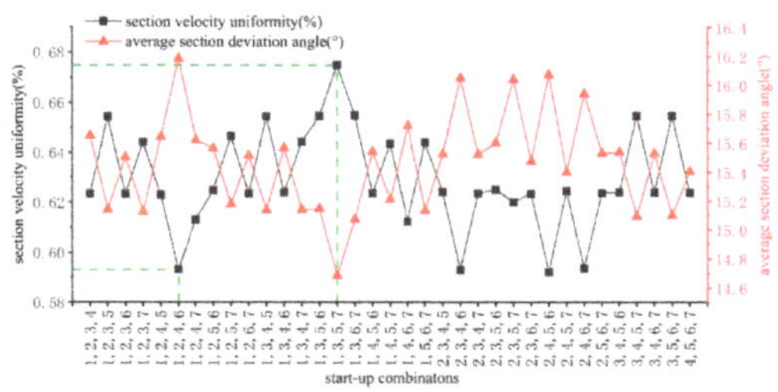

Figure 16. Uniformity of flow velocity and average section deviation angle distribution of cross-section for different start-up combination schemes. (The dotted line represents the coordinates of the point).

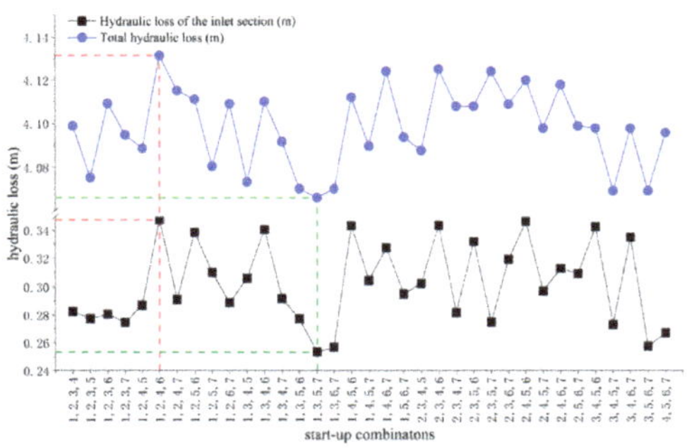

Figure 17. Hydraulic loss of the inlet section and total hydraulic loss of the pump station system for four units operating. (The dotted line represents the coordinates of the point).

Figure 18 shows the characteristic cross-section flow velocity distribution cloud plots for the worst startup combination, opening units 1, 2, 4, and 6, and the optimal startup combination, opening units 1, 3, 5, and 7. The characteristic cross-section flow velocity maps indicate the following: when the four units are operated, the corresponding flow velocity distribution area is better when the unit with the inlet runner to the left is switched on than when the unit with the inlet runner to the right is switched on. Compared with switching on three units simultaneously, no change in the flow uniformity in the characteristic section is observed when four units are simultaneously switched on. Figure 19 shows the entropy production distribution cloud diagram of the middle section of the main overflow parts (impeller, guide vane, volute) of the centrifugal pump with the optimal and worst combinations when four units are simultaneously switched on.

The entropy production of the centrifugal pump decreases when four units are simultaneously switched on compared with that when two and three units are switched on. This finding indicates that the mutual influence between the units gradually decreases at this time, the inlet conditions improve, and the energy loss is reduced. The optimal and worst combinations when four units are simultaneously switched on include unit 1. However, in the optimal combination, the entropy production of unit 1 is slightly smaller than that of the worst combination. This finding indicates that the inlet conditions of the centrifugal pumps are affected differently when varying units are switched on. Comparing units 2,

4, and 6 with units 3, 5, and 7, the high entropy production area of units 3, 5, and 7 in the optimal combination is significantly smaller than that of units 2, 4, and 6 in the worst combination. This result indicates that the internal flow of the centrifugal pumps under the conditions of the optimal combination is uniform and the energy loss is small. Compared with the worst startup combination, the entropy production value of the optimal startup combination is generally significantly reduced, and the vortex volume is also significantly smaller than the worst startup combination. This finding indicates that when opening the optimal combination of units 1, 3, 5, and 7, the fluid flow remains relatively stable, and the flow loss is small.

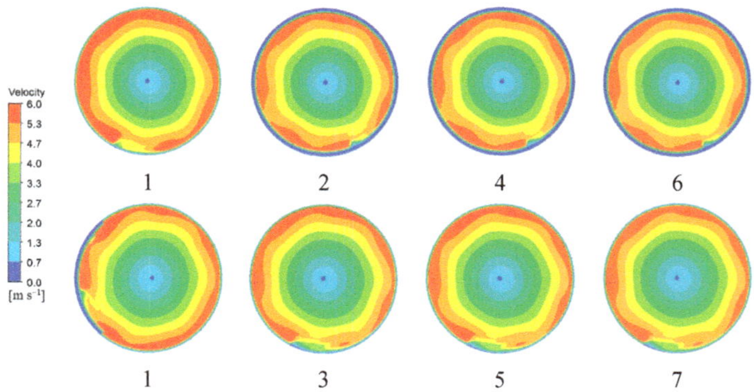

Figure 18. Velocity cloud diagram of characteristic section.

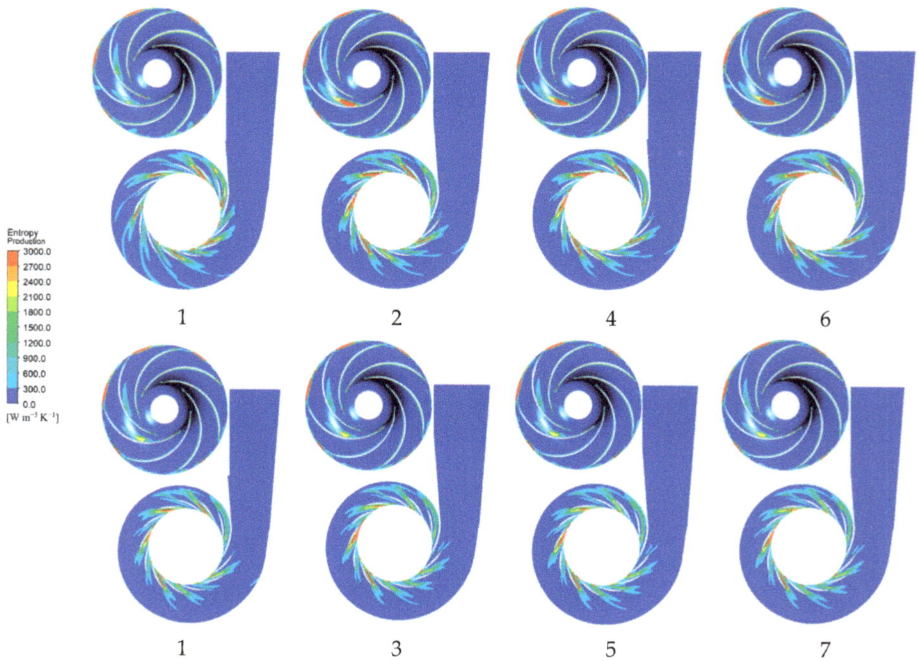

Figure 19. Entropy production distribution cloud diagram of the main flow parts of the centrifugal pump when four units are started.

Compared with the case when two and three units were simultaneously switched on, the vortices in the overflow components were reduced when four units were simultaneously switched on. This phenomenon is consistent with the distribution of the entropy production values, indicating that the flow inside the centrifugal pump was more uniform when four units were simultaneously switched on than when two and three units were simultaneously switched on. Figure 20 shows that the worst combination for high vorticity area is more evident than the optimal combination. Overall, when four units must be switched on, priority is given to switching on units 1, 3, 5, and 7.

Figure 20. Vorticity cloud diagram of main flow components of the centrifugal pump when four units are started.

4.4. Parallel Operation of Five Units

The five units were simulated in this paper for all the start combinations of operating conditions. Figure 21 shows the specific start combination scheme and the calculation results. Meanwhile, Figure 22 shows the distribution of hydraulic and total hydraulic losses in the inlet section of the pumping station during the operation of five pumps.

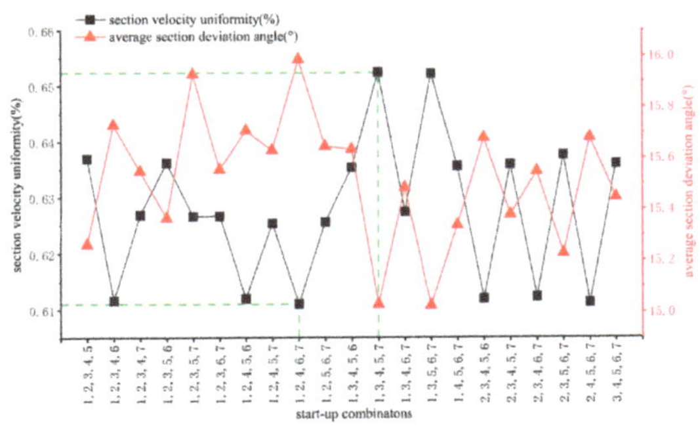

Figure 21. Uniformity of flow velocity and average section deviation angle distribution of cross-section for different start-up combination schemes. (The dotted line represents the coordinates of the point).

The calculation results reveal that for five units operating simultaneously, units 1, 3, 4, 5, and 7 are opened when the flow rate uniformity is the highest (up to 65.24%), while units 1, 2, 4, 6, and 7 are opened when the flow rate uniformity is the lowest (61.11%). The difference in maximum and minimum values of section flow rate uniformity is 3.93%, and the relative change in the average cross-section deflection angle is minimal. The difference between the maximum and minimum values is only 0.96°. Moreover, the change law of section flow rate uniformity and pump unit hydraulic loss change law is different from that of the minimum total hydraulic loss. The difference between the maximum and minimum values is only 0.96°. The rule of change of section flow uniformity and pump unit hydraulic loss change law is also different: opening units 1, 3, 4, 5, and 7 yields the smallest total hydraulic loss of 5.10 m, while opening units 1, 2, 4, 6, and 7 leads to the largest total hydraulic loss of 5.15 m. Thus, the test and numerical simulation results demonstrate the same trend. Meanwhile, units 1, 3, 4, 5, and 7 are simultaneously opened when the pumps are not in the same position. The test has the same trend as the numerical simulation results. The hydraulic loss in the inlet section of the pumping station system is the smallest when units 1, 3, 4, 5, and 7 are switched on, and the hydraulic loss in the inlet section of the pumping station system is the largest when units 1, 2, 4, 6, and 7 are switched on. The hydraulic loss in the inlet section of the pumping station system is the largest (0.451 m) when units 1, 2, 4, 6, and 7 are switched on.

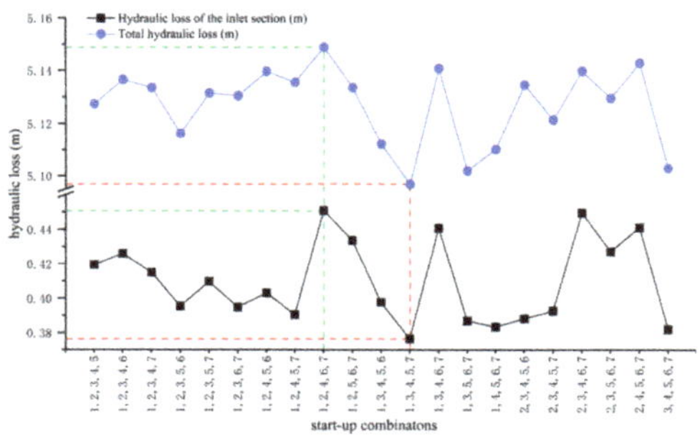

Figure 22. Hydraulic loss of the inlet section and total hydraulic loss of the pump station system for five units operating. (The dotted line represents the coordinates of the point).

Figure 23 shows the characteristic cross-section flow velocity distribution cloud diagrams for the worst startup combination (opening units 1, 2, 4, 6, and 7) and the optimal startup combination opening units (1, 3, 4, 5, and 7). It can be seen from the characteristic cross-section flow velocity maps that, under the operating conditions of five units, the corresponding flow velocity distribution area is better when the unit with the inlet runner to the left is switched on than when the unit with the inlet runner to the right is switched on. Compared with simultaneously switching on four units, the flow velocity distribution in the characteristic section of the worst combination is more uniform than that when five units are simultaneously switched on. The flow velocity uniformity in the section of the optimal combination is worse than that of switching on four units simultaneously. Therefore, when five units must be switched on, priority is given to switching on units 1, 3, 4, 5, and 7.

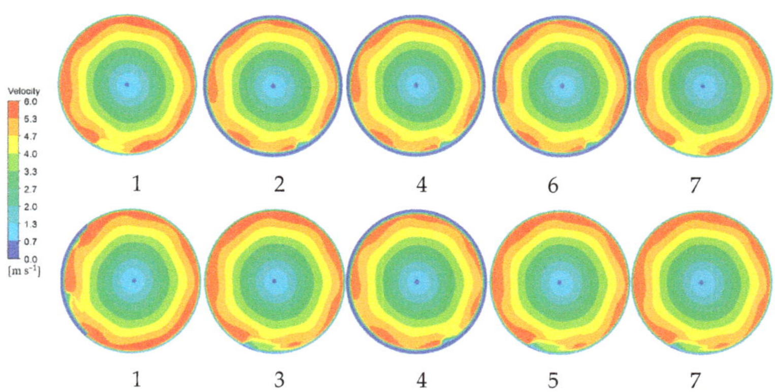

Figure 23. Velocity cloud diagram of characteristic sections.

4.5. Six Units in Parallel Operation

All the start combination conditions are simulated in this paper during the operation of six units. Figure 24 shows the specific start combination scheme and the calculation results. Meanwhile, Figure 25 shows the distribution of hydraulic losses in the inlet section of the pumping station and the total hydraulic losses during the operation of the six units.

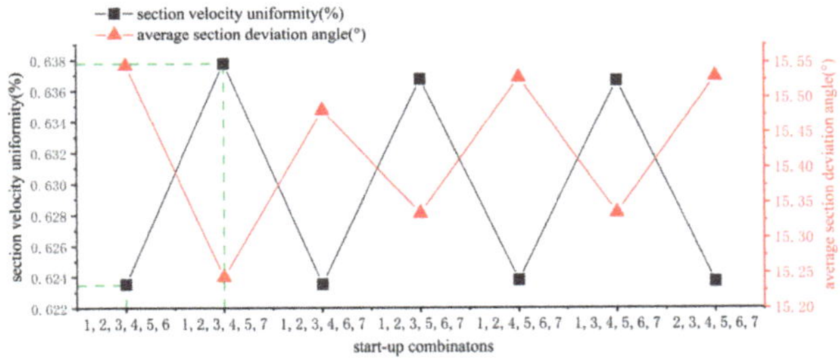

Figure 24. Uniformity of flow velocity and average section deviation angle distribution of cross-section for different start-up combination schemes. (The dotted line represents the coordinates of the point).

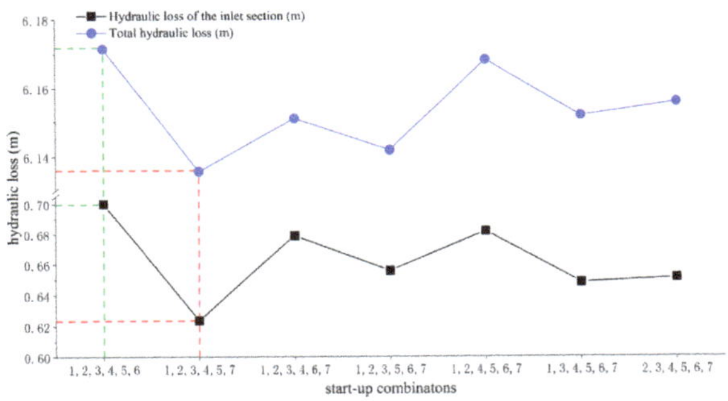

Figure 25. Hydraulic loss of the inlet section and total hydraulic loss of the pump station system for six units operating. (The dotted line represents the coordinates of the point).

The calculation results reveal that for the six units operating simultaneously, opening units 1, 2, 3, 4, 5, and 7 yields the highest flow rate uniformity of up to 63.78%, while opening units 1, 2, 3, 4, 5, and 6 results in the lowest flow rate uniformity of 62.35%. The difference between the maximum and minimum values of the section flow rate uniformity is 1.34%, and the relative change in the average cross-section deflection angle is minimal. The difference in the maximum and minimum values is only 0.3°. The change rule of the section flow rate uniformity and the pump unit hydraulic loss is different. When units 1, 2, 3, 4, 5, and 7 are switched on, the difference between the maximum and minimum values of total hydraulic loss is only 0.3°. The change rule of section flow uniformity is different from that of hydraulic loss of the pumping unit: when switching on units 1, 2, 3, 4, 5, and 7, the total hydraulic loss is the smallest (6.136 m); when switching on units 1, 2, 3, 4, 5, and 6, the total hydraulic loss is the largest (6.172 m). The test results reveal the same trend as the results of numerical simulations. When six units are simultaneously switched on, the hydraulic loss in the inlet section of the pumping station system is the smallest (0.6236 m) when units 1, 2, 3, 4, 5, and 7 are switched on, while the hydraulic loss in the inlet section of the pumping station system is the largest (0.7 m) when units 1, 2, 3, 4, 5, and 6 are switched on.

Figure 26 shows the characteristic cross-section flow velocity distribution cloud diagrams for the worst start combinations for switching on units 1, 2, 3, 4, 5, and 6 and the optimal start combinations for switching on units 1, 2, 3, 4, 5, and 7. The characteristic cross-section flow velocity map reveals that when the six units are running, compared with the previous conditions, the flow velocity uniformity of the pump inlet cross-section and the average cross-section deflection angle do not demonstrate substantial changes. The cross-section cloud map is also maintained. Overall, when switching on the six units, priority is given to switching on units 1, 2, 3, 4, 5, and 7.

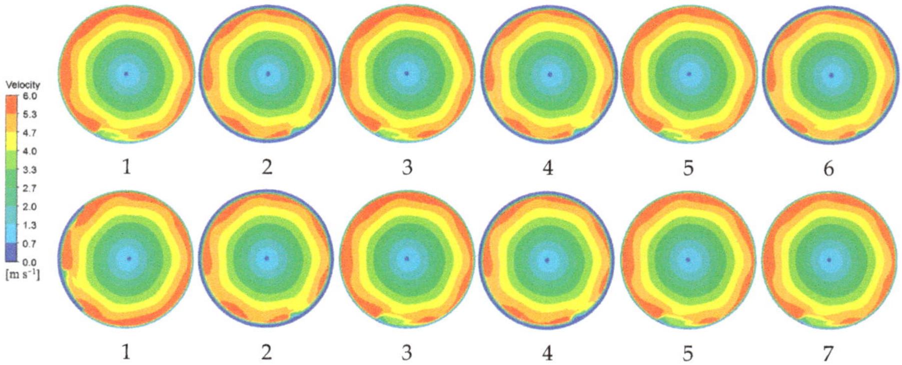

Figure 26. Velocity cloud diagram of a feature section.

5. Conclusions

In this paper, through the design of different scheme combinations under the number of starts, the simultaneous individual opening of two, three, four, five, and six units as well as other operating conditions is analyzed to calculate the axial flow velocity uniformity of the inlet section of the pump, the average section deviation angle, and the hydraulic loss. The pump is examined under different start combination conditions and pump flow characteristics to explore the opening of different numbers of units under the circumstances of economical start combinations. The following conclusions are obtained:

(1) When different numbers of units are running, the number of odd-numbered units compared to that of even-numbered units in the water conditions is superior, the section flow uniformity is high, and the average section deviation angle and hydraulic loss are small.

(2) The results of vortex and entropy maps of the impeller, guide vane, and the middle section of the volute show that the unstable flow in the centrifugal pump mainly occurs at the impeller outlet and in the guide vane runner. Therefore, the main energy loss of the centrifugal pump occurs in the vicinity of the impeller outlet and the guide vane runner. For different start combinations, the differences in the flow characteristics of the centrifugal pump are mainly reflected in the areas with the most severe undesirable flow, such as the impeller runner and the entire runner of the guide vane.

(3) Owing to the differences in the pumping station inlet conditions under different startup combinations, the overall combination of simulation calculations reveals the optimal startup scheme combination of pumping station systems as shown in Table 1: when two units are simultaneously switched on, priority is given to units 1 and 3; when three units are switched on, priority is given to units 1, 3, and 5; when four units are switched on, priority is given to units 1, 3, 5, and 7; when five units are switched on, priority is given to units 1, 2, 3, 5, and 7; when six units are simultaneously switched on, priority will be given to units 1, 2, 3, 4, 5, and 7.

Table 1. The optimal startup scheme combination of pumping station systems.

The Number of Units in Operation	The Optimal Startup Scheme
2	1, 3
3	1, 3, 5
4	1, 3, 5, 7
5	1, 2, 3, 5, 7
6	1, 2, 3, 4, 5, 7

Author Contributions: X.Z. and P.Z. conceived and supervised the study, and carried out the numerical simulations; Y.P. and W.W. conducted the experiment and wrote the first draft of the manuscript. X.Z. and Y.Z. carried out the data analysis and edited the manuscript. All authors have read and agreed to the published version of the manuscript.

Funding: This work was supported by the Joint Foundation of Shaanxi [Grant No. 2019JLP-25].

Data Availability Statement: Data are contained within the article.

Conflicts of Interest: P.Z. was employed by the Hanjiang-to-Weihe River Valley Water Diversion Project Construction Co., Ltd. The remaining authors declare that the research was conducted in the absence of any commercial or financial relationships that could be construed as a potential conflict of interest.

References

1. Zhang, R.; Cheng, J.; Zhu, H.; Yao, L.; Zhang, L. Low-head Pump Performances and Determination of Reasonable Scope for Variable Speed Operation. *J. Ineragricult. Machy* **2009**, *40*, 78–81.
2. Wu, H.; Zhou, Z.; Gao, Y.; Li, J. Research of Optimized Dispatching of Cascade Pumping Stations Based on Hydrodynamic Dispersion Model and Genetic Algorithm. *Haihe Water Resour.* **2014**, 46–49. Available online: https://www.alljournals.cn/view_abstract.aspx?pcid=5B3AB970F71A803DEACDC0559115BFCF0A068CD97DD29835&cid=3ECA06F115476E3F&jid=8D2FD083F0BFAB3C390FA3932AD11EAA&aid=34B6762159E1CD4D51FFED50B5A38286&yid=9EAD63ADE6B277ED (accessed on 1 April 2024).
3. Lin, L. Optimised Scheduling of Pumping Station System Based on Genetic Algorithm. Master's Thesis, Tianjin University, Tianjin, China, 2007.
4. Moradi-Jalal, M.; Mariño, M.A.; Afshar, A. Optimal Design and Operation of Irrigation Pumping Stations. *J. Irrig. Drain. Eng.* **2003**, *129*, 149–154. [CrossRef]
5. Pengcheng, C.; Fan, Y.; Dandan, S.; Li, Z.B.; Shen, Q.R. Analysis of flow pattern and rectification measures in lateral inlet forebay of multi-unit pumping station. *China Rural Water Conserv. Hydropower* **2021**, *12*, 229–234.
6. Xu, C.; Wang, R.; Liu, H.; Lian, H.; Wang, Y.; Wang, G. Research on the influence of start-up combinations on the flow pattern in forebay of side-in-let pumping station on sandy river. *J. Water Resour.* **2020**, *51*, 92–101.
7. Wu, P.; Zhou, Q.; Ding, H.; Wang, Z.; Xu, J. Numerical simulation of inlet flow of symmetrically distributed units in Dazhai River. *Water Conservancy Sci. Technol. Econ.* **2021**, *27*, 68–70.
8. Bagirov, A.M.; Barton, A.F.; Mala-Jetmarova, H.; Al Nuaimat, A.; Ahmed, S.T.; Sultanova, N.; Yearwood, J. Ant-Colony Optimization for Optimal Pump Scheduling. *Math. Comput. Model.* **2013**, *57*, 873–886. [CrossRef]
9. Zi, D.; Wang, B.; Wang, F.; He, C.; Xue, S. Influences of start-up pump units on the sediment concentration for the intake system of a pumping station. *J. Agric. Eng.* **2022**, *38*, 59–68.
10. Weilin, W. Research on the Optimal Operation of Terrace Pumping Station in Two Lakes Section. Master's Thesis, Jinan University, Jinan, China, 2020.
11. He, Y.; Pan, Z.; Yu, H.; Qiu, S. Optimizatin on combination of pumping unit start-up for large pumping station under complicated flow pattern in fore bay. *Water Conserv. Hydropower Technol.* **2016**, *47*, 69–74.
12. Li, Y.; Luo, W.; Li, C. An Analysis of Flow Pattern of Forebay and Units Commitment Optimization. *China Rural Water Hydropower* **2014**, 108–111. Available online: https://kns.cnki.net/kcms2/article/abstract?v=ttOPOQ75YvKtvHJwQyHY3aTJYH5e12cPUSvTuyOaLKJx6RvSfAONWbW2R2D7J0PPPh2UP5gZSN_Fqraky39xeLCEhQuuz9Ucb8T7xaQIlJfBbh5nUhCbUpp5vO7Ch43C&uniplatform=NZKPT&flag=copy (accessed on 1 April 2024).
13. Guohui, C.; Fujun, W. Numerical investigation on the flow structure and vortex behavior at a large scale pump sump. In Proceedings of the Fluids Engineering Division Summer Meeting, San Diego, CA, USA, 30 July–2 August 2007.
14. Wang, S.; Cheng, J.; Zhu, B. Optimal operation of a single unit in a pumping station based on a combination of orthogonal experiment and 0-1 integer programming algorithm. *Water Supply* **2022**, *22*, 7905–7915. [CrossRef]

15. Chen, Y.X.; Xi, B.; Chen, Z.; Shen, S. Study on the Hydraulic Characteristics of an Eccentric Tapering Outlet Pressure Box Culvert in a Pumping Station. *Processes* **2023**, *11*, 1598. [CrossRef]
16. Rahman, M.T.; Siddiqi, M.A. Study Involving the Flow Pattern Comparison of a PIV Experiment with CFD Simulation, for the Flow within a Centrifugal Pump. In Proceedings of the Fluids Engineering Division Summer Meeting, Chicago, IL, USA, 3–7 August 2014; American Society of Mechanical Engineers: New York, NY, USA, 2014; Volume 46216, p. V01AT07A006.
17. Byskov Rikke, K.; Jacobsen Christian, B.; Pedersen, N. Flow in a centrifugal pump impeller at design and off-design conditions—Part II: Large eddy simulations. *J. Fluids Eng.* **2003**, *125*, 73–83. [CrossRef]
18. Ding, H.; Li, Z.; Gong, X.; Li, M. The influence of blade outlet angle on the performance of centrifugal pump with high specific speed. *Vacuum* **2019**, *159*, 239–246. [CrossRef]
19. Zhang, K. *Principles of Fluid Machinery*; Mechanical Industry Press: South Norwalk, CN, USA, 2014.
20. Wang, F.; Tang, X.; Chen, X.; Xiao, R.; Yao, Z.; Yang, W. A review on flow analysis method for pumping stations. *J. Water Resour.* **2018**, *49*, 47–61.
21. Liu, W. Research on the Influence Law of Double Worm Shell Structure on the Radial Force of Centrifugal Pump Impeller. Master's Thesis, Lanzhou University of Technology, Lanzhou, China, 2018.
22. Zhou, S. Research on the Problem of solid-Liquid Two-Phase Flow Centrifugal Pump Erosion and Wear. Master's Thesis, Xi'an University of Technology, Xi'an, China, 2017.
23. Ming, L.; Yong, W.; Wei, X.; Erlin, W.; Houlin, L. Optimisation of geometric parameters of lateral inlet forebay of pumping station. *J. Agric. Eng.* **2022**, *38*, 69–77.

water

MDPI

Article

Comparative Analysis of the Hydrodynamic Performance of Dual Flapping Foils with In-Phase and Out-of-Phase Oscillations

Ertian Hua [1,*], Linfeng Qiu [1], Rongsheng Xie [1,2], Zhongxin Su [1] and Wenchao Zhu [1]

1 College of Mechanical Engineering, Zhejiang University of Technology, Hangzhou 310023, China; 2112102093@zjut.edu.cn (L.Q.); xrspump@163.com (R.X.); suzhongxin_66@163.com (Z.S.); 2112102394@zjut.edu.cn (W.Z.)

2 School of Mechanical & Automotive Engineering, Zhejiang University of Water Resources and Electric Power, Hangzhou 310018, China

* Correspondence: het@zjut.edu.cn; Tel.: +86-13588114369

Abstract: In the context of the plain river network, conventional water pumps suffer several drawbacks, including inadequate efficiency, poor security, and costly installation costs. In order to improve the hydrodynamic insufficiency problem and enhance the hydrodynamic performance and applicability of flapping hydrofoils, this paper proposes a bionic pumping device based on dual flapping foils. Based on the finite volume method and overlapping grid technology, the numerical simulation and experimental verification of the hydraulic performance of two typical motion modes of in-phase and out-of-phase oscillations are conducted, thereby providing a theoretical foundation for improving and optimizing the design of flapping hydrofoils. The results show that the out-of-phase oscillation has better hydraulic performance compared to the in-phase oscillation. The formation of the tail vortex structure plays a crucial role in determining the hydraulic efficiency of dual flapping foils, with in-phase oscillation forming a pair of vortex streets and out-of-phase oscillation forming two pairs of vortex streets. The pumping efficiency of the out-of-phase oscillation is significantly higher than that of the in-phase oscillation, reaching up to 38.4% at a fixed frequency of $f = 1$ Hz, which is an increase of 90.5% compared to the in-phase oscillation. The characteristic curve of the in-phase oscillation shows an "S" type unstable oscillation phenomenon, namely the hump phenomenon, while the out-of-phase oscillation does not show such a phenomenon, which can effectively expand its application range. In addition, the applicable head of the out-of-phase oscillation hydrofoil is lower, which can better meet the requirements of ultra-low head conditions.

Keywords: dual flapping foils; hydrodynamic performance; numerical simulation; test verification

Citation: Hua, E.; Qiu, L.; Xie, R.; Su, Z.; Zhu, W. Comparative Analysis of the Hydrodynamic Performance of Dual Flapping Foils with In-Phase and Out-of-Phase Oscillations. *Water* **2023**, *15*, 3275. https://doi.org/10.3390/w15183275

Academic Editors: Georg Umgiesser, Ran Tao, Changliang Ye, Kan Kan, Huixiang Chen and Yuan Zheng

Received: 20 July 2023
Revised: 1 September 2023
Accepted: 14 September 2023
Published: 15 September 2023

1. Introduction

In the region characterized by a plain river network, the presence of a mild topography and low water flow rate leads to inadequate hydrodynamics; the water pollution caused by it harms the residents' health and quality of life [1–5]. As one of the common management measures, the pump-gate joint dispatching method can enhance the hydrodynamics of river networks and efficiently enhance the quality of the water's surroundings [6,7]. However, under the ultra-low head condition, the pumping station has the problems of low operating efficiency, excessive vibration, and poor stability, which cannot adapt to the situation with almost no heads of small rivers in the plains [8–10]. Research has found that the bionic pumping device imitating the movement of the fishtail has the advantages of high propulsion efficiency, low noise, and high flow rate, which can play a beneficial role in pushing water [11,12]. The relevant studies mainly focus on single foils and dual foils need to be further studied [13,14]. In addition, studies have shown that dual foils have certain advantages over single foils, summarized by the following: Compared with single foils, dual foils are more adaptable to the river's width in application scenarios. The dual foil also has higher stability and flow field structure, as well as more robust control performance for better and more precise control of the water flow direction. Therefore, this paper proposes a bionic pumping device based on a dual flapping foil, aiming to improve the problem of insufficient river hydrodynamics and enhance the mobility of water flow.

Many scholars have studied dual foils, which are widely used in energy harvesting, underwater vehicles, and wave gliders [15–20]. In terms of motion parameters, Kinsey and Dumas investigated the energy harvesting performance of a turbine based on a tandem configuration using numerical methods and introduced the global phase parameter, and they found that the maximum efficiency could reach 64% by varying the inter-foil spacing and motion phase difference parameters [21]. Subsequently, He et al. further investigated the inter-foil spacing and analyzed the influence mechanism numerically, and concluded that the inter-foil spacing has a significant effect on the energy harvesting of the dual foils [22]; Bao et al. studied the propulsion performance of the dual foils and found that the effect of the inter-foil spacing on the propulsion performance was mainly in the wake interference [23]. Dewey et al. examined the propulsion performance of tandem hydrofoils using a vortex array model and PIV experiments, and demonstrated the wake vortex structure with varying phase differences to analyze the effect of phase difference on propulsion performance [24]. Liu et al. numerically investigated the energy harvesting efficiency of a dual foil in the presence of a 90° phase relationship in pitch amplitude, and explored the effects of frequency, amplitude, and other kinematic parameters on energy harvesting efficiency [25]. Ahmet Gungor et al. numerically investigated the effect of two oscillating foils in a side-by-side configuration on the phase difference and Strouhal number. Then, they classified the vortex patterns of the wake of the dual foil and elucidated the physical mechanism of the wake merging process [26,27]. Chao Li Ming et al. numerically investigated the thrust generation and wake structure of dual foils in tandem, parallel, and staggered arrangements, and found that it is affected by the horizontal gap and vertical gap [28]. Huerta et al. conducted an experimental study of dual foils with different phase differences in open water and found that dual foil with a side-by-side structure is more beneficial in terms of efficiency compared to a single foil [29]. Wang et al. undertook a parametric analysis of the propulsive efficiency of propulsors. Their findings revealed that a propulsive efficiency exceeding 70% could be attained by employing suitable combinations of kinematic parameters [30].

Studies have shown that dual flapping foils exhibit excellent hydrodynamic performance under appropriate motion parameters. Most of the studies have been related to energy harvesting and underwater propulsion, while relatively few studies have been focused on pumping water to enhance the water body flow. This paper proposes a bionic pumping device based on dual flapping foils to enhance water body mobility in river network areas. To further enhance the pumping performance, a comprehensive analysis of the oscillation mode is conducted. In addition, the study also pointed out that the oscillation mode of the dual flapping foils has a crucial effect on the performance of the bionic pumping device, mainly in the difference of the wake vortex structure. [24]. Therefore, this paper focuses on two typical oscillation modes, namely in-phase and out-of-phase oscillation. We conduct numerical simulations and experimental validations to compare and analyze the impact of the dual flapping foil on pump performance in these two modes. This paper sets out a guideline for optimizing the structural design of the bionic pumping device.

2. Motion Model

2.1. Motion Description

The study of hydrofoils is divided into two types: flexible hydrofoils and rigid hydrofoils. The flexible hydrofoil has more degrees of freedom, complex motion, and advantages in the higher flutter frequency region. In contrast, the motion of the rigid hydrofoil is simple, easy to control, low cost, and has more practical engineering significance, so this paper selects the rigid hydrofoil as the research object. This study chooses the NACA0012 airfoil as the subject of investigation, with its corresponding profile depicted in Figure 1.

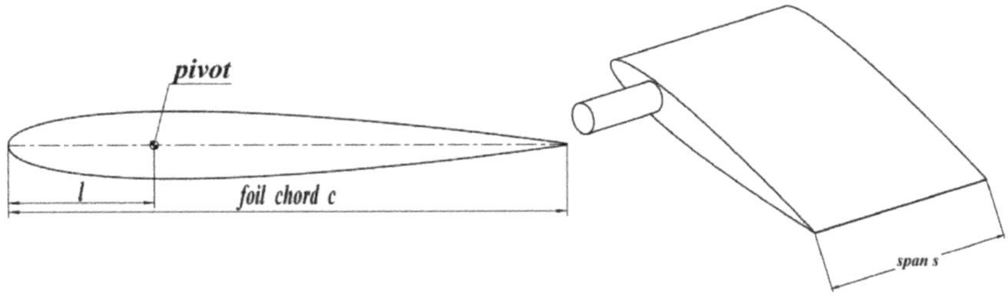

Figure 1. Schematic diagram of the hydrofoil.

In this paper, the motion is a synchronous pitch and lift motion of the dual flapping foils according to the simple harmonic law. The motion of both parts varies by sinusoidal law, as shown in Figure 2. Figure 2 illustrates the motion of the dual flapping foils, exhibiting both in-phase and out-of-phase oscillation.

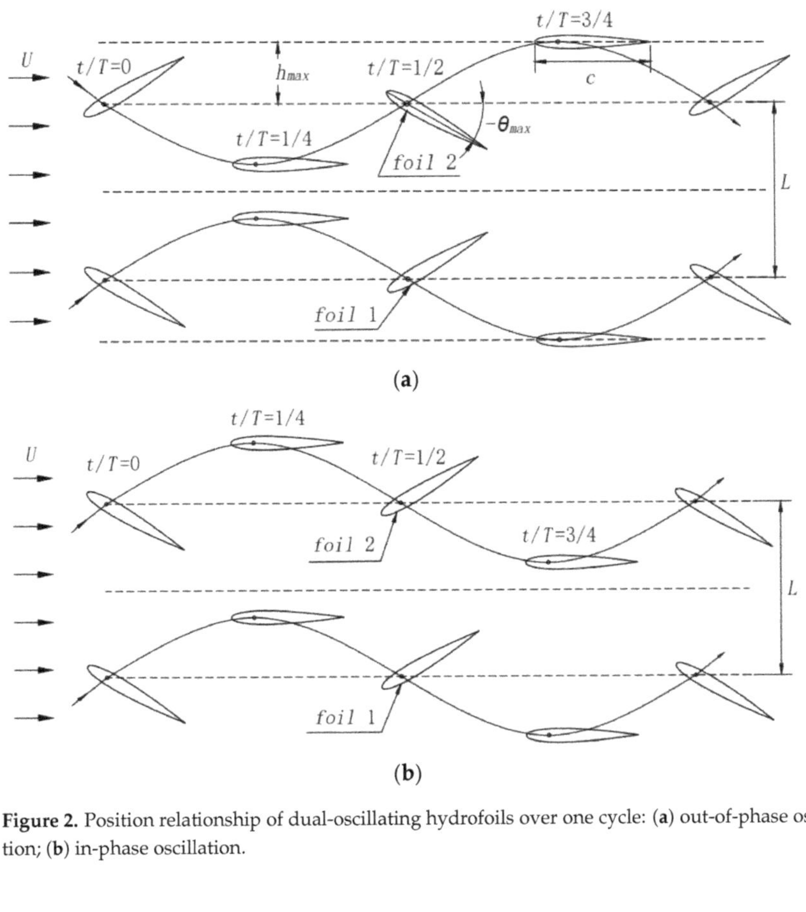

Figure 2. Position relationship of dual-oscillating hydrofoils over one cycle: (**a**) out-of-phase oscillation; (**b**) in-phase oscillation.

In Figure 2, $h_{\max}$ represents heaving amplitude and $\theta_{\max}$ represents pitching amplitude, which is 30° in this study. T represents the period of motion. L represents the distance between two foils. The equation describing the motion of dual flapping foils is given by:

$$\begin{cases} h_1(t) = h_{\max}\sin(2\pi f t) \\ \theta_1(t) = \theta_{\max}\sin(2\pi f t + \phi) \\ h_2(t) = h_{\max}\sin(2\pi f t + \varphi) \\ \theta_2(t) = \theta_{\max}\sin(2\pi f t + \phi + \varphi) \end{cases}, \tag{1}$$

where $h_1(t)$ and $h_2(t)$ are the heave displacement of foil 1 and foil 2, respectively. $\theta_1(t)$ and $\theta_2(t)$ are the pitch displacement of foil 1 and foil 2, respectively. f is the frequency of the motion, ϕ is the phase angle between heave and pitch, φ is the phase difference between the two foils. To illustrate this, $\varphi = 0$ represents the in-phase oscillation; $\varphi = \pi$ represents the out-of-phase oscillation.

The velocity equation is derived by differentiating the equation of motion provided as follows:

$$\begin{cases} \dot{h_1}(t) = 2\pi f h_{\max}\sin(2\pi f t + \varphi) \\ \dot{\theta_1}(t) = 2\pi f \theta_{\max}\sin(2\pi f t + \phi) \\ \dot{h_2}(t) = 2\pi f h_{\max}\sin(2\pi f t + \varphi) \\ \dot{\theta_2}(t) = 2\pi f \theta_{\max}\sin(2\pi f t + \phi + \varphi) \end{cases}, \tag{2}$$

where $\dot{h_1}(t)$ and $\dot{h_2}(t)$ are the heave speed of foil 1 and foil 2, respectively. $\dot{\theta_1}(t)$ and $\dot{\theta_2}(t)$ are the pitch speed of foil 1 and foil 2, respectively.

In this study, chord length c = 0.3 m, span s = 1 m, the distance between two foils $L = L'/c = 1.8$, the separation between the location of the pivot and the leading edge $l = 0.2c$, heave amplitude $h_{\max} = 0.5c$, pitch amplitude $\theta_{\max} = \pi/6$, phase angle $\phi = -\pi/2$, and the frequency of the motion f = 1 Hz. The Strouhal number $St = 2fA_{\max}/\overline{U}$, where $\overline{U}$ is the average velocity at the outflow once the flow has stabilized.

2.2. Mechanical Parameters and Pumping Indicators

The evaluation of hydrodynamic performance in flapping hydrofoils primarily relies on the measurement of two crucial parameters. The formulas are defined as follows:

The instantaneous thrust coefficient is defined as:

$$C_T(t) = \frac{2F_T(t)}{\rho \overline{U}^2 sc}, \tag{3}$$

The instantaneous lift coefficient is defined as:

$$C_L(t) = \frac{2F_L(t)}{\rho \overline{U}^2 sc}, \tag{4}$$

The average thrust coefficient is defined as:

$$\overline{C_T} = \frac{1}{T}\int_t^{t+T} C_T(t)\mathrm{d}t, \tag{5}$$

The average lift coefficient is defined as:

$$\overline{C_L} = \frac{1}{T}\int_t^{t+T} C_L(t)\mathrm{d}t, \tag{6}$$

where ρ is the fluid's density and s is the span of the foil.

The average input power of dual foils during one period of motion is defined as:

$$\overline{P_{\text{in}}} = \frac{1}{T}\left(\left|\int_t^{t+T} F_L(t)\dot{h}(t)\mathrm{d}t\right| + \left|\int_t^{t+T} M(t)\dot{\theta}(t)\mathrm{d}t\right|\right), \tag{7}$$

where $M(t)$ is the instantaneous torque around the shaft.

To facilitate a more comprehensive investigation into the pumping capabilities of the hydrofoil, a number of performance parameters were utilized for the study. The formulas are presented below:

The average flow rate of an outlet in a steady flow field is defined as:

$$\overline{Q} = \overline{U}bS, \tag{8}$$

where b is the width of the river channel, taken as 1.4 m.

The average head is defined as:

$$\overline{H} = \frac{\Delta\overline{P}}{\rho g}, \tag{9}$$

The pumping efficiency is defined as:

$$\eta = \frac{\Delta\overline{P}\cdot\overline{Q}}{\overline{P_{\text{in}}}}, \tag{10}$$

where $\Delta\overline{P}$ is the mean pressure differential between the inlet and the outlet.

3. Numerical Method and Validation

3.1. Governing Equation and Turbulence Model'

This paper utilizes the finite volume method as well as using Fluent 2022 simulation software (Ansys, Canonsburg, PA, USA) to perform numerical simulations. To describe the characteristics of a flow field that exhibits incompressibility, two equations are employed, which are shown in the previous study [14].

The equation governing motion control can be expressed as follows:

$$\frac{\partial\overline{u_i}}{\partial x_i} = 0, \tag{11}$$

$$\frac{\partial\overline{u_i}}{\partial t} + \overline{u_j}\frac{\partial\overline{u_i}}{\partial x_j} = -\frac{1}{\rho}\frac{\partial\overline{p}}{\partial x_i} + \frac{\partial}{\partial x_j}[(\gamma+\gamma_t)(\frac{\partial\overline{u_i}}{\partial x_j} + \frac{\partial\overline{u_j}}{\partial x_i})], \tag{12}$$

where $\overline{u_i}$ (i = 1, 2) is the fluid's average velocity, x_i (i = 1, 2) is the space coordinate, P is the fluid's pressure, γ is the kinematic viscosity coefficient, $\gamma_t = c_\mu k^2/\varepsilon$ is the turbulent viscosity coefficient, k is the turbulent kinetic energy, ε is the dissipation rate of the turbulent energy, and c_μ is the constant.

Throughout the process of motion, vortices are generated, moved, and shed by the hydrofoil's edges. The Realizable turbulence model is capable of effectively capturing the intricate flow field information. Additionally, the computational burden associated with this model is of moderate magnitude. Hence, the Realizable k-ε turbulence model has been chosen for this study. The relevant equations for this model can be found in the previous work [31].

3.2. Computational Domain and Mesh Generation

The utilization of computing capabilities and the mitigation of negative volume in the mesh are critical considerations in the numerical simulation of hydrofoil motion. To address these concerns, the overlapping mesh technique is employed.

In order to acquire a comprehensive wake flow field of the dual flapping foils, a fluid domain with dimensions of 5 m in length and 1.4 m in width was selected. In Figure 3, the computational domain comprises a foreground and background grid. The primary grid, referred to as the foreground grid, consists of a hybrid grid configuration. Due to the curved shape and complex geometry of the flapping foil used in this paper, the structured grid is not applicable, so a hybrid grid is chosen. In contrast, the secondary grid, known as the background grid, is constructed utilizing a structured grid format featuring a uniform global size of 0.008 m. The two sets of grids are numerically calculated by removing the area in the background grid corresponding to the foreground grid. Then, the adjacent grid cells are interpolated to transfer the data. Hence, in order to enhance the accuracy of calculations, the global dimensions of both grids are configured to be identical.

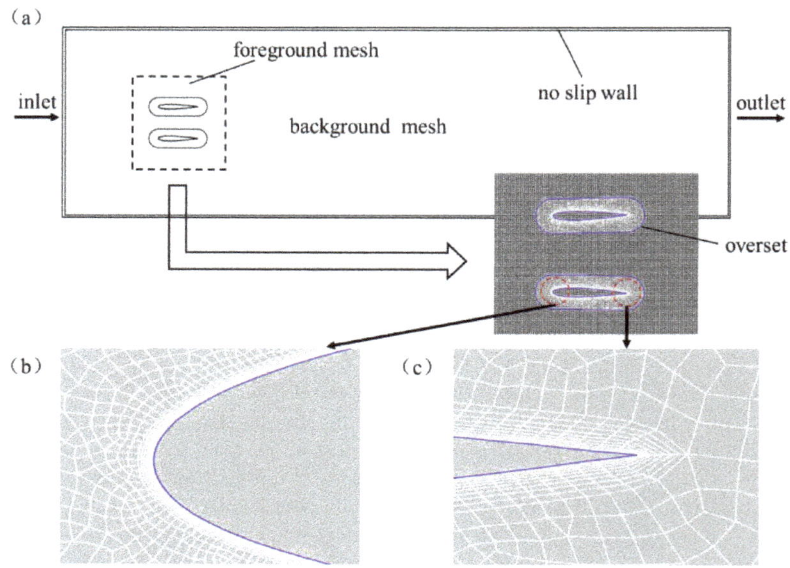

Figure 3. Computational domain mesh and boundary conditions: (**a**) grid overview; (**b**) leading edge grid; (**c**) trailing-edge grid.

Different turbulence models and different wall functions require different grid Y^+ values. To enhance the accurate representation of the dynamic wall characteristics and mitigate numerical dispersion, it is imperative to build boundary layers on both sides of the flow channel and hydrofoil wall. Hence, the initial mesh size of the boundary layer adjacent to the hydrofoil surface is established at 0.03 mm, satisfying $Y^+ < 1$ and a growth rate of 1.2.

3.3. Boundary Conditions and Parameter Settings

In this paper, in the context of boundary condition configuration, the left boundary is assigned the pressure-inlet condition, while the right boundary is assigned the pressure-outlet condition. The hydrofoil boundary and the walls flanking the flow channel are designated as no-slip walls. The outermost boundary line of the foreground grid is designated as an overset boundary, while the overlapping region between the foreground grid and the background grid is designated as a grid overlap region, as shown in Figure 3.

The oscillation mode of the hydrofoil is given via the User Defined Functions (UDF). Within the solution settings, it is necessary to choose a transient solver. The corresponding parameter settings are given in the previous study [32].

3.4. Verification of the Irrelevance of the Time Steps and Grid Number

In the context of computational fluid dynamics (CFD), the accuracy and efficiency of numerical computations are influenced by the sparsity of the grid and the time step [33]. Hence, it is imperative to examine the correlation between the grid number, the time step, and the computational results. By verifying the independence of these factors, we can enhance the precision and dependability of the computed outcomes. The validation conditions are: heave amplitude, $h_{\max} = 0.15$ m; pitch amplitude, $\theta_{\max} = 30°$; flapping frequency, f = 1 Hz; and inlet flow rate, $U = 0.4$ m/s.

To validate the irrelevance of the time steps, we chose three sets of data, which are 0.1 s, 0.01 s, and 0.005 s. As shown in Figure 4a, through comparison it was found that when the time step was 0.1 s, the instantaneous thrust coefficient obtained from the simulation was more different than the other two, so 0.01 s was ultimately chosen for the following calculation considering the calculation accuracy and efficiency.

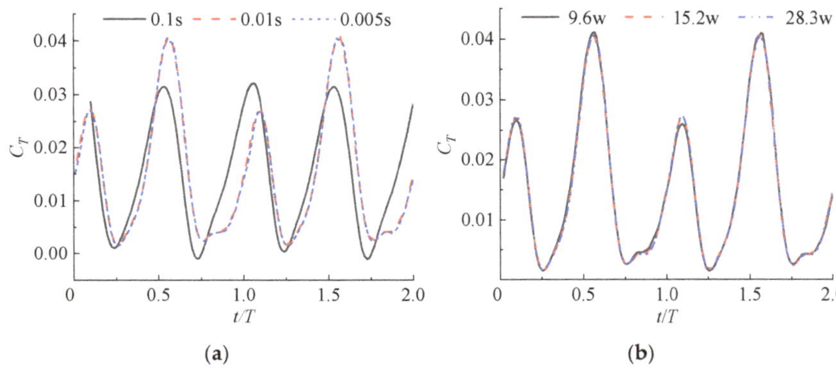

Figure 4. Verification of independence: (**a**) time steps; (**b**) grid number.

To validate the irrelevance of the grid number, three sets of grid numbers, 96,000, 152,000, and 283,000, were selected for numerical calculations. As shown in Figure 4b, through comparison it was found that the results obtained from the three sets of grids are similar, so the grid of 152,000 was ultimately chosen for the subsequent calculation considering the calculation accuracy and efficiency.

3.5. Validation

To authenticate the validity of the computational simulation method, a comparative analysis was conducted using the experimental data obtained from the towed tank laboratory at MIT (Cambridge, MA, USA) [33]. The computational model uses the NACA0012 hydrofoil with chord length $c = 0.1$ m. The simulation parameters were set in accordance with the experimental working parameters of the previous study [33]: The domain's capacity was $20c \times 15c$; inlet velocity $U = 0.4$ m/s; phase angle $\phi = 90°$; heave amplitude and chord length $h_{\max} = c = 0.1$ m; maximum angle of attack $\alpha_{\max} = 35°$; Strouhal number $St = 2fh_{\max}/U$; and the frequency of the motion $f = 1$ Hz.

The findings are depicted in Figure 5. The outcomes of the simulations correspond to the observed patterns of variation in the experimental data from the previous study. Hence, the effectiveness of the computational method utilized has been shown.

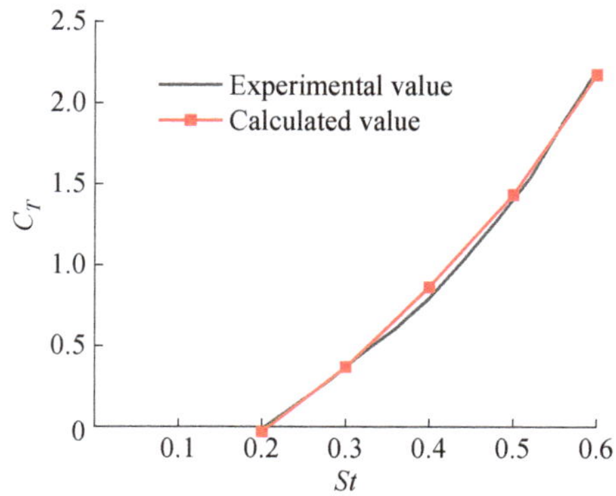

Figure 5. Comparison between the numerical simulation outcomes and the experimental data in the previous study [33].

4. Experimental Setup

4.1. Bionic Pumping Device

To verify the effects of the dual foil's in-phase and out-of-phase oscillations on its pumping performance and to evaluate the accuracy of the numerical calculations, a bionic pumping device incorporating dual flapping foils was developed. The experimental apparatus was created using two sets of synchronous belt linear modules and motors, both motor-driven and controlled by actuators, to meet the demand for the compound harmonic motion of the dual flapping foils. Figure 6 illustrates the schematic diagram of the dual flapping foils device in three dimensions.

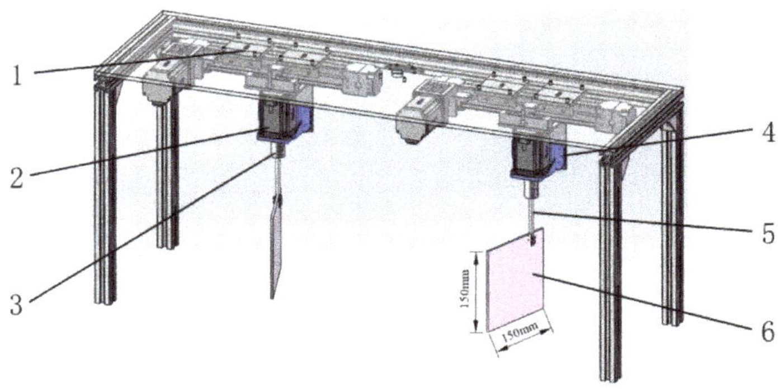

Figure 6. Three-dimensional schematic diagram of flapping hydrofoil device: 1. synchronous belt linear module; 2. motor; 3. coupling; 4. support frame; 5. connecting rod; 6. foil.

In this study, the dual flapping foils bionic pump included passage and water flapping foils. The hydrofoil's chord length was 150 mm and clearance between the hydrofoils and the top and bottom sides of the passage was 5 mm. Figure 7 shows a dual flapping foil device test bench. The test set was built on a 2.5 m × 0.7 m × 0.16 m rectangular passage, as illustrated in Figure 7a. In addition, Figure 7b illustrates the internal schematic diagram of the experimental flow passage.

This study employs the WIM-@ADV acoustic Doppler flowmeter (Beijing, China) for the purpose of experimental data acquisition. The flowmeter is utilized to monitor the average flow velocity at the outlet section. It has a flowmeter velocity measurement range from 0 to 3 m/s, with a measurement accuracy of 0.005 m/s, and a maximum sampling frequency of 50 Hz. The differential pressure sensor (Jiangsu, China) was set on the flow passage's inlet and outlet side panels, with a measuring range up to 300 Pa and a minimal scale of 0.2 Pa. In addition, the input power of the power supply side of the motor was measured via the electrical measurement method using a clamp power meter (Shanghai, China).

(a) (b)

Figure 7. (**a**) Dual flapping foil device test bench; (**b**) internal schematic diagram of the experimental flow passage.

4.2. Uncertainty Analysis

The synthesis error of the test includes systematic and random errors, which are calculated as follows:

$$E_\eta = \pm\sqrt{E_{\eta\cdot S}^2 + E_{\eta\cdot R}^2}, \tag{13}$$

$$E_{\eta,S} = \pm\sqrt{E_{Q\cdot S}^2 + E_{H\cdot S}^2 + E_{P\cdot S}^2}, \tag{14}$$

$$E_{\eta,R} = \pm\sqrt{E_{Q\cdot R}^2 + E_{H\cdot R}^2 + E_{P\cdot R}^2}, \tag{15}$$

where $E_{\eta\cdot S}$ and $E_{\eta\cdot R}$ are the systematic errors and the random errors; $E_{Q\cdot S}$, $E_{H\cdot S}$, and $E_{P\cdot S}$ are the systematic error of the Doppler flowmeter, differential pressure sensor, and clamp power meter, respectively; $E_{Q\cdot R}$, $E_{H\cdot R}$, and $E_{P\cdot R}$ are the random error of flow testing, head testing, and torque speed testing. These systematic error parameters are shown in Table 1.

Table 1. The main equipment of the experimental test system.

Terms	Equipment	Type	Systematic Error
Flow	Doppler flowmeter	WIM-@ADV	±1%
Head	Differential pressure sensor	3051	±0.2%
Current	Clamp power meter	VC6412D	±2.5%

Through the model test data, the total random error is counted using probabilistic statistical methods, and finally the total uncertainty of the experiment is calculated, for which the E_η was ±4.05%.

To ensure the test bench's reliability, the bionic pumping device underwent multiple tests in the out-of-phase oscillation mode. These tests were conducted at a flapping frequency ranging from 0.1 to 0.8 Hz, using consistent test methods and working conditions. The findings, presented in Table 2, demonstrate that the flow rate pattern of the device remains consistent across the three tests. Moreover, the information gathered under comparable workplace circumstances is extremely similar, confirming that the test results are reliable.

Table 2. Repeatability test.

	f/Hz	0.1	0.2	0.3	0.4	0.5	0.6	0.7	0.8
$v/m \cdot s^{-1}$	Expeiment 1	0.020	0.047	0.070	0.097	0.120	0.152	0.172	0.201
	Expeiment 2	0.021	0.045	0.075	0.095	0.125	0.151	0.170	0.207
	Expeiment 3	0.025	0.050	0.068	0.096	0.123	0.147	0.175	0.205

5. Results and Analysis

5.1. Influence of Two Oscillation Modes on Mechanical Properties

To investigate the effect of the two oscillation modes on the mechanical characteristics of the foils, the corresponding instantaneous thrust and lift coefficients were calculated and plotted for the oscillation frequency f = 1 Hz. Both were analyzed for foil 1; Figure 8 shows the results.

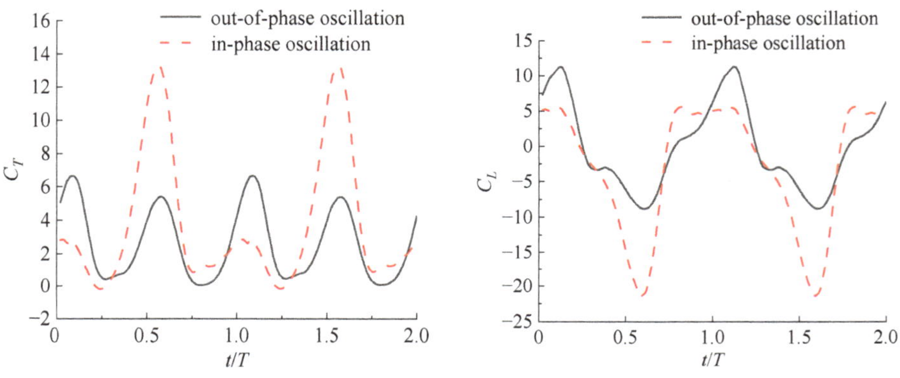

Figure 8. Variation in the instantaneous thrust and lift coefficients for two oscillation modes: (**a**) instantaneous thrust coefficients; (**b**) instantaneous lift coefficients.

To enable a comprehensive comparison, the horizontal coordinate was established by taking into account the values of the dimensionless time for each corresponding period. Figure 8a shows the instantaneous thrust coefficient curves corresponding to the two oscillation modes. Figure 8a shows that the curves are periodic, with two peaks and two valleys per motion cycle. In the in-phase oscillation mode, when combined with the relative motion form of the dual flapping foils, the motion direction of the upper and lower foil is identical. In contrast, the motion is in the opposite direction in the out-of-phase oscillation mode. For the out-of-phase oscillation mode, when the foil starts to flap upward from the initial position, the thrust gradually increases and then decreases, and reaches a valley when the motion reaches the maximum heave amplitude, and when the foil starts to flap downward in the opposite direction, the thrust keeps increasing and reaches a peak after reaching the initial position, and then keeps decreasing and reaches a second valley in the amplitude position. For the in-phase oscillation mode, the thrust decreases gradually and reaches a valley when the foil begins to flap upward from the original position, then flap downward, and the thrust increases and reaches a peak after crossing the initial position.

The comparison reveals a distinct difference between the peaks and valleys of the waves for both oscillations. The thrust peaks and valleys fluctuate more in the in-phase oscillation than in the out-of-phase oscillation. In addition, the instantaneous thrust coefficient produces negative values in the in-phase oscillation mode, indicating that the oscillating motion can at times impede the motion of the water body during a motion cycle. In contrast, the instantaneous thrust coefficient for the out-of-phase oscillation mode is always positive, indicating that the oscillating motion can propel the motion of the water body throughout the entire motion cycle.

Figure 8b shows the instantaneous lift coefficient curves corresponding to the two oscillation modes. As shown in Figure 8b, in the out-of-phase oscillation mode, the instantaneous lift coefficient curve has two peaks and two valleys in one motion cycle, both near the equilibrium position, and the curve is symmetrically distributed along the zero scale. In contrast, the instantaneous lift coefficient curve fluctuates more in the in-phase oscillation mode.

Through comparison it is found that the average lift coefficient of the out-of-phase oscillation is almost zero. In contrast, the average lift coefficient of the in-phase oscillation is negative. The instantaneous lift coefficient curve of the out-of-phase oscillation is more symmetrical than the in-phase oscillation, indicating that the flow field is more uniform.

5.2. Influence of Two Oscillation Modes on the Flow Field

Since the propulsion effect of various oscillation modes on the water body is different, the velocity nephograms of the two oscillation modes are contrasted and examined to ascertain the disparities. Figure 9 depicts a comparison and analysis of the velocity nephograms of the two oscillation modes with the oscillation frequency f = 1 Hz.

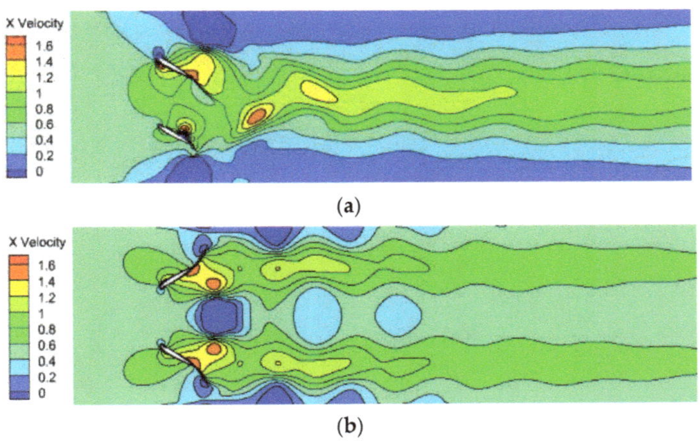

Figure 9. Velocity nephograms of two oscillation modes: (**a**) in-phase oscillation; (**b**) out-of-phase oscillation.

Figure 9 presents the velocity nephograms of the two oscillation modes. Upon conducting a comparative analysis, it has been observed that the distribution patterns of the two flow fields exhibit noticeable differences. The hydrofoil's left-side inlet exhibited a consistent and steady inflow. When the incoming flow enters, the in-phase oscillation mode moves toward the centerline of the wake between the dual flapping foils due to the interaction between the vortex pairs, forming a straight jet; at the same time, the out-of-phase oscillation mode has symmetry due to the opposite direction of the vortex formed by the two foils, forming two straight jets. In addition, the low-velocity region in the flow field near the wall area on both sides of the in-phase oscillation mode accounts for a more significant proportion of the whole region than that of the out-of-phase oscillation mode.

In the out-of-phase oscillation mode, the flow velocity in the middle region is reduced because, as depicted in Figure 10, when the flapping foils move to both sides, a low-pressure region is formed in the middle region, which generates a Karman vortex street, thereby creating a specific obstruction to the water flow. Figure 10 depicts the characteristics of the instantaneous flow field in the out-of-phase oscillation mode ($t/T = 3/4$). At this moment, two distinct vortices are formed in the trailing-edge trailing area of the dual flapping foil.

The reason for this is that the upper and lower flapping foils move away from each other to produce a low-pressure region, thus forming a trailing-edge backflow.

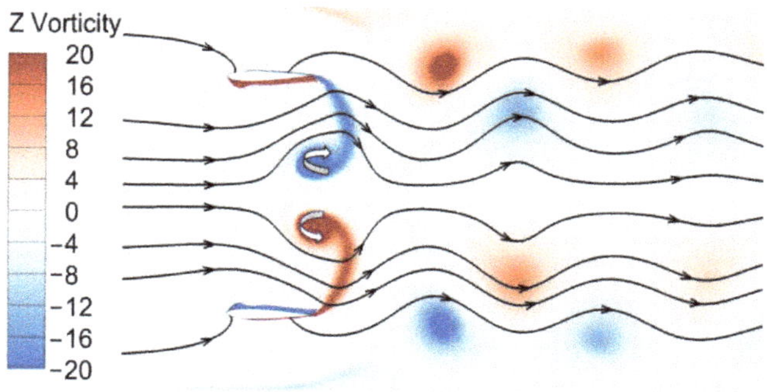

Figure 10. Characteristic diagram of the instantaneous flow field in the out-of-phase oscillation mode ($t/T = 3/4$).

To further investigate the effect of the two oscillation modes on the flow field structure, the vorticity nephogram of the two oscillation modes was analyzed by taking the oscillation frequency f = 1 Hz, as shown in Figure 11.

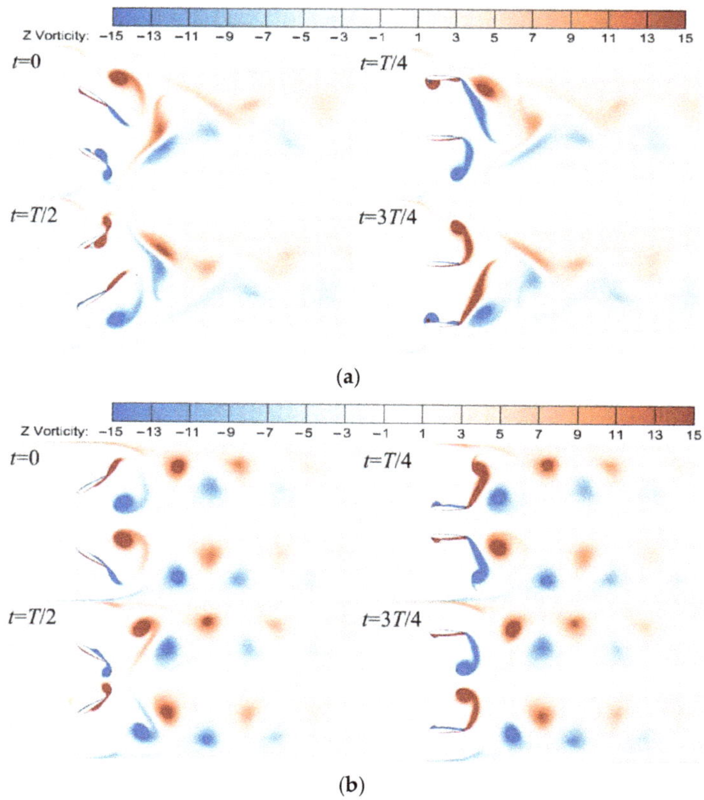

Figure 11. Vorticity nephogram of two oscillation modes during one cycle: (**a**) in-phase oscillation; (**b**) out-of-phase oscillation.

To characterize the rotational flow in the wake of the hydrofoil, we have elected to calculate the z-direction component of spin, that is, the average angular velocity of the fluid micro-cluster rotating around the rotation axis (parallel to the z-axis), ω_z, is used as the characteristic quantity.

$$\omega_z = \frac{1}{2}\left(\frac{\partial u_y}{\partial x} - \frac{\partial u_x}{\partial y}\right), \tag{16}$$

where ω_z is the angular velocity component of rotation about the z-axis and u_x and u_y are the velocity components in the x-axis and y-axis directions.

Figure 11 shows the vorticity nephogram for the two oscillation modes during one cycle. From Figure 11a, it can be seen that in the in-phase oscillation mode, a downward-deflected top vortex pair and an upward-deflected bottom vortex pair are formed during the motion, and these vortex pairs will move along the trailing-edge centerline and eventually merge to form a pair of vortex streets. Throughout the entire cycle, both the leading and trailing edges of the dual oscillating foils generate vortices, with the leading edge vortices moving along the hydrofoil surface toward the trailing edge, combining with and shedding the trailing-edge vortices. During the movement, the vortices of the same type formed by the dual flaps combine to create an enormous vortex and stagger above and below the centerline of the trailing edge to form a pair of vortex streets. In addition, the vortices formed by the up-flap and down-flap motions are flattened at $T/4$ and $3T/4$ cycles. This is due to the different strengths of the wake vortices produced by the up-flap and down-flap motions, as well as the mutual influence of the walls. When these vortices are close to each other, they will become flattened.

Figure 11b shows that in the out-of-phase oscillation mode, two pairs of vortices are formed during the motion, with the bottom vortex street and the top vortex street parallel to each other. Taking Foil 1 as an example, when the foil moves from the initial position to the $T/4$ cycle, the negative vortex formed at the trailing edge of the flapping foil is shed along the upper surface of the foil, and moves to the $T/2$ cycle, at which time the positive vortex formed at the leading edge moves along the lower surface of the flapping foil to the trailing edge, and when it moves to the $3T/4$ cycle, the positive vortex formed in the previous process is shed along the lower surface of the hydrofoil, and the previous negative vortex forms a pair of rotationally opposite and staggered anti-Karman vortex streets. The Foil 1 and Foil 2 transition processes for the tail vortex are identical.

Figures 8–11 illustrate the close relationship between the instantaneous thrust coefficient and the tail vortex structure. In the in-phase oscillation mode, the tail vortex of the dual foils is deflected due to the interaction between vortex pairs, which suppresses the negative vortex formed by the upper surface of the upper flap and the positive vortex formed by the lower surface of the lower flap, forming a flattened vortex. Since the strength of the positive and negative vortices formed by a flapping foil is different, the force of the vortex on the hydrofoil during upward swing is different from that of the vortex on the hydrofoil during downward swing, which leads to a significant difference between the two peaks of the instantaneous thrust coefficients in the in-phase oscillation mode. In addition, the vortices formed by dual flapping foils are of comparable strength in the out-of-phase oscillation mode. Hence, the difference between the two peak instantaneous thrust values in the out-of-phase oscillation is smaller.

5.3. *Influence of Two Oscillation Modes on the Pumping Performance*

Due to their various oscillations, the dual flapping foils have different effects on the pumping performance of the water body, as shown in Figure 12.

Figure 12 presents the pumping efficiency curves. Figure 12 demonstrates that pumping efficiency gradually increases as the oscillation frequency f increases, but the rate of increase gradually decreases.

Under the same operating conditions, the pumping efficiency of the out-of-phase oscillation mode is greater than that of the in-phase oscillation mode, mainly due to the following two reasons. First, the instantaneous thrust coefficient diagram shows that the out-of-phase oscillation has smoother thrust coefficient operating characteristics compared to the in-phase oscillation, with less energy loss. Hence, the pumping efficiency is relatively higher.

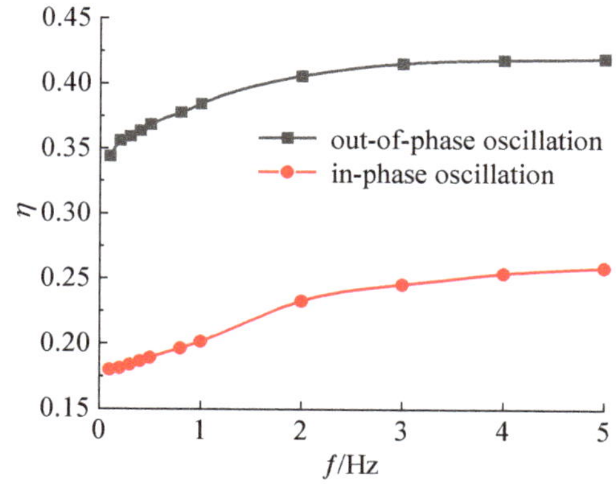

Figure 12. Pump efficiency curves of two oscillation modes.

Secondly, according to the flow field structure analysis, the interaction between the vortex pairs generated by the in-phase oscillation mode is more substantial, and the direction of the vortex pairs is severely tilted, resulting in a stronger lateral jet in the front part and a rapid dissipation of the vortex energy in the wake. In contrast, the interaction between the vortex pairs in the out-of-phase oscillation mode is relatively weak, which reduces the vortex energy dissipation and improves the pumping efficiency.

Figure 13 presents the flow and head change curves of the two oscillation modes. The figure illustrates that, when the oscillation frequency f is the same, the average flow and head of the out-of-phase oscillation mode are more significant than that of the in-phase oscillation mode. Furthermore, as the oscillation frequency f increases, the average flow demonstrates a direct proportionality to the oscillation frequency, while the average head exhibits a quadratic proportionality to the oscillation frequency.

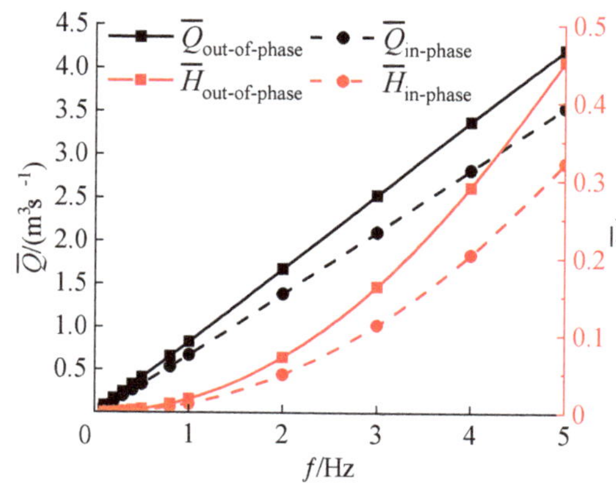

Figure 13. Variation curves of flow and head in two oscillation modes.

In order to investigate the effects of the two oscillation modes on the characteristic curves, the inlet boundary is set as the velocity inlet, and the velocity variation range is set from 0 to 1 m/s, with a variation of 0.1 m/s at each interval. In addition, the oscillation frequency f = 1 Hz is defined, as shown in Figure 14.

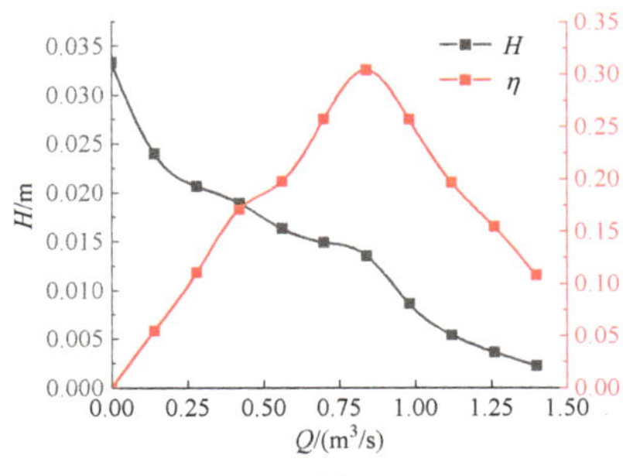

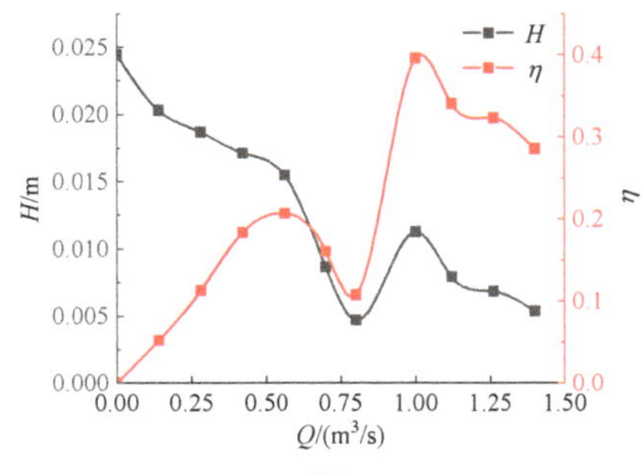

Figure 14. Pump efficiency and head characteristic curves of two oscillation modes: (**a**) out-of-phase oscillation; (**b**) in-phase oscillation.

Figure 14 presents the characteristic curves of pump efficiency versus head for the two oscillation modes. Figure 14 illustrates that the characteristic curve of dual flapping foils exhibits a comparable trend of change compared to that of the conventional pump. As the flow rate increases, there is a gradual decrease in the head, while the pumping efficiency initially increases and subsequently decreases. Both variables exhibit an optimal efficiency point. The rationale behind this phenomenon lies in the fact that at low flow rates, the flapping foil assumes a propulsive function with respect to the surrounding water body. Nevertheless, it can be observed that there is a gradual decrease in the thrust exerted by the flapping foil on the water body as the flow rate increases. The propulsive function undergoes a gradual transition into an obstructive function, leading to a decline in pumping efficiency.

In addition, Figure 14b shows that the in-phase oscillation mode has a noticeable "S" type unstable oscillation phenomenon between the flow rate of 0.64~1 m^3/s, namely the hump phenomenon. As shown in Figure 15, the graph shows that at a flow rate of 0.84 m^3/s, the internal flow field under the corresponding operating conditions shows strong instability, which poses a significant threat to the safety, stability, and reliability of the bionic pumping device. Previous studies show that the humping phenomenon leads to the instability of the device [34–37]. Hence, the pumping device with the out-of-phase oscillation is more reliable compared with the in-phase oscillation. At the same time, the maximum head is smaller when the dual flapping foil is used to push water in the out-of-phase oscillation, which can meet the demand of ultra-low head conditions.

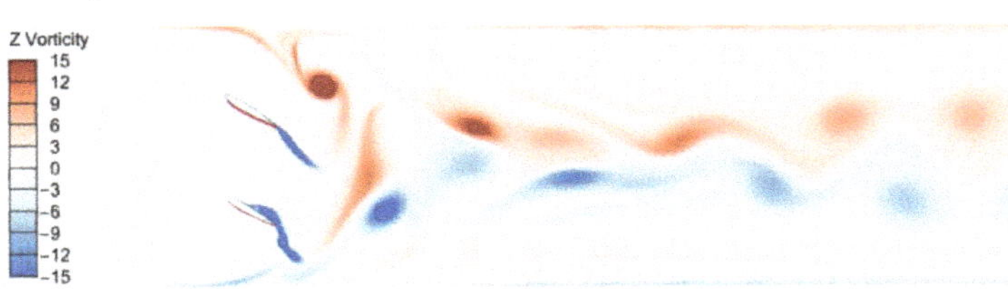

Figure 15. Vorticity nephogram of in-phase oscillation mode at a flow rate of 0.84 m^3/s ($t/T = 0$).

5.4. Performance Test

Experiments were conducted on dual flapping foils with various motion frequencies for the distinct oscillating modes. Experiments were conducted in both of the two oscillation modes for comparison, with the oscillation frequency of 0.1~0.8 Hz which changed in 0.1 Hz increments. The flow rate measurement requires six distinct sites at the exit of the rectangular flow channel for each group after 60 cycles of wing flapping, with the average value of the six data groups representing the exit flow rate under the given conditions. The results are shown in Figure 16.

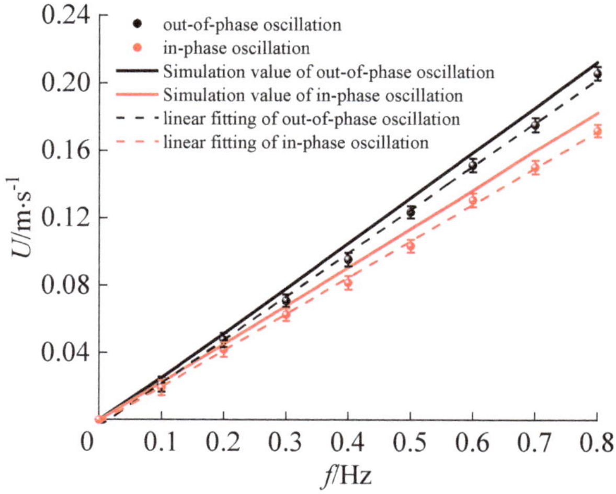

Figure 16. Comparison of flow velocity between experimental testing and simulation.

Figure 16 contrasts the results and simulation values for the flow velocity of the dual flapping foils with the two oscillation modes. Figure 16 demonstrates that the out-of-phase oscillation mode has a greater exit flow velocity than the in-phase oscillation mode. With the oscillation frequency f = 0.8 Hz, the in-phase oscillation has an exit flow velocity of 0.174 m/s, and the out-of-phase oscillation has an exit flow velocity of 0.203 m/s, which is 16.6% greater. In addition, the outlet flow velocity was proportional to the frequency of hydrofoil motion.

Moreover, the numerical calculation results exhibit a similar variation trend as the experimental results. However, the flow velocity in the test is slightly low, which is primarily due to two factors: on the one hand, the linkage bar between the plate and the device has an effect on the flow field structure; and on the other hand, due to the limited computing power of the computing equipment, the numerical simulation is a two-dimensional calculation, and the three-dimensional outcome generated in the test has some effect on the results.

6. Conclusions

This paper establishes the motion model of dual flapping foils with in-phase and out-of-phase oscillations. The mechanical characteristics, flow field structure, and pumping performance are analyzed through simulation and experimentation. The findings indicate that the out-of-phase oscillation exhibits superior hydraulic performance compared to the in-phase oscillation, as evidenced by the following results:

(1) The mechanical properties of the two oscillation modes exhibit notable distinctions. Out-of-phase oscillation consistently produces thrust throughout a motion cycle, whereas in-phase oscillation generates both thrust and drag forces. Furthermore, under the instantaneous lift coefficient curve, the out-of-phase oscillation is more symmetrical than the in-phase oscillation, indicating that the out-of-phase oscillation's flow field is more uniform.

(2) The form of the tail vortex structure is a crucial determinant affecting the hydraulic performance of the dual flapping foil. There is a significant difference in the tail vortex structure between the two oscillation modes, with in-phase oscillation forming a pair of vortex streets and out-of-phase oscillation forming two pairs of vortex streets. Furthermore, it influences the flow field, whereby in-phase oscillation results in the formation of a single straight jet, while out-of-phase oscillation led to the formation of two parallel straight jets.

(3) Both oscillations have propulsive effects on the water body. The pumping efficiency of the out-of-phase oscillation is greater than that of the in-phase oscillation. Specifically, with the oscillation frequency $f = 1$ Hz, the pumping efficiency of the out-of-phase oscillation reaches 38.4%, which is 90.5% greater than that of the in-phase oscillation. Furthermore, it should be noted that the out-of-phase oscillation results in a greater outlet flow, a more uniform flow field structure, and a superior pumping effect. Experimental verification has demonstrated that the out-of-phase oscillation yields a greater outlet flow rate when compared to the in-phase oscillation.

(4) The calculation results show that the dual flapping foil and conventional pumps have similar characteristic curves. However, in the flow rate range of 0.64~1 m^3/s, the characteristic curve of the dual flapping foils with in-phase oscillation reveals an "S" type unstable oscillation phenomenon, namely the hump phenomenon, which will lead to the instability of the device. In contrast, out-of-phase oscillation does not exhibit this phenomenon, effectively extending its application range. In addition, the out-of-phase oscillating hydrofoil has a reduced applicable head, allowing it to better meet the requirements of ultra-low head conditions.

Author Contributions: E.H. presented the main idea of analyzing the hydrodynamic performance of dual flapping foils; L.Q. contributed to the overall composition and writing of the manuscript; R.X. conducted numerical theory research; Z.S. reviewed the manuscript; W.Z. provided writing ideas and revised the manuscript. All authors have read and agreed to the published version of the manuscript.

Funding: This research was funded by the National Natural Science Foundation of China (Grant No. 51976202, 61772469), the Zhejiang Provincial Key Research and Development Project (Grant No. 2021C03019), and the Zhejiang Provincial Key Laboratory of Rural Water Conservancy and Hydropower Resources Allocation and Control Key Technologies Open Fund Project (Grant No. UZJWEU-RWM-20200304B).

Data Availability Statement: Not applicable.

Conflicts of Interest: The authors declare no conflict of interest.

References

1. Geng, J.; Wang, Y.; Hu, S.G. WQI based water quality assessment and spatial-temporal change in plain river network area of Jiashan county. *Environ. Eng.* **2023**, *41*, 187–193+209.
2. Liu, J.; Ding, R.; Chen, Y.; Zhu, W.; Fan, Z.; Ma, J. Assessment and Optimization of Water Resources Regulation for River Networks in the Tidal Plain-A Case Study of the Qingsong Area in Shanghai. *Water* **2022**, *14*, 3523. [CrossRef]
3. Wang, H.; Shen, Z.; Zeng, Y.; Yan, H.; Li, Y.; Yuan, W. Connection between Anthropogenic Water Diversion and Hydrodynamic Condition in Plain River Network. *Water* **2021**, *13*, 3596. [CrossRef]
4. Ma, Y.Q.; Cao, W.; Zhao, Y.M. Eutrophication characteristics, cause analysis and control strategies in a typical plain river network region. *Acta Sci. Circumst.* **2022**, *42*, 174–183.
5. Yuan, X.Z.; Xu, X.F.; Yu, L.L. Simulation analysis of sewage discharge in plain river network based on coupled hydrodynamic model. *China Rural Water Hydropower* **2022**, *12*, 39–46.
6. Xiao, X.M.; Chen, D.Y.; Hu, H.P. Semi-closed Water Flowing Project Based on Joint Scheduling of Sluice & Pumping Station. *China Water Wastewater* **2022**, *38*, 113–116.
7. Li, X.; Tang, H.W.; Wang, L.L. Simulation of water environment under joint operation of gates and pumps in plain river network area. *J. Hohai Univ. Nat. Sci.* **2016**, *44*, 393–399.
8. Cao, W.; Li, W. Study on the Performance Improvement of Axial Flow Pump's Saddle Zone by Using a Double Inlet Nozzle. *Water* **2020**, *12*, 1493. [CrossRef]
9. Mu, T.; Zhang, R.; Xu, H.; Zheng, Y.; Fei, Z.; Li, J. Study on improvement of hydraulic performance and internal flow pattern of the axial flow pump by groove flow control technology. *Renew. Energy* **2020**, *160*, 756–769. [CrossRef]
10. Zheng, Y.; Chen, Y.J.; Zhang, R. Analysis on unsteady stall flow characteristics of axial-flow pump. *Trans. Chin. Soc. Agric. Mach.* **2017**, *48*, 127–135.
11. Wu, X.; Zhang, X.; Tian, X.; Li, X.; Lu, W. A review on fluid dynamics of flapping foils. *Ocean Eng.* **2020**, *195*, 106712.
12. Feng, Y.K.; Su, Y.M.; Su, Y.Y. Study on hydrodynamic performance of oscillating wing under simulated fish swimming mode. *J. Huazhong Univ. Sci. Technol. Nat. Sci. Ed.* **2019**, *12*, 38–43.
13. Hua, E.R.; Chen, W.Q.; Tang, S.W.; Xie, R.S.; Guo, X.M.; Xu, G.H. Study on the characteristics of pushing flow of flapping hydrofoil device in small channel. *J. Agric. Mach.* **2022**, *53*, 9.

14. Hua, E.R.; Luo, H.T.; Xie, R.S.; Chen, W.Q.; Tang, S.W.; Jin, D.Y. Investigation on the Influence of Flow Passage Structure on the Performance of Bionic Pumps. *Processes* **2022**, *10*, 2569. [CrossRef]
15. Ma, P.; Wang, Y.; Xie, Y.; Liu, G. Behaviors of two semi-passive oscillating hydrofoils with a tandem configuration. *Energy* **2021**, *214*, 118908. [CrossRef]
16. Moreira, D.; Mathias, N.; Morais, T. Dual flapping foil system for propulsion and harnessing wave energy: A 2D parametric study for unaligned foil configurations. *Ocean Eng.* **2020**, *215*, 107875. [CrossRef]
17. Xu, W.; Xu, G.; Duan, W.; Song, Z.; Lei, J. Experimental and numerical study of a hydrokinetic turbine based on tandem flapping hydrofoils. *Energy* **2019**, *174*, 375–385. [CrossRef]
18. Mannam, N.P.B.; Krishnankutty, P.; Vijayakumaran, H.; Sunny, R.C. Experimental and Numerical Study of Penguin Mode Flapping Foil Propulsion System for Ships. *J. Bionic Eng.* **2017**, *14*, 770–780. [CrossRef]
19. Wang, Y.L.; Jiang, W.; Xie, Y.H. Numerical Investigation Into the Effects of Motion Parameters on Energy Extraction of the Parallel Foils. *J. Fluids Eng.-Trans. ASME* **2019**, *141*, 061104. [CrossRef]
20. Dahmani, F.; Sohn, C.H. Effect of convergent duct geometry on the energy extraction performance of tandem oscillating hydrofoils system. *J. Fluids Struct.* **2020**, *95*, 102949. [CrossRef]
21. Kinsey, T.; Dumas, G. Optimal Tandem Configuration for Oscillating-Foils Hydrokinetic Turbine. *J. Fluids Eng.-Trans. ASME* **2012**, *134*, 031103. [CrossRef]
22. He, G.; Yang, H.; Mo, W.; Zhao, Z.; Wang, W.; Ghassemi, H. Influence of inter-foil spacing on energy extraction of tandem oscillating hydrofoils. *Ocean Eng.* **2022**, *259*, 111953. [CrossRef]
23. Bao, Y.; Zhou, D.; Tao, J.J.; Peng, Z.; Zhu, H.B.; Sun, Z.L.; Tong, H.L. Dynamic interference of two anti-phase flapping foils in side-by-side arrangement in an incompressible flow. *Phys. Fluids* **2017**, *29*, 033601. [CrossRef]
24. Dewey, P.A.; Quinn, D.B.; Boschitsch, B.M.; Smits, A.J. Propulsive performance of unsteady tandem hydrofoils in a side-by-side configuration. *Phys. Fluids* **2014**, *26*, 041903. [CrossRef]
25. Liu, H.B.; Wang, Y.; Ma, P.L. Coupling analysis of parameters based on parallel-oscillating hydrofoils hydrokinetic turbine. *J. Zhejiang Univ. Eng. Sci.* **2017**, *51*, 153–159.
26. Gungor, A.; Hemmati, A. Wake symmetry impacts the performance of tandem hydrofoils during in-phase and out-of-phase oscillations differently. *Phys. Rev. E* **2020**, *102*, 043104. [CrossRef]
27. Gungor, A.; Khalid, M.S.U.; Hemmati, A. Classification of vortex patterns of oscillating foils in side-by-side configurations. *J. Fluid Mech.* **2022**, *951*, A37. [CrossRef]
28. Chao, L.-M.; Pan, G.; Zhang, D.; Yan, G.-X. On the thrust generation and wake structures of two travelling-wavy foils. *Ocean Eng.* **2019**, *183*, 167–174. [CrossRef]
29. Huera-Huarte, F.J. Propulsive performance of a pair of pitching foils in staggered configurations. *J. Fluids Struct.* **2018**, *81*, 1–13. [CrossRef]
30. Wang, J.; Liu, P.; Chin, C.; He, G.; Song, W. Parametric study on hydro-elasticity characteristics of auto-pitch wing-in-ground effect oscillating foil propulsors. *Ocean Eng.* **2020**, *201*, 107115. [CrossRef]
31. Du, X.X.; Zhang, Z.D. Numerical analysis of the influence of four flapping modes on the propulsive performance of underwater flapping wing. *Eng. Mech.* **2018**, *35*, 8.
32. Hua, E.R.; Zhu, W.; Xie, R.; Su, Z.; Luo, H.; Qiu, L. Comparative Analysis of the Hydrodynamic Performance of Arc and Linear Flapping Hydrofoils. *Processes* **2023**, *11*, 1579. [CrossRef]
33. Read, D.A.; Hover, F.S.; Triantafyllou, M.S. Forces on oscillating foils for propulsion and maneuvering. *J. Fluids Struct.* **2003**, *17*, 163–183. [CrossRef]
34. Wang, K.J.; Zhao, Y.; Wang, S.H. Analysis of pressure fluctuation and internal flow characteristics of axial flow pump under off design condition. *J. Hydroelectric. Eng.* **2023**, *42*, 86–96.
35. Xie, R.S.; Hua, E.T.; Xu, G.H. Research on Double Hump Phenomenon of Axial Flow Pump. *Trans. Chin. Soc. Agri. Mach.* **2022**, *53*, 178–185.
36. He, N.C.; Tang, M.G.; Liu, H.L. Test and analysis on pressure pulsation and hydraulic performance of saddle zone in axial flow pump. *J. Drain. Irri. Mach. Eng.* **2018**, *36*, 118–123.
37. Ye, C.; An, D.; Huang, W.; Heng, Y.; Zheng, Y. Investigation on Stall Characteristics of Centrifugal Pump with Guide Vanes. *Water* **2023**, *15*, 21. [CrossRef]

water

Article

Study on Impeller Optimization and Operation Method of Variable Speed Centrifugal Pump with Large Flow and Wide Head Variation

Yang Zheng [1,*], Long Meng [1], Guang Zhang [2], Peng Xue [1], Xin Wang [1], Chiye Zhang [1] and Yajuan Tian [1]

[1] China Institute of Water Resources and Hydropower Research, Beijing 100083, China
[2] Harbin Electric Machinery Company Limited, Harbin 150040, China
* Correspondence: 18246178676@163.com

Abstract: The benefits of variable speed centrifugal pumps include high stability, a broad operating range, and adjustable input power. In water distribution systems, the pump units are increasingly using variable speed technology. The energy-saving features and operational stability of the pump station are directly impacted by the hydraulic performance and the operation strategy. In this study, CFD numerical analysis and model tests were adopted to design and evaluate the hydraulic performance of the variable speed centrifugal pump with large flow and wide head variation in Liyuzhou Pump Station. Under the premise of ensuring the wide head variation, the optimized centrifugal pump met the requirements of hump margin and efficiency in the high head zone and the cavitation margin in the low head zone. The test results demonstrated that the operational range of the variable speed centrifugal pump was successfully widened by reasonable hydraulic parameters selection and impeller optimization. The safe and efficient operational range of the variable speed unit was determined by means of taking the performance requirements of the pump's maximum input shaft power, cavitation characteristics and pressure fluctuation into consideration. The scientific and reasonable operational path to meet the various operation needs was also investigated and determined for the pump station's actual operation needs. A high efficiency, safe operation, and a simplified control logic were achieved by using the operational path, which makes it a reasonable potential guide for hydraulic design and operational optimization of variable speed centrifugal pumps with large flow and wide head range.

Keywords: variable speed centrifugal pump; large flow; wide head variation; hydraulic design; variable speed operation range; operation path

Citation: Zheng, Y.; Meng, L.; Zhang, G.; Xue, P.; Wang, X.; Zhang, C.; Tian, Y. Study on Impeller Optimization and Operation Method of Variable Speed Centrifugal Pump with Large Flow and Wide Head Variation. *Water* **2024**, *16*, 812. https://doi.org/10.3390/w16060812

Academic Editors: Helena M. Ramos and Bommanna Krishnappan

Received: 29 January 2024
Revised: 24 February 2024
Accepted: 7 March 2024
Published: 9 March 2024

1. Introduction

Large centrifugal pump stations play a critical role in long-distance water transfer projects and large-scale irrigation and drainage systems [1]. The performance of pumps, which are the core power equipment of pump stations, directly affects the energy-saving characteristics and operational stability of pump stations [2]. When pump stations require high hydraulic and energetic characteristics, larger flow rate range, larger pump station head range, it may be difficult for conventional fixed speed centrifugal pump to meet the all the operating conditions due to its relatively narrow efficiency range. To guarantee the stability and safety of the pump station during operation, the large-scale speed regulation operation strategy of centrifugal pumps is employed. The variable speed centrifugal pump offers several advantages over fixed speed centrifugal pump, including excellent stability, wide operating range, high hydraulic efficiency, and adjustable input power. Variable speed drives can increase the centrifugal pump's stability and operational efficiency while allowing for control of the pump's input power. However, within the variable speed operation range of the centrifugal pump, unstable hump characteristics and cavitation phenomena often occur, shortening the pump's service life and reducing the reliability of its operation due to the generation of vibration and noise. Consequently, there are some restrictive technical problems involved in addressing the hump instability characteristic [3–7] and the cavitation characteristic [8–13] in the optimization design of the variable speed centrifugal pump.

Numerical estimation and performance analysis based on the CFD method have rapidly become the primary methods for designers due to the rapid development of computer fluid dynamics technology [14–18]. At the same time, many scholars have conducted comprehensive research on optimization methods [19,20], as well as on the monitoring and predicting of safe unit operation [21]. The flow characteristics of the centrifugal pump were simulated and analyzed using CFD, and the hydraulic performance was confirmed through the model test. Numerous domestic and international studies have been conducted in this field to optimize the design of centrifugal pumps. Cao [22] integrated CFD numerical simulation with response surface method to optimize guide vane centrifugal pumps. The Box–Behnken experiment design method was used to analyze 46 sets of hydraulic performance data, and the optimized model demonstrated a significant improvement in hydraulic performance compared to the original model. Yang [23] proposed a matching optimization on hydraulic components. Various approximate models were utilized to establish the relationship between the optimization design variables and the objective function. Reducing the number of layers and vanes of the diffuser and enhancing the matching of hydraulic components can dramatically improve the hydraulic performance and the hump instability characteristic of large vertical centrifugal pumps. Yuan [24] combined numerical simulation with prototype experiments and employed orthogonal experiment and the GA_PSO algorithm to improve the head variation and efficiency of the centrifugal pump with an elbow inlet. Sun [25] analyzed the influence of different blade numbers and structural designs on the cavitation performance and efficiency of the pump. The blade performance is evaluated comprehensively based on pressure distribution, flow conditions, cavitation, and the velocity diagram under different blade numbers. It is concluded that the efficiency of the four-blade pump was higher. Sun [26] combined Taguchi design and numerical calculations to enhance the efficiency and operational stability of an LVTCP, thereby reducing energy consumption. Based on the efficient house theory, a set of design schemes with the most extensive range of high-efficiency zones was calculated. These schemes improved the overall operating efficiency and stability of the storage pump under both design and non-design conditions. Hu [27] proposed an effective optimization design process for wide-output pump-turbines, highlighting the relationship between runner geometric characteristics and performance. The results indicate that the present approach can significantly improve the efficiency and stability.

However, the studies mentioned above mainly focused on fixed-speed centrifugal pumps. There are few papers studying the design methods of variable speed centrifugal pumps, and few applications of variable speed centrifugal pumps were reported in the literature. The design of a variable speed centrifugal pump can be completed by combining the design of a fixed speed centrifugal pump with the verification of variable speed operating conditions.

The operational stability and operating method of hydraulic mechanics have always been the key research topics of many scholars [28–31]. Merciera [32] studied the stability of pump operations of variable speed units under different control strategies. Theoretical and simulation analysis indicates that power control by speed can ensure the stability of the system. The research on the operation methods of variable speed units [33–37] has gradually become the prominent topic in the field of hydraulic mechanics. The advantages [38–41] and response characteristics [42] of variable speed operations for the hydraulic characteristics were also discussed by many researchers. However, there are few related studies in the literature on parameter selection, and there is currently no relevant research on methods to expand the operating area of centrifugal pumps.

This article intends to explore how to widen the operational range of the centrifugal pumps by optimizing the impeller. Firstly, this study involves the rational selection of operating parameters for the centrifugal pump. Secondly, it employs a fixed-speed optimization method to broaden its high efficiency area based on the selected centrifugal pump. Finally, the overall optimization of the centrifugal pump is carried out according to the variable speed operating range, considering the different characteristic requirements of the small flow and large flow areas. The effectiveness of this design approach and process has been verified in the Liyuzhou pump station in China. The research provides technical support for the application of centrifugal pump stations with large flow and wide head variation.

2. Hydraulic Design

2.1. Pump Station Parameter Analysis

In this article, the centrifugal pump with large flow and wide head range was used as the research object. The pump station has eight installed variable centrifugal pump units with a total flow rate of 80 m^3/s, and is the centrifugal pump station with the largest single design flow in China. The pump station's overall flow varies between 20 m^3/s and 80 m^3/s. The maximum head of the centrifugal pump is 48.0 m, and the minimum head is 16.3 m. Although the pump station head is not high, the relative amplitude of the head is enormous, and the ratio of the maximum head to the minimum head reaches $K = 2.945$. Table 1 presented the main parameters of the variable centrifugal pump unit. The specific speed, i.e., $n_s = 202.5$, and the formula are shown in Equation (1).

$$n_s = 3.65 \times \frac{n \times \sqrt{Q}}{H^{\frac{3}{4}}} \quad (1)$$

where n_s denotes the specific speed of the pump, n is the rotating speed (r/min), Q represents the mass flow rate of design point (m^3/s), and H stands for the pump head (m).

Table 1. Main parameters of the variable centrifugal pump unit.

Physical Quantity	Measure Value	Physical Quantity	Measure Value
Maximum head H_{max} (m)	48	Design flow, Q (m^3/s)	13.5
Design head H_r (m)	42.2	Rated speed, n (r/min)	250
Minimum head H_{min} (m)	16.3	Head variation ratio	2.945
Hump margin	≥2%	Variable speed range	(0.65~1.04)n

According to the analysis of the main technical parameters of the pump station, the project required a wide range of speed regulation operation to accommodate the wide range operation capacity of the pump and to ensure the safe and stable operation of the pump station. The variable speed operating range and power regulation capability of a centrifugal pump are closely related to its parameter selection and hydraulic design. Due to the relatively small high-efficiency operating region, fixed speed centrifugal pumps are unable to fully meet the performance demands of pump stations with wide head variation. At present, there is no large centrifugal pump in China that can meet the performance requirement of the pump station implemented in this project.

2.2. Impeller Optimization Design

Based on the complex hydraulic conditions, such as large flow rates, wide head ranges, and high stability, the parameters selection and calibration method are systematically analyzed initially. Aiming at examining the pump station's wide head variation characteristics, the hydraulic design must not only satisfy the requirements for high efficiency levels and hump margin but also regulate cavitation performance within a reasonable range. A variable speed centrifugal pump has the characteristics of flexible operation. As shown in Figure 1, the centrifugal pump can operate at different head conditions at variable speed, such as point A, point A′ and point A″, which are similar operating points; conversely, the input power of the pump can be changed by changing the rotational speed, such as point

A and point B, in which case the head maintained stable while the flow rate and input power changed.

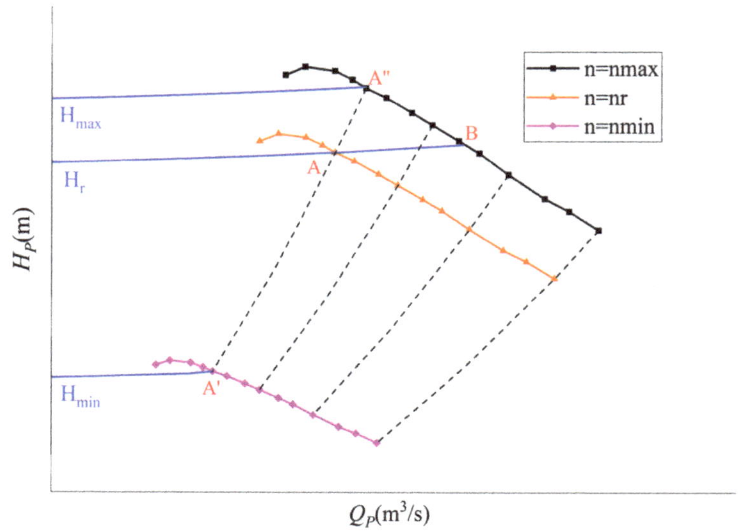

Figure 1. Centrifugal pump flow-head curves at different rotational speeds.

According to the operation characteristics and performance requirements of the centrifugal pump, the whole channel is designed based on parameter selection. The draft tube, impeller, guide vane, and volute are the four primary flow components of a centrifugal pump. The geometric parameters of draft tube and volute should be adjusted to the pumping station's specifications before conducting CFD calculation and performance studies. To minimize the hydraulic loss of the guide vane, it is important to optimize not only the airfoil profile but also the relative position connection between the guide vane and the impeller, as well as the appropriate size of the vaneless area. Additionally, maintaining the width of the impeller in proportion to the guide vane aperture is crucial.

The hydraulic design of the impeller, which is the core energy conversion equipment of the centrifugal pump, directly affects the overall performance of the centrifugal pump. The pump station belongs to the medium–high specific speed centrifugal pump. During the preliminary model selection process, the hydraulic model with a similar specific speed is selected as the initial impeller. The main geometric parameters of the initial impeller are shown in Table 2. The rated rotation speed characteristic curve of the initial impeller prototype centrifugal pump (Test) are shown Figure 2.

Table 2. Main geometry parameters of the initial impeller.

Physical Quantity	Measure Value	Physical Quantity	Measure Value
Impeller blade number	7	High-pressure edge diameter (mm)	427
Inlet liquid flow angle (°)	15.1	Low-pressure edge diameter (mm)	300
Inlet liquid flow angle (°)	17.8	Guide vane high (mm)	74

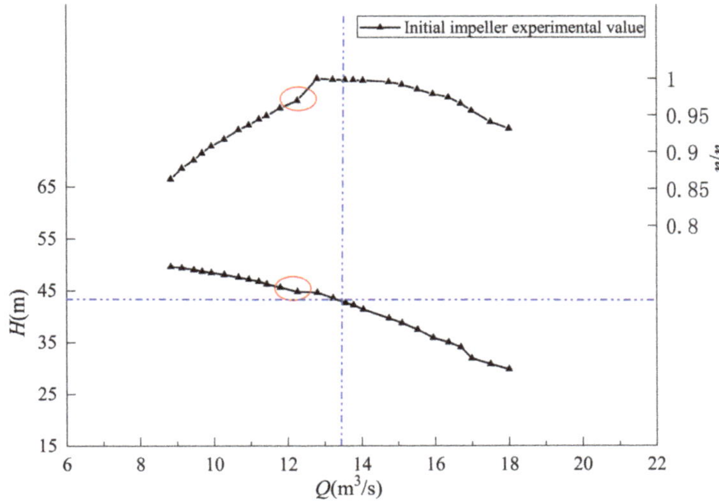

Figure 2. Rated rotation speed characteristic curve of the initial impeller prototype centrifugal pump (Test).

In Figure 2, the intersection point of the blue dashed lines is the design condition point of centrifugal pump. The hump margin of centrifugal pump is less than 2% at the location of red circles in the operating area of pump station. Analyzing the initial impeller based on the model test indicated that the cavitation performance and hump margin are difficult to meet the performance requirements. Therefore, the initial impeller, the impeller geometry, blade profile, exit edge, wrap angle, and inlet blade angle for the centrifugal pump to be developed were all improved to enhance cavitation performance and characteristics of hump instability. The targeted hydraulic design is determined by considering the operational features of the pump station and then combining the reasonable requirements for hydraulic performance between the minimum head and the maximum head. Firstly, during the impeller hydraulic design process, the small flow hump margin should be guaranteed to meet the requirements of the centrifugal pump running at low speed, thereby enhancing the water supply guarantee rate and ensuring the unit's excellent hydraulic properties. Secondly, the cavitation characteristics are improved to ensure the safe and stable operation of the pump station within its operational range.

The cavitation performance is improved by optimizing the shape of the blade inlet edge and the placement angle of the blade inlet. To improve the pressure fluctuation characteristics, the matching relationship between the impeller and guide vane, as well as the vaneless zone width was optimized. By decreasing the guide vane height and strengthening the internal flow characteristics between the guide vane and impeller under small flow conditions, the hump margin was also increased. Numerical simulation results indicate that the optimized impeller has better energy characteristics, cavitation characteristics, and hump margin than the initial impeller. Following repeated calculations, analyses, and adjustments, the hydraulic performance meets the operational requirements of the pump station. The rated speed characteristic comparison curves of the initial impeller and optimized impeller prototype centrifugal pumps (CFD and Test) are shown Figures 3 and 4, respectively.

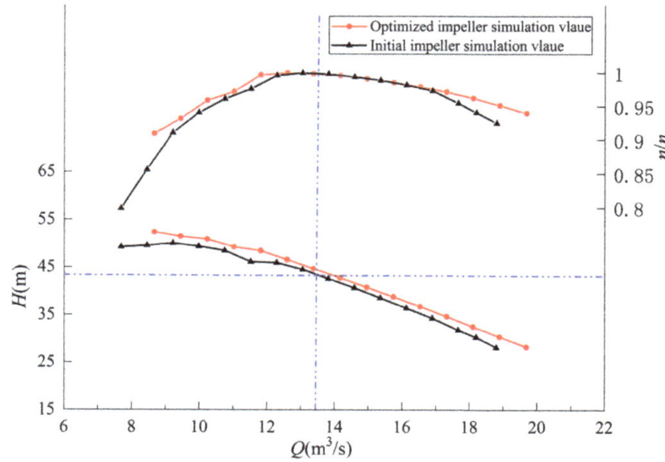

Figure 3. Rated speed characteristic comparison curve of initial impeller and optimized impeller prototype centrifugal pumps (CFD).

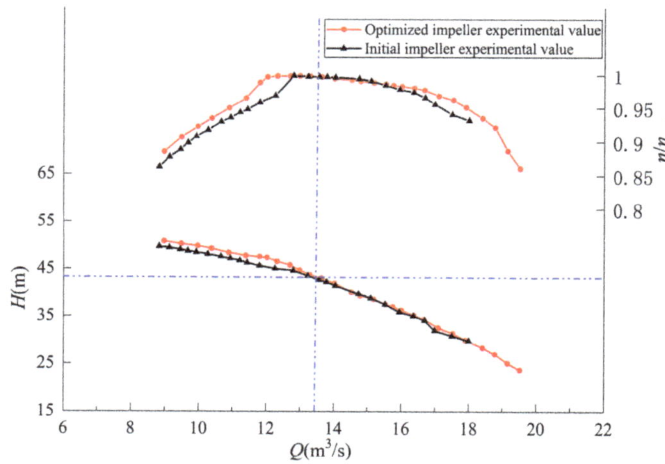

Figure 4. Rated speed characteristic comparison curve of initial impeller and optimized impeller prototype centrifugal pumps (Test).

The main geometric parameters of the optimized impeller are shown in Table 3, and the comparison of model impeller meridian flow channel are shown in Figure 5. The comparison of model impeller blades are also shown Figure 6, where the green part is the initial impeller and the red part is the optimized impeller.

Table 3. Main geometry parameters of optimized impeller.

Physical Quantity	Measure Value	Physical Quantity	Measure Value
Impeller blade number	7	High pressure edge diameter (mm)	427
Inlet liquid flow Angle (°)	16.5	High pressure edge diameter (mm)	300
Inlet liquid flow Angle (°)	18.7	Guide vane high (mm)	68

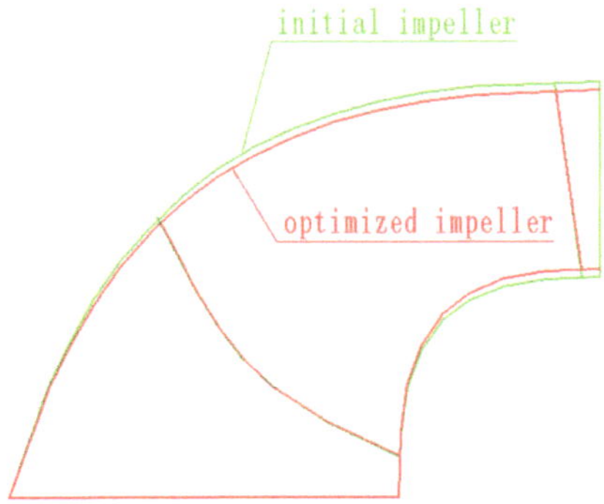

Figure 5. The meridian flow channel of initial impeller and optimized impeller.

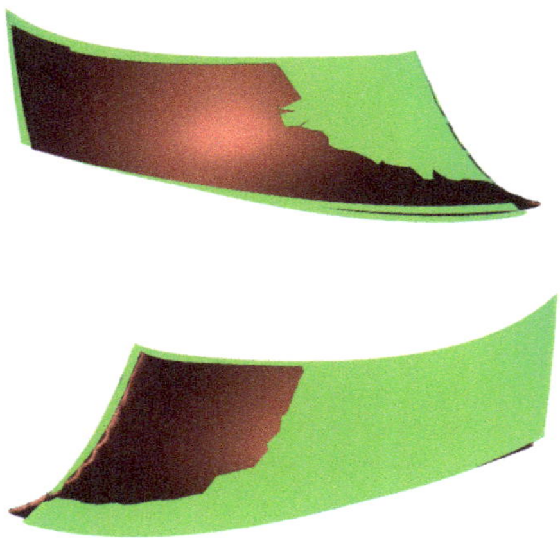

Figure 6. The comparison of model impeller blades with initial impeller and optimized impeller.

3. Performance Estimation Method

3.1. CFD Numerical Calculation Method

In this article, the simulation computation of the whole flow channel is conducted for a large centrifugal pump. Flow characteristics and pressure distribution in the internal flow field are obtained to provide essential data for the performance analysis and optimization of the centrifugal pump.

Pre-processing is essential for numerical simulation, including solid modeling and mesh division. The overall calculation domain of centrifugal pumps is divided into three main parts: the volute and guide vane domain, the impeller domain, and the draft tube domain. The calculation domain model of centrifugal pumps is shown in Figure 7.

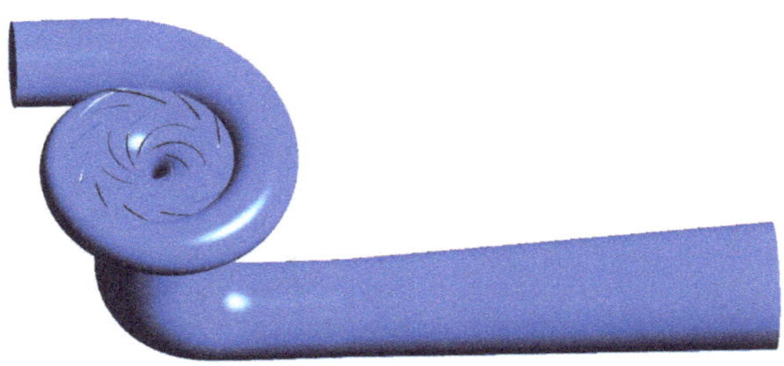

Figure 7. Calculation domain model of centrifugal pumps.

The computational grid for the three calculation domains were generated respectively, and the division of the regional mesh was refined to ensure precise representation. To improve the calculation speed and accuracy of iterative calculations in the simulation, a grid independence check was carried to determine the number of grids for final calculation. According to the numerical calculation results of centrifugal pump efficiency and head, when the grid number of grid elements exceeds 5.71×10^6, the head and hydraulic efficiency tends to be stable. The number of grid elements for volute and guide vanes domain, impeller domain, and draft tube domain were 2.2 million, 2.4 million, and 1.1 million, respectively. The grids for each part are shown in Figure 8.

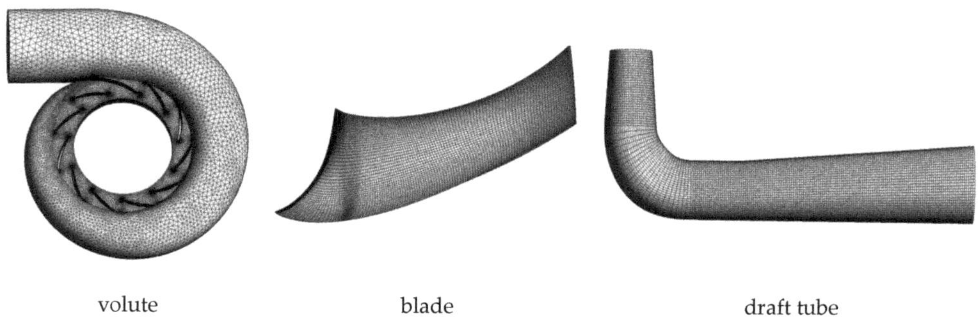

Figure 8. Grid of model calculation domain.

To obtain the flow characteristics and pressure distribution in the internal flow field of centrifugal pump, this article adopts the software SST (shear stress transport) turbulence model to carry out numerical simulation of the computational model. The SST turbulence model consists of the turbulent kinetic energy transport equation and the turbulent dissipation rate transport equation, providing accurate predictions for flow separation under reverse pressure gradients.

For boundary conditions, the mass flow rate is used as the inlet boundary condition for the draft tube, the outlet condition for the volute is defined by setting the mean static pressure to 0 pa, the standard wall function is used in the near wall area, and the wall is without any slip boundary conditions.

For the impeller and the guide vane, the static interface adopts the stage interface transfer model. To ensure consistency with the model test and facilitate data analysis and further optimization design, the model rotated speed is set to 1200 r/min. To keep the boundary conditions similar to the real conditions, the reference pressure was set as 101.325 kPa. The solver is set to the higher order solution mode, and the standard of convergence is the residual of the continuity and momentum equations with a value of 1×10^{-4}.

3.2. Model Test Validation Method

The hydraulic performance of the variable-speed centrifugal pump developed by the numerical calculation method is further validated through the model test method. The model test was carried out at a hydraulic mechanical model test rig, as shown in Figure 9.

A model test rig is a hydraulic mechanical test rig with high parameters and high precision, primarily for pump and pump turbine testing. The uncertainty in efficiency is less than ±0.2%.

The model test includes examining energy characteristics, cavitation characteristics, runaway speed characteristics, pressure pulsation characteristics, water thrust, and other test items. All test items are conducted on the same test rig and model device.

Efficiency testing primarily focuses on measuring the efficiency, flow rate, and head of centrifugal pump in clear water. It includes efficiency characteristics of the expected frequency change range and frequency (speed) within the operating range. The reference surface of the plant cavitation is the centerline of the guide vane, and the cavitation tests are conducted in the whole operating range of the centrifugal pump. Lastly, the incipient cavitation coefficient, σi, and the critical cavitation coefficient, σ1, corresponding to the different operating conditions are determined.

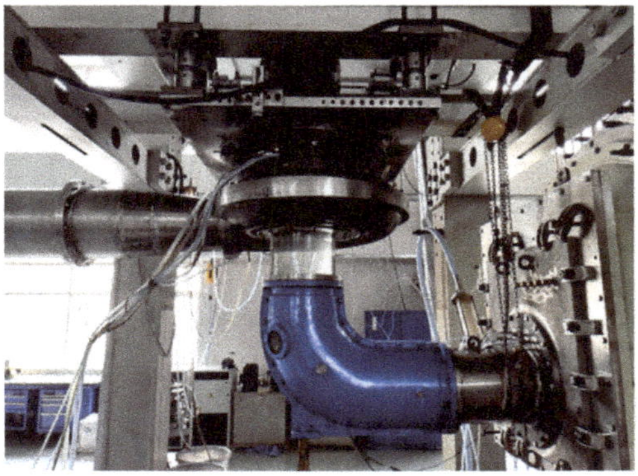

Figure 9. Centrifugal pump model test rig.

4. Performance Evaluation and Result Analysis

4.1. Centrifugal Pump Performance Evaluation

The centrifugal pump should be capable of reducing flow while maintaining good operating efficiency in the low head zone when the pump station is close to the rated head. It is necessary to maintain the same flow under various head situations. Therefore, the centrifugal pump must possess the following characteristics: a reasonably wide high-efficiency zone and a cavitation-free zone of operation; its wide-range speed strategy can match the operational requirements. The flow characteristic and pressure distribution of the centrifugal pump are analyzed, and the impeller's overall performance level is estimated using the CFD numerical simulation. To validate the effectiveness of performance compliance, tests are conducted on the designed impeller.

This article adopts the model centrifugal pump as the numerical calculation model. It focuses on analyzing the internal flow field of the maximum head, design head, and the minimum head frequency conversion conditions (i.e., operating 65% of the rated speed to maintain the designed flow of the pump). The main parameters of the calculation condition point of the centrifugal pump are shown in Table 4.

Table 4. Main parameters of calculation condition points of the centrifugal pump.

Condition Point	Rated Speed n (r/min)	Mass Flow Q (m^3/s)	Head H (m)
T1 Prototype (H_{max})	250	11.48	48.48
T1 Model (H_{max})	1200	0.29	33.79
T2 Prototype (H_r)	250	13.86	42.47
T2 Model (H_r)	1200	0.35	29.60
T3 Prototype (H_{min})	162.5	10.29	16.6
T3 Model (H_{min})	780	0.26	11.57

The impeller internal pressure distribution and streamline under maximum head conditions are depicted in Figures 10 and 11. It can be observed that internal pressure within the impeller is uniform, with static pressure gradually increasing from the impeller's inlet to outlet. Additionally, the internal streamline within the impeller is uniformly distributed, with fluid velocity gradually increasing.

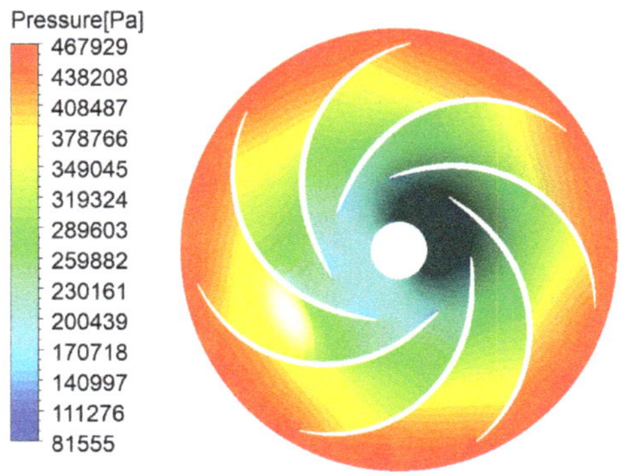

Figure 10. T1 pressure distribution of the impeller on span = 0.5.

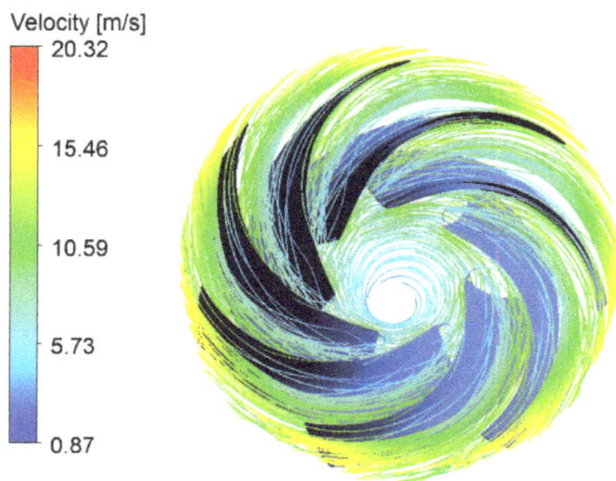

Figure 11. T1 streamline distribution inside the impeller.

The pressure distribution of the pressure surface and suction surface is uniform based on Figures 12 and 13, indicating that there is no prominent low pressure local area. This suggests that the energy conversion on the blade is uniform.

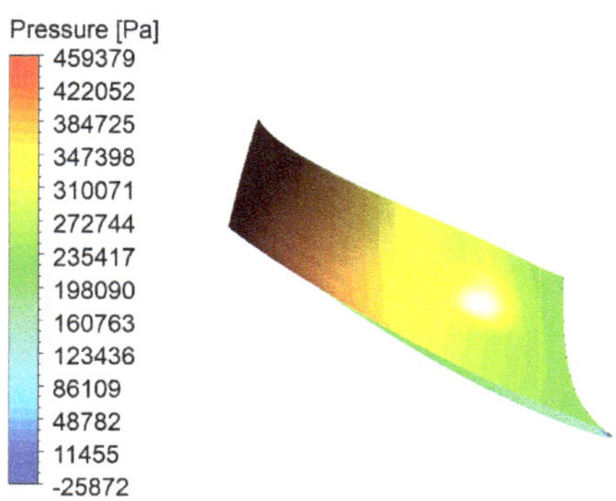

Figure 12. T1 pressure distribution on blade pressure side.

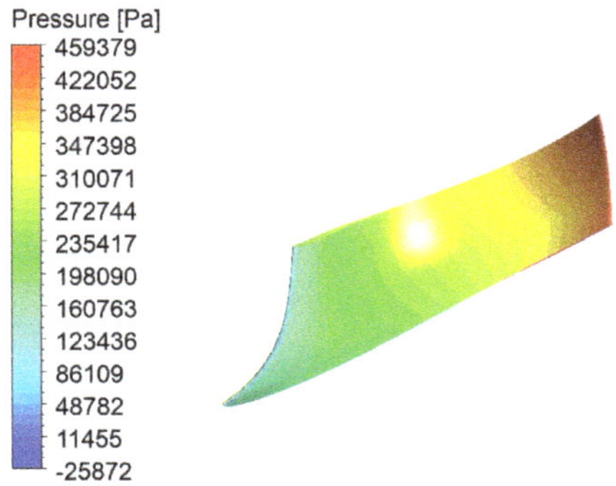

Figure 13. T1 pressure distribution on blade suction side.

Figures 14 and 15 show the distribution of the internal pressure and streamline of the impeller under the design head condition. The figures show that the internal pressure gradient of the impeller changes uniformly, and the internal streamline of the impeller is evenly distributed without local turbulence and vortex flow.

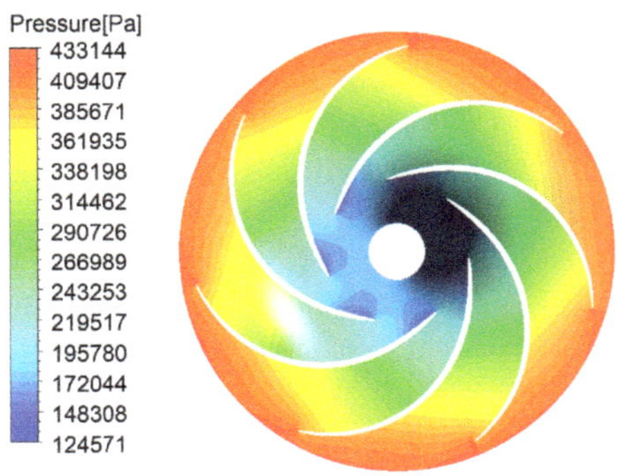

Figure 14. T2 pressure distribution of the impeller on span = 0.5.

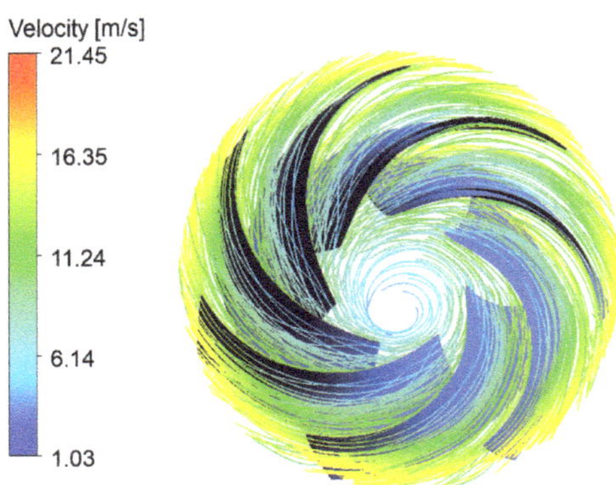

Figure 15. T2 streamline distribution inside the impeller.

Figures 16 and 17 indicate that there is no obvious local low-pressure area on the blade surface, with uniform pressure distribution observed on both the pressure surface and the suction surface of the blade; this indicates that the energy conversion on the blade is uniform. Following similar rules, the centrifugal pump has good internal flow field distribution characteristics at similar working conditions at different speeds.

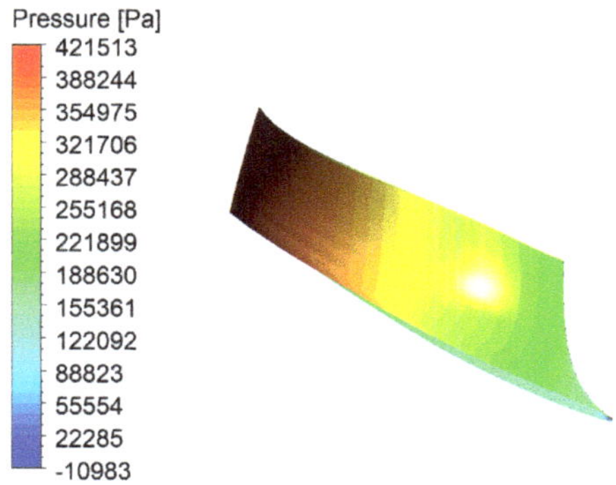

Figure 16. T2 pressure distribution on blade pressure side.

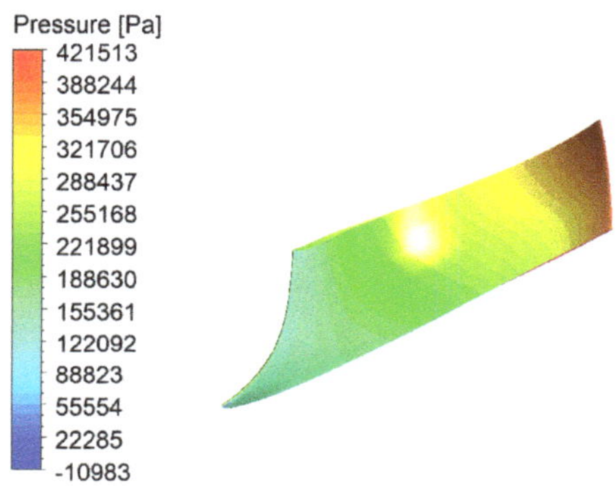

Figure 17. T2 pressure distribution on blade suction side.

Figures 18–21 show the CFD numerical calculation results of the minimum head frequency conversion condition (65% rated speed). Given that the minimum head of the pump is low, the frequency conversion adjustment is adopted to ensure the efficient and stable operation of the pump. So that the pump can still maintain a good running condition at the minimum head. It can be seen from the figure that the internal pressure variation with the impeller is uniform, and the streamline distribution inside the impeller is reasonable. The pressure gradient of impeller blade is reasonably distributed, with no obvious uneven pressure distribution.

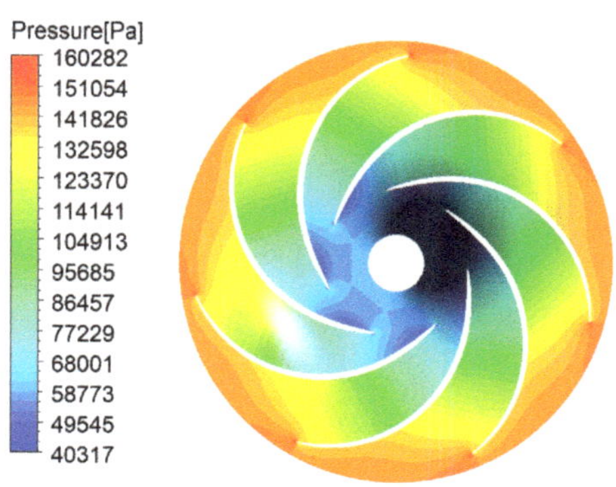

Figure 18. T3 pressure distribution of the impeller on span = 0.5.

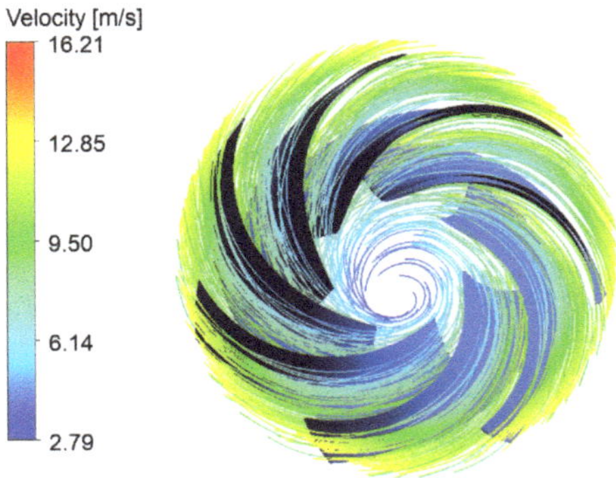

Figure 19. T3 streamline distribution inside the impeller.

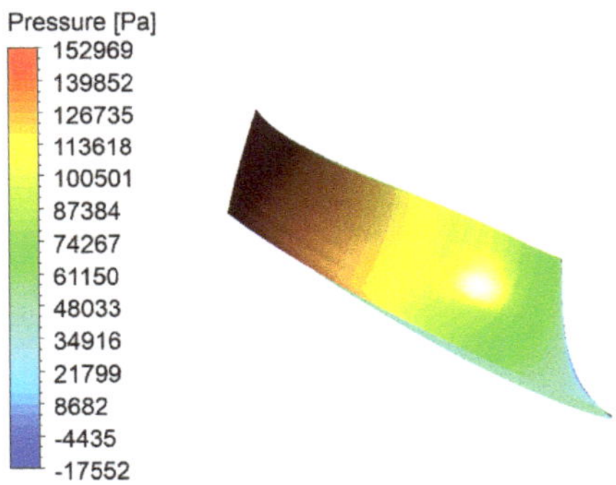

Figure 20. T3 pressure distribution on blade pressure side.

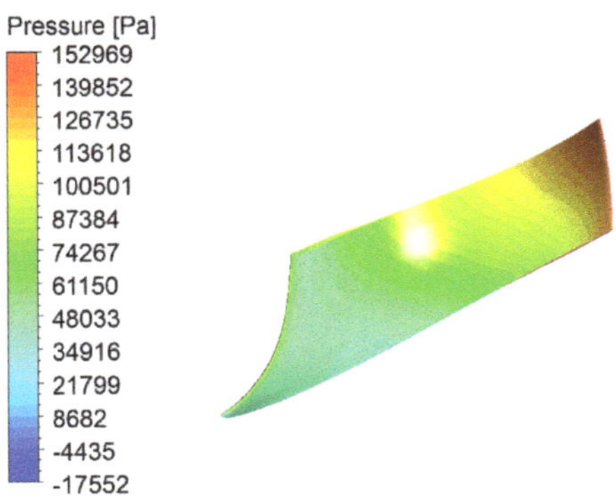

Figure 21. T3 pressure distribution on blade suction side.

To facilitate comparison, the model head, flow, efficiency, and other parameters are transformed into prototype values using the similarity law. The numerical calculation and experimental values of the prototype energy characteristic parameters for the rated speed condition are contrasted in Figure 22. The horizontal axis represents the flow Q (m^3/s), and the vertical axis represents the head H (m) and the relative efficiency of the centrifugal pump η/η_{opt}, respectively. The intersection point of the blue dashed lines is the design head and design flow and the red circle is the optimum operating condition point. The centrifugal pump has great operating efficiency at design conditions, as demonstrated by the red lines in Figure 22. It maintains this level of efficiency throughout a wide operating region (11 m^3/s–17 m^3/s) near the rated flow rate ($\eta/\eta opt > 0.975$).

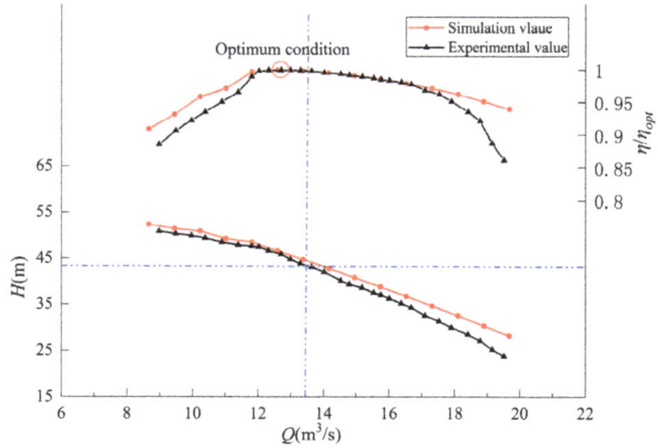

Figure 22. Rated speed characteristic comparison curve of prototype centrifugal pump (CFD and Test).

The black lines represent the test results. Under design conditions, the test results are consistent with the CFD numerical results in predicting the head and efficiency, thus verifying the accuracy of the numerical simulations. Near the design condition, there is a high coincidence and trend within test results and the CFD numerical results, which indicates that CFD calculations accurately predict the hydraulic characteristics of the centrifugal pump.

4.2. Variable Speed Operating Characteristics Analysis

The energy characteristic curve obtained from the model test at the rated speed is converted into the prototype energy characteristic curve by the similarity rule. Figure 23 illustrates the variable speed characteristic curves of the prototype pump at different rotated speeds. In Figure 23, point A, point A′ and point A″ are the optimal operating points of the centrifugal pump under the design head, minimum head and maximum head conditions, respectively.

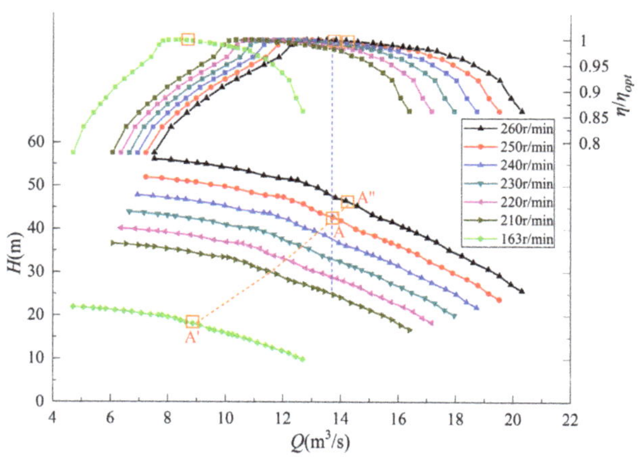

Figure 23. Variable speed characteristic curves of prototype pump (Test).

From the flow, head, and efficiency parameters in Figure 23, the maximum input shaft power curve of the centrifugal pump at different rotational speeds can be calculated. Figure 24 illustrates the correlation between the rated speed and the maximum input shaft power at each operating condition. Considering the motor capacity of the pump station has been determined and the allowance is required, the maximum rotation speed of the centrifugal pump is determined to be 260 r/min.

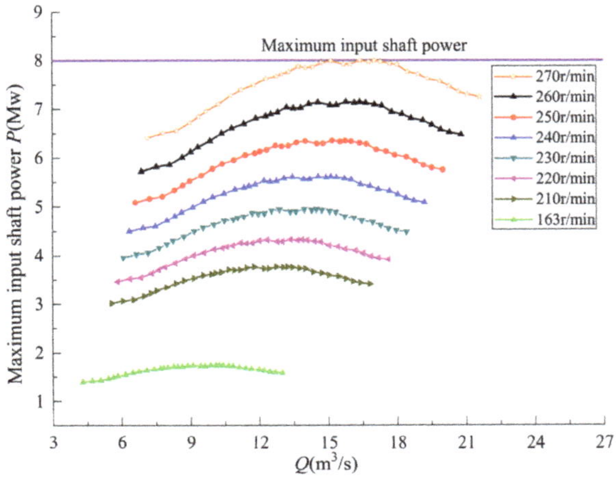

Figure 24. The variable speed curves based on maximum input shaft power characteristics.

The incipient cavitation curve coefficient is displayed in Figure 25 for various operating conditions. When the plant cavitation coefficient is high, the head is low, and the downstream level is high. When the plant cavitation coefficient is low, the head is high, and the downstream level is low. Figure 25 illustrates that when the pump is operating at high speeds, the incipient cavitation coefficient is consistently smaller than the plant cavitation coefficient. This indicates that, under appropriate operating conditions, the centrifugal pump can operate cavitation-free.

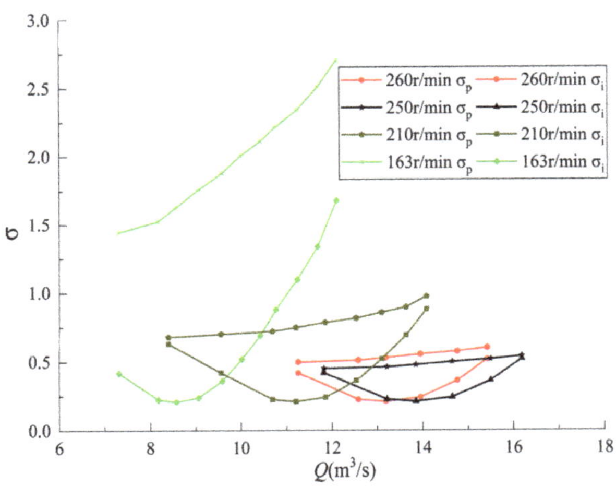

Figure 25. Pump incipient cavitation coefficient performance curve (Test), where σp denotes the plant cavitation coefficient and σi represents the incipient cavitation coefficient (IEC60193).

The variable speed pump unit operates within a range which is limited by the rated speed, maximum input shaft power, hydraulic efficiency, hump margin, pressure pulsation, and the cavitation limit in the low head.

Based on the above analysis, the pump operating zone shall be determined by multiple boundaries: the maximum rotational speed and maximum input shaft power line pump curve (B′–C′), the cavitation line (C–D′), the hump margin line (A′–B′), the minimum variable speed limit (A′–E′) and the relatively high efficiency line (D′–E′). If the boundary lines intersect, the part within the intersection point is taken as the operating area. The final safe and efficient variable speed operating zone is shown as the blue area in Figure 26. The variable speed operation of the centrifugal pump can ensure efficient and stable operation without cavitation in this zone.

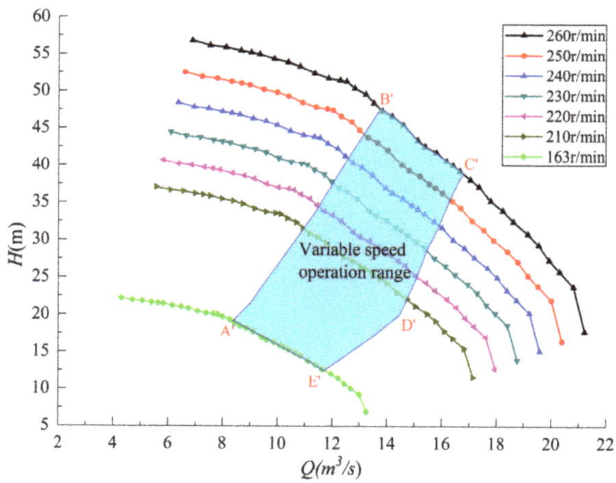

Figure 26. The variable range of the prototype centrifugal pump.

To meet the actual requirements of the pump station and facilitate the actual operational management, the operation path can be further established. This path aims to ensure the simplicity of operation and maintenance of the pump station while enhancing the efficiency and stability.

When the head required by the pump station ranges from 25.5 m to 48.0 m, the appropriate speed can be selected to keep the rated flow unchanged, as shown in Figure 27. The intersection point of the blue dashed line (Q = 13.5 m^3/s) and the head curve of different speeds (210 r/min–260 r/min) represents the operating condition under different heads. From the position of the intersection point between the blue dashed line and the efficiency curve, we can see that the efficiency points for these operating conditions are concentrated and high.

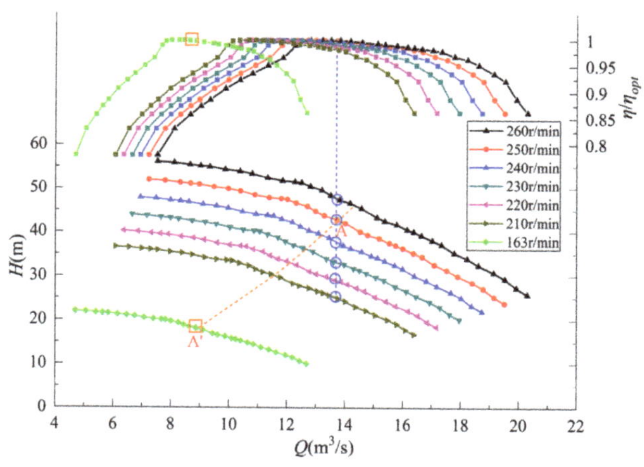

Figure 27. Variable speed operation curves of prototype pump.

If the pump station flow is maintained at the rated flow when its head approaches the minimum head of 16.3 m, hydraulic efficiency will further decline. Meanwhile, reducing the flow rate should ensure the high level of efficiency. The rated speed design point A in Figure 1 is comparable to A′, and the orange square in the high efficiency zone represents the corresponding efficiency of point A′. Consequently, by decreasing the rotation speed (163 r/min–210 r/min), the pump station can operate along the orange dotted line when the pump station head ranges from 16.3 m to 25.5 m.

In summary, the characteristics of small flow hump and large flow cavitation can be improved to achieve wide operating range of variable speed centrifugal pump. For specific pumping stations, appropriate operation paths can be selected according to the actual operation requirements to simplify the operation and ensure the efficiency and stability within the operation range.

5. Conclusions

This study investigates and optimizes a variable speed centrifugal pump through a combination of numerical simulation and model testing. The analysis aims to establish optimal design rationality between the maximum head hump margin and the minimum head cavitation characteristics under the large flow and wide head variation conditions by adjusting the shape of the blade inlet edge and the blade inlet airfoil, improve the small flow hump characteristics through optimizing the matching relationship between the guide blade and the impeller, and improve the pressure pulsation characteristics by optimizing the width of the vaneless area. The centrifugal pump that can satisfy the project's needs is obtained by continuously modifying the geometric flow channel and impeller parameters. The project achieves the goal of expanding the variable speed centrifugal pump's wide operating range. We can effectively increase the centrifugal pump's operating range and safety stability in the high efficiency area by using numerical optimization technology and model testing methods. The study also offers a technical guarantee for the pump unit's wide range of variable speed operations.

To achieve high efficiency and stability across the whole operating range, this article focused on analyzing the pump station's suitable operation path in accordance with its actual operation needs. The operation strategy of the variable speed centrifugal pump is proposed for various operating conditions, creating a combined operation mode that maintains an efficient operation in low head area while maintaining steady flow rate in the high head area. The research results can function as a reference for the optimal design of large flow and wide range variable speed centrifugal pumps and reasonable operation of units in the future.

Author Contributions: Methodology, G.Z.; software, P.X. and Y.T.; validation, C.Z. and X.W.; writing—original draft, Y.Z.; supervision, L.M. All authors have read and agreed to the published version of the manuscript.

Funding: This research was funded by the IWHR Research & Development Support Program (HM0145B012021).

Data Availability Statement: The data presented in this study are available on request from the corresponding author. The data are not publicly available due to engineering technical requirements.

Conflicts of Interest: Author Guang Zhang was employed by the company Harbin Electric Machinery Company Limited. The remaining authors declare that the research was conducted in the absence of any commercial or financial relationships that could be construed as a potential conflict of interest.

References

1. Shi, X.; Lu, J.; Zhao, L. Investigations on the influence of tandem blades on inner flow and performance characteristics of centrifugal pump. *Proc. Inst. Mech. Eng. Part E J. Process Mech. Eng.* **2020**, *234*, 46–55. [CrossRef]
2. Torii, D.; Nagahara, T.; Okihara, T. Suppression of the secondary flow in a suction channel of a large centrifugal pump. *IOP Conf. Ser. Mater. Sci. Eng.* **2013**, *52*, 032005. [CrossRef]

3. Ramirez, R.; Avila, E.; Lopez, L.; Bula, A.; Forero, J.D. CFD characterization and optimization of the cavitation phenomenon in dredging centrifugal pumps. *Alex. Eng. J.* **2020**, *59*, 291–309. [CrossRef]
4. Al-Obaidi, A.R.; Towsyfyan, H. An experimental study on vibration signatures for detecting incipient cavitation in centrifugal pumps based on envelope spectrum analysis. *J. Appl. Fluid Mech.* **2019**, *12*, 2057–2067.
5. Wang, D.; Liu, Z.; Han, W. Study on improving cavitation performance of centrifugal Pump by Perforation at the Front Cover Plate. *Int. J. Fluid Mach. Syst.* **2020**, *13*, 668.
6. Hesamedin, K.; Abdolah, S.; Siamak, H. Improvement of the centrifugal pump performance by restricting the cavitation phenomenon. *Chem. Eng. Trans.* **2018**, *71*, 1369–1374.
7. Song, P.; Wei, Z.; Zhen, H.; Liu, M.; Ren, J. Effects of pre-whirl and blade profile on the hydraulic and cavitation performance of a centrifugal pump. *Int. J. Multiph. Flow* **2022**, *157*, 104261. [CrossRef]
8. Yang, G.; Shen, X.; Shi, L.; Zhang, D.; Zhao, X.; van Esch, B.P.M. Numerical investigation of hump characteristic improvement in a large vertical centrifugal pump with special emphasis on energy loss mechanism. *Energy* **2023**, *273*, 127163. [CrossRef]
9. Ye, C.; Wang, C.; Yan, H.; Wang, F.; Zheng, Y.; van Esch, B.P.M. Investigation on transition characteristics of laminar separation bubble on a hydrofoil. *Phys. Fluids* **2023**, *35*, 105154. [CrossRef]
10. Ye, C.; Tang, Y.; An, D.; Wang, F.; Zheng, Y.; van Esch, B.P.M. Investigation on stall characteristics of marine centrifugal pump considering transition effect. *Ocean Eng.* **2023**, *280*, 114823. [CrossRef]
11. Feng, J.; Ge, Z.; Yang, H.; Zhu, G.; Li, C.; Luo, X. Rotating stall characteristics in the vaned diffuser of a centrifugal pump. *Ocean Eng.* **2021**, *229*, 108955. [CrossRef]
12. Wang, Y.; Ding, Z. Optimization design of hump phenomenon of low specific speed centrifugal pump based on CFD and orthogonal test. *Sci. Rep.* **2022**, *12*, 12121.
13. Li, P.; Li, N.; Tao, R.; Lu, Z.; Zhu, D.; Liu, W.; Yao, Z.; Xiao, R. Quantitative investigation of the head-hump of large-scale vaned-voluted centrifugal pump. *Proc. Inst. Mech. Eng. Part C J. Mech. Eng. Sci.* **2022**, *236*, 10779–10791. [CrossRef]
14. De Donno, R.; Ghidoni, A.; Noventa, G.; Rebay, S. Shape optimization of the ERCOFTAC centrifugal pump impeller using open-source software. *Optim. Eng.* **2019**, *20*, 929–953. [CrossRef]
15. Wang, Y.; Zhou, L.; Han, M.; Shen, L. Performance Prediction of an Optimized Centrifugal Pump with High Efficiency. *Fluid Dyn. Mater. Process.* **2023**, *19*, 2215–2228. [CrossRef]
16. Huang, R.; Zhang, Z.; Zhang, W.; Mou, J.; Zhou, P.; Wang, Y. Energy performance prediction of the centrifugal pumps by using a hybrid neural network. *Energy* **2020**, *213*, 119005. [CrossRef]
17. Ramadhan, A.A.; Ali, Q. Effect of outlet impeller diameter on performance prediction of centrifugal pump under single-phase and cavitation flow conditions. *Int. J. Nonlinear Sci. Numer. Simul.* **2022**, *23*, 1203–1229.
18. Dong, J.; Li, W. Numerical simulation of centrifugal pump cavitation based on ANSYS. *J. Phys. Conf. Ser.* **2023**, *2450*, 012031. [CrossRef]
19. Mahdi, M.T.; Amin, D.A.; Ahmad, H.; Mohamed BB, H.; Mohsen, I. MLP and optimized FCM-ANFIS models pro-posed for inlet turbulent flow under ultrasonic vibration. *J. Therm. Anal. Calorim.* **2023**, *148*, 13995–14009.
20. Morteza, E.; Amin, D.A.; Ali, J. Optimization of ultrasonic-excited double-pipe heat exchanger with machine learning and PSO. *Int. Commun. Heat Mass Transf.* **2023**, *147*, 106985.
21. Li, L.; Xu, W.; Tan, Y.; Yang, Y.; Yang, J.; Tan, D. Fluid-induced vibration evolution mechanism of multiphase free sink vortex and the multi-source vibration sensing method. *Mech. Syst. Signal Process.* **2023**, *189*, 110058. [CrossRef]
22. Cao, W.; Wang, H.; Yang, X.; Leng, X. Optimization of Guide Vane Centrifugal Pumps Based on Response Surface Methodology and Study of Internal Flow Characteristics. *J. Mar. Sci. Eng.* **2023**, *11*, 1917. [CrossRef]
23. Yang, G.; Zhao, X.; Zhang, D.; Geng, L.; Yang, X.; Gao, X. Hydraulic components' matching optimization design and entropy production analysis in a large vertical centrifugal pump. *J. Mech. Sci. Technol.* **2021**, *35*, 5033–5048. [CrossRef]
24. Yuan, Y.; Jin, R.; Tang, L.; Lin, Y. Optimization Design for the Centrifugal Pump under Non-Uniform Elbow Inflow Based on Orthogonal Test and GA_PSO. *Processes* **2022**, *10*, 1254. [CrossRef]
25. Sun, X.; Zhu, R.; Wang, J.; Li, Y. Effect analysis of centrifugal pump parameters optimization design on performance. *J. Phys. Conf. Ser.* **2023**, *2599*, 012034. [CrossRef]
26. Sun, J.; Pei, J.; Wang, W. Effect of impeller and diffuser matching optimization on broadening operating range of storage pump. *J. Energy Storage* **2023**, *72*, 108737. [CrossRef]
27. Hu, Z.; Cheng, Y.; Liu, D.; Chen, H.; Ji, B.; Ding, J. Broadening the operating range of pump-turbine to deep-part load by runner optimization. *Renew. Energy* **2023**, *207*, 73–88. [CrossRef]
28. Iliev, I.; Trivedi, C.; Dahlhaug, O.G. Variable-speed operation of Francis turbines: A review of the perspectives and challenges. *Renew. Sustain. Energy Rev.* **2019**, *103*, 109–121. [CrossRef]
29. Badshah, M.; Badshah, S.; Jan, S. Comparison of computational fluid dynamics and fluid structure interaction models for the performance prediction of tidal current turbines. *J. Ocean Eng. Sci.* **2020**, *5*, 164–172. [CrossRef]
30. Nasir, U.; Iqbal, Z.; Rasheed, M.; Minxiao, H. Active and reactive power control of a variable speed pumped storage system. In Proceedings of the 2015 IEEE 15th International Conference on Environment and Electrical Engineering (EEEIC), Rome, Italy, 10–13 June 2015; pp. 6–11.
31. Li, Y.; Yang, W.; Huang, Y.; Ma, W.; Zhao, Z.; Yang, J.; Yang, J. Variation law of pressure pulsation during variable speed operation of pumped storage units. *IOP Conf. Ser. Earth Environ. Sci.* **2022**, *1079*, 012111. [CrossRef]

32. Mercier, T.; Hardy, C.; Van Tichelen, P.; Olivier, M.; De Jaeger, E. Control of variable-speed pumps used as turbines for flexible grid-connected power generation. *Electr. Power Syst. Res.* **2019**, *176*, 105962. [CrossRef]
33. Salome, D.M.; Meirelles, G.L.; Melo, B.B.; Bezerra, D.B. Optimal pump selection for variable speed operation in water distribution network. *RBRH-Rev. Bras. Recur. Hidreicos* **2020**, *25*. [CrossRef]
34. Gevorkov, L.; Rassõlkin, A.; Kallaste, A.; Vaimann, T. Simulation study of mixed pressure and flow control systems for optimal operation of centrifugal pumping plants. *Electr. Control. Commun. Eng.* **2018**, *14*, 89–94. [CrossRef]
35. Liu, X.; Liu, S.; Wang, Y. Optimal control study of solar-air source heat pump coupled heating system operation. *J. Phys. Conf. Ser.* **2023**, *2474*, 012015. [CrossRef]
36. Amul, G.; Atmaram, K.; Lal, P.B.; Sailesh, C.; Prasad, H.N. Numerical analysis of francis turbine being operated in variable speed from sediment erosion perspective. *J. Phys. Conf. Ser.* **2023**, *2629*, 012006.
37. Wang, H.; Ma, Z. Regulation Characteristics and Load Optimization of Pump-Turbine in Variable-Speed Operation. *Energies* **2021**, *14*, 8484. [CrossRef]
38. Delgado, J.; Ferreira, J.P.; Covas DI, C.; Avellan, F. Variable speed operation of centrifugal pumps running as tur-bines. Experimental investigation. *Renew. Energy* **2019**, *142*, 437–450. [CrossRef]
39. Selçuk; Arslan, S. Comparison of energy efficiencies of a small centrifugal pump at constant and variable speed operations. *Tarim Bilim. Derg.* **2016**, *22*, 444–454. [CrossRef]
40. Edson, B.; Zulcy, S.; Augusto, V.; Helcio, V.; Ângelo, R.; Roberto, S.; Rafael, B.; José, B. The benefits of variable speed operation in hydropower plants driven by francis turbines. *Energies* **2019**, *12*, 3719.
41. Becker, M.R.; Torres TC, J.; Matías, D. Adaptive control of M3C-based variable speed drive for multiple permanent-magnet-synchronous-motor-drive centrifugal pumps. *Machines* **2023**, *11*, 884. [CrossRef]
42. Mercier, T.; Olivier, M.; Dejaeger, E. Operation ranges and dynamic capabilities of variable-speed pumped-storage hydropower. *J. Physics Conf. Ser.* **2017**, *813*, 012004. [CrossRef]

water

MDPI

Article

Investigation on Stall Characteristics of Centrifugal Pump with Guide Vanes

Changliang Ye [1,*], Dongsen An [2], Wanru Huang [3], Yaguang Heng [4] and Yuan Zheng [1]

1 College of Energy and Electrical Engineering, Hohai University, Nanjing 211100, China
2 Beijing Aerospace Propulsion Institute, China Aerospace Science and Technology Corporation, Beijing 100176, China
3 National Marine Hazard Mitigation Service, Beijing 100194, China
4 Key Laboratory of Fluid and Power Machinery of Ministry of Education, Xihua University, Chengdu 610039, China
* Correspondence: yechangliang@hhu.edu.cn

Abstract: Stall usually occurs in the hump area of the head curve, which will block the channel and aggravate the pump vibration. For centrifugal pumps with guide vanes usually have a clocking effect, the stall characteristic at different clocking positions should be focused. In this paper, the flow field of the centrifugal pump under stall conditions is numerically simulated, and the rotor–stator interaction effects of the centrifugal pump under stall conditions are studied. The double-hump characteristic is found in the head curve by using SAS (Scale Adaptive Simulation) model. The hump area close to the optimal working condition is caused by hydraulic loss, while the hump area far away from the optimal working condition point is caused by the combined action of Euler's head and hydraulic loss. The SAS model can accurately calculate the wall friction loss, thus predicting the double-hump phenomenon. The pressure fluctuation and head characteristics at different clocking positions under stall conditions are obtained. It is found that when the guide vanes outlet in line with the volute tongue, the corresponding head is the highest, and the pressure fluctuation is the lowest. The mechanism of the clocking effect in the centrifugal pump with guide vanes is obtained by simplifying the hydrofoil. It is found that when the downstream hydrofoil leading edge is always interfered with by the upstream hydrofoil wake, the wake with low energy mixes the boundary layer with low energy, which causes small-pressure pulsation. The results could be used for the operation of centrifugal pumps with guide vanes.

Keywords: pressure fluctuation; SAS; clocking effect; hydrofoil; wake

Citation: Ye, C.; An, D.; Huang, W.; Heng, Y.; Zheng, Y. Investigation on Stall Characteristics of Centrifugal Pump with Guide Vanes. *Water* **2023**, *15*, 21. https://doi.org/10.3390/w15010021

Academic Editor: Giuseppe Pezzinga

Received: 8 December 2022
Revised: 17 December 2022
Accepted: 19 December 2022
Published: 21 December 2022

1. Introduction

The stall of a centrifugal pump is an unstable flow phenomenon when it operates at low flow rate [1–4]. The stall phenomenon was first discovered by Stenning in a centrifugal compressor test [5], and in-depth research was carried out in this field [6,7]. In the mid-1980s, the stall problem in centrifugal pumps began to receive widespread attention [8]. Centrifugal pumps often suffer from severe pressure pulsations when operating under stall conditions, which further aggravates the vibration and noise of the pump and, in severe cases, causes fatigue damage to the blades [9,10]. Especially in recent years, with the continuous construction of large pump stations, higher requirements have been put forward for the safe and stable operation of pumps, so it is of great significance to understand the stall problem and then control it.

CFD is an important way to reveal the internal flow mechanism [11,12]. For the stall of centrifugal pumps, accurate prediction of stall phenomenon is the basis of research. The stall of centrifugal pumps is a complex rotational separated flow. At present, the calculation of three-dimensional separated flow mainly includes Reynolds average method (RANS), large eddy simulation method (LES), and mixed URANS/LES method [13]. Among them,

LES model can analyze the turbulent vortex structure, turbulence fluctuation, and other information of complex flow in more detail, but it is difficult to apply to the calculation of existing engineering problems due to its huge amount of calculation [14]. At present, RANS and URANS/LES methods are the most commonly used methods. The stall in the guide vane is calculated by the standard k-ε model [15], but there is a large difference between the experimental data and the simulated data. Braun [16] used the SST k-ω model to simulate a pump and found the model can predict the phenomenon of rotating stall, but the flow at the stall point and the propagation speed of the stall cell are larger than those in the test. Lucius [17] and Zhang [18] adopted the URANS/LES model to calculate a centrifugal pump that was tested at Magdeburg University in Germany. It was found that the predicted head and rotating stall vortex frequency were close with experiment, but the predicted peak values were different. Therefore, it is necessary to select turbulence models according to specific problems.

Many scholars have carried out a great deal of research on the factors affecting the stall of centrifugal pumps. The main influencing factors are pump flow [19–23], impeller speed [24–26], number of blades [27,28], and rotor–stator interaction [29–31]. For the guide-vane centrifugal pump, the clocking position of the guide vane is also an important factor affecting the internal flow characteristics of the centrifugal pump. Clocking effect refers to the influence of changing the circumferential relative position between the blades on turbomachinery [32]. Benigni [33] and Wang [34,35] studied the clocking effect of an annular volute centrifugal pump. The results show that the clocking effect has a greater impact on the pressure fluctuation in the pump, especially on the pressure fluctuation intensity of the volute. Tan [36] studied the clocking effect of the high head multistage centrifugal pump and found that the clocking effect had a significant impact on the internal pressure and radial force of the pump. However, the influence of clocking effect on the internal flow characteristics of centrifugal pumps at stall conditions is rarely reported.

In this paper, the turbulence models are compared and analyzed in order to accurately predict the flow characteristics in the pump, and then, the influence of the clocking effect on the pump flow at stall conditions is further analyzed. Meanwhile, due to the complex structure of the centrifugal pump, in order to better explore the clocking effect of the guide-vane centrifugal pump, this paper establishes a simplified model based on the structural characteristics of the rotor and stator in the centrifugal pump in order to deeply analyze the mechanism of clocking effect in the centrifugal pump.

2. Research Object and Numerical Calculation Method

The geometric structure of the centrifugal pump with guide vane selected as the calculation object in this paper is shown in Figure 1. Combined with the internal flow structure of the pump, the pressure fluctuations in the guide vane and impeller regions are analyzed, respectively. Figure 1 is the axial sectional view of the guide vane centrifugal pump. The six flow channels are named A-F. Three monitoring points are set on the pressure surface and suction surface of the blades in the two flow channels of the impeller along the direction from the inlet to the outlet. Similarly, three monitoring points are set in two adjacent flow channels of the guide vane. During the calculation, the pressure fluctuation of these points is recorded with time.

The hexahedral structure grid is used to divide the computational domain. With the purpose of ensuring the reliability of the calculation, taking the near-wall grid into account is necessary. To do this, the first layer of the grid is validated. After testing, $y+ < 30$, which means the mesh is suitable for turbulence models, as shown in Figure 2. In order to verify the independence of grid, different grid amounts are calculated in the rated condition. The pump head is selected as the target parameter for grid independence verification, and its variation with the number of grids is observed. The number of grids is added at the tongue region of volute and blade surface. The calculation domain with different grid numbers is simulated by applying the SST k-ω turbulence model. The head rises when grids elements increase from 2.34 million to 4.17 million. The variation of heads is scarce when grid elements increase to 5.27 million. In conclusion, the elements of grids make a small difference in the results when they continue to increase. Therefore, the elements of grids of the pump will be controlled at about 5.27 million, and the corresponding number of nodes is 4.1 million.

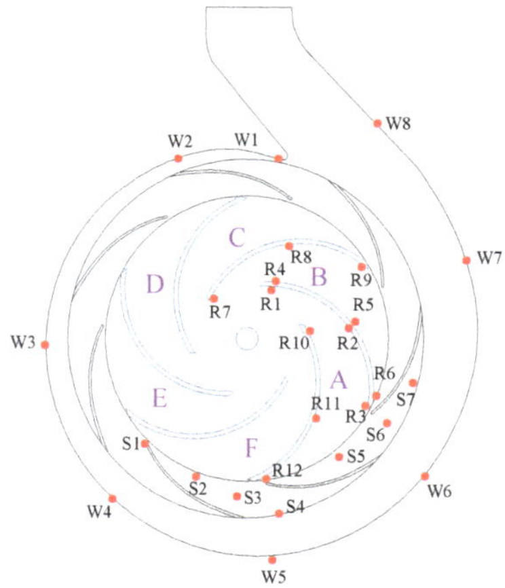

Figure 1. Schematic Diagram of Guide-vane Centrifugal Pump.

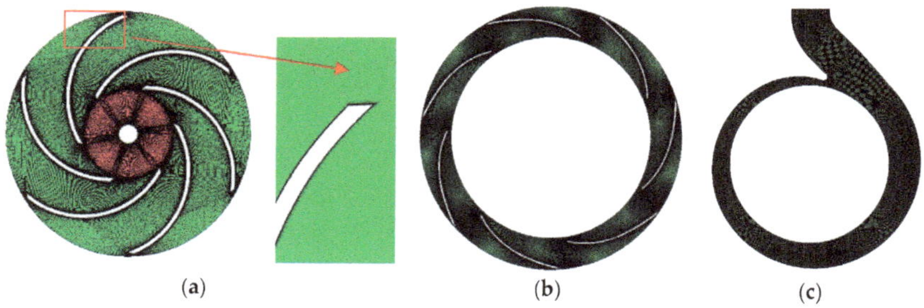

Figure 2. Grid of the centrifugal pump: (**a**) impeller; (**b**) guide vane; (**c**) volute.

The velocity inlet and pressure outlet are adopted. The rotor–stator interface adopts the transient frozen rotor technology, and the impeller speed is set as 725 r/min to keep consistent with the test. The no-slip condition is adopted in the wall boundary. The governing equations are discretized by the finite volume method in space and the second-order implicit scheme in the time domain. The time step is 2.3×10^{-4} s, which is 1/360 of the rotation period. The convergence residual is set to 1.0×10^{-5}. This calculation is carried out on the rack server of Hohai University. The servers all use 64-bit high-performance processors and 12 dual computing nodes, including 24 Intel E5-2600 eight-core processors and 192 CPU cores. Each node contains 64 GB of memory, and the total system memory is 678 GB.

When the centrifugal pump impeller stall occurs, the particularity and complexity of the internal flow put forward higher requirements for the numerical simulation method. The numerical results calculated by the SST k-ω and SAS model are compared.

The turbulent kinetic energy and dissipation rate in the SST k-ω model can be expressed as follows [37]:

$$\frac{\partial \rho k}{\partial t} + \frac{\partial(\rho U_j k)}{\partial x_j} = P_k - D_k + \frac{\partial}{\partial x_j}\left((\mu + \sigma_{k3}\mu_t)\frac{\partial k}{\partial x_j}\right) \tag{1}$$

$$\frac{\partial \rho \omega}{\partial t} + \frac{\partial(\rho U_j \omega)}{\partial x_j} = P_\omega - D_\omega + \frac{\partial}{\partial x_j}\left((\mu + \sigma_{\omega 3}\mu_t)\frac{\partial \omega}{\partial x_j}\right) + 2\rho(1 - F_1)\frac{\sigma_{\omega 2}}{\omega}\frac{\partial k}{x_j}\frac{\partial \omega}{x_j} \tag{2}$$

This model has become the standard in many industrial applications. The coefficients in the formulas are given in reference [38].

The "scale-adaptive" feature of SAS model is mainly embodied in its definition of a second order velocity ladder [39]. Many scholars have applied the model in the simulation of rotating machinery and proved the reliability of the model. The construction method of the SAS model is based on the SST model by adding the source term Q_{SAS} to the ω equation:

$$\frac{\partial(\rho k)}{\partial t} + \frac{\partial(\rho U_i k)}{\partial x_i} = P_k - \beta^* \rho k \omega + \frac{\partial}{\partial x_i}\left|(\mu + \sigma_k \mu_t)\frac{\partial k}{\partial x_i}\right| \tag{3}$$

$$\frac{\partial(\rho \omega)}{\partial t} + \frac{\partial(\rho U_i \omega)}{\partial x_i} = \alpha \rho S^2 - \beta \rho \omega^2 + \frac{\partial}{\partial x_i}\left|(\mu + \sigma_\omega \mu_t)\frac{\partial \omega}{\partial x_i}\right| + 2(1 - F_1)\rho \sigma_{\omega_2}\frac{1}{\omega} \cdot \frac{\partial k}{\partial x_i} \cdot \frac{\partial \omega}{\partial x_i} + Q_{SAS} \tag{4}$$

$$Q_{SAS} = \rho \cdot \max\left[\zeta_2 \kappa S^2 \left(\frac{L}{L'_{vK}}\right)^2 - C\frac{2k}{\sigma_\Phi} \cdot \max\left(\frac{1}{\omega^2} \cdot \frac{\partial \omega}{\partial x_i} \cdot \frac{\partial \omega}{\partial x_i}, \frac{1}{k^2} \cdot \frac{\partial k}{\partial x_i} \cdot \frac{\partial k}{\partial x_i}\right), 0\right] \tag{5}$$

$$L = \sqrt{k}/\left(c_\mu^{1/4} \cdot \omega\right), c_\mu = 0.09 \tag{6}$$

The coefficients in the formulas are given in reference [39].

3. Prediction of Centrifugal Pump Performance Curve

The internal flow of the centrifugal pump with guide vanes is complex. In order to systematically analyze the internal stall flow characteristics of the pump, the full operating ranges are calculated. The research object of this paper is the experimental centrifugal pump of Technical University of Denmark (DTU). In the experiment, PIV and LDV methods are used to collect the data of single-stage impeller so as to obtain the information on average flow field. The impeller inlet flow is changed by adjusting the diameter of the hole on the resistance plate. The adjustment range is (15–100%) Q_d, where Q_d is the flow under design conditions. In order to stabilize the flow field at the pump inlet, a rectifier is placed in the pipe section at the inlet of the resistance plate and impeller [40].

The comparison results of the numerical simulation of centrifugal pumps at different working conditions are shown in Figure 3. The head of centrifugal pumps without guide vanes calculated by the two models are close to the experimental values, indicating that the numerical model method can accurately predict the energy characteristics of centrifugal pumps. In the guide-vane centrifugal pump, the existence of the guide vane increases the mechanical loss and changes the head. In terms of head prediction performance, the two models both predict the hump area at 0.2 Q_d–0.3 Q_d, in which the hump peak point is 0.3 Q_d, and the wave trough is 0.2 Q_d. The difference is that the SAS model observed the occurrence of the hump phenomenon at 0.5 Q_d–0.6 Q_d, in which the peak point is 0.6 Q_d, and the wave trough point is 0.5 Q_d. The double-hump phenomenon of guide vane centrifugal pump predicted by the SAS model is also observed in the pump turbine [41].

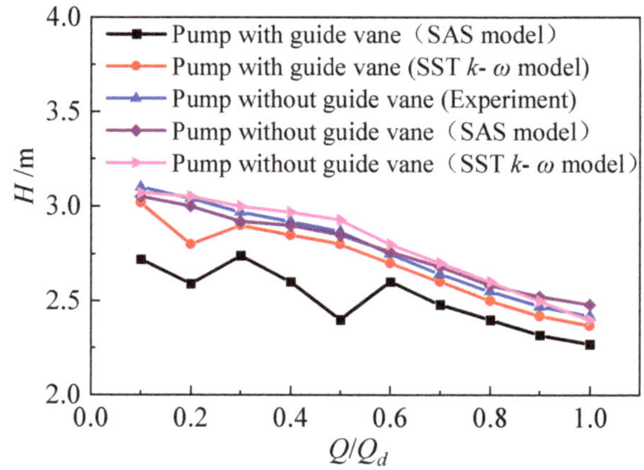

Figure 3. Centrifugal Pump Head Performance Curve.

4. Characteristic Analysis of Double Humps

Through the numerical simulation analysis in the previous section, the guide vane centrifugal pump has a double-hump characteristic, and the emergence of the hump area must be accompanied by the unstable flow. In order to further analyze the causes of this phenomenon, this section focuses on the hydraulic loss of each component.

The head characteristic of the guide-vane centrifugal pump is caused by the combined action of Euler head and hydraulic losses of flow passage components. Euler energy ($\Delta Cu \cdot U$) is the difference between impeller outlet velocity moment ($Cu_2 \cdot U_2$) and impeller inlet velocity moment ($Cu_1 \cdot U_1$). In the process of numerical simulation, Euler energy and Euler head can be obtained through equations [41]:

$$\Delta Cu \cdot U = Cu_2 \cdot U_2 - Cu_1 \cdot U_1 \approx \frac{\Gamma \cdot \omega}{Q} \tag{7}$$

$$H_{\text{Euler}} = \frac{\Delta Cu \cdot U}{g} \tag{8}$$

Figure 4 shows the results of Euler head calculated by the two models. It can be seen that the Euler head factor increases with the decrease of flow rates. In the Euler head calculated by SAS, when the flow rate changes from 1.0 Q_d to 0.5 Q_d, the Euler head decreases slowly; when the flow decreases from 0.5 Q_d to 0.4 Q_d, the Euler head decreases rapidly. For the Euler head calculated by SST k-ω, when the flow rate changes from 1.0 Q_d to 0.6 Q_d, the Euler head decreases slowly; when the flow decreases from 0.6 Q_d to 0.5 Q_d, the Euler head decreases rapidly. The hump area is observed at 0.2 Q_d–0.3 Q_d of Euler head calculated by the two models, where the peak point is 0.3 Q_d, and the trough is 0.2 Q_d.

In the hydraulic loss analysis of flow passage components, the loss size of a section of the area can be obtained through the traditional differential pressure analysis method. Figure 5 shows the calculation of the total head loss of the guide-vane centrifugal pump in the full flow channel. In the Euler head calculated by the SAS model, when the flow changes from 1.0 Q_d, to 0.6 Q_d, the head loss decreases slowly; when the flow decreases from 0.6 Q_d, to 0.5 Q_d, the hydraulic loss increases rapidly. For the Euler head calculated by SST k-ω, when the flow rate changes from 1.0 Q_d to 0.5 Q_d, the Euler head decreases slowly; when the flow decreases from 0.5 Q_d, to 0.4 Q_d, the hydraulic loss increases rapidly. Under each working condition, the hydraulic loss calculated by SAS model is greater than SST k-ω. Due to the entropy, generation of the wall area is calculated more accurately by the SAS model, so the hydraulic loss obtained is greater than SST k-ω.

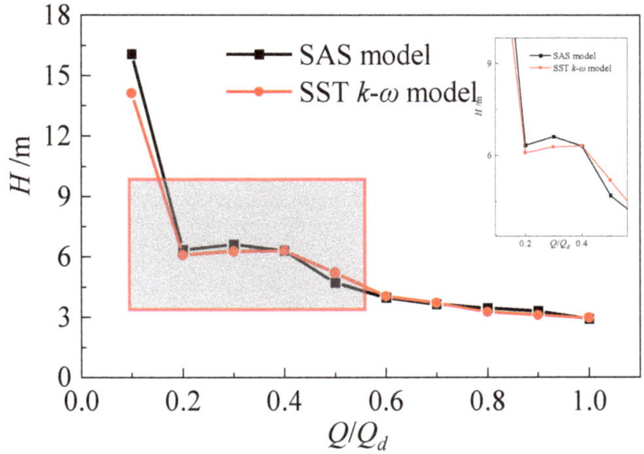

Figure 4. Comparison Diagram of Euler Head of Guide Vane Centrifugal Pump.

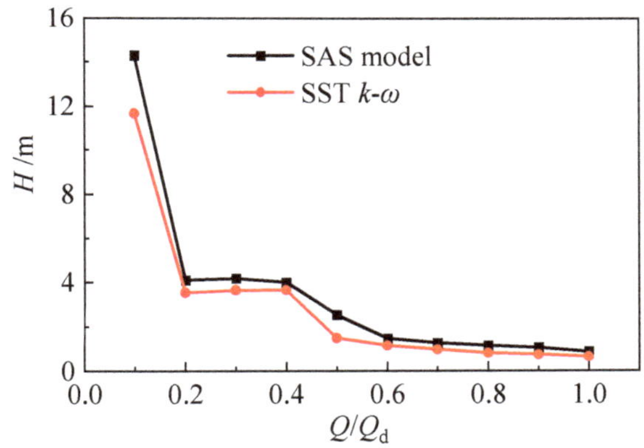

Figure 5. Calculation of Total Head Loss of Guide-Vane Centrifugal Pump in Full Flow Channel.

Figure 6 shows the head loss calculation of each component under different flow rates. The hydraulic loss of each component is the minimum at the design condition. Hydraulic loss of guide-vane centrifugal pump mainly comes from guide vane and impeller. For the hydraulic loss calculated by the SAS model, the head curve corresponds to the 0.5 Q_d–0.6 Q_d hump area close to the design working condition. From the peak point (0.6 Q_d) to the trough point (0.5 Q_d) in the hump area, the hydraulic loss of each component suddenly increases, in which the impeller and guide vane increase greatly, while the corresponding Euler head changes little. Therefore, the hump here is mainly caused by hydraulic loss. During the 0.5 Q_d–0.6 Q_d operating condition, there is no hump area calculated by the SST k-ω model because the hydraulic loss of components does not change much. When the flow rates decrease from 0.5 Q_d to 0.4 Q_d, the hydraulic loss increases rapidly, but the calculated Euler head increases at this time, offsetting the occurrence of the hump area. In the hump area of 0.2 Q_d–0.3 Q_d, the Euler head calculated by the two models decreases, and the hydraulic loss increases. Therefore, the hump area is the result of the joint action of Euler head and hydraulic loss.

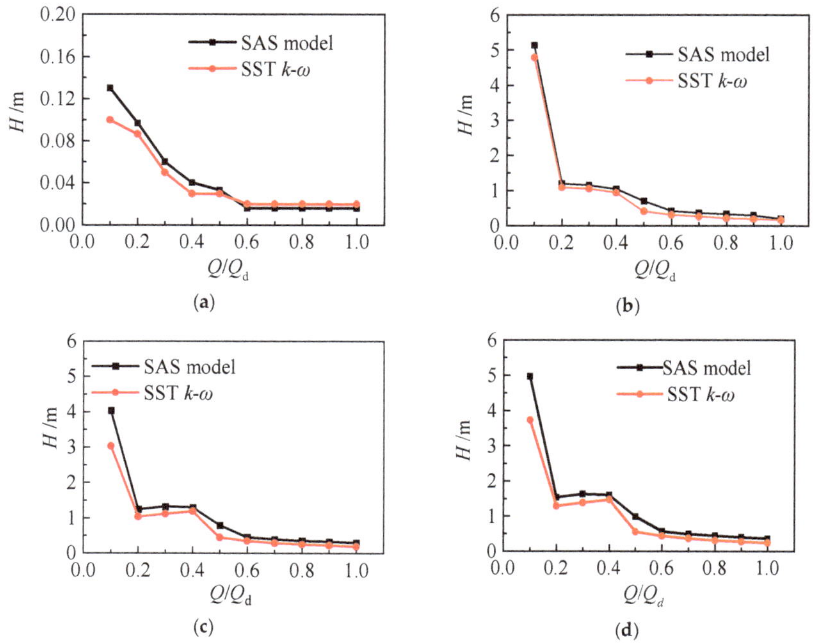

Figure 6. Calculation of Head Loss of Each Component of Guide-Vane Centrifugal Pump. (**a**) Inlet pipe; (**b**) Impeller; (**c**) Guide vane; (**d**) Volute.

As shown in Figures 7 and 8, the flow patterns inside the impeller at different working conditions are compared. At different working conditions, the influence of the guide vane on the internal flow of the impeller is different. At 0.7 Q_d, the stall passage of the impeller calculated by the two models does not change much. However, in the guide-vane passage, a separation vortex at the back of the guide vane is observed in the flow field calculated by the SAS model, which indicates that the loss of the guide vane begins to increase at this time. At 0.5 Q_d, the stall vortex is obvious in the impeller calculated by the two models. There are relatively large stall cells calculated by the SST k-ω model in channels A, C, and E, and they are all located on the suction surface of the impeller blade head and in the middle of the impeller pressure surface. There are also stall cells calculated by the SAS model in the flow channels in A, C, and E channels, which is similar to SST k-ω. The difference is that the previous two small vortices become a larger one, which is located near the suction surface of the impeller inlet, and the size of the stall cells almost blocks the entire flow passage. At 0.3 Q_d, the flow pattern of the impeller calculated by the two models becomes worse. From the results calculated by SST k-ω, the relatively large stall vortices are observed in channels A, C, and E, while the flow in the other three channels is relatively stable. Stall cells appear in all six channels from the results calculated by the SAS model, which is consistent with the phenomenon observed in the external characteristic curve in Figure 3. When the pump operates at 0.3 Q_d, the head curve is at the trough of the wave, which indicates that the hump in the head flow curve is largely caused by the stall in the impeller flow passage.

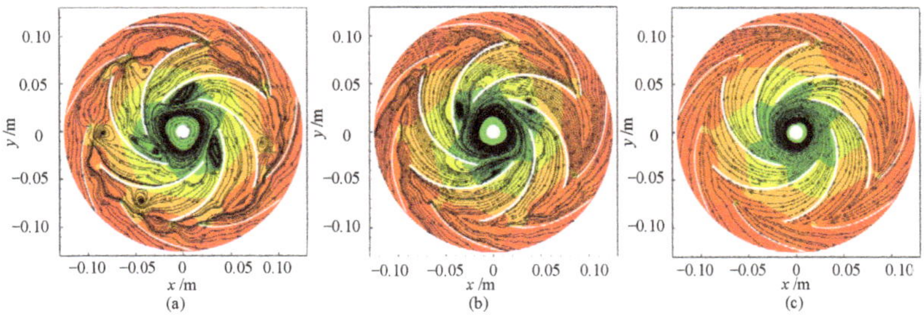

Figure 7. Pressure Contour and Streamline of Centrifugal Pump Calculated by SAS model. (a) 0.3 Q_d; (b) 0.5 Q_d; (c) 0.7 Q_d.

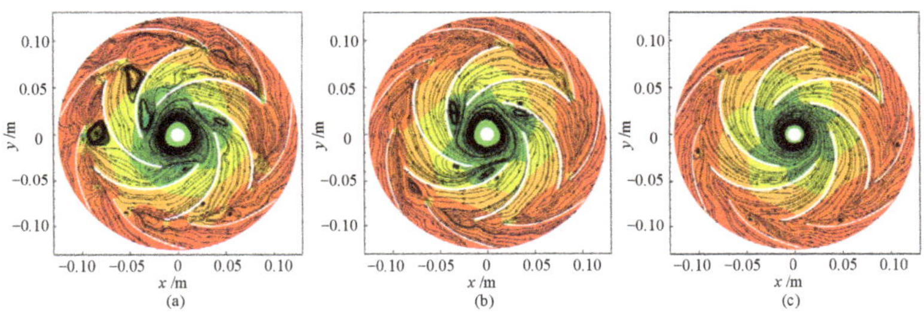

Figure 8. Pressure Contour and Streamline of Centrifugal Pump Calculated by SST k-ω model. (a) 0.3 Q_d; (b) 0.5 Q_d; (c) 0.7 Q_d.

5. Clocking Effect of Guide-Vane Centrifugal Pump under Stall Characteristics

Figure 9 shows the general assembly drawing of the centrifugal pump with a guide vane. In order to study the influence of the clocking position of the guide vane on the pump performance and the pressure fluctuation on the volute wall, four guide-vane installation positions are selected by rotating the guide vane, as shown in Figure 9. The guide-vane positions 2, 3, and 4 are obtained by rotating the guide-vane position 1 clockwise by 12.5°. The included angle between two adjacent vanes of the guide vane is 51.43°. For clocking position 1, the guide vanes outlet in line with the volute tongue. For clocking positions 2, 3, and 4, the position of the guide-vane outlet and volute tongue are staggered by 12.5°, 25°, and 37.5°, respectively.

The head curve of the centrifugal pump with guide vanes at four clocking positions is compared, as shown in Figure 10. When the guide vane is installed at clocking position 1, the head is maximum at large flow rate conditions. When the guide vane is installed at clocking position 3, the head is the minimum at all flow rate conditions. Different clocking positions have a great impact on pump performance at low flow rate conditions, and the maximum head difference is 0.5 m. The double humps occur at all four clocking positions, and the humps are observed at 0.2 Q_d–0.3 Q_d and 0.5 Q_d–0.6 Q_d.

The comparison of the streamline distribution at four clocking positions at 0.3 Q_d is shown in Figure 11. The clocking position has a great influence on the flow structure of the guide vane, and different sizes of reflux areas are generated in the flow passage, which is due to the flow separation occurring at the pressure side of the guide vane. Meanwhile, part of the fluid in the guide-vane flow passage flows out of the volute outlet, and part of the fluid enters the volute flow passage. In addition, the velocity direction on the right side of the volute outlet and the vortex structure on the left side is obviously affected by the guide-vane position. In the head curve, the head difference is large at 0.3 Q_d because the flow structure at the outlet of the guide vane close to the volute outlet is greatly affected by the position of the guide vane, which results in flow separation at the volute outlet, causing hydraulic loss.

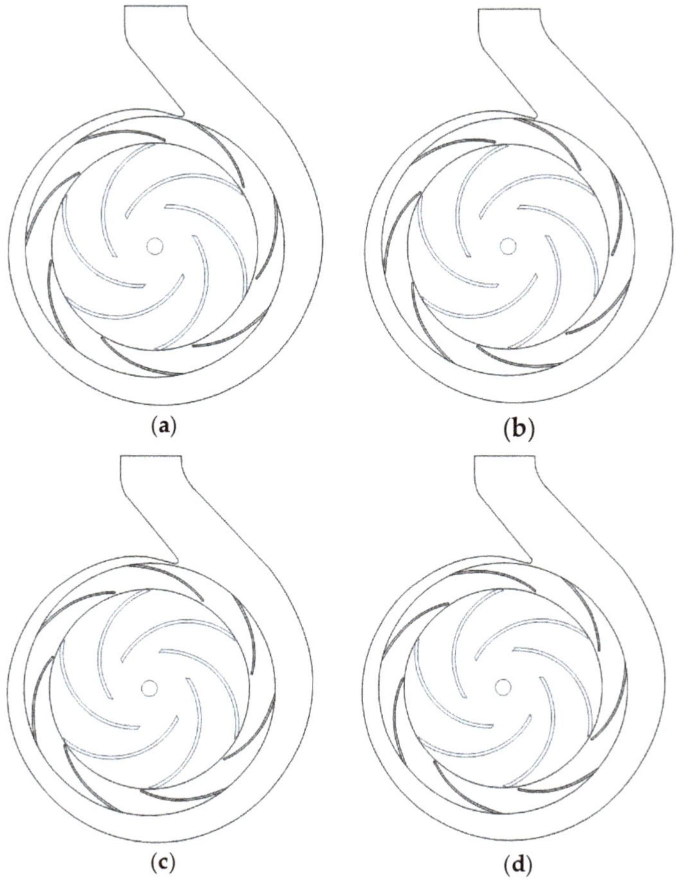

Figure 9. Clocking Position of the Pump. (**a**) Clocking position 1; (**b**) Clocking position 2; (**c**) Clocking position 3; (**d**) Clocking position 4.

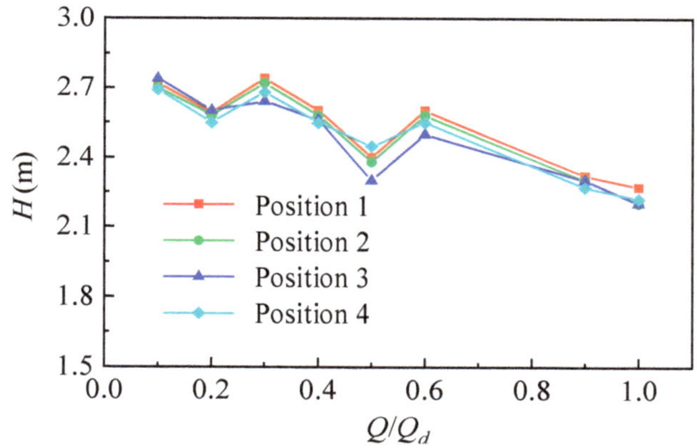

Figure 10. Centrifugal Pump Head Flow Curve at Different Clocking Positions.

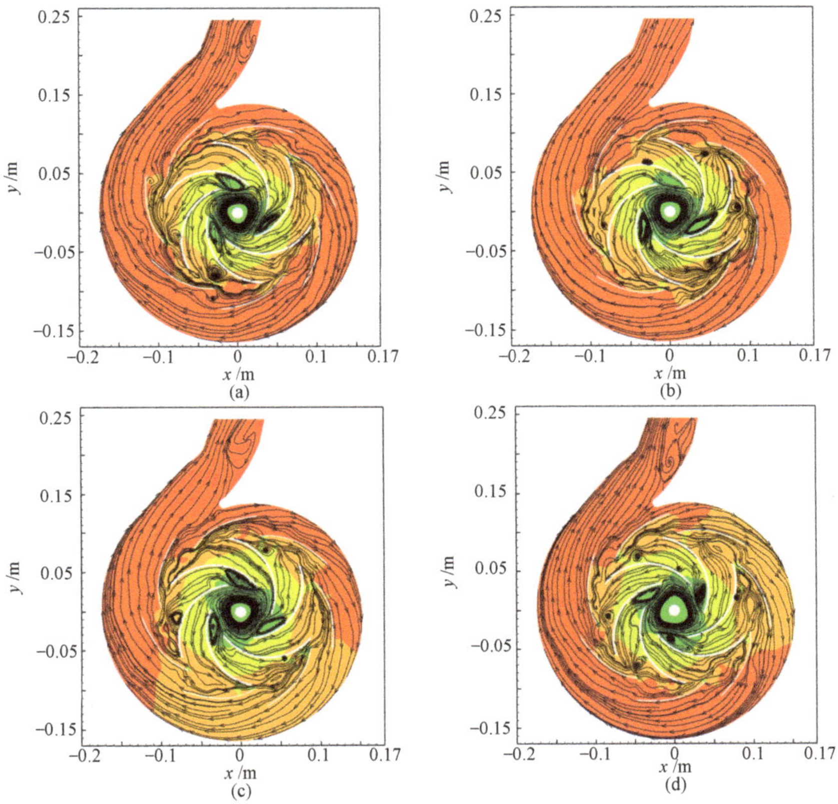

Figure 11. Velocity Streamline at Different Clocking Positions. (**a**) Clocking position 1; (**b**) Clocking position 2; (**c**) Clocking position 3; (**d**) Clocking position 4.

Figure 12 shows the pressure fluctuation of monitoring points in the impeller and guide vane at 0.3 Q_d. In the middle of the impeller, Figure 12a shows that the guide-vane position at stall condition has a significant impact on the instability characteristics of pressure fluctuation caused by rotor–stator interaction. When the guide vane is installed at position 1, the amplitude of the blade is at a relatively stable minimum; when the guide vane is at position 3, the pressure fluctuation amplitude of the blade has strong instability, which is greatly affected by the rotating position of the impeller. Due to the existence of the vortex at the inlet of the blade suction side, the pressure fluctuation amplitude is enhanced, and the peak-to-peak value is large. In the vaneless region, which is the gap between the impeller and the guide vanes, the flow interference is strong, and the pressure fluctuation signals are relatively greater, and the peak-to-peak value of the pressure fluctuation is the largest. Therefore, when the stall cell moves to the monitoring point, the pressure decreases rapidly, and after the stall cell falls off, the pressure value rises again. Stall cells periodically form, develop, and fall off, inducing low-frequency pressure fluctuations.

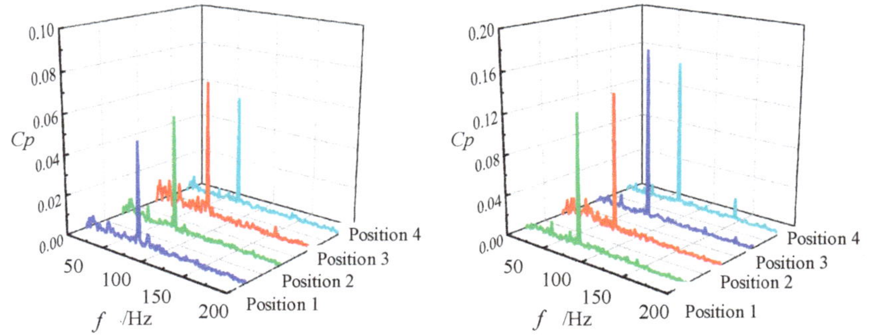

(a) (b)

Figure 12. Pressure Fluctuation of Impeller and Guide Vane. (**a**) Monitor point R2; (**b**) Monitor point S2.

Figure 13 shows the pressure fluctuation of the volute at 0.3 Q_d. The clocking position of the guide vane has a significant impact on the dominant frequency and corresponding amplitude of the volute. In the four clocking positions, the dominant frequency at the volute tongue is blade frequency (75 Hz), and the secondary frequency is low frequency. When the guide vane is installed at clocking position 1, the fluctuation amplitude of the dominant frequency is maximum, and when it is installed at clocking position 3, the fluctuation amplitude of the dominant frequency is minimum. The amplitude of dominant frequency fluctuation at clocking position 3 is 1.3 times of that at clocking position 1. At the outlet of the volute, one time of the dominant frequency corresponding to clocking position 1 is the blade frequency (75 Hz), and the dominant frequency at other clocking positions is the low frequency. From Figure 11, it can be seen that the streamline at clocking positions 2, 3, and 4 is twisted at the outlet of the volute, while the streamline at the outlet corresponding to timing position 1 is relatively smooth. Therefore, the timing position affects the flow at the outlet of the volute so that the volute will produce a low-frequency fluctuation.

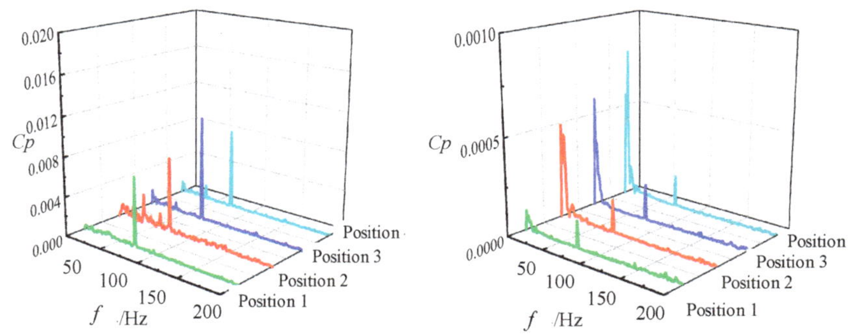

(a) (b)

Figure 13. Pressure Fluctuation of Volute. (**a**) Monitor point W1 near tongue; (**b**) Monitor point W8 near outlet.

In order to further study the clocking effect, the guide vane and volute tongue are simplified as two static hydrofoils [42]. The flow is controlled by the upstream valve angle for hydraulic characteristics research at different operating conditions. The clocking position is changed by changing the vertical position of the first hydrofoil, as shown in Figure 14.

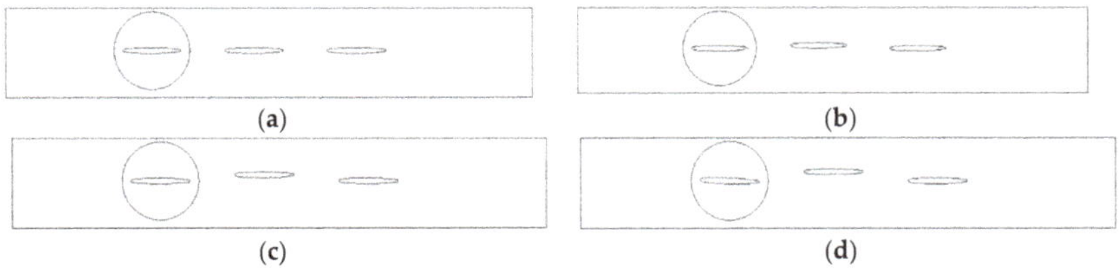

Figure 14. Simplified Hydrofoil. (**a**) Position 1; (**b**) Position 2; (**c**) Position 3; (**d**) Position 4.

Figure 15 shows the change of lift and drag coefficient of the secondary hydrofoil. It can be seen that with the decrease of flow rate, the secondary hydrofoil at the four clocking positions shows the characteristics of first stabilization and then oscillation. The difference is that at clocking position 1, the secondary hydrofoil starts to oscillate at a lower flow rate, and the amplitude of oscillation is much smaller than that at the other three positions.

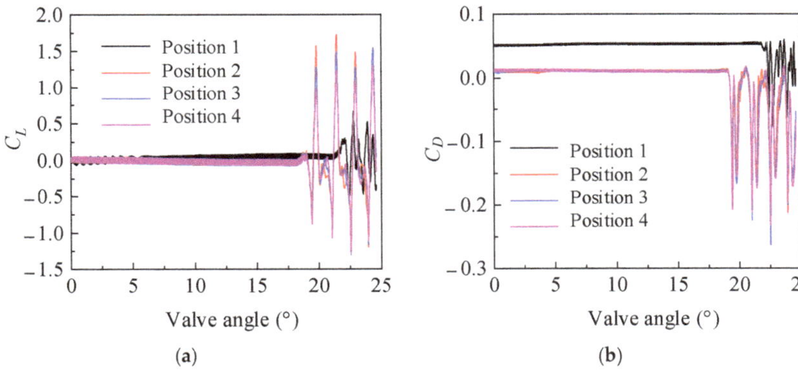

Figure 15. Variation of Lift, Drag Coefficient of Secondary Hydrofoil. (**a**) Lift; (**b**) Drag.

Figure 16 shows the arrangement of pressure fluctuation measuring points. The monitoring points, P1~P4, are arranged at different positions upstream and downstream of the secondary hydrofoil center. P1 and P2 are located at 0.1 L and 0.5 L at the tail of the first hydrofoil, and P3 and P4 are located at 0.1 L and 0.5 L at the tail of the second hydrofoil.

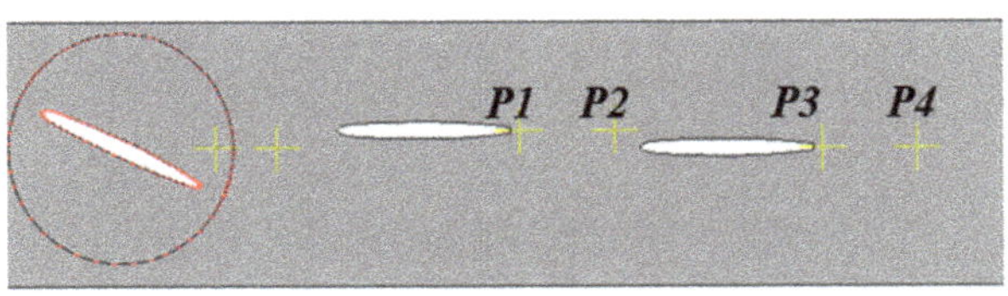

Figure 16. Arrangement of Pressure Fluctuation Measuring Points.

Figure 17 shows the pressure distribution of different monitoring points. It can be seen from the figure that for the monitoring points P1 and P2 at upstream of the secondary hydrofoil, the pressure changes with time tend to stabilize first and then increase. The pressure changes at the four clocking positions are 20°, 19°, 18°, and 17°, respectively. With the continuous rotation of the valve plate, the clearance between the valve plate and the pipe wall gradually changes, the flow rate decreases, and the pressure increases. Among the four clocking positions, the pressure fluctuation range of clocking position 1 is obviously smaller than that of other clocking positions.

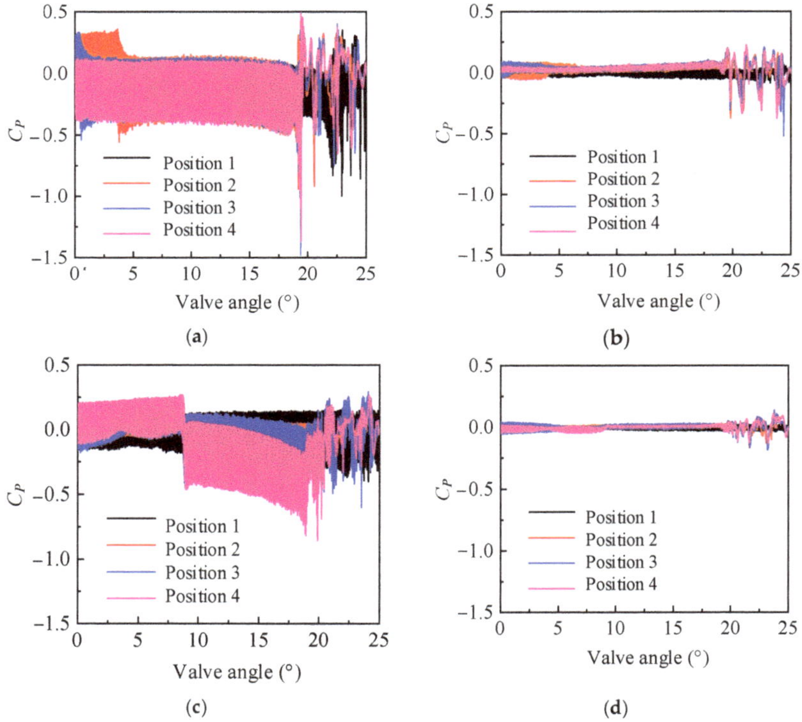

Figure 17. Pressure Distribution of Different Monitoring Points with Valve Opening. (**a**) Monitoring point P1; (**b**) Monitoring point P2; (**c**) Monitoring point P3; (**d**) Monitoring point P4.

Figure 18 shows the vorticity distribution at different clocking positions. It can be clearly observed that under the four positions, the wake of the first stage hydrofoil has a greater impact on the secondary hydrofoil. At clocking position 4, the wake of the first hydrofoil almost dissipates when it reaches the secondary hydrofoil passage and cannot be observed. At the clocking positions 1, 2, and 3, the first hydrofoil wake is mixed with the mainstream region and boundary layer of the secondary hydrofoils. Since the mainstream belongs to high-energy fluid, and the wake belongs to low-energy fluid, its mixing loss is obviously greater than that of the boundary layer of the same low-energy fluid, resulting in a vibration difference.

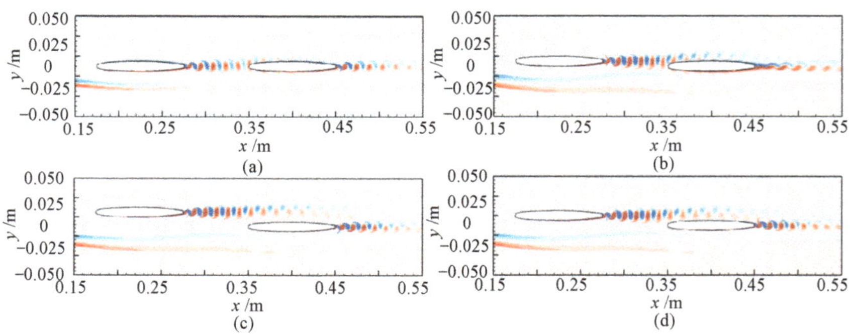

Figure 18. Vorticity Distribution at Different Time Sequence Positions. (**a**) Position 1; (**b**) Position 2; (**c**) Position 3; (**d**) Position 4.

Figure 19 shows the change of the shape factor of the suction surface of the secondary hydrofoil. It can be seen that under the condition of clocking position 1, no matter how the valve opening changes, the shape factor of the whole hydrofoil surface is less than 1.5, which reflects that the whole hydrofoil area is turbulent. On the suction surface, the momentum of the fluid in the boundary layer of the first hydrofoil wake and the secondary-stage hydrofoil is exchanged, and the momentum of the boundary layer increases compared with that without the influence of the wake. With the introduction of high energy, the boundary layer becomes thinner, and the separation area decreases, thus reducing the flow loss of the boundary layer. The high turbulence wake generated by the upstream hydrofoil will induce the transition of the downstream hydrofoil boundary layer. For clocking positions 3 and 4, the position where the transition occurs is also different, which indicates that with the change of clocking position, the wake begins to affect the boundary layer on the suction surface of the blade at a certain position downstream of the leading edge of the blade, and the wake induces the boundary layer to start transition before boundary layer separation.

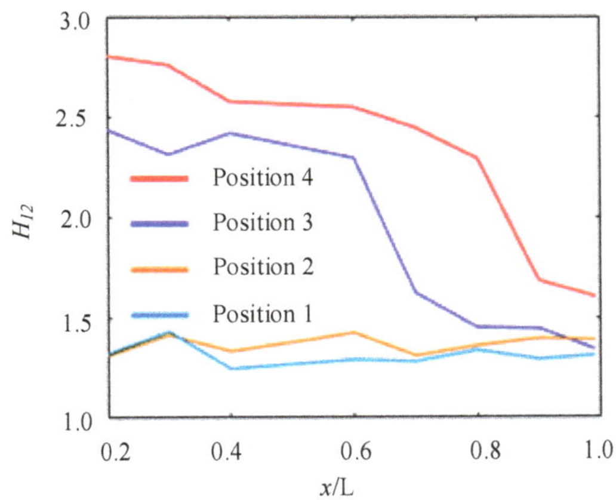

Figure 19. Change of Shape Factor of Suction Surface of Secondary Hydrofoil.

6. Conclusions

In this article, the stall of centrifugal pump with guide vanes is studied by numerical simulation, and the following conclusions were obtained:

(1) The double-hump characteristic was found in the head discharge curve by using the SAS model. Comparing the flow field characteristics at different flow rate conditions, it was found that the hump area close to the optimal working condition is caused by hydraulic loss, and the hump area far away from the optimal working condition point is caused by the combined action of Euler head and hydraulic loss. The SAS model can accurately calculate the wall friction loss, thus predicting the double hump phenomenon.

(2) The pressure fluctuation and head characteristics at different clocking positions under stall conditions were obtained. It was found that when the guide vanes outlet in line with the volute tongue, the flow pattern of the volute and guide vane is good, so the head is high due to small hydraulic loss, and the pressure fluctuation is low.

(3) The mechanism of clocking effect in the centrifugal pump with guide vanes was obtained by simplifying the hydrofoil. Based on the simplified hydrofoil, it can be found that the disturbance of wake to the boundary layer will affect the boundary layer transition and then affect the friction stress of the blade, resulting in the change of flow field pressure amplitude. When the downstream hydrofoil head area is always interfered with by the upstream hydrofoil wake, the wake of the low-energy fluid is mixed with the boundary layer of the same low-energy fluid. At this time, the boundary layer is in a turbulent state to avoid laminar flow separation, causing small

vibration of the downstream hydrofoil, so the position where the guide vane's outlet is in line with the volute tongue is most recommended.

Author Contributions: Software, D.A.; Validation, W.H.; Writing—original draft, C.Y.; Supervision, Y.H. and Y.Z. All authors have read and agreed to the published version of the manuscript.

Funding: This research was funded by the National Natural Science Foundation of China (grant number: 52209109, No.52271275) and the Open Research Subject of Key Laboratory of Fluid and Power Machinery (Xihua University), Ministry of Education (grant number: LTDL-2022005).

Data Availability Statement: Not applicable.

Acknowledgments: The authors would like to acknowledge the financial support given by the National Natural Science Foundation of China (No.52209109, No.52271275) and the Open Research Subject of Key Laboratory of Fluid and Power Machinery (Xihua University), Ministry of Education (grant number LTDL-2022005).

Conflicts of Interest: The authors declare no conflict of interest.

Nomenclature

Q_d	Design flow	SAS	Scale adaptive simulation
Cu	Circumferential component of absolute velocity	LES	Large eddy simulation method
U	Circumferential velocity	RANS	Reynolds average method
H_{Enter}	Euler head	C_p	Pressure coefficient
$\Delta Cu \cdot U$	Euler energy	ω	Angular velocity

References

1. Wang, C.; Wang, F.; Li, C.; Chen, W.; Wang, H.; Lu, L. Investigation on energy conversion instability of pump mode in hydro-pneumatic energy storage system. *J. Energy Storage* **2022**, *53*, 105079. [CrossRef]
2. Zhou, P.; Wang, F.; Mou, J. Investigation of rotating stall characteristics in a centrifugal pump impeller at low flow rates. *Eng. Comput.* **2017**, *34*, 1989–2000. [CrossRef]
3. Feng, J.; Ge, Z.; Yang, H.; Zhu, G.; Li, C.; Luo, X. Rotating stall characteristics in the vaned diffuser of a centrifugal pump. *Ocean Eng.* **2021**, *229*, 108955. [CrossRef]
4. Ye, C.L.; Wang, C.Y.; Zi, D.; Tang, Y.; van Esch, B.; Wang, F.J. Improvement of the SST γ-$Re_{\theta t}$ transition model for flows along a curved hydrofoil. *J. Hydrodyn.* **2021**, *33*, 520–533. [CrossRef]
5. Stenning, A.H.; Kriebel, A.R. Stall propagation in a cascade of airfoils. *Trans. Am. Soc. Mech. Eng.* **1958**, *80*, 777–789. [CrossRef]
6. Emmons, H.W.; Pearson, C.E.; Grant, H.P. Compressor surge and stall propagation. *Trans. Am. Soc. Mech. Eng.* **1955**, *77*, 455–467. [CrossRef]
7. Day, I.J. Stall, surge and 75 years of research. *ASME J. Turbomach.* **2016**, *138*, 011001. [CrossRef]
8. Zhang, N.; Gao, B.; Ni, D.; Liu, X. Coherence analysis to detect unsteady rotating stall phenomenon based on pressure pulsation signals of a centrifugal pump. *Mech. Syst. Signal Process.* **2021**, *148*, 107161. [CrossRef]
9. Huang, X.-B.; Liu, Z.-Q.; Li, Y.-J.; Yang, W.; Guo, Q. Study of the internal characteristics of the stall in a centrifugal pump with a cubic non-linear SGS model. *J. Hydrodyn.* **2019**, *31*, 788–799. [CrossRef]
10. Wang, C.; Wang, F.; Xie, L.; Wang, B.; Yao, Z.; Xiao, R. On the Vortical Characteristics of Horn-Like Vortices in Stator Corner Separation Flow in an Axial Flow Pump. *J. Fluids Eng.* **2021**, *143*, 061201. [CrossRef]
11. Abusorrah, A.M.; Mebarek-Oudina, F.; Ahmadian, A.; Baleanu, D. Modeling of a MED-TVC desalination system by considering the effects of nanoparticles: Energetic and exergetic analysis. *J. Therm. Anal. Calorim.* **2021**, *144*, 2675–2687. [CrossRef]
12. Hassan, M.; Mebarek-Oudina, F.; Faisal, A.; Ghafar, A.; Ismail, A. Thermal energy and mass transport of shear thinning fluid under effects of low to high shear rate viscosity. *Int. J. Thermofluids* **2022**, *15*, 100176. [CrossRef]
13. Wang, C.; Wang, F.; Li, C.; Ye, C.; Yan, T.; Zou, Z. A modified STRUCT model for efficient engineering computations of turbulent flows in hydro-energy machinery. *Int. J. Heat Fluid Flow* **2020**, *85*, 108628. [CrossRef]
14. Posa, A. LES investigation on the dependence of the flow through a centrifugal pump on the diffuser geometry. *Int. J. Heat Fluid Flow* **2021**, *87*, 108750. [CrossRef]
15. Sano, T.; Yoshida, Y.; Tsujimoto, Y.; Nakamura, Y.; Matsushima, T. Numerical Study of Rotating Stall in a Pump Vaned Diffuser. *J. Fluids Eng.* **2002**, *124*, 363–370. [CrossRef]
16. Braun, O. Part Load Flow in Radial Centrifugal Pumps. Ph.D. Thesis, EPFL, Lausanne, Switzerland, 2009.
17. Lucius, A.; Brenner, G. Unsteady CFD simulations of a pump in part load conditions using scale-adaptive simulation. *Int. J. Heat Fluid Flow* **2010**, *31*, 1113–1118. [CrossRef]

18. Zhang, N.; Jiang, J.; Gao, B.; Liu, X. DDES analysis of unsteady flow evolution and pressure pulsation at off-design condition of a centrifugal pump. *Renew. Energy* **2020**, *153*, 193–204. [CrossRef]
19. Ji, L.; Li, W.; Shi, W.; Tian, F.; Agarwal, R. Effect of blade thickness on rotating stall of mixed-flow pump using entropy generation analysis. *Energy* **2021**, *236*, 121381. [CrossRef]
20. Kan, K.; Zheng, Y.; Chen, Y.; Xie, Z.; Yang, G.; Yang, C. Numerical study on the internal flow characteristics of an axial-flow pump under stall conditions. *J. Mech. Sci. Technol.* **2018**, *32*, 4683–4695. [CrossRef]
21. Krause, N.; Pap, E.R.; The´ venin, D. Influence of the blade geometry on flow instabilities in a radial pump elucidated by time-resolved particle-image velocimetry. In Proceedings of the Turbo Expo: Power for Land, Sea, and Air, Montreal, Canada, 14–17 May 2007; ASME: New York, NY, USA, 2007; Volume 47950, pp. 1659–1668.
22. Berten, S.; Dupont, P.; Fabre, L.; Kayal, M.; Avellan, F.; Farhat, M. Experimental investigation of flow instabilities and rotating stall in a high-energy centrifugal pump stage. In Proceedings of the Fluids Engineering Division Summer Meeting, Vali, CO, USA, 2–6 August 2009; Volume 43727, pp. 505–513.
23. Chudina, M. Noise as an indicator of cavitation in a centrifugal pump. *Acoust. Phys.* **2003**, *49*, 463–474. [CrossRef]
24. Johnson, D.A.; Pedersen, N.; Jacobsen, C.B. Measurements of rotating stall inside a centrifugal pump impeller. In Proceedings of the Fluids Engineering Division Summer Meeting, Houston, TX, USA, 19–23 June 2005; Volume 41987, pp. 1281–1288.
25. Ye, W.; Huang, R.; Jiang, Z.; Li, X.; Zhu, Z.; Luo, X. Instability analysis under part-load conditions in centrifugal pump. *J. Mech. Sci. Technol.* **2019**, *33*, 269–278. [CrossRef]
26. Ren, X.; Fan, H.; Xie, Z.; Liu, B. Stationary stall phenomenon and pressure fluctuation in a centrifugal pump at partial load condition. *Heat Mass Transf.* **2019**, *55*, 2277–2288. [CrossRef]
27. Takamine, T.; Furukawa, D.; Watanabe, S.; Watanabe, H.; Miyagawa, K. Experimental Analysis of Diffuser Rotating Stallin a Three-Stage Centrifugal Pump. *Int. J. Fluid Mach. Syst.* **2018**, *11*, 77–84. [CrossRef]
28. Liu, X.-D.; Li, Y.-J.; Liu, Z.-Q.; Yang, W.; Tao, R. Dynamic evolution process of rotating stall vortex based on high-frequency PIV system in centrifugal impeller. *Ocean Eng.* **2022**, *259*, 111944. [CrossRef]
29. Pavesi, G.; Cavazzini, G.; Ardizzon, G. Time–frequency characterization of the unsteady phenomena in a centrifugal pump. *Int. J. Heat Fluid Flow* **2008**, *29*, 1527–1540. [CrossRef]
30. Takao, S.; Konno, S.; Ejiri, S.; Miyabe, M. Suppression of Diffuser Rotating Stall in A Centrifugal Pump by Use of Slit Vane. In Proceedings of the Fluids Engineering Division Summer Meeting, Online, 10–12 August 2021; American Society of Mechanical Engineers: New York, NY, USA, 2021; Volume 85291, p. V002T03A007.
31. Yan, H.; Heng, Y.; Zheng, Y.; Tao, R.; Ye, C. Investigation on Pressure Fluctuation of the Impellers of a Double-Entry Two-Stage Double Suction Centrifugal Pump. *Water* **2022**, *14*, 4065. [CrossRef]
32. Griffini, D.; Insinna, M.; Salvadori, S.; Martelli, F. Clocking Effects of Inlet Nonuniformities in a Fully Cooled High-Pressure Vane: A Conjugate Heat Transfer Analysis. *J. Turbomach.* **2015**, *138*, 021006. [CrossRef]
33. Benigni, H.; Jaberg, H.; Yeung, H.; Salisbury, T.; Berry, O.; Collins, T. Numerical Simulation of Low Specific Speed American Petroleum Institute Pumps in Part-Load Operation and Comparison with Test Rig Results. *J. Fluids Eng.* **2012**, *134*, 024501. [CrossRef]
34. Gu, Y.; Pei, J.; Yuan, S.; Wang, W.; Zhang, F.; Wang, P.; Liu, Y. Clocking effect of vaned diffuser on hydraulic performance of high-power pump by using the numerical flow loss visualization method. *Energy* **2019**, *170*, 986–997. [CrossRef]
35. Wang, W.; Pei, J.; Yuan, S.; Yin, T. Experimental investigation on clocking effect of vaned diffuser on performance characteristics and pressure pulsations in a centrifugal pump. *Exp. Therm. Fluid Sci.* **2018**, *90*, 286–298. [CrossRef]
36. Tan, M.; Lian, Y.; Wu, X.; Liu, H. Numerical investigation of clocking effect of impellers on a multistage pump. *Eng. Comput.* **2019**, *36*, 1469–1482. [CrossRef]
37. Menter, F.R. *Improved Two-Equation k-Turbulence Models for Aerodynamic Flows*; NASA Technical Memorandum, 103975(1_), 3t; NASA: Washington, DC, USA, 1992.
38. Hellsten, A. Some improvements in Menter's k-omega SST turbulence model. In Proceedings of the 29th AIAA, Fluid Dynamics Conference, Albuquerque, NM, USA, 15–18 June 1998; p. 2554.
39. Menter, F.R.; Egorov, Y. SAS turbulence modelling of technical flows. In *Direct and Large-Eddy Simulation VI*; Springer: Dordrecht, The Netherlands, 2006; pp. 687–694.
40. Pedersen, N. Experimental Investigation of Flow Structures in a Centrifugal Pump Impeller Using Particle Image Velocimetry. Ph.D. Thesis, Technical University of Denmark, Kongens Lyngby, Denmark, 2000.
41. Li, D.; Song, Y.; Lin, S.; Wang, H.; Qin, Y.; Wei, X. Effect mechanism of cavitation on the hump characteristic of a pump-turbine. *Renew. Energy* **2020**, *167*, 369–383. [CrossRef]
42. Yan, H.; Zhang, H.; Zhou, L.; Liu, Z.; Zeng, Y. Optimization design of the unsmooth bionic structure of a hydrofoil leading edge based on the Grey–Taguchi method. *Proc. Inst. Mech. Eng. Part M J. sEng. Marit. Environ.* **2022**, 14750902221128140. [CrossRef]

water

MDPI

Article

Analysis of Stress Characteristics of a Vertical Centrifugal Pump Based on Fluid-Structure Interaction

Siwei Li [1], Yongsha Tu [1], Changliang Ye [2,*], Hongyeyu Yan [2], Jin Dai [3], Mengfan Dang [1], Chunxia Yang [2], Yuan Zheng [2] and Yongbiao Li [3]

1 Hunan Water Resources and Hydropower Survey, Design, Planning and Research, Co., Ltd., Changsha 410007, China; siwei0055@126.com (S.L.); tyongsha@foxmail.com (Y.T.); 18229835971@163.com (M.D.)
2 College of Energy and Electrical Engineering, Hohai University, Nanjing 211100, China; 2020020215@hhu.edu.cn (H.Y.); yangchunxia@hhu.edu.cn (C.Y.); zhengyuan@hhu.edu.cn (Y.Z.)
3 Hunan Provincial Water Resources Development and Investment Co., Ltd., Changsha 410007, China; chc_dj@163.com (J.D.); liyong_biao@163.com (Y.L.)
* Correspondence: yechangliang@hhu.edu.cn

Abstract: Vertical centrifugal pumps play a crucial role in numerous water conservancy projects. However, their continuous operation can lead to the development of cracks or even fractures in some centrifugal pump blades, resulting in a substantial adverse impact on the operation of the pumping station unit and jeopardizing safe production. This study employs the fluid-structure interaction method to comprehensively investigate the modal characteristics of the impeller, both in an air environment and immersed in water. Furthermore, the analysis of static and dynamic stress attributes is conducted. The natural frequency of the impeller when submerged in water is significantly lower than its frequency in an air medium, typically accounting for approximately 0.35 to 0.46 of the air-based natural frequency. There are conspicuous stress concentrations at specific locations within the system, specifically at the rounded corners of the blade back exit edge, the impeller front cover, the middle of the blade inlet edge, and the junction where the blade interfaces with the front and back cover. It is crucial to underscore that when the system operates under high-flow or low-flow conditions, there is a pronounced stress concentration at the interface between the impeller and the rear cover plate. Any deviation from the intended design conditions results in an escalation of equivalent stress levels. Through dynamic stress calculations during a single rotational cycle of the impeller, it is discerned that the cyclic nature of stress at the point of maximum stress is primarily influenced by the number of blades and the rotational velocity of impeller. This research carries significant implications for effectively mitigating blade fractures and cyclic fatigue damage, thereby enhancing the operational reliability of vertical centrifugal pumps in water conservancy applications.

Keywords: pump; modal analysis; static stress; dynamic stress; numerical simulation; fluid-structure interaction

Citation: Li, S.; Tu, Y.; Ye, C.; Yan, H.; Dai, J.; Dang, M.; Yang, C.; Zheng, Y.; Li, Y. Analysis of Stress Characteristics of a Vertical Centrifugal Pump Based on Fluid-Structure Interaction. *Water* **2023**, *15*, 4269. https://doi.org/10.3390/w15244269

Academic Editor: Ling Zhou

Received: 13 November 2023
Revised: 8 December 2023
Accepted: 11 December 2023
Published: 13 December 2023

1. Introduction

The increasing prevalence of large centrifugal pumps can be attributed to the successive implementation of cross-basin water transfer projects, extensive irrigation and drainage initiatives, and the development of nuclear power plants. With the increasing size of the centrifugal pump impellers, the pump blade cracking problem has become particularly prominent, for example, the blade cracking results in certain economic losses in the Dongjiang-Shenzhen Water Supply Project of China as well as the Grand Coulee Hydroelectric Power Station project of America [1,2].

Many studies have shown that the causes of blade fatigue fracture are the residual stresses, static stresses and vibration alternating variable stresses acting on the blade, and the expansion of microcracks and cracks caused by the cumulative damage of the impeller under its action [3–5]. Therefore, stress characterization of large centrifugal pumps is of great significance for the prevention of blade cracks.

The process of stress characterization encompasses the interplay between fluid and solid domains, typically employing a coupled fluid-structure methodology. Currently, more studies use the coupling of Computational Fluid Dynamics (CFD) and Finite Element Analysis (FEA) for fluid-structure interaction [6–8]. CFD is used to simulate the flow inside the fluid, while the FEA method is used to analyze the structural response of the pump. The combination of these two methodologies facilitates the dynamic interaction between the fluid and the structure. Researchers usually use commercial software or self-developed coupled solvers to realize it. This approach allows flow-induced vibrations, stress distributions, and the effect of pump structural deformation on flow performance to be considered in a single simulation [9–11].

The intrinsic frequency and intrinsic mode of vibration of an undamped system are called the intrinsic modes of the system [12,13]. The investigation of impeller mechanical modal issues holds considerable importance. When the water excitation frequency such as Karman vortex is equal or similar to a certain order of the impeller mechanical intrinsic frequency, resonance occurs, which results in unit vibration and rotor blade cracking problems [14–16]. Krzemianowski [17,18] used a three-dimensional finite element method with full fluid-structure interaction to investigate the modal features of a mixed-flow rotor in aqueous medium. The vibration characteristics of the runner in water, such as self-oscillation frequency and vibration mode, were obtained. The results show primary inherent frequency of the runner when submerged in water has a resemblance to the rotational frequency of the blades, and the Carmen vortex frequency at rated condition is close to the higher-order intrinsic frequency of the runner. This proximity significantly increases the likelihood of inducing resonance in the structure, perhaps resulting in the formation of cracks. Zhou [19–21] adopted the displacement-velocity potential format of the finite element method without considering the mesh variation and used the Galerkin method of the weighted residual method to investigate the vibration characteristics of a coupled runner in a water medium. Galerkin's method to perform sequential coupling calculations for the coupled system. A comprehensive analysis is conducted to examine how the flow velocity field influences the vibrational properties of the blades, and the method of calculating the added mass and the factors affecting the added mass are analyzed. B Hübner [22,23] calculated the intrinsic frequency and intrinsic vibration pattern of a hydraulic turbine in water using the acoustic-hydraulic equations, and examined the impact of the vibration pattern caused by the flow, including the flow within the intricate and unstable flow regime. To accomplish this, he utilized the partitioned fluid-structure interaction method. By coupling the fluid flow, the nonlinear solids as a whole calculate the dynamic stresses applied to the solids and analyze the hydro elastic instability of elastomers.

The bearer of the hydro-mechanical runner, water, produces a complex hydraulic excitation on its action, and people have been actively seeking methodologies to determine the stress properties of runner blades resulting from hydraulic excitation. Nowadays, ANSYS software is one of the more commonly used software for calculating fluid structure [24–26]. Kan [27] used the flow field and structure to compute and examine the flow field of the axial pump impeller, as well as the strength characteristics of the runner body, and the distribution of stresses was obtained. The research findings indicate a precise alignment between the regions of stress concentration and the specific sites where fracture formation occurs. Quan [28] analyzed the static stress and vibration pattern of the inducer on the basis of the results of three-dimensional numerical calculations of the full flow channel of the high-speed centrifugal pump, the results show that, when subjected to water pressure, the blade deformation reaches its highest displacement in close proximity to the outside edge of the blade outflow edge. Additionally, the region where the blade and the hub are connected demonstrates the utmost magnitude of comparable stress. It is worth noting that blade deformation significantly impacts the vibration characteristics. Egusquiza [29,30] employed the sequential fluid-structure coupling technique to assess the stress characteristics of the runner blade across different operational scenarios. The findings indicate a predominantly linear relationship between the maximum static stress of the runner blade and the turbine power in the majority of working conditions. The maximum static stress of the runner blade under various water heads in the construction of a Francis turbine was determined by Chen [31]. The analysis of the runner static stress distribution at various water heads during rated output reveals that the highest stress occurs at the junction between the blade outlet edge and the upper crown. Furthermore, it indicates a negative correlation between the water head and the highest static stress value. Menéndez [32] carried out static stress and deformation calculations of a pump-turbine and found that the highest static stress is observed on the impeller pressure surface at the junction of the blades and the hub. Additionally, the maximum deformation of the impeller increases as the blade radius increases. Li [33] obtained the stress strain of the impeller structure, and the pressure and flow velocity distribution of the flow field through the simulation of the coupling of the impeller blade and the internal flow field. Birajdar [34] employed the idea of sequential fluid-structure coupling to investigate a vertical turbine pump. They utilized the internal non-constant flow field information as mechanical boundary conditions and employed the finite element method to conduct transient analysis. The utilization of the finite element method was also employed in order to investigate the impeller of a large double-suction centrifugal pump. The findings indicate that the dynamic stress distribution pattern on the impeller surface remains largely consistent across various flow conditions. The highest stress point is observed in the root region adjacent to the front cover plate, either at the inlet or outlet side. Furthermore, the dynamic stress reaches its peak under conditions of low flow. Shi [35] carried out the fluid-structure interaction calculation on the runner of a pump and found that the larger stress area was basically the same as that in which the actual cracks appeared. Then the transient dynamic analysis of the runner under dynamic load was carried out, and the Newmark algorithm was used to derive the stress distribution when the guide vane opening was changed, and when the frequency of dynamic load change was large, the dynamic stress was calculated to be roughly around 15–20 MPa, which was closer to the empirical value. The primary cause of blade cracking is attributed to the elevated levels of dynamic stress.

In this paper, the simulation computation of the whole flow channel is conducted for a particular large centrifugal pump. Fine water pressure distribution maps are obtained to provide important basic data for the performance analysis and optimization of the centrifugal pump. Further, a method known as sequential flow-solid coupling is employed to integrate the flow field data collected earlier with the impeller structure for simulation purposes. The objective is to analyze the distribution of structural stress within the impeller under various operating situations. By comprehensively analyzing the stress distributions under multiple operating conditions, the evaluation of the stresses exerted on the impeller is conducted under various operating situations, and possible stress concentration areas are identified.

2. Computational Model and Boundary Conditions

2.1. Computational Models

Pre-processing is the premise of numerical simulation, including solid modeling and mesh division. The centrifugal pump impeller is subjected to modal analysis under the condition of fluid-structure interaction, so in order to accurately simulate the behavior of an impeller and its surrounding water body, it is imperative to incorporate both components into the computational model simultaneously and subsequently partition the mesh accordingly.

Firstly, the centrifugal pump blade is modeled. According to the wood mold drawing, the cylindrical coordinates of each wood mold section control point are extracted and transformed into Cartesian coordinates, the blade is firstly drawn in BladeGen, then the blade is imported into UG 3D modeling software, and the blade solid is generated by sewing the surface. After that, based on the device layout diagram, the impeller front and rear cover plates are observed to be created subsequent to rotation, as indicated by the axial section. Finally, the front and rear cover plates and blades are summed up to generate the impeller entity. The impeller diameter is D = 400 mm, the number of blades is Z = 6, the rated speed is n = 1480 r/min. The impeller solid model and mesh division are shown in Figure 1. The average y+ on the wall is within 30.

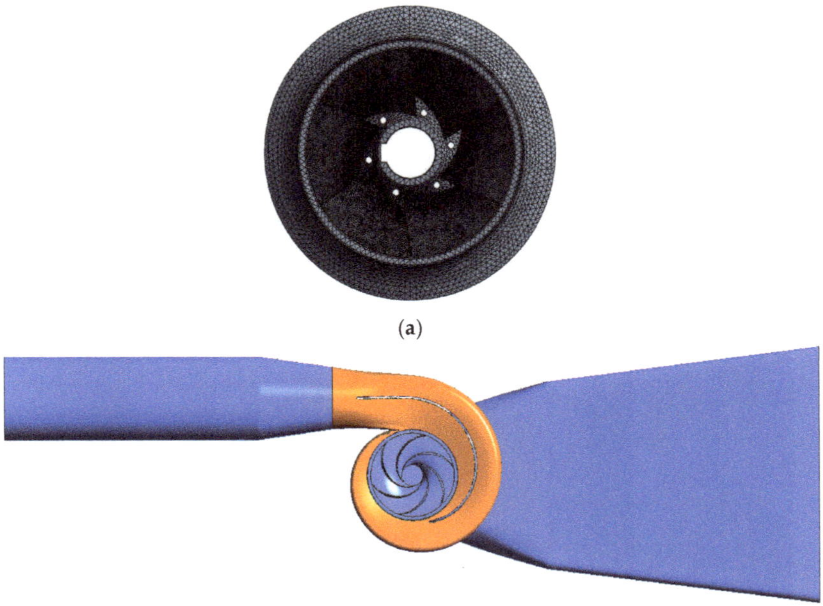

(a)

(b)

Figure 1. Computational domain. (**a**) Finite element model of impeller; (**b**) Fluid domain.

2.2. Boundary Conditions

The given boundary condition for the computation of the flow field involves the intake, which is determined by the mass flow rate. On the other hand, the exit condition for the volute is defined by setting the mean static pressure to zero. By employing CFD analysis to evaluate the entire flow channel, a more precise internal flow field may be computed. This enhanced accuracy facilitates the determination of blade surface pressure loads, which in turn contributes to more accurate calculations in the subsequent structural field analysis [36–38]. The focus of the structural field calculation is on an individual impeller within a centrifugal pump. Since stress concentration often occurs in the root region, the impeller modeling of the blade and the impeller before and after the cover plate of the transition between the rounded corners of the impeller is also accurately simulated, and in this sensitive area of the mesh refinement, to avoid the calculation process due to the mesh division of the cause of the stress concentration phenomenon that should not be generated [39].

In the structural field calculation, the loads mainly include displacement constraints, water pressure on the blade surface and centrifugal force. The displacement constraint is the fixed hub end face, the gravity and centrifugal force are applied as long as they are defined in ANSYS Workbench [40], and the water pressure at the interface between the fluid and solid is determined by analyzing the outcomes of the comprehensive computation of the flow field in the channel. For the modal analysis of the impeller in water, the fluid-structure interaction needs to be considered, so in addition to the 3D modeling of the impeller, it is also necessary to construct the water region around it. We can visualize this as placing the impeller in a cylindrical body of water, which is to completely submerge the impeller. In order to form a fluid-structure interaction integral finite element model of the impeller submerged in water, we can realize this by constructing a rectangular fluid outside the impeller that is slightly larger than the size of the impeller. When meshing, it is necessary to ensure that the mesh sizes at the fluid-structure interaction interface are consistent. Figure 2 illustrates the comprehensive finite element model of the fluid-solid connection, while the mesh division is visually presented in the same figure. The cases run in the High-Performance Computing Center of Hohai University, using 256 GB of memory and 128 cores for parallel computing. The numerical calculation is carried out on ANSYS CFX 2020.

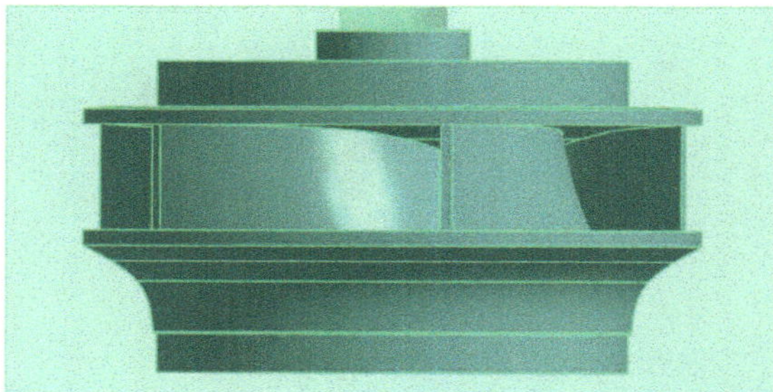

Figure 2. Finite element model of impeller in water.

2.3. Numerical Model

(1) Turbulence Model

The calculation of the flow field adopts SST k-ω turbulence model. The model consists of turbulent kinetic energy transport equation and turbulent dissipation rate transport equation, which has good predictive ability for flow separation under reverse pressure gradient.

$$\frac{\partial \rho k}{\partial t} + \frac{\partial(\rho \overline{u}_j k)}{\partial x_j} = P_k - D_k + \frac{\partial}{\partial x_j}\left((\mu + \sigma_{k3}\mu_t)\frac{\partial k}{\partial x_j}\right) \tag{1}$$

$$\frac{\partial \rho \omega}{\partial t} + \frac{\partial(\rho \overline{u}_j \omega)}{\partial x_j} = P_\omega - D_\omega + \frac{\partial}{\partial x_j}\left((\mu + \sigma_{\omega 3}\mu_t)\frac{\partial \omega}{\partial x_j}\right) + 2\rho(1 - F_1)\frac{\sigma_{\omega 2}}{\omega}\frac{\partial k}{x_j}\frac{\partial \omega}{x_j} \tag{2}$$

Specific parameters can be referenced in reference [38].

(2) Modal Analysis Methods

Equation (3) is the expression of the linear differential equation of structural dynamics:

$$[M]\{\ddot{u}\} + [C]\{\dot{u}\} + [K]\{u\} = \{F(t)\} \tag{3}$$

where, $[M]$; $[C]$; $[K]$ are the structural mass matrix, damping matrix, and stiffness matrix, respectively. $\{\ddot{u}\}$; $\{\dot{u}\}$; $\{u\}$ are the acceleration vector, velocity vector, and displacement of each node of the structure; $\{F(t)\}$ is the external structure load vector.

Modal analysis is a commonly used dynamic analysis method in the field of fluid machinery simulation. It is mainly used to analyze the natural frequency and mode shape of a structure. The natural frequency of a structure is related to many factors, such as the material used by the structure, operating state, the magnitude of the load, etc. When the operating frequency of the centrifugal pump is the same as the natural frequency of the structure, large structural deformation will occur due to structural resonance, which will affect the safety and stability of pump operation in severe cases. Hence, obtaining pump natural frequency and its mode shape by modal analysis can provide a reference for the design of the centrifugal pump.

The linear differential equation of structural dynamics is the same as Equation (3). For modal analysis, the overall load matrix of the structure is $\{F(t)\} = 0$, and the damping of the structure can be ignored, that is, $[C] = 0$. Therefore, the structural dynamic linear differential equations can be expressed by Equation (4):

$$[M]\{\ddot{u}\} + [K]\{u\} = \{0\} \tag{4}$$

From the free vibration of the structure as simple harmonic vibration, the displacement function Equation (5) can be obtained:

$$x = x\sin(\omega t) \tag{5}$$

the dynamic equation of modal analysis can be obtained as Equation (6):

$$\left([K] - \omega^2[M]\right)\{x\} = \{0\} \tag{6}$$

To solve Equation (6), the natural frequency and mode shape of the structure are the eigenvalues and eigenvectors, and the structure must be linear. Specific parameters can be referenced in reference [41].

2.4. Validation

The test was conducted at the China Institute of Water Resources and Hydropower Research test rig, as shown in Figure 3. The maximum test head of the test bench is 150 m, the maximum flow rate is 2.2 m^3/s, and the efficiency measurement uncertainty is less than $\pm 0.2\%$. The test bench is equipped with a hydraulic mechanical energy performance test.

Figure 3. Test rig.

Figure 4 compares the calculated and experimental values of the energy characteristic parameters under each working condition. Under the design conditions, the calculated value is more accurate in predicting the head, and the calculated value exceeds the experimental one. The shaft power value obtained through calculation is 0.65 percent greater than that of another value, while the calculated efficiency value is 0.14 percentage points lower than the experimental value. In the case of a low flowrate, the computed values of head, shaft power, and efficiency exhibit a decrease. In the case of a large flowrate, the computed values of head and power exceed the experimental values, while the calculated value of efficiency falls short of the experimental value. Overall, the errors are all within 5%, and the numerical calculation has high reliability.

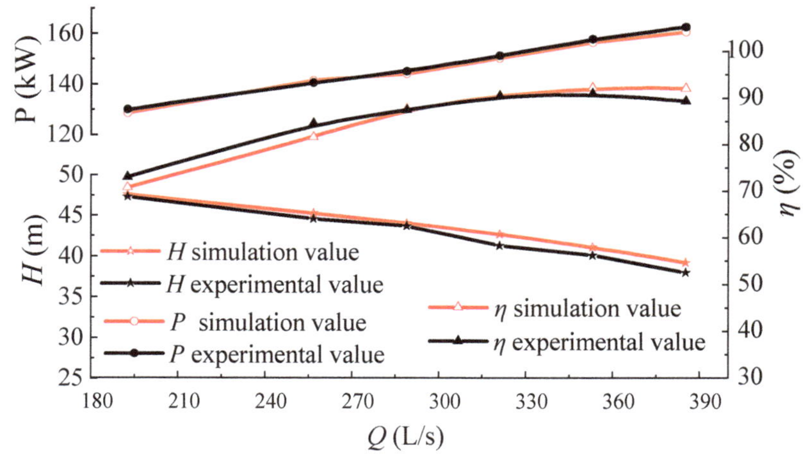

Figure 4. Comparison of energy characteristics.

3. Impeller Modal Analysis

3.1. Modal Analysis of Impeller in Air

Table 1 presents the material properties and meshing details of the impeller. Modal analysis is performed to ascertain the inherent frequencies and mode shapes of a structure's unconstrained oscillations, so only displacement constraints need to be defined and other external loads can be neglected. The full displacement constraint is defined at the end face of the shaft bore of the impeller. According to the modal extraction method, the determination of the modal characteristics of the impeller in an air environment is often performed by employing a mass matrix that is consistent throughout the analysis. So the Block Lanczos algorithm is used in the calculation.

Table 1. Material properties and meshing of impeller.

Parameters	Value
Modulus of elasticity E (Gpa)	206
Poisson ratio μ	0.288
Material density ρ (kg/m^3)	7700
Meshing	8-node tetrahedral solid unit solid 45

3.2. Modal Analysis of Impeller in Water

The pressure of the Fluid 30 unit is obtained by solving the wave equation. In situations where the pressure distribution of the liquid needs to be observed, using the Fluid 30 unit as the simulation unit is a more appropriate choice [42]. For the modal analysis immersed in water, 8-node tetrahedral solid unit Solid 45 and acoustic unit Fluid 30 are selected for the impeller and the water body, respectively, and the mechanical properties of the water include underwater sound velocity and density. In the actual analysis, the water body is divided into two major categories, one is not in contact with the impeller blades of the unit, and the nodes of these units have only pressure degrees of freedom, only need to apply the surface pressure, where it is a free liquid surface, the pressure is 0; the other is in contact with the impeller surface of the unit, that is, the fluid-structure interaction interface, these units have both pressure degrees of freedom, and displacement degrees of freedom. For the modal analysis in the water, the same only needs to define the full displacement constraints on the end face of the shaft hole of the impeller. According to the theoretical analysis, the unsymmetric method has been selected as the analytical approach to assess the modal characteristics of the impeller immersed in water.

Figure 5 shows the graphs of the first 10 orders of the intrinsic frequency of the impeller in air and water, and Table 2 shows the comparison of the first 10 orders of the intrinsic frequency of the impeller. Where f_a denotes the intrinsic frequency in air, f_w denotes the intrinsic frequency in water, and $\eta = f_w/f_a$ denotes the effect of the water body on the intrinsic frequency of the impeller.

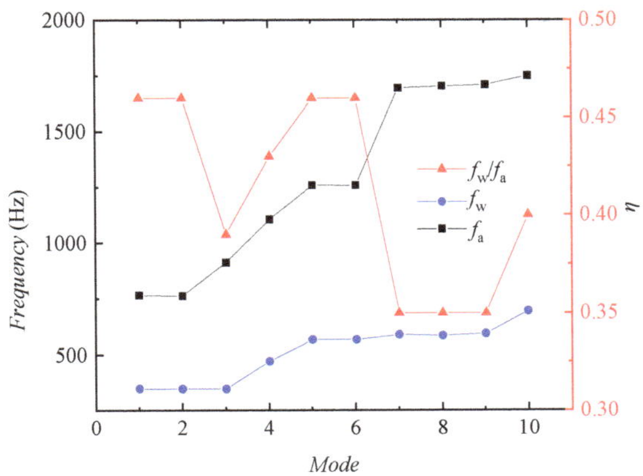

Figure 5. The natural frequencies of the impeller at different stages in air and water.

Table 2. Comparison of impeller intrinsic frequencies of each order in air and water.

Mode Order	f_a (Hz)	f_w (Hz)	$\eta = f_w/f_a$
1	767.54	352.84	0.46
2	768.19	354.42	0.46
3	917.33	354.43	0.39
4	1105.8	477.61	0.43
5	1259.2	573.66	0.46
6	1260.6	574.01	0.46
7	1696.7	591.56	0.35
8	1704.7	591.56	0.35
9	1708.9	604.96	0.35
10	1747.3	695.88	0.40

Figures 6 and 7 show the displacement and vibration patterns of each order mode of the impeller in air and water, respectively. The figures show that the displacements and vibration patterns of each order mode of the impeller in water and air are very similar, with the maximum displacements occurring at the outlet of the impeller rear cover plate. The intrinsic frequencies and vibration patterns of the impeller in air and water usually appear in pairs, mainly due to the periodical symmetric structure of the impeller. The 1st and 2nd order vibration pattern are shown as the impeller with the z-axis as the center point, up and down vibration in the axial plane; the 3rd and 4th order vibration pattern is shown as the torsional vibration of the impeller around the z-axis; and the 5th and 6th order vibration pattern is shown as the unilateral bending deformation of the impeller in the axial plane. The impeller's intrinsic frequency in water is lower than its intrinsic frequency in air due to the extra mass of the water body, but the reduction of the intrinsic frequency of each order is different, as shown in the influence factor, is fluctuating between 0.35 and 0.46.

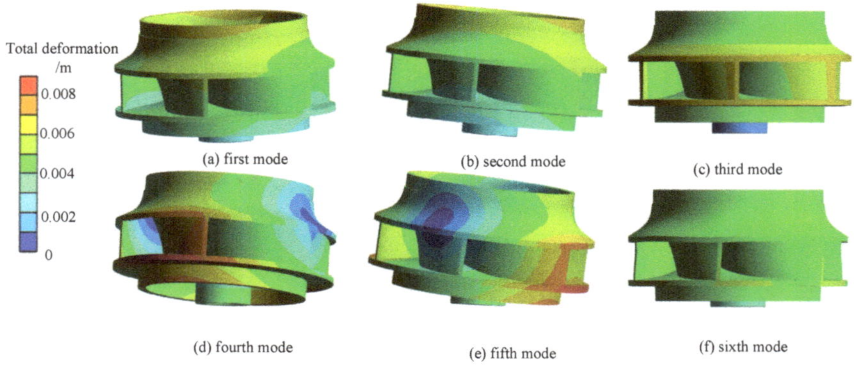

Figure 6. Modal diagrams of each order of the impeller in air.

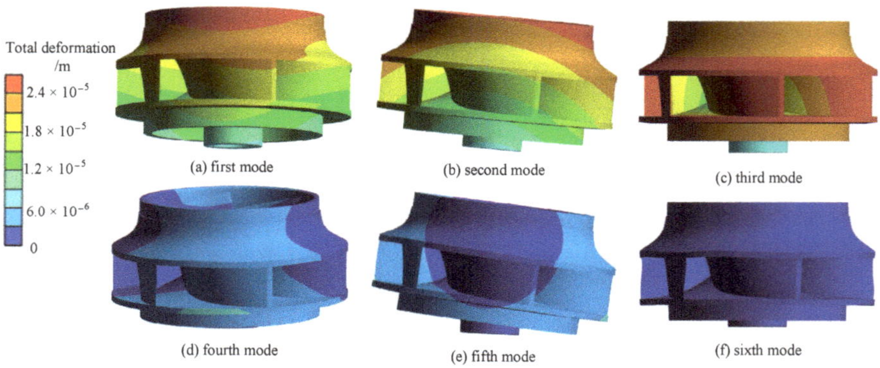

Figure 7. Modal diagrams of each order of the impeller in water.

4. Centrifugal Pump Impeller Full-Condition Stress Characterization

4.1. Computational Modeling and Meshing

The structural field calculation object is a separate centrifugal pump impeller, and the impeller material is martensitic mild steel (13Cr4Ni), the specific characteristics are displayed in Table 3. The impeller modeling is crucial in precisely simulating the transition between the blade and the impeller front and rear cover plate, since stress concentration frequently occurs in the blade root region. Additionally, the division of the regional mesh is encrypted to ensure precise representation, to avoid the calculation process due to the mesh division of the cause of the stress concentration phenomenon that should not occur. Due to the complex structure of the impeller, tetrahedral cells are used for meshing.

Table 3. Physical parameters of impeller materials.

Physical Quantity	Measure Value	Physical Quantity	Measure Value
Density ρ (kg/m^3)	7700	Tensile strength σ_b (MPa)	580
Modulus of elasticity E (GPa)	206	Yield limit σ_s (MPa)	950
Poisson's ratio μ	0.288	Permissible stress $[\sigma]$, (MPa)	560

This work aims to mitigate the impact of mesh quantity on the structural field calculation of the impeller. To achieve this, the mesh division is conducted using varying numbers, and the static stress calculation of the impeller is carried out for a certain same working condition ($Q = 900\ m^3/h$). In order to facilitate observation, Figure 8 illustrates the relationship between the static stress and deformation values at various mesh counts. It is evident that the number of meshes employed in the impeller design has a negligible impact on the deformation. Additionally, the maximum equivalent stress variation stabilizes as the number of meshes reaches 1,600,000, and considering the computational efficiency, this research utilizes a mesh count of 1,600,000 to conduct computations and analyze static stresses on the impeller.

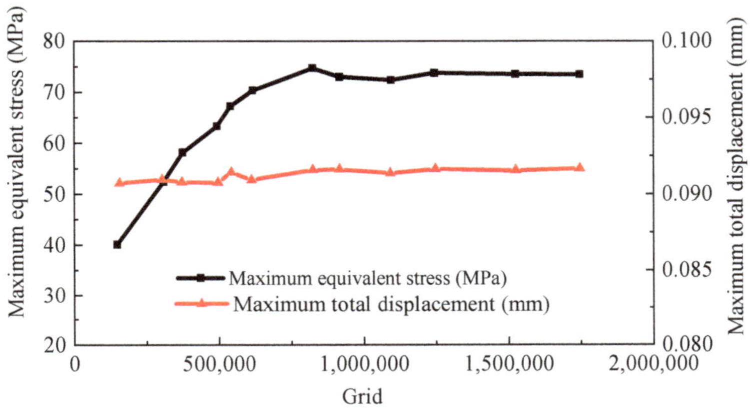

Figure 8. Grid independence check.

4.2. Boundary Conditions for Structural Field Calculations

Constraints: the displacement constraints are fixed impeller shaft hole end face. Fluid-structure interaction boundary: the part of the impeller in contact with the flow field. Load: mainly includes centrifugal force, gravity and water pressure on the blade surface. The impeller is a rotating part, which is realized by applying centrifugal force load to the impeller area. The applied z-axis direction gravity acceleration is $9.8\ m/s^2$. The water pressure at the interface of the fluid-structure interaction is determined by analyzing the outcomes of the flow field computation.

4.3. Calculated Field Results in Structures Analyzed

By solving the impeller with finite elements, Figure 9 illustrates the maximum equivalent stress and the maximum total deformation across several operational scenarios. With the climb of flow, the maximum stress value and the total deformation are first reduced and then gradually increased, and reached the minimum value near the design condition point, the maximum stress condition appeared at low flow rates, especially near the 0 flow condition, the equivalent force of the impeller reaches 181 MPa, about 3 times of the maximum equivalent force value of the design condition, analyze the reason, mostly attributed to the elevated head at modest flow rates. This leads to bigger pressure differentials on both the front and rear surfaces of the vane. Notably, the maximum deformation experienced is approximately three times greater than what was initially anticipated during the design phase, the primary reason for this phenomenon can be attributed to the fact that in low flow conditions, the centrifugal pump exhibits a higher head, resulting in a bigger pressure differential between the leading and trailing edges of the vane, which leads to the larger equivalent force on the impeller. In the vicinity of 0 flow condition, the total deformation also reaches the maximum value of 0.26 mm. The permissible stress of the impeller material is 560 MPa. The static stress of the centrifugal pump meets the requirements of the material.

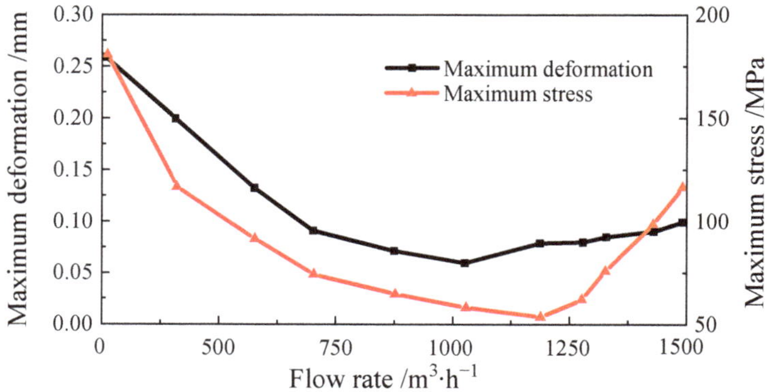

Figure 9. Maximum equivalent stress and maximum deformation under various operational situations.

The graphical representation in Figure 10 illustrates the correlation between the maximum equivalent force and the shaft power at each different operating condition. With the increase of shaft power, the observed trend in the maximum equivalent force exerted by the impeller exhibits a pattern of initial decrease followed by subsequent increase. After the design condition point, the maximum equivalent force of the impeller shows an approximately linear relationship with the power, which indicates that at this point, the maximum equivalent force is related to the torque.

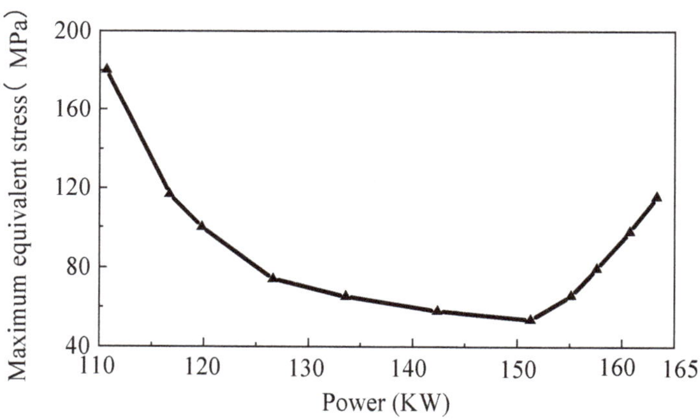

Figure 10. Maximum equivalent force of impeller vs. shaft power of centrifugal pumps.

Figures 11 and 12 show the displacement and deformation distribution and equivalent force diagrams under three typical conditions, respectively. Through observation, the maximum deformations all occur at the outer edge of the back cover plate outlet, which is because the impeller has a lower stiffness here and the fluid pressure here is higher, and stiffness damage is easy to occur in these places. The impeller experiences its highest equivalent stresses at the curved corner where the intersection of the back exit edge of the blade and the front cover plate takes place, as shown in Figure 11. Here the blade outlet pressure is higher, and here the thickness is thin, and it is also the intersection round corner of the front cover plate and the blade, which makes it easy to produce stress concentration. Significant stress concentrations are seen near the midpoint of the intake side of the blade, as well as at the junction where the blade intersects with both the front and rear cover plates.

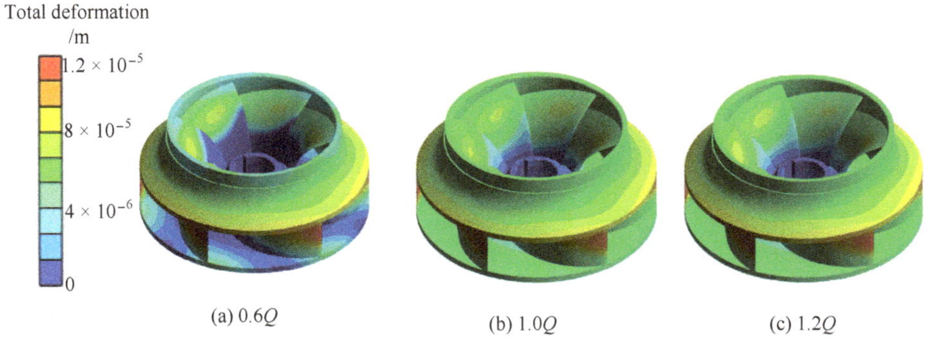

Figure 11. Distribution of impeller displacement and deformation.

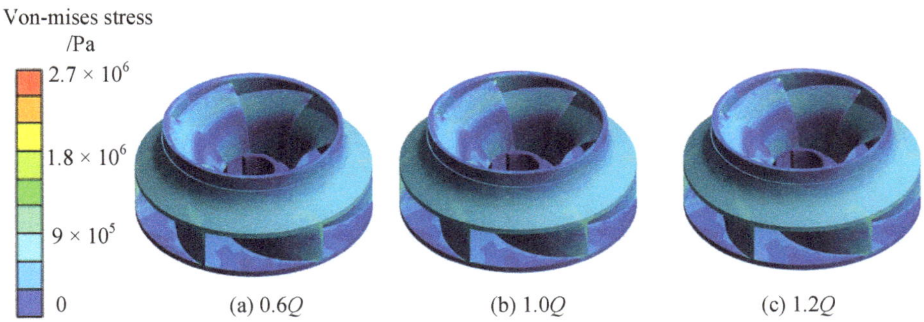

Figure 12. Equivalent force distribution of impeller.

4.4. Stress Characterization of Impeller at Different Moments

The rated rotational speed of the centrifugal pump is $n = 1450$ r/min, and the rotational period is T = 0.04 s. To conduct an analysis of the stress experienced by the impeller at various time points throughout a rotational cycle, the structural field and the impeller field of the centrifugal pump are rotated sequentially by 5° until the end of the rotational period (360°), and, at the same time, the positional relationship between the vane and the tongue of the volute will likewise undergo alteration. The results of the rotated flow field are sequentially loaded into the corresponding impeller structure field, and the stresses of the impeller at different moments are calculated. Figure 13 illustrates the temporal distribution of the greatest equivalent stresses experienced by the impeller during three representative operational scenarios at different moments in one rotation cycle (T = 0.04 s). It can be seen that, under each condition, the stress experienced at the maximum stress point of the impeller during one rotation cycle exhibits a cyclic pattern, characterized by six peaks and valleys. This discovery is consistent with the relationship between the number of impeller blades and the frequency of maximum stress, indicating that both the number of impeller blades and the rotational speed have a significant influence on the frequency of maximum stress. Furthermore, it is evident that the maximum equivalent stress amplitude significantly increases when subjected to high flow rates. The periodic variations in the internal flow field characteristics of a centrifugal pump are caused by the dynamic and static interference between the impeller and the volute casing, as well as the changing relative positions of the impeller blades and the volute tongue. Consequently, these factors also impact the distribution of equivalent forces acting on the impeller. Although the maximum stress is very small compared with the ultimate destructive stress of the material, it is necessary to prevent the fatigue damage of the vane because its change shows periodicity.

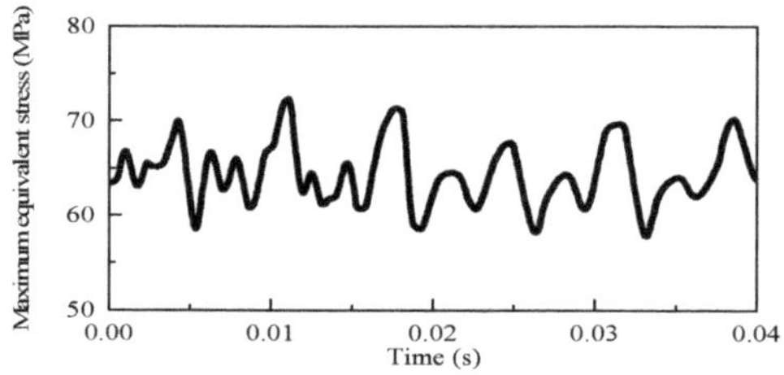

(**a**) 0.6Q

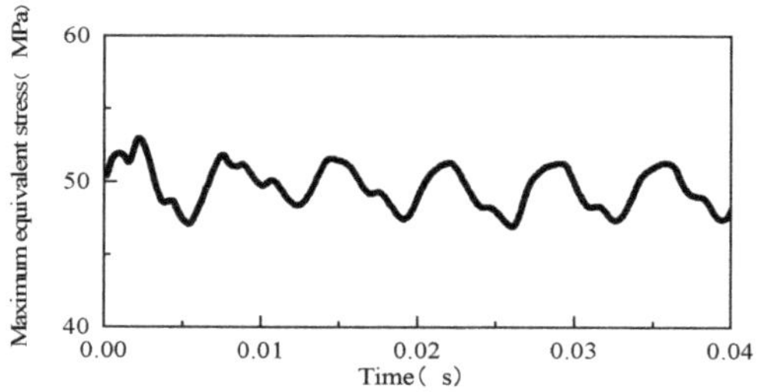

(**b**) 1.0Q

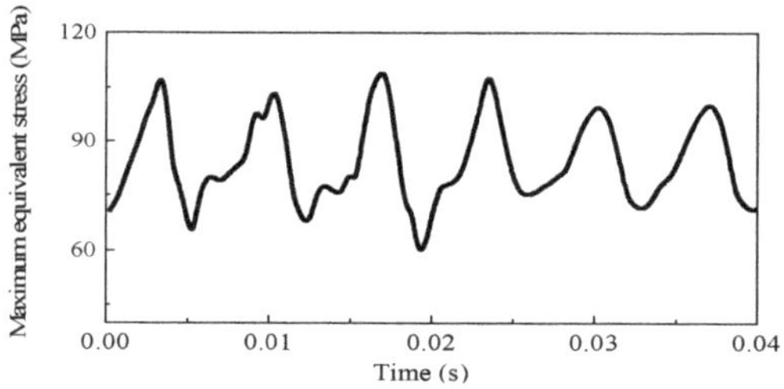

(**c**) 1.2Q

Figure 13. Time-domain diagram of stress at the maximum stress point.

Calculation results show that, as the flow rate increases, the impeller undergoes variations in both maximum equivalent stress and deformation. Initially, these values decrease and then increase, reaching their minimum near the design condition point. The impeller exhibits its highest stress and deformation levels under low flow conditions, particularly at zero flow where the maximum equivalent stress reaches 181 MPa and the maximum total deformation is 0.26 mm. To guarantee the secure and dependable operation of the pump and pumping station system, centrifugal pumps should be avoided to operate under 0–0.2 Q flow conditions. The impeller undergoes its most significant distortion at the exit of the rear cover plate. The maximum equivalent stress of the impeller occurs at the corner where the outlet edge of the back of the blade intersects with the front cover plate of the impeller. There is a significant stress concentration in the middle of the inlet edge of the blade and at the junction with the front and rear cover plates.

During a single rotation cycle, the stress experienced at the maximum stress point of the impeller exhibits periodic fluctuations, characterized by six peaks and valleys. This pattern aligns with the quantity of impeller blades, suggesting that the frequency of stress at the maximum stress point of the blade is primarily influenced by both the quantity of impeller blades and the rotational speed. Furthermore, under conditions of high flow rate, the maximum equivalent stress amplitude demonstrates a substantial increase. The primary reason for the cyclic variations in the internal flow characteristics of the centrifugal pump is the rotor-stator interference effect that occurs between the impeller and the volute. The alterations are determined by the relative positioning of the blade and the tongue of the volute. Consequently, the impact on the impeller's equivalent stress distribution is likewise influenced. Despite the impeller's maximum stress being significantly lower than the material's ultimate destructive stress, it is imperative to mitigate fatigue damage to the blades due to the cyclic nature of stress fluctuations.

5. Conclusions

This study centers on an analysis of a large-scale centrifugal pump using the finite element method. The research adopts a one-way fluid-structure interaction approach to systematically evaluate the flow dynamics within the pump. Subsequently, this analysis facilitates the determination of the inherent modes of the impeller, both in air and water environments. Following this, a comprehensive static stress analysis of the impeller is conducted. The results of the calculations presented in this research unveil the following key insights:

The natural frequency of the impeller when submerged in water exhibits a noticeable reduction in comparison to its natural frequency in an air environment. This reduction varies across different modes, with higher-order modes experiencing a more pronounced decrease, suggesting a heightened impact of the surrounding medium on the intrinsic frequency of the impeller. A detailed examination of the first ten order displacement and vibration pattern diagrams of the impeller in both air and water reveals a similarity, with paired patterns evident in both environments. This observed symmetry stems from the impeller's inherent periodic structural characteristics.

Structural stress field calculations reveal a pattern: as the flow rate increases, there is an initial decrease followed by a subsequent increase in both maximum equivalent stress and deformation. These values reach their minimum near the design condition point. Notably, the lowest flow condition, particularly at 0 flow, exhibits the highest levels of stress and deformation. At its peak, the maximum equivalent stress registers at 181 MPa, while the overall deformation reaches a maximum of 0.26 mm.

The most pronounced distortion occurs at the exit of the rear cover plate, with the maximum equivalent force on the impeller observed at the curved corner where the outlet side of the blade's back meets the front cover plate of the impeller. Furthermore, there is a conspicuous concentration of stress both in the middle of the blade's inlet side and at its junction with both the front and rear cover plates. The periodic alterations in the internal flow field characteristics of the centrifugal pump stem from the dynamic and static interference effects arising from the interaction between the impeller and the worm case. These variations are further influenced by the changing relative positions of the blade and the spacer's tongue. Consequently, the maximum equivalent stress also exhibits periodic fluctuations.

In light of these findings, ensuring the secure and reliable operation of pumps and pumping station systems necessitates avoiding the utilization of centrifugal pumps under flow conditions falling within the range of 0 to 0.2 Q.

Author Contributions: Methodology, S.L.; Software, Y.T. and Y.L.; Validation, J.D. and M.D.; Writing—original draft, H.Y.; C.Y. (Changliang Ye); Supervision, C.Y. (Chunxia Yang) and Y.Z. All authors have read and agreed to the published version of the manuscript.

Funding: This research was funded by the National Natural Science Foundation of China (Grant number: 52209109, No.52271275, No.52379086) and the Scientific Research Start-up Funding, Hohai University (1047/423160).

Data Availability Statement: The data presented in this study are available on request from the corresponding author.

Conflicts of Interest: Author Siwei Li, Yongsha Tu and Mengfan Dang were employed by Hunan Water Resources and Hydropower Survey, Design, Planning and Research, Co., Ltd. Author Jin Dai and Yongbiao Li were employed by Hunan Provincial Water Resources Development and Investment Co., Ltd. The remaining authors declare that the research was conducted in the absence of any commercial or financial relationships that could be construed as a potential conflict of interest.

References

1. Adams, R. What Constitutes' High Energy' In Centrifugal Pumps? In Proceedings of the 30th International Pump Users Symposium, Houston, TX, USA, 23–25 September 2014; Turbomachinery Laboratories, Texas A&M Engineering Experiment Station: College Station, TX, USA, 2014.
2. Chan, E.Y.Y.; Ho, J.Y.E. Urban water and health issues in Hong Kong. In *Urban Drought: Emerging Water Challenges in Asia*; Springer Nature: Singapore, 2019; pp. 241–262.
3. Roy, A.; Palit, P.; Das, S.; Mukhyopadyay, G. Investigation of torsional fatigue failure of a centrifugal pump shaft. *Eng. Fail. Anal.* **2020**, *112*, 104511. [CrossRef]
4. Ye, C.; Tang, Y.; An, D.; Wang, F.; Zheng, Y.; van Esch, B.P.M. Investigation on stall characteristics of marine centrifugal pump considering transition effect. *Ocean Eng.* **2023**, *280*, 114823. [CrossRef]
5. Huang, X.; Fang, T.; Pang, K.; Guo, Q.; Qiu, B.; Lu, J. Air-entrained vortex in open intake: Time–frequency analysis and the interaction with subsurface vortices. *Phys. Fluids* **2022**, *34*, 113313. [CrossRef]
6. Zheng, L.; Chen, X.; Dou, H.-S.; Zhang, W.; Zhu, Z.; Cheng, X. Effects of clearance flow on the characteristics of centrifugal pump under low flow rate. *J. Mech. Sci. Technol.* **2020**, *34*, 189–200. [CrossRef]
7. Hoseini, S.S.; Najafi, G.; Ghobadian, B.; Akbarzadeh, A.H. Impeller shape-optimization of stirred-tank reactor: CFD and fluid structure interaction analyses. *Chem. Eng. J.* **2021**, *413*, 127497. [CrossRef]
8. Hirschhorn, M.; Tchantchaleishvili, V.; Stevens, R.; Rossano, J.; Throckmorton, A. Fluid–structure interaction modeling in cardiovascular medicine—A systematic review 2017–2019. *Med. Eng. Phys.* **2020**, *78*, 1–13. [CrossRef]
9. Cuamatzi-Meléndez, R.; Flores-Cuamatzi, E. Modelling fluid-structure interaction of water recirculating flow to predict damage and/or failure in a jet-pump assembly of a nuclear boiling water reactor. *Eng. Struct.* **2020**, *206*, 110155. [CrossRef]
10. De Donno, R.; Ghidoni, A.; Noventa, G.; Rebay, S. Shape optimization of the ERCOFTAC centrifugal pump impeller using open-source software. *Optim. Eng.* **2019**, *20*, 929–953. [CrossRef]
11. Olimstad, G.; Østby, P.T.K. Failure and redesign of a high-speed pump with respect to rotor-stator interaction. *Eng. Fail. Anal.* **2019**, *104*, 704–713. [CrossRef]
12. Wang, W.; Zhou, L.; Tao, R.; Song, X.; Wang, Z. Numerical simulation of dynamic characteristics of hydrofoil structure under cavitation conditions. *Ocean Eng.* **2023**, *280*, 114937. [CrossRef]
13. Griffith, B.E.; Patankar, N.A. Immersed Methods for Fluid–Structure Interaction. *Annu. Rev. Fluid Mech.* **2020**, *52*, 421–448. [CrossRef] [PubMed]
14. Mo, J.-O.; Lee, Y.-H. Performance Prediction and Flow Characteristics of a Hydraulic Pump for ABS and ESC Systems Using FSI Simulation. *Int. J. Automot. Technol.* **2020**, *21*, 1419–1429. [CrossRef]
15. Ye, C.; Wang, C.; Yan, H.; Wang, F.; Zheng, Y.; van Esch, B.P.M. Investigation on transition characteristics of laminar separation bubble on a hydrofoil. *Phys. Fluids* **2023**, *35*, 105154. [CrossRef]
16. Luo, H.; Zhou, P.; Shu, L.; Mou, J.; Zheng, H.; Jiang, C.; Wang, Y. Energy Performance Curves Prediction of Centrifugal Pumps Based on Constrained PSO-SVR Model. *Energies* **2022**, *15*, 3309. [CrossRef]
17. Krzemianowski, Z.; Steller, J. High specific speed Francis turbine for small hydro purposes-Design methodology based on solving the inverse problem in fluid mechanics and the cavitation test experience. *Renew. Energy* **2021**, *169*, 1210–1228. [CrossRef]
18. Krella, A.; Maurin, A.; Krzemianowski, Z. Degradation of Armco iron caused by cavitation: Part II—Correlation with stress analysis. *Eng. Fail. Anal.* **2021**, *128*, 105621. [CrossRef]
19. He, L.; Zhou, L.; Ahn, S.-H.; Wang, Z.; Nakahara, Y.; Kurosawa, S. Evaluation of gap influence on the dynamic response behavior of pump-turbine runner. *Eng. Comput.* **2019**, *36*, 491–508. [CrossRef]
20. Xia, X.; Zhou, L.; Lv, Y.; Wang, Z. Numerical investigation of two degree-of-freedom galloping oscillation of a cylinder attached with fixed fairing device. *Ocean Eng.* **2021**, *240*, 109971. [CrossRef]
21. Yang, J.; Zhou, L.-J.; Wang, Z.-W.; Jiang, X.-Y.; Zhou, X.-J.; Ding, J.-H.; Han, W.-F. Numerical investigation of the cavitation dynamic parameters in a Francis turbine draft tube with columnar vortex rope. *J. Hydrodyn.* **2019**, *31*, 931–939. [CrossRef]
22. Hübner, B.; Silva, D.R. Advanced simulation of coupled physics in thrust bearings. In *IOP Conference Series: Earth and Environmental Science*; IOP Publishing: Bristol, UK, 2019; Volume 240, p. 062009.
23. Hübner, B.; Seidel, U.; Roth, S. Application of fluid-structure coupling to predict the dynamic behavior of turbine components. In *IOP Conference Series: Earth and Environmental Science*; IOP Publishing: Bristol, UK, 2010; Volume 12, p. 012009.
24. Trivedi, C.; Cervantes, M.J. Fluid-structure interactions in Francis turbines: A perspective review. *Renew. Sustain. Energy Rev.* **2017**, *68*, 87–101. [CrossRef]
25. Badshah, M.; Badshah, S.; Jan, S. Comparison of computational fluid dynamics and fluid structure interaction models for the performance prediction of tidal current turbines. *J. Ocean Eng. Sci.* **2020**, *5*, 164–172. [CrossRef]
26. Zanetti, G.; Cavazzini, G.; Santolin, A. Effect of the von Karman Shedding Frequency on the Hydrodynamics of a Francis Turbine Operating at Nominal Load. *Int. J. Turbomach. Propuls. Power* **2023**, *8*, 27. [CrossRef]
27. Kan, K.; Zheng, Y.; Fu, S.; Liu, H.; Yang, C.; Zhang, X. Dynamic stress of impeller blade of shaft extension tubular pump device based on bidirectional fluid-structure interaction. *J. Mech. Sci. Technol.* **2017**, *31*, 1561–1568. [CrossRef]
28. Quan, H.; Cheng, J.; Guo, Y.; Kang, L.; Peng, G. Influence of Screw Centrifugal Inducer on Internal Flow Structure of Vortex Pump. *J. Fluids Eng.* **2020**, *142*, 091203. [CrossRef]

29. Presas, A.; Valentin, D.; Zhao, W.; Egusquiza, M.; Valero, C.; Egusquiza, E. On the use of neural networks for dynamic stress prediction in Francis turbines by means of stationary sensors. *Renew. Energy* **2021**, *170*, 652–660. [CrossRef]
30. Valentín, D.; Presas, A.; Valero, C.; Egusquiza, M.; Egusquiza, E.; Gomes, J.; Avellan, F. Transposition of the mechanical behavior from model to prototype of Francis turbines. *Renew. Energy* **2020**, *152*, 1011–1023. [CrossRef]
31. Chen, X.; Lai, X.; Gou, Q.; Song, D. Effect of guide vane profile on the hydraulic performance of moderate low-specific-speed Francis turbine. *J. Mech. Sci. Technol.* **2023**, *37*, 1289–1300. [CrossRef]
32. Menéndez -Blanco, A.; Oro, J.M.F.; Meana-Fernández, A. Unsteady three-dimensional modeling of the Flu-id-Structure Interaction in the check valves of diaphragm volumetric pumps. *J. Fluids Struct.* **2019**, *90*, 432–449. [CrossRef]
33. Li, W.; Ji, L.; Shi, W.; Zhou, L.; Jiang, X.; Zhang, Y. Fluid-structure interaction study of a mixed-flow pump impeller during startup. *Eng. Comput.* **2018**, *35*, 18–34. [CrossRef]
34. Birajdar, R.; Keste, A. Prediction of Flow-Induced Vibrations due to Impeller Hydraulic Unbalance in Vertical Turbine Pumps Using One-Way Fluid-Structure Interaction. *J. Vib. Eng. Technol.* **2020**, *8*, 417–430. [CrossRef]
35. Shi, L.; Zhu, J.; Wang, L.; Chu, S.; Tang, F.; Jin, Y. Comparative Analysis of Strength and Modal Characteristics of a Full Tubular Pump and an Axial Flow Pump Impellers Based on Fluid-Structure Interaction. *Energies* **2021**, *14*, 6395. [CrossRef]
36. Wang, H.; Wang, F.; Wang, B.; Wu, J.; Lu, H.; Wang, C. Partial flow separation in guide-vane region of large-capacity/low-head pumped hydro energy storage system with horizontal shaft. *J. Energy Storage* **2023**, *71*, 108173. [CrossRef]
37. Ye, C.; An, D.; Huang, W.; Heng, Y.; Zheng, Y. Investigation on Stall Characteristics of Centrifugal Pump with Guide Vanes. *Water* **2023**, *15*, 21. [CrossRef]
38. Yan, H.; Heng, Y.; Zheng, Y.; Tao, R.; Ye, C. Investigation on Pressure Fluctuation of the Impellers of a Double-Entry Two-Stage Double Suction Centrifugal Pump. *Water* **2022**, *14*, 4065. [CrossRef]
39. Wang, C.; Wang, F.; Li, C.; Chen, W.; Wang, H.; Lu, L. Investigation on energy conversion instability of pump mode in hydro-pneumatic energy storage system. *J. Energy Storage* **2022**, *53*, 105079. [CrossRef]
40. Zeng, Y.; Yao, Z.; Huang, B.; Wu, Q.; Wang, F. Experimental investigation of the hydrodynamic damping of a vibrating hydrofoil in cavitating flow. *Ocean Eng.* **2022**, *266*, 112734. [CrossRef]
41. Tickoo, S. *ANSYS Workbench 2021 R1: A Tutorial Approach*; CADCIM Technologies: Schererville, IN, USA, 2021.
42. Zhou, W.; Zhou, P.; Xiang, C.; Wang, Y.; Mou, J.; Cui, J. A Review of Bionic Structures in Control of Aerodynamic Noise of Centrifugal Fans. *Energies* **2023**, *16*, 4331. [CrossRef]

water

MDPI

Article

Investigation of Structural Strength and Fatigue Life of Rotor System of a Vertical Axial-Flow Pump under Full Operating Conditions

Haoyu Li [1], Zhizhou Cai [2], Yuan Zheng [1], Jiangang Feng [3], Hui Xu [3,4], Huixiang Chen [3,4], Maxime Binama [4,5] and Kan Kan [1,4,*]

1 College of Energy and Electrical Engineering, Hohai University, Nanjing 211100, China; lihaoyu@hhu.edu.cn (H.L.); zhengyuan@hhu.edu.cn (Y.Z.)
2 Shanghai Nuclear Engineering Research & Design Institute, Shanghai 200233, China; caizhizhou2023@163.com
3 College of Agricultural Science and Engineering, Hohai University, Nanjing 211100, China; jgfeng@hhu.edu.cn (J.F.); hxu@hhu.edu.cn (H.X.); chenhuixiang@hhu.edu.cn (H.C.)
4 College of Water Conservancy and Hydropower Engineering, Hohai University, Nanjing 210098, China; binamamaxime@lnpu.edu.cn
5 College of Science and Technology (MEE Department), University of Rwanda, Kigali 3900, Rwanda
* Correspondence: kankan@hhu.edu.cn

Abstract: Axial-flow pumps consider both the conventional pump mode and the pump as turbine (PAT) mode operation and put forward higher requirements for long-term operation stability and structural strength; therefore, it is of great engineering significance to evaluate the structural strength and fatigue life of the rotor under full operating conditions. In this study, based on computational fluid dynamics and the one-way fluid-structure interaction algorithm, the structural strength and fatigue life of the rotor system of a large vertical axial-flow pump under full operating conditions were evaluated and studied. The results show that blade deformation and equivalent stress are generally higher in the PAT mode than in the pump mode. The maximum deformation in both modes occurs at the tip of the blade, while the area of stress concentration is at the root of the blade. Both the deformation and the equivalent stress increase with increasing flow rate. The minimum safety factor occurs at the blade root in both modes, and the safety factor in the PAT mode is relatively smaller than that in pump mode. Therefore, when designing and manufacturing axial flow pumps for turbine duties, priority should be given to material strength at the blade root during PAT mode operation to ensure safe and stable operation. The aim of this study is to provide technical references and theoretical foundations for evaluating the service cycle of axial-flow pumps and the influence on pump life under different operation modes.

Keywords: axial-flow pump; rotor system; structural strength; fatigue life; pump as turbine; fluid-structure interaction

Citation: Li, H.; Cai, Z.; Zheng, Y.; Feng, J.; Xu, H.; Chen, H.; Binama, M.; Kan, K. Investigation of Structural Strength and Fatigue Life of Rotor System of a Vertical Axial-Flow Pump under Full Operating Conditions. *Water* **2023**, *15*, 3041. https://doi.org/10.3390/w15173041

Academic Editor: Tammo Steenhuis

Received: 27 June 2023
Revised: 21 August 2023
Accepted: 22 August 2023
Published: 24 August 2023

1. Introduction

Pumping stations can be used for cross-regional water resource planning to achieve rational distribution of water resources, and they also perform functions such as flood control, drainage, and urban water supply, making it one of the key construction and research projects in various water conservancy projects worldwide [1–4]. The pumping device is the heart of the pumping station technology and the basic equipment for converting mechanical energy into gravitational potential energy of the liquid [5,6]. Axial-flow pump devices are capable of transporting larger liquid flow rates due to their axial geometric structural characteristics and are a mature and widely used pumping device solution for low-head pumping stations [7,8].

The widespread application of axial-flow pumps in pumping station construction worldwide has led to increasing demands on their long-term stability and reliability, driven by the continuously increasing number and capacity of the systems. However, during operation, axial-flow pump units face potential hazards in the form of vibration and blade cracking caused by water pressure loads, which seriously threaten the safety and stable operation of the units [9]. The rotor system, with the impeller as the main component, is the most vulnerable link to shock loads and one of the main causes of pump failure accidents. Therefore, the structural strength of axial-flow pump impellers has received increasing attention from researchers in related fields in recent years.

The traditional application of experimental methods to study the above problems is undoubtedly associated with high costs [10]. With the advancement of algorithms and computational power, fluid-structure interaction (FSI) technology based on computational fluid dynamics (CFD) and finite element method (FEM) offers a more convenient, faster and less expensive solution for researchers to investigate the structural strength properties of rotating hydraulic machines such as axial-flow pumps that have been investigated in recent years [11–13] and that have been able to achieve certain results. Pei et al. [14] applied the FSI method to evaluate and quantitatively analyze the deformation and stress of the blades of an axial-flow pump unit with a bi-directional channel. The results showed that the maximum deformation occurred at the blade rim, and the deformation gradually decreased along the edge from the leading edge (LE) to the trailing edge (TE). Liu et al. [15] analyzed the stress and deformation of the blades of a vertical bi-directional axial-flow pump based on the flow field pressure distribution and found that the stress concentration occurred at the center of the connection between the blade root and the hub and the maximum stress value at stress concentration decreased with increasing flow rate. Shi et al. [16] compared the equivalent stress distributions of an axial-flow pump and a full tubular pump and found that the maximum equivalent stress of an axial-flow pump was lower than that of full tubular pump under different flow conditions. The axial-flow pump also has a smaller stress concentration area compared to the full tubular pump. Under the influence of periodic loading, impeller blades may experience some fatigue. The numerical simulation results from Zhang et al. [17] revealed that the interaction between the rotors and stators of axial-flow pumps is an important factor in the periodic action of the blade, and that the maximum equivalent stress of the blade exhibits periodic oscillations within the rotation cycle with a frequency that coincides with the passing frequency of the blade. The study by Gao et al. [18] showed that the stress distribution of the axial-flow pump impeller was not evenly distributed, but the dynamic stress distribution characteristics were basically similar under different flow conditions. The researchers also found that the dynamic loading on the impeller varied periodically as the impeller rotated. Zhang et al. [19] conducted a comprehensive analysis of the axial-flow pump blade equivalent stress using methods such as fast Fourier transform (FFT) and showed that the stress vibration amplitude of the impeller blade was significantly higher than that of the guide vane. The maximum vibration amplitude of the equivalent stress of the impeller occurred in the center position between LE and TE of the blade hub.

So far, the structure of the rotor system of an axial flow pump in conventional operating mode has been discussed in detail. However, it is important to note that in certain regions of the world with typical monsoon climates, pumping stations are tasked with seasonal water intake [20]. At certain time intervals, pumping stations must divert upstream flooding to downstream areas. To use this water resource more efficiently, pump devices are often designed and operated with their impellers rotating backwards as turbines, referred to as the pump as turbine (PAT) mode [21]. This mode provides a more stable and cleaner power source for the power grid compared to wind and solar energy [22]. Due to their large flow rate, relatively simple structure and easy installation, axial-flow pumps offer good economy [23,24]. In fact, they are even utilized as simple turbine units in certain developing regions and remote districts, and are a popular form of reverse power generation with various types of pumps [25]. Equally important is the investigation of the structural strength of the rotor when operating the axial flow pump in the PAT mode, which has already attracted the interest of some researchers. Meng et al. [26] found that the maximum equivalent stress and total deformation of the impeller blades occurred at 0.75 times the design flow rate in pump mode, while in the PAT mode they occurred at 0.65 times the design flow rate, with the equivalent stress increasing with increasing flow rate. Bai et al. [27] stated that the dynamic stress frequency characteristics were different in the pump mode and the PAT mode. In the pump mode, the primary frequency of the dynamic stress was the rotational frequency of the impeller, with no significant secondary frequencies observed. However, in the PAT mode, blade stress was affected by the combined effects of the adjacent blades, and the primary frequency was a multiple of the rotation frequency, which was highly dependent on the number of blades. Overall, there is still relatively little research on the structural strength and fatigue life of the rotor system of axial-flow pumps operating in both the conventional pump mode and the PAT mode under all operating conditions.

Therefore, this paper utilizes the commercial software interface provided by ANSYS to conduct a study on the rotor system and structural strength of a large vertical axial-flow pump, including the pump mode and the PAT mode, under all operating conditions using the Reynolds-averaged Navier–Stokes (RANS) and one-way FSI methods. In the following sections, Section 2 presents the numerical methods used in the simulation, Section 3 provides the computational settings for the numerical simulation, including the results of the mesh discretization into fluid and solid domain. Section 4 discusses and analyzes the results of the simulation, while the specific main conclusions are presented in Section 5.

2. Methodology

2.1. Fluid Governing Equations

The internal flow of an axial-flow pump is a complex turbulent flow with fluid viscosity, and the compressibility of the fluid is generally neglected [28]. In this study, the RANS algorithm was used to model the turbulence internal the pump device; thus, the continuity equation can be expressed as

$$\frac{\partial \bar{u}_j}{\partial x_j} = 0, \tag{1}$$

and the momentum conservation equation is written as

$$\frac{\partial(\rho \bar{u}_i)}{\partial t} + \frac{\partial}{\partial x_j}(\rho \bar{u}_i \bar{u}_j) = -\frac{\partial \bar{p}}{\partial x_i} + \frac{\partial}{\partial x_j}\left(\mu \frac{\partial \bar{u}_i}{\partial x_j} - \rho \overline{u_i' u_j'}\right) + f_i, \tag{2}$$

where the overline symbol ($\bar{}$) represents the time-averaged operation, u_i and u_j stand for the velocity components in x, y and z directions, the sub-scripts are the directional indices. t is the physical time, ρ is the fluid density and μ is the dynamic viscosity of the fluid. Additionally, f denotes the external force. The two-way Re-Normalization Group (RNG) $k-\epsilon$ turbulence model is used to close the governing equations [29–32]. The transport equations of turbulence kinetic energy k and turbulence dissipation rate ϵ are calculated, respectively, as

$$\frac{\partial(\rho k)}{\partial t} + \frac{\partial}{\partial x_j}(\rho k u_i) = \frac{\partial}{\partial x_j}\left[\left(\mu + \frac{\mu_t}{\sigma_k}\right)\frac{\partial k}{\partial x_j}\right] + P_k - \rho\varepsilon, \tag{3}$$

$$\frac{\partial(\rho\varepsilon)}{\partial t} + \frac{\partial}{\partial x_j}(\rho\varepsilon u_j) = \frac{\partial}{\partial x_j}\left[\left(\mu + \frac{\mu_t}{\sigma_\varepsilon}\right)\frac{\partial \varepsilon}{\partial x_j}\right] + \frac{\varepsilon}{k}(C_{\varepsilon 1\mathrm{RNG}}P_k - C_{\varepsilon 2}\varepsilon), \tag{4}$$

where ν_t stands for the turbulent viscosity and P_k is the turbulence production rate due to viscous forces. $C_{\epsilon 1\mathrm{RNG}}$, $C_{\epsilon 2}$, σ_k and σ_ϵ are all the constants of the RNG model.

2.2. Structural Governing Equations

Considering the significant difference in magnitude between the expected deformation of the axial-flow pump and the scale of the flow field, the influence of the deformation on the flow field can be essentially neglected [33,34]. Therefore, in this study, a one-way FSI method was adopted for investigation. The governing equation of blade structure dynamics is expressed as

$$\boldsymbol{M}\ddot{\boldsymbol{q}}_t(t) + \boldsymbol{C}\dot{\boldsymbol{q}}_t(t) + \boldsymbol{K}\boldsymbol{q}_t(t) = \boldsymbol{Q}(t). \quad (5)$$

Here, $\boldsymbol{M}$, $\boldsymbol{C}$ and $\boldsymbol{K}$ refer to the matrix of structure mass, structural damping, and structural rigidity, respectively. $\ddot{\boldsymbol{q}}$, $\dot{\boldsymbol{q}}$ and $\boldsymbol{q}$ denote the nodal acceleration vector, the velocity vector, and the displacement vector, and $\boldsymbol{Q}$ represents the fluid load vector. The solution flow chart of one-way FSI adopted in this paper is shown in Figure 1.

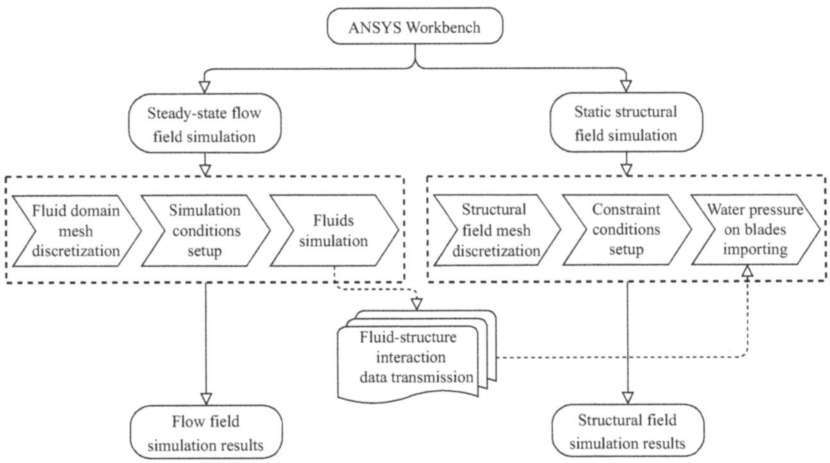

Figure 1. Solution flow chart of fluid–structure interaction.

3. Numerical Settings

3.1. Geometry of Pump Device

This paper conducts numerical simulation and theoretical research on a large vertical axial-flow pump in a certain pumping station. Figure 2 shows the schematic diagram of the pump device, the entire pump device consists of an open inlet channel, an elbow-shaped inlet passage, a pump section, a low hump-shaped outlet passage, and an open outlet channel. The pump section includes an impeller part composed of five blades and a guide vane part composed of eight blades. The basic parameters of the pump section are listed in Table 1.

The axial-flow pump device has two different modes. One is the conventional pump mode that uses the rotation of the impeller to transport water. The other mode utilizes the elevation difference of the water flow to make the pump impeller rotate in the opposite direction and drive the generator to generate electricity. This mode is usually called the reverse power generation mode of the pump or the PAT mode. Both modes are also indicated in Figure 2. In addition, the axial-flow pump blades are made of $ZG_0Cr13Ni_4CuMo$ stainless steel, with a yield strength of 668 MPa, an ultimate strength of 800 MPa and a fatigue limit of 210 MPa.

Table 1. Main parameters of axial-flow pump section.

Impeller Diameter D/(mm)	Rated Rotational Speed n_r/(r/min)	Rated Flow Rate Q_r/(m^3/s)	Number of Impeller Blades	Number of Guide Vanes
2350	166.7	16.67	5	8

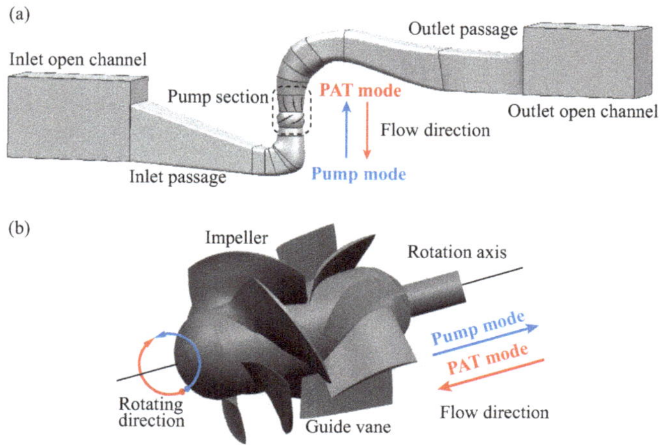

Figure 2. Schematic diagram of (**a**) integral flow channel and (**b**) pump section of axial-flow pump device.

3.2. *Fluid Field*

3.2.1. Spatial Discretization

In this study, ANSYS ICEM was used to discretize the fluid domain of the entire axial-flow pump device. The discretization was carried out using unstructured grids, while structured grids were used to discretize the inlet and outlet channels. In order to ensure accurate flow solutions, local refinements of the grid were performed in critical areas such as the impeller and guide vanes, where complex flows exist. Initially, eight grid configuration schemes were established, and the head and efficiency of the axial flow pump under each scheme were computed. Figure 3 shows the verification curve of grid independence of the fluid domain. Both efficiency and head increase with the increase in the number of grids, indicating gradual convergence. When the number of grids exceeds 11.83 million, the degree of change in head and efficiency with the number of grids becomes very small.

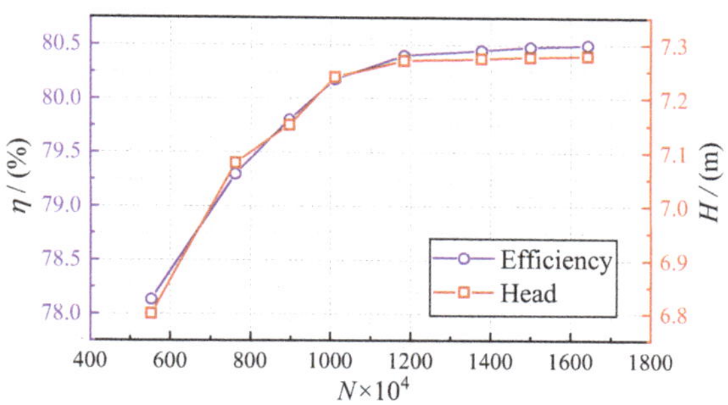

Figure 3. Verification curve of fluid domain grid independence.

Therefore, after considering the comprehensive factors of efficiency and computational accuracy, the final total number of grids was determined to be 11.83 million grid cells, of which the inlet passage accounted for 3.4 million, the impeller part accounted for 2.83 million, the guide vane part accounted for 2.8 million, and the outlet passage part accounted for 2.8 million. The grid quality reached 0.32, which met the requirements for accurate simulation. A schematic diagram of the meshing result of the pump device is shown in Figure 4.

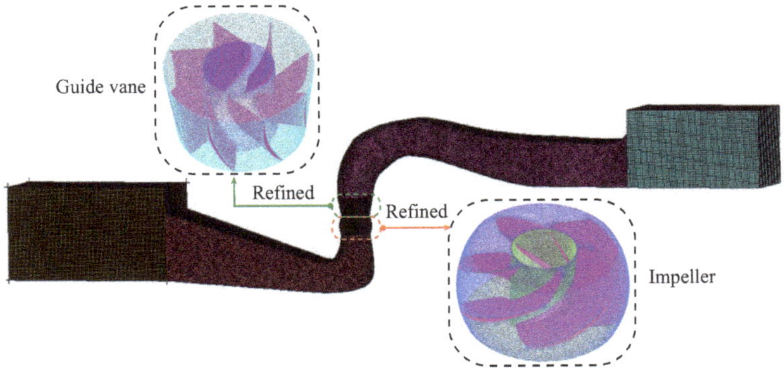

Figure 4. Computational domain division of fluid field.

3.2.2. Solution Method and Boundary Conditions

The steady-state flow fields of an axial-flow pump were simulated under six different operating conditions, including low flow rate ($Q = 0.8Q_{BEP}$), optimal flow rate ($Q = 1.0Q_{BEP}$), and high flow rate ($Q = 1.2Q_{BEP}$), in both the pump mode and the PAT mode. Here, Q_{BEP} refers to the respective optimal operating point flow rate of the two modes. The optimal guide vane angle was selected for each operating condition. The numerical simulations of the flow fields were conducted using the FLUENT 19.2 commercial software. The RNG $k-\epsilon$ turbulence model was selected, and the SIMPLEC algorithm was used to couple the velocity and pressure fields. A first-order implicit scheme was used for time discretization, while a second-order upwind scheme was used for the convective terms of the momentum equation and other spatial terms. The rotating and stationary regions were connected through interfaces. The inlet boundary condition was set as a mass flow inlet, while the outlet boundary condition was set as a pressure outlet and the solid wall was set as a no-slip wall. The number of iterations was designated as 3000.

3.2.3. Validation of External Characteristics

To verify the reliability of the numerical simulation results, model tests were conducted for the pump mode and the PAT mode of the axial-flow pump at a rated speed in this study. The experiments were carried out on the hydraulic machinery multi-functional model test bench of Hohai University [35], as shown in Figure 5. The total uncertainty of the experiment was less than 0.4%. The diameter of the model pump impeller was D = 300 mm, and the experimental rotation speed was 1305.8 r/min. The experimental results were converted to the values of the prototype axial-flow pump according to the similarity law and compared with the simulated efficiency and head. The comparison results are shown in Figure 6. The trend of efficiency and head with flow rate in a numerical simulation is consistent with the experimental results, but the numerical simulation overestimates the efficiency and head slightly. This is partly due to the large roughness and leakage losses of the model pump in the experiment. The maximum error at each flow rate condition did not exceed 3%; thus, the numerical calculation results can be considered acceptable.

Figure 5. Hydraulic machinery multi-functional model test bench of Hohai University.

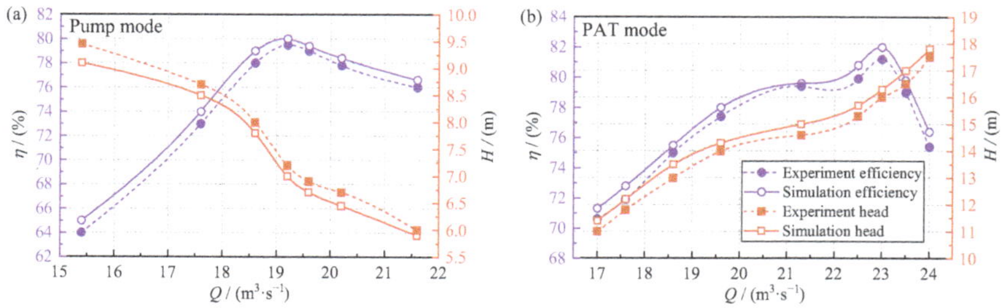

Figure 6. Comparison between simulation and experiment values of head and efficiency for axial-flow pump under (**a**) Pump mode and (**b**) PAT mode.

3.3. Structural Field

In this study, the Mesh module interface provided by the ANSYS Workbench platform was used to perform mesh generation on the solid domain of the blade [36]. To ensure the accuracy and reliability of the mesh generation, four mesh generation schemes were established, and the independence of the solid field mesh was verified. The blade region was divided into four different sizes of 45 mm, 35 mm, 25 mm, and 15 mm, respectively, and local refinement was performed on the blade. The grid division schemes and the maximum stress and maximum displacement values under each scheme are shown in Table 2.

Table 2. Solid grid division scheme.

Scheme	Grid Size at Blade/(mm)	Number of Units	Number of Nodes	Ultimate Stress/(Mpa)	Maximum Offset/(mm)
I	45	45,557	83,243	38.87	0.398
II	35	74,511	133,740	43.86	0.412
III	25	145,001	255,622	47.92	0.413
IV	15	322,237	557,320	49.93	0.408

Among the four schemes, as the grid size of the blade region decreased and the number of elements and nodes increased, both the maximum stress value and the maximum displacement value showed a monotonic increase. Therefore, Scheme IV was selected as the mesh scheme for subsequent computations. In this scheme, the number of elements reached 322,237, the number of nodes reached 557,320, and the final details of the mesh can be found in Figure 7a.

When performing a one-way FSI computation on the axial-flow pump device, it is necessary to add constraint conditions in various directions of the blade. Figure 7b shows the constraint conditions added to the blades, where A represents the gravity constraint (Standard Earth Gravity), and the acceleration due to gravity is set to 9.8066 m/s^2. B represents the circumferential velocity (Rotational Velocity), and the rotating centrifugal force acts on the structural part of the blade during rotation. The rotational speed of the impeller is set to 166.7 r/min. C and D represent cylindrical constraints (Cylindrical Support), which prevent displacement of the rotating part of the impeller. E represents the pressure-bearing surface (Imported Pressure), where the water pressure on the surface of the blade computed from the steady-state flow field is imported to this pressure-bearing surface.

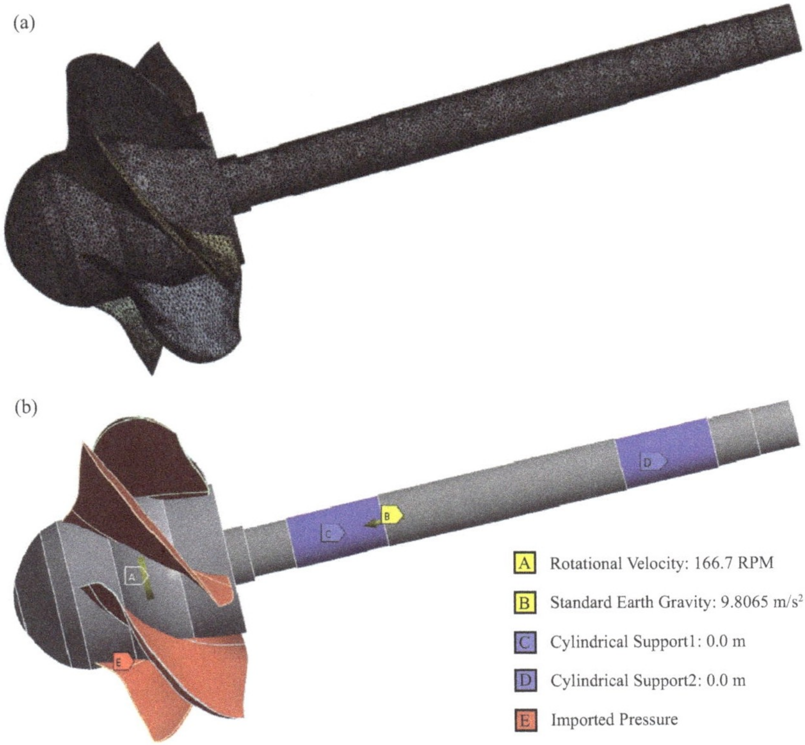

Figure 7. Details of structural field setting including (**a**) grid division of structural solid domain and (**b**) structural field constraint.

4. Results and Discussion

4.1. Analysis of Deformation and Stress of Blades

4.1.1. Total Deformation of Blades

Figure 8 shows the solid total deformation of the blade of the axial-flow pump device operating in pump mode at $0.8Q_{BEP}$, $1.0Q_{BEP}$, and $1.2Q_{BEP}$ flow rates. The maximum deformation of the blade is $\delta = 0.158$ mm at $0.8Q_{BEP}$, $\delta = 0.245$ mm at $1.0Q_{BEP}$, and $\delta = 0.407$ mm at $1.2Q_{BEP}$. As the flow rate increases, the maximum deformation on the blade surface gradually increases. The main deformation distribution of the blade is increasing gradually from the hub along the wheel rim, while there is almost no deformation at the hub. The deformation gradient near the hub is relatively small, and the increase in radial deformation is slow. The maximum deformation occurs at the water-inlet edge of the blade pressure surface (PS), and the deformation gradually decreases along the circumferential direction towards the water-outlet edge. This is because the thickness of the blade is the thinnest at the water-inlet edge and the thickest at the hub.

The distribution of solid total deformation of the blades of the axial-flow pump device operating in the PAT mode at $0.8Q_{BEP}$, $1.0Q_{BEP}$, and $1.2Q_{BEP}$ flow rates is shown in Figure 9. At $0.8Q_{BEP}$, the maximum deformation of the blade is $\delta = 0.167$mm, at $1.0Q_{BEP}$

it is $\delta = 0.256$ mm, and at $1.2Q_{BEP}$ it is $\delta = 0.410$ mm. The total deformation of the blade increases with the flow rate, and the area of deformation of the blade gradually increases. The maximum deformation values of the PS and suction surface (SS) can be found at the top wheel rim of the blade, and the thinner top of the blade is more susceptible to deformation due to the impact of the water flow, which can have a certain impact on the safe use of the blade. Similar to the pump mode, the deformation at the thicker hub can be ignored. In the PAT mode, the deformation of the blade is slightly increased compared to the pump mode. This is because during reverse power generation, the flow state is more turbulent, and the fluid does more work on the blade, which intensifies the deformation of the blade and thus has a certain impact on the safety and stability of the pump device.

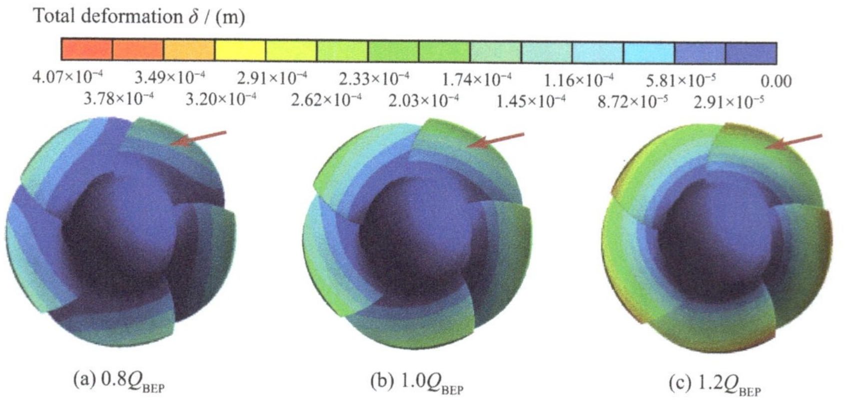

Figure 8. Total deformation of blades under the flow rate of (**a**) $0.8Q_{BEP}$, (**b**) $1.0Q_{BEP}$ and (**c**) $1.2Q_{BEP}$ in pump mode.

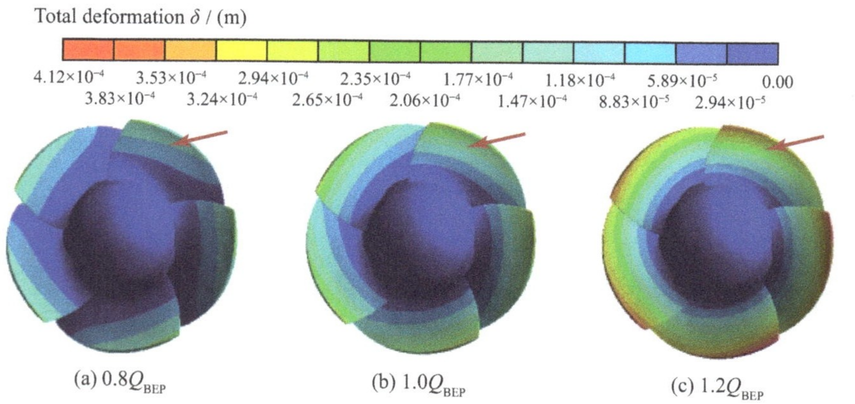

Figure 9. Total deformation of blades under the flow rate of (**a**) $0.8Q_{BEP}$, (**b**) $1.0Q_{BEP}$ and (**c**) $1.2Q_{BEP}$ in PAT mode.

4.1.2. Equivalent Stress of Blades

In this study, the equivalent stress σ of the rotor system is calculated according to the fourth strength theory, and the calculation equation is

$$\sigma = \sqrt{\frac{1}{2}\left[(\sigma_1 - \sigma_2)^2 + (\sigma_1 - \sigma_3)^2 + (\sigma_3 - \sigma_2)^2\right]}, \quad (6)$$

where σ_1, σ_2 and σ_3 are principle stresses. The distribution of the equivalent stress of the blade of the axial-flow pump device operating in the pump mode and the PAT mode at the above-mentioned three different flow rates is shown in Figures 10 and 11. Stress concentration occurs at the root of the blade in both modes, which may cause fatigue cracks and static load fracture in the blade. The maximum equivalent stress is at the root of the blade on the SS, and the stress on the blade decreases gradually from the root near the hub to the rim, which is caused by the shape of the blade. It can be observed that the back of the blade is more prone to flow separation near the position close to the hub, leading to a significant difference in equivalent stress between the front and back of the blade.

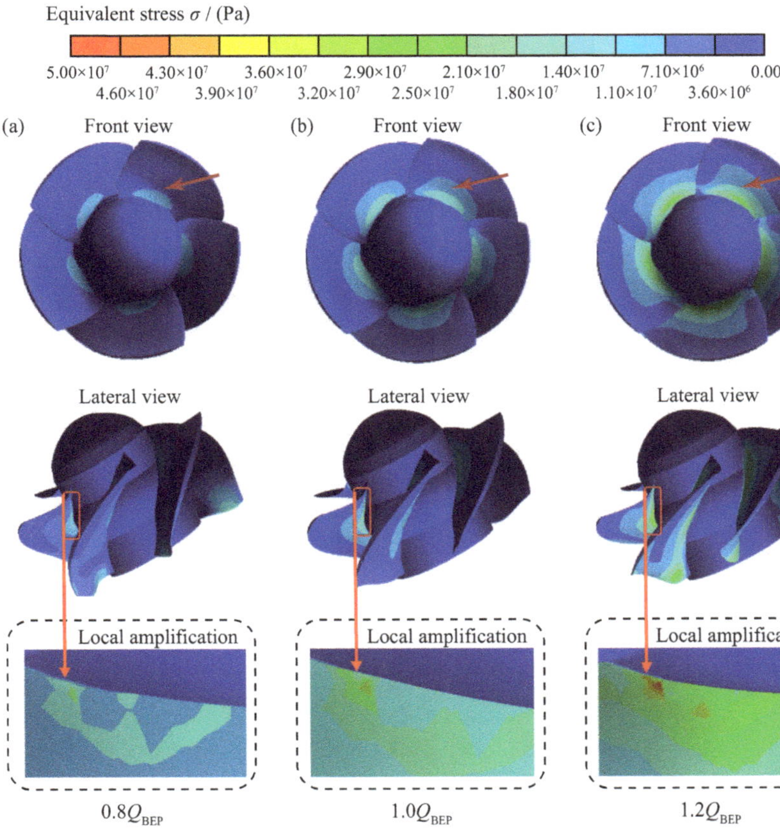

Figure 10. Equivalent stress of blades under the flow rate of (**a**) $0.8Q_{BEP}$, (**b**) $1.0Q_{BEP}$ and (**c**) $1.2Q_{BEP}$ in pump mode.

When operating in the pump mode, the equivalent stress of the blade increases with the increase in the flow rate. At $0.8Q_{BEP}$, the maximum equivalent stress is 20.34 MPa, while at $1.0Q_{BEP}$ it reaches 31.13 MPa, and at $1.2Q_{BEP}$ it reaches up to 47.95 MPa. The corresponding trend between flow rate and equivalent stress in the PAT mode is consistent with that in the pump mode. At $0.8Q_{BEP}$, the maximum stress value of the blade is 22.23 MPa, while at $1.0Q_{BEP}$, it reaches 33.12 MPa, and at $1.2Q_{BEP}$, it reaches up to 49.93 MPa. The maximum equivalent stress in the PAT mode is higher than that in the pump mode, and the high-stress distribution area is larger at $0.8Q_{BEP}$.

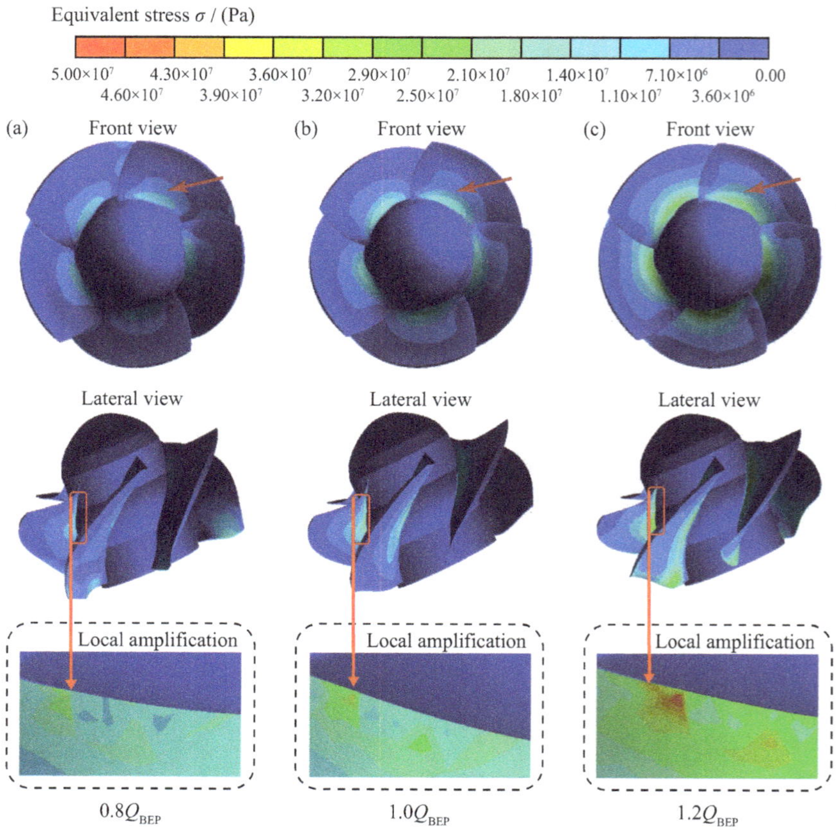

Figure 11. Equivalent stress of blades under the flow rate of (**a**) $0.8Q_{\mathrm{BEP}}$, (**b**) $1.0Q_{\mathrm{BEP}}$ and (**c**) $1.2Q_{\mathrm{BEP}}$ in PAT mode.

4.2. Analysis of Fatigue Life of Impeller

According to the fatigue loss theory, the fatigue strength is checked by calculating the safety factor n_σ, and its theoretical calculation equation is

$$n_\sigma = \frac{\sigma_{-1}}{K_{\sigma D}\sigma_a} \geq [n]. \tag{7}$$

Here, σ_{-1} is the fatigue limit, $K_{\sigma D}$ is the fatigue reduction coefficient which equals 2.2 and σ_a is the stress amplitude. Using the maximum stress value for checking, the corresponding safety allowance coefficient $\geq [n]$ is 1.5. In this study, the Fatigue Tools module in ANSYS Workbench is called to evaluate the fatigue life of axial flow pump. The Goodman model is used to modify the average value of alternating stress.

4.2.1. Number of Stress Cycles

The minimum blade cycle times in all operating conditions for both modes exceeded 10^6 cycles, indicating that the blade usage time and usage cycles are close to infinite cycling. This proves that when operating in the pump mode and the PAT mode, the blade usage time and usage cycles of the axial-flow pump device are within the safe allowable range, meeting the safety requirements of the material and ensuring structural strength for long-term safe and stable operation.

4.2.2. Blade Safety Factor

Figure 12 depicts the distribution of blade safety factors of the axial-flow pump device in the pump mode at $0.8Q_{BEP}$, $1.0Q_{BEP}$, and $1.2Q_{BEP}$ flow rates. In all operating conditions, the region where the minimum safety factor of the blade is located is at the connection between the blade root and the hub, while the safety factor is significantly higher at the blade tip. The amplified area in the figure shows the region where the minimum safety factor is located, and the distribution characteristics under each operating condition exhibit a noticeable regular pattern. The minimum safety factor at $0.8Q_{BEP}$ is $n_\sigma = 4.24$, at $1.0Q_{BEP}$ it is $n_\sigma = 2.77$, and at $1.2Q_{BEP}$ it is $n_\sigma = 1.79$. As the flow rate increases, the safety factor of the blade root decreases, making it more susceptible to damage, which is consistent with the pattern observed at the maximum equivalent stress of the blade. Additionally, Figure 13 compares the blade safety factors of the impeller obtained from numerical simulation and theoretical analysis for the axial-flow pump device operating in the pump mode at the three different flow rates mentioned above, and the relative error is provided. The relative error between the simulated minimum safety factor and the theoretical value increases with the flow rate. At a $0.8Q_{BEP}$ flow rate, the relative error of the minimum safety factor is 2.17%, while it is 3.75% at $1.0Q_{BEP}$. The maximum error is 6.55% at a $1.2Q_{BEP}$ flow rate because as the flow rate increases, the relative error in the solution denominator decreases due to the decrease in the safety factor. The absolute error is within the acceptable range, which verifies the accuracy and reliability of the numerical simulation results for the safety factor.

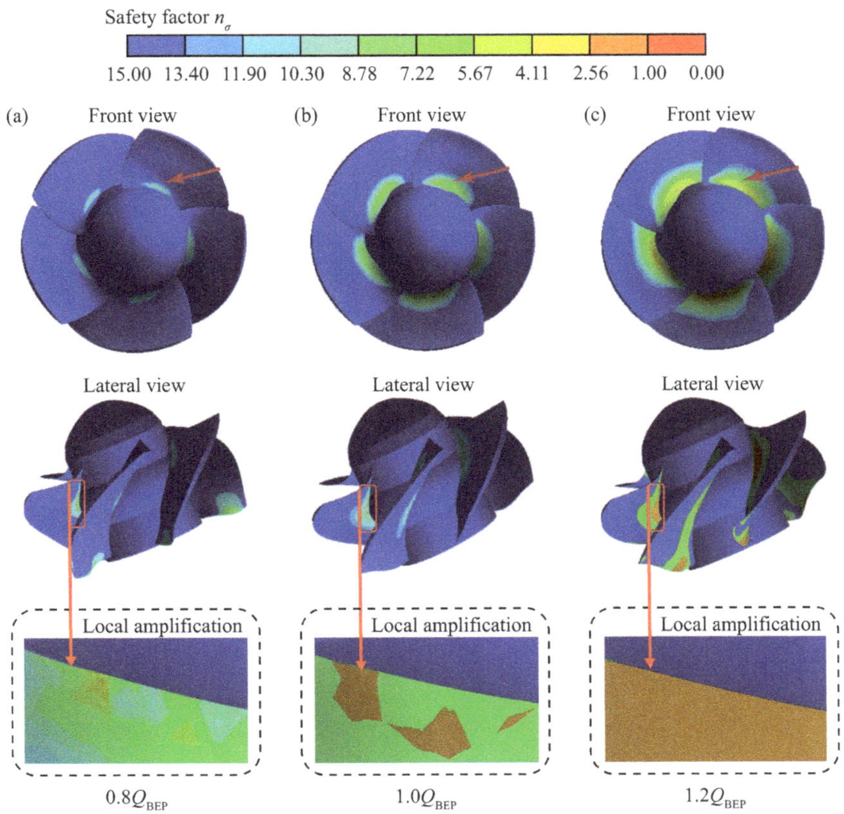

Figure 12. Blade safety factor under the flow rate of (**a**) $0.8Q_{BEP}$, (**b**) $1.0Q_{BEP}$ and (**c**) $1.2Q_{BEP}$ in pump mode.

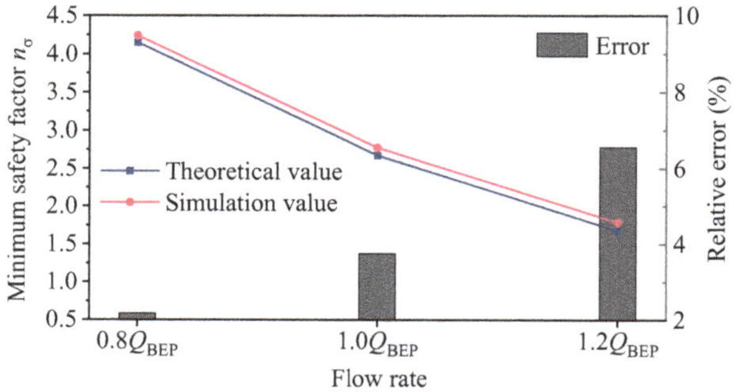

Figure 13. Comparison between theoretical value and numerical simulation value of minimum safety factor of impeller fatigue in pump mode.

Figure 14 presents the distribution of blade safety factors for the axial-flow pump device operating in the PAT mode at $0.8Q_{BEP}$, $1.0Q_{BEP}$, and $1.2Q_{BEP}$ flow rates. Similar to the operation in the pump mode, the minimum safety factor of the blade mainly concentrates at the blade root where it connects to the hub, while the safety factor at the blade edge is relatively larger for different flow rates. The figure also has local amplification on the region where a small safety factor distribution exists. At a $0.8Q_{BEP}$ flow rate, the minimum safety factor is $n_\sigma = 3.86$, while it is $n_\sigma = 2.60$ at $1.0Q_{BEP}$, and $n_\sigma = 1.73$ at $1.2Q_{BEP}$. As the flow rate increases, the safety factor of the blade root decreases, which corresponds to the pattern observed at the maximum equivalent stress of the blade. Figure 15 compares the blade safety factors obtained from numerical simulation and theoretical analysis for the axial-flow pump device operating in the PAT mode at the three different flow rates. The trend of the minimum safety factor error is similar to that under the pump mode. The relative error between the simulated and theoretical values increases with the flow rate. At a $0.8Q_{BEP}$ flow rate, the relative error of the minimum safety factor is 4.61%, while it is 7.44% at $1.0Q_{BEP}$ and 8.81% at $1.2Q_{BEP}$. It can be observed that the error between the simulated and theoretical values under the PAT mode is generally higher than that under the pump mode, which can be attributed to the more complex internal flow field under the PAT mode. Considering that the value of the safety factor is small, it leads to a small denominator and a large relative error when solving the relative error. However, the maximum error is still less than 10% under all operating conditions, so the simulation results can be considered accurate.

In summary, for the axial-flow pump device operating in both pump and PAT modes, the minimum safety factor of the blade is mainly concentrated at the blade root. Therefore, the blade root needs to be reinforced during the manufacturing process to ensure the material safety during operation. Compared with the operation under the pump mode, the minimum safety factor of the blade under the PAT mode is smaller at a low flow rate of $0.8Q_{BEP}$ and an optimal flow rate of $1.0Q_{BEP}$. This may be attributed to the more turbulent and poorer flow state of the water flow during the PAT mode, which results in more work performed by the water flow on the blade and greater impact on the blade. Under these conditions, the blade is more susceptible to damage. Therefore, the material strength and safety stability of the blade under PAT mode should be given special consideration during the design and manufacturing process.

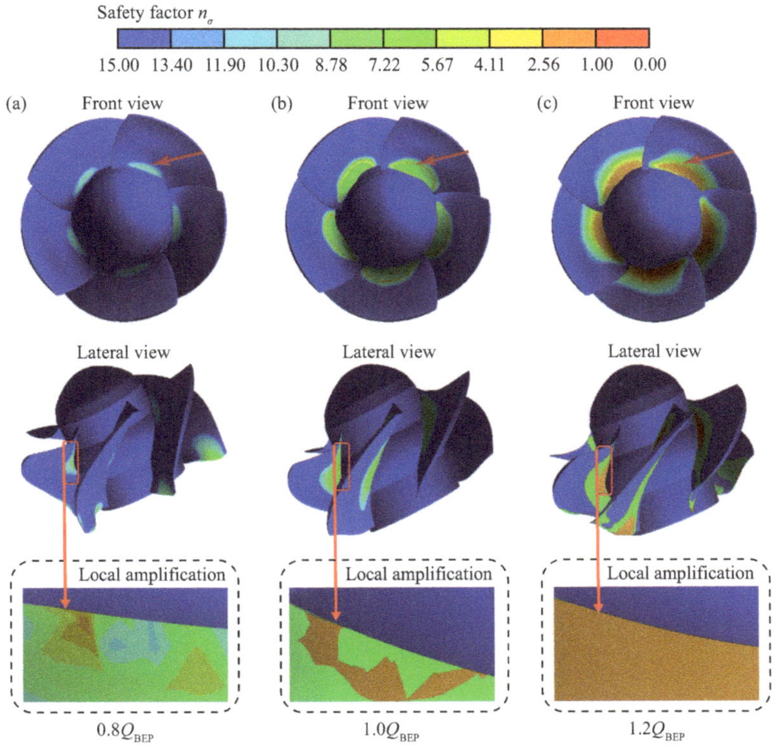

Figure 14. Blade safety factor under the flow rate of (**a**) 0.8Q_{BEP}, (**b**) 1.0Q_{BEP} and (**c**) 1.2Q_{BEP} in PAT mode.

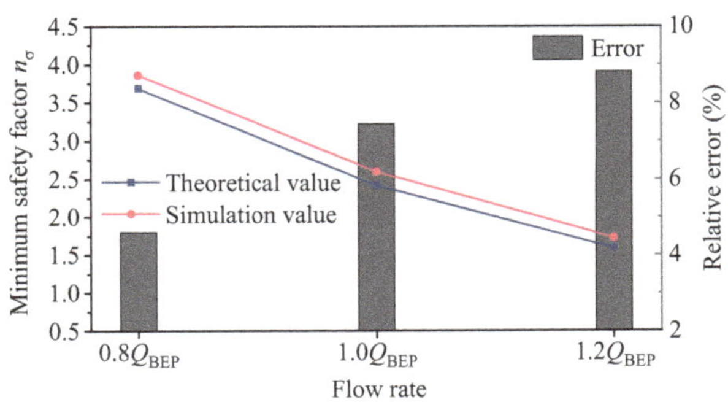

Figure 15. Comparison between theoretical value and numerical simulation value of minimum safety factor of impeller fatigue in PAT mode.

5. Conclusions

In this paper, commercial software ANSYS and the one-way FSI algorithm are used to study the equivalent stress, blade deformation, and fatigue life characteristics of the rotor system of a large vertical axial-flow pump operating in both pump and PAT modes at their respective three different flow rates: 0.8Q_{BEP}, 1.0Q_{BEP}, and 1.2Q_{BEP}. The main conclusions of this study are summarized as follows:

(1) When the axial-flow pump device operates in both pump mode and PAT mode, the maximum blade deformation increases with the increasing flow rate. The deformation at the blade root can be neglected and the radial deflection gradient is small, while

the deflection gradient increases gradually near the top of the blade. In comparison, the maximum deformation in the PAT mode is generally higher than that in the pump mode at all flow rates.

(2) Under all operating conditions, the stress concentration phenomenon mainly occurs at the blade root, with the maximum equivalent stress at the impeller root occurring at the blade suction surface, and the stress gradually decreasing from the blade root to the blade edge. Comparing different flow rate conditions, the maximum equivalent stress of both modes occurs at a flow rate of $1.2Q_{BEP}$. Under the same flow rate conditions, the equivalent stress in the PAT mode is relatively higher than in the pump mode.

(3) The number of cycles of the impeller exceeds 10^6 under all operating conditions, indicating that the load carrying capacity of the impeller is within the safe allowable range when the axial-flow pump device operates in both modes. The simulated impeller safety factor in the pump mode and the pump-turbine mode is slightly higher than the theoretical calculation value, which is reasonable. The minimum safety factor appears at the blade root, and the safety factor is smaller when operating in the PAT mode. Therefore, the blade root of the blade needs to be strengthened during processing to ensure the safe and stable operation of the pump device.

The above conclusions can be used to evaluate the service cycle of axial-flow pumps and the influence on pump life under different operation modes. At present, we only analyze the influence of water pressure on the structural strength of the rotor system under ideal conditions. In practical engineering, the impeller may have cracks in the long-term operation process. Therefore, in future work, more research will be conducted on the influence of cracks on the life of the axial-flow pump under full operating conditions and the comprehensive evaluation of the PAT generation economy and the influence on the life of the pump.

Author Contributions: Conceptualization, K.K.; methodology, K.K. and Z.C.; software, H.L. and Z.C.; validation, H.L., Z.C. and K.K.; formal analysis, H.L. and Z.C.; investigation, H.L. and Z.C.; resources, K.K. and H.C.; data curation, H.L. and Z.C.; writing—original draft preparation, H.L., Z.C.; writing—review and editing, K.K., H.C. and M.B.; visualization, H.L. and Z.C.; supervision, Y.Z., J.F. and H.X.; project administration, Y.Z., J.F. and H.X.; funding acquisition, K.K. and H.L. All authors have read and agreed to the published version of the manuscript.

Funding: This research was funded by the National Natural Science Foundation of China (52379086; 52009033), the Natural Science Foundation of Jiangsu Province (BK20200509), the Postdoctoral Research Foundation of China (2022T150185; 2022M711021), and the Project on Excellent Post-graduate Dissertation of Hohai University (422003515).

Data Availability Statement: Not applicable.

Conflicts of Interest: The authors declare no conflict of interest.

Abbreviations

CFD	Computational fluid dynamics
FEM	Finite element method
FSI	Fluid–structure interaction
FFT	Fast Fourier transform
PAT	Pump as turbine
RNG	Re-normalization group
RANS	Reynolds-averaged Navier–Stokes
BEP	Best efficiency point
PS	Pressure surface
SS	Suction surface
LE	Leading edge
TE	Trailing edge

u_i and u_j	Velocity components in x, y and z direction
i and j	Directional indices for x, y and z
t	Physical time
ρ	Fluid density
μ	Fluid dynamic viscosity
f_i	External force
k	Turbulence kinetic energy
ϵ	Turbulence dissipation rate
$C_{\epsilon 1RNG}$, $C_{\epsilon 2}$, σ_k and σ_ϵ	Constants of the RNG model
$\boldsymbol{M}$	Matrix of structure mass
$\boldsymbol{C}$	Matrix of structural damping
$\boldsymbol{K}$	Matrix of structural rigidity
$\ddot{q}$	Nodal acceleration vector
$\dot{q}$	Nodal velocity vector
q	Nodal displacement vector
$\boldsymbol{Q}$	Fluid load vector
η	Efficiency
H	Head
D	Impeller diameter
Q	Flow rate
Q_r	Rated flow rate
n	Rotation speed
n_r	Rated rotation speed
δ	Deformation
σ	Equivalent stress
σ_1, σ_2 and σ_3	Principle stresses
n_σ	Safety factor
σ_{-1}	Fatigue limit
$K_{\sigma D}$	Fatigue reduction coefficient
σ_a	Stress amplitude
$[n]$	Safety allowance coefficient

References

1. Feng, J.; Luo, X.; Guo, P.; Wu, G. Influence of tip clearance on pressure fluctuations in an axial flow pump. *J. Mech. Sci. Technol.* **2016**, *30*, 1603–1610. [CrossRef]
2. Shi, L.; Zhang, W.; Jiao, H.; Tang, F.; Wang, L.; Sun, D.; Shi, W. Numerical simulation and experimental study on the comparison of the hydraulic characteristics of an axial-flow pump and a full tubular pump. *Renew. Energy* **2020**, *153*, 1455–1464. [CrossRef]
3. Zhang, X.; Tang, F.; Chen, Y.; Huang, C.; Chen, Y.; Wang, L.; Shi, L. Experimental study on the internal pressure pulsation characteristics of a bidirectional axial flow pump operating in forward and reverse directions. *Machines* **2022**, *10*, 167. [CrossRef]
4. Abera, F.F.; Asfaw, D.H.; Engida, A.N.; Melesse, A.M. Optimal Operation of Hydropower Reservoirs under Climate Change: The Case of Tekeze Reservoir, Eastern Nile. *Water* **2018**, *10*, 273. [CrossRef]
5. Xi, S.; Desheng, Z.; Bin, X.; Weidong, S.; van Esch, B.B. Experimental and numerical investigation on the effect of tip leakage vortex induced cavitating flow on pressure fluctuation in an axial flow pump. *Renew. Energy* **2021**, *163*, 1195–1209. [CrossRef]
6. Kan, K.; Yang, Z.; Lyu, P.; Zheng, Y.; Shen, L. Numerical study of turbulent flow past a rotating axial-flow pump based on a level-set immersed boundary method. *Renew. Energy* **2021**, *168*, 960–971. [CrossRef]
7. Shen, S.; Qian, Z.; Ji, B.; Agarwal, R.K. Numerical investigation of tip flow dynamics and main flow characteristics with varying tip clearance widths for an axial-flow pump. *Proc. Inst. Mech. Eng. Part A J. Power Energy* **2019**, *233*, 476–488. [CrossRef]
8. Wang, C.; Wang, F.; Tang, Y.; Zi, D.; Xie, L.; He, C.; Zhu, Q.; Huang, C. Investigation into the phenomenon of flow deviation in the S-shaped discharge passage of a slanted axial-flow pumping system. *J. Fluids Eng.* **2020**, *142*, 041205. [CrossRef]
9. Zhang, D.S.; Shi, W.D.; Bin, C.; Guan, X.F. Unsteady flow analysis and experimental investigation of axial-flow pump. *J. Hydrodyn. Ser. B* **2010**, *22*, 35–43. [CrossRef]
10. Zhang, D.; Shi, W.; Van Esch, B.B.; Shi, L.; Dubuisson, M. Numerical and experimental investigation of tip leakage vortex trajectory and dynamics in an axial flow pump. *Comput. Fluids* **2015**, *112*, 61–71. [CrossRef]
11. Keck, H.; Sick, M. Thirty years of numerical flow simulation in hydraulic turbomachines. *Acta Mech.* **2008**, *201*, 211–229. [CrossRef]

12. Hübner, B.; Weber, W.; Seidel, U. The role of fluid-structure interaction for safety and life time prediction in hydraulic machinery. In *Proceedings of the IOP Conference Series: Earth and Environmental Science*; IOP Publishing: Bristol, UK 2016; Volume 49, p. 072007.
13. Trivedi, C.; Cervantes, M.J. Fluid-structure interactions in Francis turbines: A perspective review. *Renew. Sustain. Energy Rev.* **2017**, *68*, 87–101. [CrossRef]
14. Pei, J.; Meng, F.; Li, Y.; Yuan, S.; Chen, J. Fluid–structure coupling analysis of deformation and stress in impeller of an axial-flow pump with two-way passage. *Adv. Mech. Eng.* **2016**, *8*, 1687814016646266. [CrossRef]
15. Liu, X.; Xu, F.; Cheng, L.; Pan, W.; Jiao, W. Stress Characteristics Analysis of Vertical Bi-Directional Flow Channel Axial Pump Blades Based on Fluid–Structure Coupling. *Machines* **2022**, *10*, 368. [CrossRef]
16. Shi, L.; Zhu, J.; Wang, L.; Chu, S.; Tang, F.; Jin, Y. Comparative Analysis of Strength and Modal Characteristics of a Full Tubular Pump and an Axial Flow Pump Impellers Based on Fluid-Structure Interaction. *Energies* **2021**, *14*, 6395. [CrossRef]
17. Zhang, D.; Pan, D.; Xu, Y.; Shao, P.; Wang, G. Numerical investigation of blade dynamic characteristics in an axial flow pump. *Therm. Sci.* **2013**, *17*, 1511–1514. [CrossRef]
18. Gao, J.; Hou, Y.; Xi, S.; Cai, Z.; Yao, P.; Shi, H. Prediction of flow-induced dynamic stress in an axial pump impeller using FEM. In *Proceedings of the IOP Conference Series: Materials Science and Engineering*; IOP Publishing: Bristol, UK 2013; Volume 52, p. 022041.
19. Zhang, L.; Wang, S.; Yin, G.; Guan, C. Fluid–structure interaction analysis of fluid pressure pulsation and structural vibration features in a vertical axial pump. *Adv. Mech. Eng.* **2019**, *11*, 1687814019828585. [CrossRef]
20. Kan, K.; Xu, Z.; Chen, H.; Xu, H.; Zheng, Y.; Zhou, D.; Muhirwa, A.; Maxime, B. Energy loss mechanisms of transition from pump mode to turbine mode of an axial-flow pump under bidirectional conditions. *Energy* **2022**, *257*, 124630. [CrossRef]
21. Kan, K.; Zhang, Q.; Xu, Z.; Zheng, Y.; Gao, Q.; Shen, L. Energy loss mechanism due to tip leakage flow of axial flow pump as turbine under various operating conditions. *Energy* **2022**, *255*, 124532. [CrossRef]
22. Singh, V.K.; Nath, T. Energy generation by small hydro power plant under different operating condition. *Int. J. Hydromechatron.* **2021**, *4*, 331–349. [CrossRef]
23. Asomani, S.N.; Yuan, J.; Wang, L.; Appiah, D.; Zhang, F. Geometrical effects on performance and inner flow characteristics of a pump-as-turbine: A review. *Adv. Mech. Eng.* **2020**, *12*, 1687814020912149. [CrossRef]
24. Kan, K.; Binama, M.; Chen, H.; Zheng, Y.; Zhou, D.; Su, W.; Muhirwa, A. Pump as turbine cavitation performance for both conventional and reverse operating modes: A review. *Renew. Sustain. Energy Rev.* **2022**, *168*, 112786. [CrossRef]
25. Nautiyal, H.; Kumar, A. Reverse running pumps analytical, experimental and computational study: A review. *Renew. Sustain. Energy Rev.* **2010**, *14*, 2059–2067. [CrossRef]
26. Meng, F.; Yuan, S.; Li, Y. Fluid–structure coupling analysis of impeller in unstable region for a reversible axial-flow pump device. *Adv. Mech. Eng.* **2018**, *10*, 1687814017751762. [CrossRef]
27. Bai, Y.; Wu, D. Study on Fatigue Characteristics of Axial-Flow Pump Based on Two-Way Fluid–Structure Coupling. *Energies* **2022**, *15*, 8965. [CrossRef]
28. Kan, K.; Chen, H.; Zheng, Y.; Zhou, D.; Binama, M.; Dai, J. Transient characteristics during power-off process in a shaft extension tubular pump by using a suitable numerical model. *Renew. Energy* **2021**, *164*, 109–121. [CrossRef]
29. Li, Y.; Wang, F. Numerical investigation of performance of an axial-flow pump with inducer. *J. Hydrodyn.* **2007**, *19*, 705–711. [CrossRef]
30. Shi, W.; Zhang, D.; Guan, X.; Leng, H. Numerical and experimental investigation of high-efficiency axial-flow pump. *Chin. J. Mech. Eng.* **2010**, *23*, 38–44. [CrossRef]
31. Zhang, R.; Chen, H. Numerical analysis of cavitation within slanted axial-flow pump. *J. Hydrodyn.* **2013**, *25*, 663–672. [CrossRef]
32. Shi, L.; Yuan, Y.; Jiao, H.; Tang, F.; Cheng, L.; Yang, F.; Jin, Y.; Zhu, J. Numerical investigation and experiment on pressure pulsation characteristics in a full tubular pump. *Renew. Energy* **2021**, *163*, 987–1000. [CrossRef]
33. Birajdar, R.; Keste, A. Prediction of flow-induced vibrations due to impeller hydraulic unbalance in vertical turbine pumps using one-way fluid- structure interaction. *J. Vib. Eng. Technol.* **2020**, *8*, 417–430. [CrossRef]
34. Pei, J.; Yuan, S.; Yuan, J. Fluid-structure coupling effects on periodically transient flow of a single-blade sewage centrifugal pump. *J. Mech. Sci. Technol.* **2013**, *27*, 2015–2023. [CrossRef]
35. Chen, X.; Zheng, Y.; Xu, J.; Zhang, Y.; Fernandez-Rodriguez, E.; Li, C.; Zhou, Y.; Jiang, T. Fatigue life study of francis pump under reverse generation condition based on fluid solid coupling. *Water* **2020**, *12*, 1162. [CrossRef]
36. Balakrishna, A.; Mishra, P. Modelling and analysis of static and modal responses of leaf spring used in automobiles. *Int. J. Hydromechatron.* **2021**, *4*, 350–367. [CrossRef]

water

MDPI

Article

Variable Speed Control in PATs: Theoretical, Experimental and Numerical Modelling

Frank A. Plua [1,2], Francisco-Javier Sánchez-Romero [3], Victor Hidalgo [4], Petra Amparo López-Jiménez [1] and Modesto Pérez-Sánchez [1,*]

[1] Hydraulic and Enviromental Engineering Department, Universitat Politècnica de València, 46022 Valencia, Spain; fapluagu@doctor.upv.es (F.A.P.); palopez@upv.es (P.A.L.-J.)
[2] Carrera de Ingeniería Civil, Facultad de Ingeniería y Ciencias Aplicadas, Universidad Central del Ecuador, Quito 170129, Ecuador
[3] Rural and Agrifood Engineering Department, Universitat Politècnica de València, 46022 Valencia, Spain; fcosanro@agf.upv.es
[4] Centro de Investigación en Mecatrónica y Sistemas Interactivos—MIST, Universidad Indoamérica, Av. Machala y Sabanilla, Quito 170103, Ecuador; vhhidalgo@uce.edu.ec
* Correspondence: mopesan1@upv.es

Abstract: The selection of pumps as turbines (PATs) for their respective use in energy optimisation systems is a complicated task, because manufacturers do not provide the characteristic curves. For this reason, some research has been carried out to predict them with computational fluid dynamics (CFD) and mathematical models. The purpose of this study is to validate these two prediction methodologies of flow (Q) vs. head (H) curves through numerical modelling using the computational package OpenFOAM, together with a comparison with the experimental data obtained from a PAT for the case in which the nominal rotation speed of the machine varies. Depending on the configuration and working conditions of the PAT, the simulation performed with OpenFOAM was validated by calibrating it with the nominal curve of the pump and with another simulation performed with CFD workbench SOLIDWORKS FloEFD. Subsequently, the second methodology related to the analyses and mathematical models proposed to predict the Q vs. H curves were also validated with new models in OpenFOAM and the experimental data. The results show that these prediction methods are effective when a machine's operating point is close to the BEP (best efficient point). The absolute error ranges obtained with these two prediction methodologies for rotation speeds of 880 rpm, 1020 rpm, 1200 rpm, and 1500 rpm are between 5 and 24%, 2 and 17%, 0 and 12%, and 1 and 24%, respectively.

Keywords: pumps as turbines (PAT); computational fluid dynamics; variable rotational nominal speed; OpenFOAM

Citation: Plua, F.A.; Sánchez-Romero, F.-J.; Hidalgo, V.; López-Jiménez, P.A.; Pérez-Sánchez, M. Variable Speed Control in PATs: Theoretical, Experimental and Numerical Modelling. *Water* **2023**, *15*, 1928. https://doi.org/10.3390/w15101928

Academic Editor: Yuan Zheng

Received: 27 April 2023
Revised: 15 May 2023
Accepted: 17 May 2023
Published: 19 May 2023

1. Introduction

The availability of water resources at the global level has significantly decreased. Among the main agents that have caused this situation are climate change, environmental pollution, human activities, and failures in hydraulic structures, among others. It is becoming increasingly complicated to access appropriate sources that meet the quality and quantity of the resource. Despite this, water loss due to leaks in pressurised distribution systems still manages considerable values, with losses of 8 to 24% in developed countries [1]. Considering that the need for water is increasing, there is an urgent need to implement sustainable projects that allow the user to carry them out efficiently [2]. This type of project requires the use and development of new technologies that are easy to implement and apply [3].

In the case of sustainable water systems, there are some approaches from which improvements can be proposed. Among these approaches are the determination of water quality parameters [4], optimisation of energy efficiency [5,6] reduction of water leaks [7,8], mathematical modelling of management, and optimisation of systems [9,10], among others.

Water distribution systems are not energy-efficient, because they depend on pressure demands that can generate leaks, increasing energy costs [11]. One of the elements that has a negative effect from the point of view of energy efficiency but is necessary for the hydraulic operation of systems is the pressure-reducing valve (PRV) [12]. PRVs are used to reduce the pressure at one point by regulating the flow passage. An alternative to the use of these devices, to reduce the dependence on non-renewable energy [13] and take advantage of the excess energy of these systems [14], is the use of PATs (pumps working as turbines). In addition, PATs have been used as energy-generating devices in micro-hydroelectric power plants as a sustainable solution in the water industry [15]. For this reason, it has become a trend to study the use of PATs to optimise different water systems to improve their sustainability [16–18].

PATs are pumps that work in reverse mode to generate energy. This machine's cost is cheaper than a conventional turbine of the same size [19], although they have lower hydraulic efficiencies in ranges between 0.6 and 0.7 [20]. When all electromechanical equipment is considered, the overall efficiency decreases to values between 0.5 to 0.6 [21]. The use of pumps operating as turbines (PATs) has increased due to their application, availability, and cost advantages [6,22–27]. For example, Novara et al. [28] concluded that an installation with PATs could be 5 to 15 times cheaper than a conventional installation with turbines.

The study of PATs began with Thoma and Kittredge [29], who accidentally found that pumps can operate efficiently as turbines when trying to evaluate the complete characteristics of pumps. In 1957, Stepanoff [30] reported several modes of operation of the pumps on performance curves plotted in quadrants. Once it was discovered that PATs could be applied in the chemical industry and the supply of drinking water, different researchers developed some techniques to predict the operation of this type of machine. In 1962, Childs [31] carried out comparative studies between efficiencies in devices working in both modes (pump–turbine). Subsequently, the first studies were carried out to predict the performance values in turbine mode and discover the best efficient point (BEP) through linear equations. The study of PATs has been developed using different approaches, such as in water distribution systems, where Jain [32] researched placing PATs in distribution systems. Fecarotta [33] and Morani [11] proposed an analysis regarding the proper location of PATs; the latter focused the research to look for cost reduction and the maximisation of production and energy savings. Moazeni [34] investigated the optimal number and location of PATs through mixed nonlinear programming models. Macias [18] established a methodology that was applied in an irrigation project in a rural area in the province of Valencia (Spain) that focused on optimising the location and selection of PATs based on the influence of leaks. The same author [13] developed a new methodology for self-calibration of leaks to learn the injected flow rate and the volume consumed in water networks. This methodology was applied in the city of Manta, Ecuador.

Since the performance curves are not available in pumps that work in turbine mode [27], different studies and methodologies have been carried out to obtain them and to select the appropriate machine depending on the type of working conditions required. Rossi [35] proposed a general method to predict PAT performance using artificial neural networks (ANN). Based on the datasheets provided by the pump manufacturers, the author obtained the BEP and off-design performance using the ANN methodology. In addition, the resulting predictions were compared with experimental data not used in the training process, which resulted in a high degree of compatibility. The study concluded that the BEP flow rate increased in reverse mode while the specific speed in BEP decreased slightly and also recommended the use of this tool to choose the proper PAT. To estimate the BEP and the characteristic curves of PATs, Perez-Sánchez [20] proposed new approach equations from an experimental base of 181 machines. Additionally, Plua [21] presented new empirical expressions to estimate the head, efficiency, and power curves for PATs with variable speeds. These equations allow the application of various operation strategies in hydraulic simulation tools (e.g., Epanet and WaterGEMS).

Micro-hydroelectric power plants (MHP) have become very effective solutions for rural sectors with powers of 5–100 kW. The big problem with these facilities is the high turbine cost concerning the entire project [27]. In the case of MHP, the price of these elements can be higher than 60–70% [36]. One possibility to reduce this cost is to use PATs instead of a conventional turbine [37], which would favour the expansion of MHP and the reduction of greenhouse gas emissions [28]. In 2012, Pascoa [38] proposed a new approach for a hydroelectric plant with PAT with a constant flow. Rossi [39] suggested the economic feasibility of placing PATs in the Merano aqueduct, which resulted in the production of 338 kWh of daily electricity and power of 19.18 kW. Table 1 [27] shows different PAT installations in power generation projects.

Table 1. PAT installations in MHP [27].

Location	The Capacity of the Plant (kW)	Year of Installation
Sainyabulli Province, Laos	2	2008
Thima, Kenya	2.2	2001
Mae Wei Village, Thailand	3	2008
West Java, Indonesia	4.5	1992
Kinko village, Tanzania	10	2006
Fazenda Boa Esperanca, Brazil	45	2007
Ambotia Micro-hydro project, India	50	2004
British Columbia, Canada	200	-
Vysni Lhoty, Czech Republic	332	2008

CFD techniques have been widely used to predict characteristic curves and the performance of pumps in direct and reverse modes and proven to be an effective solution in PAT approaches [40–43]. Additionally, an experimental investigation is fundamental for obtaining reliable results for PATs under different optimisation stages [14,21,44,45]. Different types of machines, such as axial, mixed, and radial PATs with horizontal and vertical axes that single- and multistage [46], have been studied using CFD simulations for fixed and variable speeds [24–26]. However, very few studies related to numerical modelling in PATs of variable rotational speeds have been executed, so it is imperative to establish equations and laws that predict their behaviour [47]. The numerical simulations were carried out to define the performance of the pump [48], analyse the flow in turbine mode [49], predict and extrapolate the characteristic curves [50], etc.

Plua [40] presented research in which the main parameters and techniques that have been simulated for PATs through CFD are shown and which main simulation ranges are mentioned: specific speed: 0.8–306, rotational speed: 250–3900 rpm, flow rate: 2.9–300 L/s, and mass flow: 13–17.8 kg/s. Concerning the numerical simulations, the principal turbulence models used were Reynolds Average Navier–Stokes (RANS) and Unsteady Reynolds Average Navier–Stokes (URANS). The most used closure model was k-ε, followed by k-ω and k-ω-SST, among others. Regarding packages, ANSYS-CFX was the most used, followed by CFD Code Fluent and OpenFOAM. With respect to the mesh generation, the number of cells was 1×10^6 to 4.2×10^6, with hexahedral, tetrahedral, mixed blocks, and pyramids. Depending on each situation, boundary conditions such as the total pressure, mass flow rate, stagnation pressure, constant total pressure, static pressure, and volumetric flow were placed at the inlet and outlet of the model. In conclusion, it was established that the CFD methodology to predict the performance of a pump working as a turbine presented adequate accuracy based on the comparison of the results with the experimental tests. However, numerous errors were also reported in some studies. The authors assumed that the reported errors were due to the geometries between the tests and the simulations not being identical; the loss estimation was not exact, and more experience in computational analysis is required for modelling this type of phenomenon. Finally, the same author [47] evaluated the application of numerical CFD simulation in PATs in comparison with experimental results and obtained conclusions for future numerical analyses. As a result, it was evidenced that there have been a few simulated cases where a flow with variable speed was simulated and that the number of studies with free code computational packages is minimal, and their use should be promoted due to their outstanding capabilities.

Therefore, the present study is focused on a numerical simulation in the OpenFOAM 3D free code package of PATs that have experimental data to validate the use of the new empirical expressions proposed for machines with different rotational speeds. The particularity in the modelling is that the study of a rotating PAT at different speeds will be carried out, and comparisons will be made with the experimental results obtained on a test bench to calibrate the model.

2. Materials and Methods

Figure 1 depicts the main tasks performed to determine the validity of new expressions obtained by Plua [21] to predict the behaviour of PATs with variable speeds. The figure comprises three very defined stages: the first shows the activities that are carried out before the numerical simulation, stage two refers to the specific work that is executed during the simulation, and in part three, the post-modelling works are shown, in which the results of the numerical modelling for the analysis case are compared with the experimental data and with the new expressions.

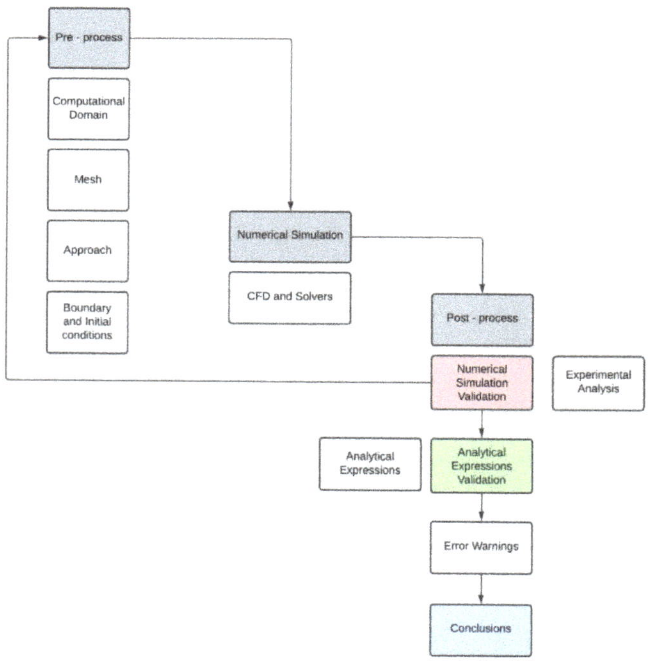

Figure 1. Methodology flowchart.

2.1. Preprocess

2.1.1. Computational Domain

The PAT model presented in this study was taken from research conducted by Pérez Sánchez [51] and experimentally tested at the CERIS-Hydraulic Lab of the Instituto Superior of Lisbon, Portugal. The geometry corresponds to an installation of a PAT in a laboratory that allows experiments where the flow, pressure conditions, and rotational speed can be varied. The hydraulic facility consists of a 1 m^3 air vessel tank, a 50 mm HDPE pipe, a KSB radial impeller centrifugal pump (model Etarnom 232) that operates in turbine mode, a regulating tank, pressure transducers, valves, and a flow recirculation pump. The air vessel tank sends water to reach the PAT, which discharges to the open free surface tank and then incorporates it into the system through the recirculation pump. The 3D model was built in the SOLIDWORKS CAD system from which the following drawing view was extracted (see Figure 2). This figure shows the geometry that will be entered into the CFD package and from which the results will be compared with the experimental data and with the new expressions. The interactions of the other elements that comprise the installation of the PAT in the laboratory, such as valves, tank, and pump, are placed in the model through the boundary conditions.

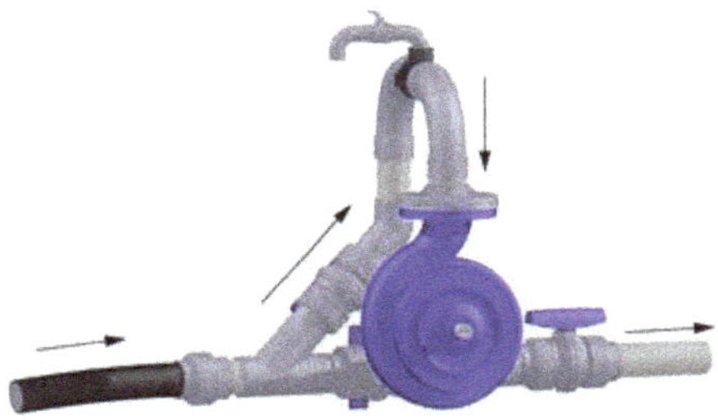

Figure 2. PAT 3D model.

Considering the complexity of this modelling, both due to the geometry and the operation required for the PATs, the computational domain must be divided into parts, which will each be meshed with different meshing levels, with an emphasis on the details of interest and on which of their boundary conditions will be determined individually, to then be configured as a total domain that delivers the results of the whole set. The computational domain consists of four parts: the inlet pipe, starting in the inlet section and reaching the pump's runner; the rotating part that is the impeller of the PAT, the rotating part of the domain; the casing, the stationary part of the pump; and the outlet pipe, which corresponds to the discharge of the pump to the outlet. The original geometry was redefined according to the configuration of the control volumes to obtain the optimal meshing (see Figure 3).

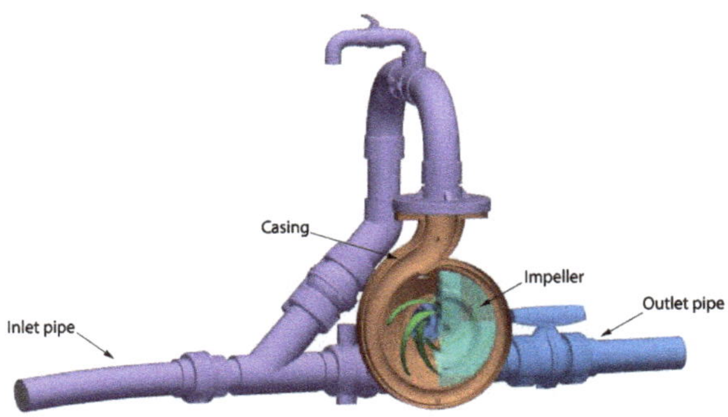

Figure 3. Domains of the case.

Depending on the actual geometry and its characteristics, the .stl files were modified with Autodesk Inventor software (https://www.autodesk.com/products/inventor/overview?term=1-YEAR&tab=subscription) to achieve a better-quality mesh. The areas of meshing interest were prioritised: the casing, the impeller, and the blades. Each has different elements and details with simultaneously different levels and definition angles. It allows the surfaces to stick more to the edges, bringing the mesh's geometry closer to the actual configuration. The geometries modelled were the volute, the discharge pipe, the inlet pipe, and the impeller. In the case of the impeller, it was divided into three parts, as seen in Figure 3. The impeller is composed of the lower and upper parts and the blades. These elements are treated independently to improve the mesh quality and then facilitate the visualisation of results at the post-processing stage. In addition, six blades were configured inside the impeller, which allowed a better study of the phenomenon presented in the PAT.

2.1.2. Mesh

The mesh was created with snappyHexMesh, an automatic mesh generator that adjusts to the surface to obtain the required mesh. First, the 3D model was exported to format .stl using Autodesk® Inventor® software. Later, with the help of HELIX-OS, the BlockMeshDict file was created to generate, using the BlockMesh utility, orthogonal mesh elements for the casing, inlet pipe, impeller, and outlet, respectively. Once the block meshing was ready, the domain geometries were admitted into the snappyHexMeshDict file. The local refinement was defined using castellatedMesh, and the internal points within the closed domain were entered. Finally, it was necessary to use the topoSet tool to generate zones with movable cells for the runner and merge the meshes with the mergeMeshes utility. The mesh characteristics are presented in Table 2, and the generated mesh is shown in Figure 4. In this figure, the different levels of meshing applied to the subdomains can be seen. In addition, it is observed that their configuration is very close to the original geometry. On the other hand, the model looks appropriately balanced, a situation that is confirmed later.

Table 2. Mesh characteristics.

Parameter	Value/Characteristic
Element type	Hexahedra, Polyhedra, Prism
Number of Elements	827,578
Hexahedral	639,704
Prism	28,238
Polyhedra	159,612
Number of Nodes	1,203,219
Number of Patches	8
Max. Aspect Ratio	14.68619
Min. Surface Area	6.19213^{-9}
Min. Volume	1.39587^{-11}
Max. Skewness	12.918596

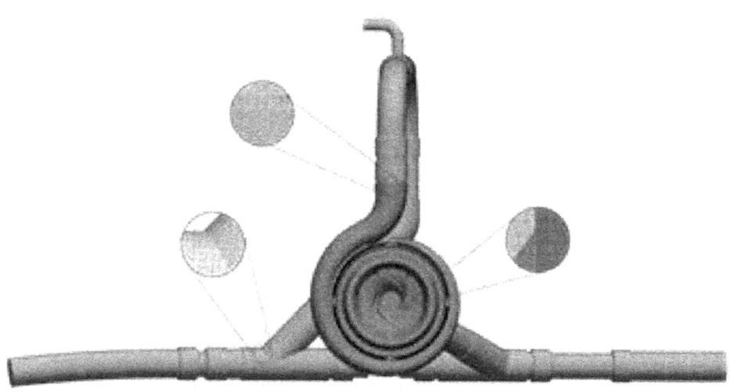

Figure 4. Generated mesh.

2.1.3. Approach

The MRF technique (multiple reference frame) was the technique used for modelling rotation in CFD in this case. This methodology establishes a separate reference frame for each region of the domain, for both rotational and static [49]. It is based on the creation of a local region around the rotating object where the relative velocity is determined for each point. First, Navier–Stokes equations are built, taking into account the centrifugal and Coriolis forces, and then, a set of equations for the stationary and rotational regions are created. This technique can accurately capture instantaneous local flows, which depend on the relative position of the rotative element vs. static geometry. In the MRF approach, the Navier–Stokes equations are solved in terms of the global/inertial velocity. Since, in this case, there is a separation between the impeller and the scroll, the AMI approach is not applicable. For that reason, a set was used that allows a simulation of these elements in the MRF approach.

2.1.4. Boundary and Initial Conditions

Initial and boundary conditions should be applied when solving the Navier–Stokes and continuity equations. Table 3 summarises the initial conditions related to the turbulence models used in this research. For the calibration of the mathematical model, the κ-ε turbulence model was used (the same one used by [51]). κ is turbulent kinetic energy, and ε is turbulent dissipation rate. The κ-ω (specific turbulent dissipation rate)-SST turbulence model was used to analyse the experimental data, the nominal rotational speed curve, and the results of the new expressions contained in [21].

Table 3. Initial conditions.

Initial Conditions	Value
Turbulent Kinetic Energy (κ)	0.032856 (m^2/s^2)
Turbulent Dissipation Rate (ε)	0.320573 (m^2/s^3)
Specific turbulent Dissipation Rate (ω)	108.4104 (s^{-1})
Turbulent kinematic viscosity (nut)	3.03×10^{-4} (m^2/s)

The turbulent kinematic viscosity value "nut" represents the roughness in the walls confirming the domain. Regarding boundary conditions, a constant velocity input condition and a static pressure output condition were used. The boundary conditions of the computational domain are detailed in Table 4.

2.2. Numerical Simulation

2.2.1. CFD

The Navier–Stokes equations were solved using CFD methods based on a continuum mechanics approach for fluid mechanics to define the fluid behaviour in the PATs [50]. For that, two equations were considered that obtain the values of velocity and pressure that allow for defining the average behaviour of the flows. The equations correspond to the conservation of mass and linear momentum and are indicated in a tensor with the following expressions [47]:

$$\frac{\partial \rho}{\partial t} + \overline{u}_j \frac{\partial \rho}{\partial x_j} = 0 \tag{1}$$

$$\frac{\partial(\rho \overline{u}_i)}{\partial t} + \frac{\partial(\rho \overline{u}_i \overline{u}_j)}{\partial x_j} = -\frac{\partial \overline{P}}{\partial x_i} + \frac{\partial\left(\overline{\tau}_{ij} - \tau'_{ij}\right)}{\partial x_j} \tag{2}$$

where i and j are subscripts for the three axes of space, respectively; $\overline{u}$ is the filtered velocity magnitude; $\overline{P}$ is the filtered pressure; the subgrid stress tensor is $\overline{\tau}_{ij}$; and τ'_{ij} is the filtered viscous stress tensor.

Table 4. Boundary conditions.

	Runner1	**Runner**	**RunnerIn**	**Volute**	**Pipe—Inlet**	**Pipe—Outlet**	**Inlet**	**Outlet**
Velocity (u-m/s)	movingWallVelocity uniform (0 0 0)	movingWallVelocity uniform (0 0 0)	movingWallVelocity uniform (0 0 0)	fixedValue uniform (0 0 0)	fixedValue uniform (0 0 0)	fixedValue uniform (0 0 0)v	flowRateInletVelocity volumetricFlowRate constant 0.0045	inletOutlet valueuniform (0 0 0)
Static Pressure (p-m^2/s^2)	zeroGradient	zeroGradient	zeroGradient	zeroGradient	zeroGradient	zeroGradient	zeroGradient	uniform 115,198 (810) 116,694 (930) 112,472 (1050) 112,909 (1170) 115,756 (1275) 110,971 (1500)

2.2.2. CFD and Solvers

The CFD package used is the CFD OpenFOAM 9, which models multiphysics simulations applicable to computational fluid dynamics for incompressible and compressible flows with applications in dynamic mesh management to make rotating reference frames with adaptable mesh refinements as required. OpenFOAM uses a directory structure to solve the cases, where the case is the name of the analysis case; the system sets the numerical control to run time and solver; the constant contains the physical properties, modelling, and mesh information; and 0 has the edge conditions, as well as the beginning to the modelling and time directories that correspond to the solutions and derived cases.

Regarding meshing, OpenFOAM has some mesh utilities, such as BlockMesh, snappyHexMesh, foamyHexMesh, and foamyQuadmesh. OpenFOAM also allows the mesh to be generated with other packages, since mesh conversion utilities are compatible with popular mesh formats (Gmsh, Fluent, Ideas, and Netgen, among others). As stated above, snappyHexMesh generated the mesh. The snappyHexMesh utility is an automatic hybrid mesh that divides, refines, and adjusts to the analysed surface, attaching the mesh with complex details of the geometry [52].

For the calibration of the model, the solver simpleFoam was applied to a steady-state incompressible flow based on the SIMPLE (Semi-Implicit Method for Pressure Linked Equations) algorithm for pressure velocity coupling [53], with applications in turbulent and transient flows in pipes.

3. Results

3.1. Numerical Simulation Validation

3.1.1. Mesh Quality

The checkMesh tool was used to evaluate the mesh quality, giving the mesh stats, the overall number of cells of each type, topology, geometry, and conclusions concerning the mesh. Two parameters were used to verify the quality of the mesh; one of them was Ω, which corresponds to the following expression $\Omega = NE/ND$, where ND is the number of nodes and NE is the number of elements. Ω indicating the homogeneity of the mesh, a good mesh quality will present Ω values close to 1, and values close to 2 have very dispersed meshes. For this case study, the calculated Ω value was 0.69, which is acceptable. The other value was the so-called y^+, which verifies the acceptable range of values for the turbulence model. If this value is less than 1, it is considered that the quality of the mesh is good. In this study, it was found that the average y^+ values in all the simulations of the mesh were less than 1.

3.1.2. Calibration

For the CFD simulation validation, two calibrations were performed concerning the Pérez-Sánchez study [51]. The first concerned the mathematical model made with SOLIDWORKS FloEFD, and the second concerned experimental research. In the Pérez-Sánchez CFD model, the simulated global variables were the head (H), the output hydraulic torque (T), the discharge (Q), and the rotational speed (N). Within the simulations, the absolute static pressure contours were obtained for a flow rate of 4.5 L/s and rotation speeds of 810 rpm, 930 rpm, 1050 rpm, 1170 rpm, 1275 rpm, and 1500 rpm. The results showed that the pressure decreased from upstream to downstream as the fluid flowed within the domains and along the impeller, from the inner to the outer region, as the energy was transmitted to the shaft. On the other hand, it was found that the higher the speed, the lower the pressure value downstream of the impeller.

The results of the simulation performed with OpenFOAM in this study are shown in Figures 5 and 6. As can be seen, the pressure decreased from upstream to downstream, and the lowest pressure value occurred at point D (before the first elbow of the volute outlet) for the maximum speed. A comparison with the original work [51] showed a remarkable similarity between the two. It was observed that, in all cases, the passage of the fluid through the PAT showed a similar behaviour, the pressure difference increased as the speed in the PAT increased. The error of this simulation concerning the original work varied in ranges from 0.014 to 14.297% at points A, B, C, and F of the model (see Table 5).

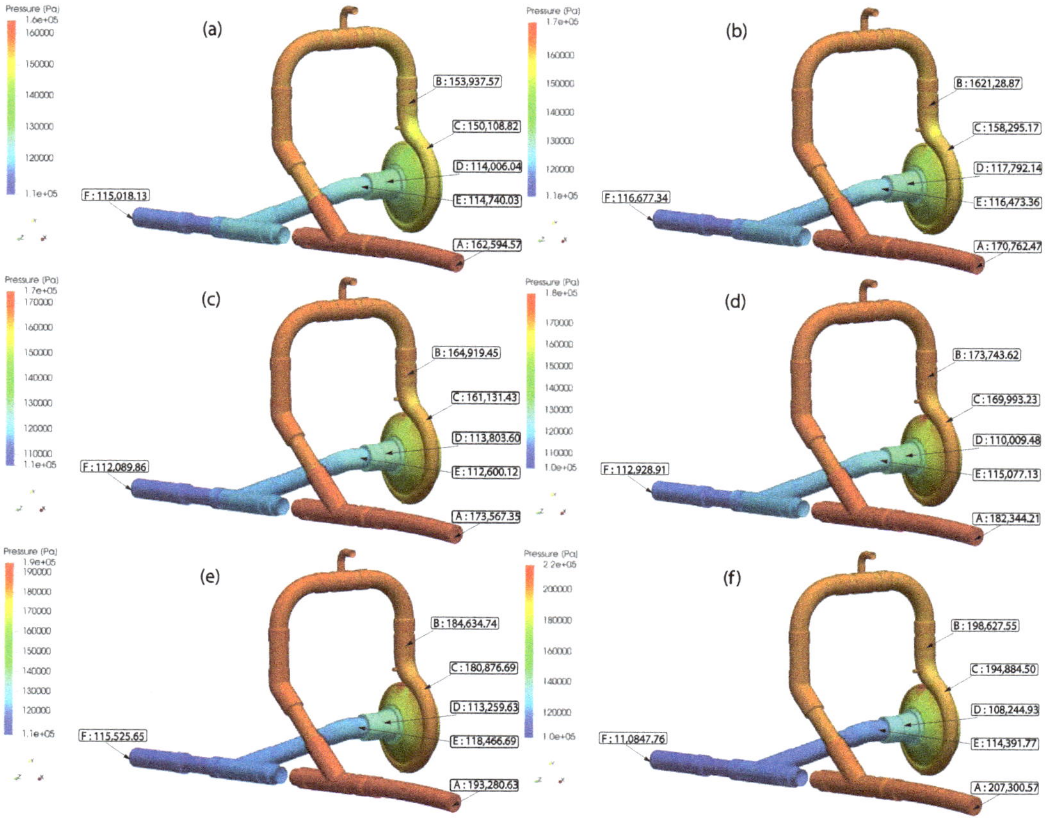

Figure 5. Absolute static pressure contours for Q = 4.50 l/s: (**a**) N = 810 rpm; (**b**) N = 930 rpm; (**c**) N = 1050 rpm; (**d**) N = 1170; (**e**) N = 1275 rpm; (**f**) N = 1500 rpm.

Regarding the calibration of the mathematical simulation with the experimental data, a sensitivity analysis was performed to identify which turbulence model produced the best results. Simulations were executed on the machine's best efficient point (BEP) tested in [51] when operating in turbine mode (Q_{BEP} = 3.6 L/s) for speeds of 200, 600, 880, 1020, 1200, and 1500 rpm using the κ-ε κ-ω-SST models. The results obtained for both simulations are shown in Table 6. As can be seen, the simulations produced errors of similar magnitude. Still, for the nominal rotational speed of 1020 rpm, the κ-ω-SST model was the one with the lowest error. An error index analysis was performed to define the turbulence model with which the cases of experimental data, nominal rotational speed curve, and the results of the new expressions [21] were simulated. Considering that, in all cases, the error indices closest to zero were those that had a better fit and data compatibility, it was observed that the κ-ω-SST turbulence model presented the best fit in all cases. However, it was verified that the order of magnitude of both turbulence models was close, so they proved their validity when they were applied. Figure 7 shows that, in all cases, the κ-ω-SST model had a better performance. Therefore, the κ-ω-SST turbulence model was adopted for the rest of the cases.

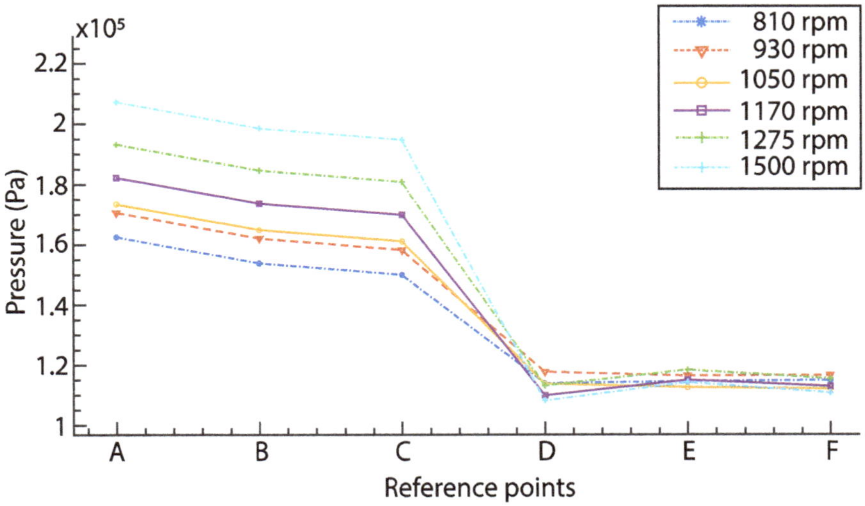

Figure 6. Absolute static pressure vs. referenced sections.

Table 5. Calibration results at points A, B, C, and F compared to [51].

	% Error					
Referenced Sections	**810**	**930**	**1050**	**1170**	**1275**	**1500**
A	8.724%	14.297%	8.218%	0.035%	12.881%	14.042%
B	4.455%	10.425%	4.286%	5.324%	9.068%	13.066%
C	5.979%	12.040%	5.643%	3.999%	10.389%	11.936%
F	0.156%	0.014%	0.340%	0.018%	0.199%	0.111%

Table 6. Sensitivity analysis for κ-ε vs. κ-ω-SST.

	Experimental	**Simulation**			
		κ-ε		**κ-ω-SST**	
n (rpm)	**H (mca)**	**H (mca)**	**% Error**	**H (mca)**	**% Error**
200	3.27	2.28	30.23	2.39	27.00
600	3.66	2.90	20.74	3.02	17.58
880	4.68	4.21	10.10	4.27	8.73
1020	5.22	5.03	3.67	5.08	2.70
1200	6.22	6.21	0.12	6.14	1.30
1500	7.86	8.60	9.35	8.77	11.52

Once the mathematical model has been validated concerning the results obtained in the numerical modelling and experimental works in [51], the curve of the machine working at nominal speed is contrasted, as seen in Figure 8. In both cases, the increasing trend is shown as a function of the increase in flow rate. In this figure, it is observed that the best results are shown near the volumetric flow value equal to 3.6 L/s, which is precisely the QBEP.

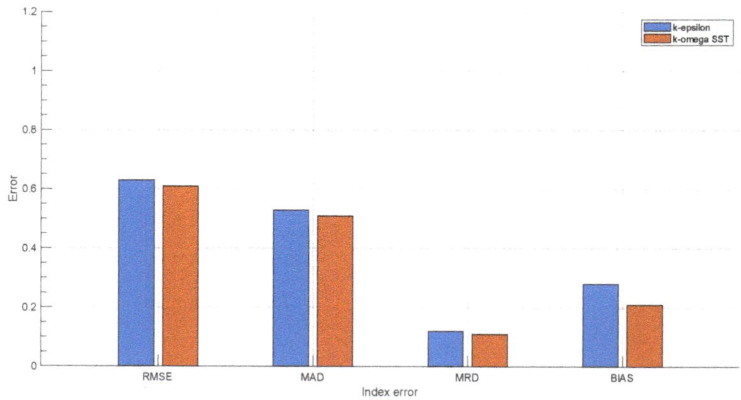

Figure 7. Index error analysis for the sensitivity analysis.

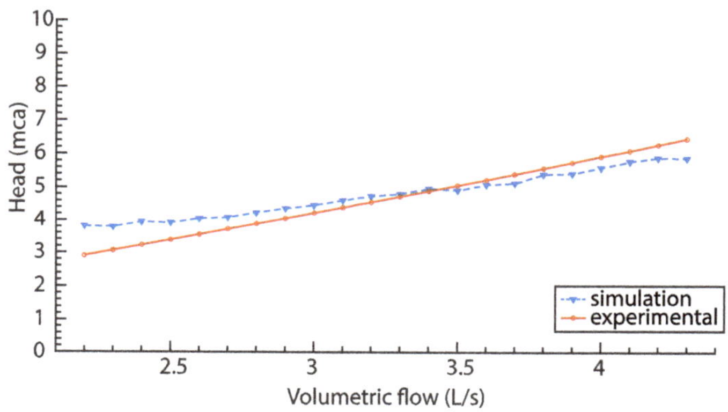

Figure 8. The nominal curve obtained with CFD OpenFOAM vs. the nominal curve in [51].

The stability of the simulation is related to the convergence, which can be seen in Figure 9, where the residuals of the velocity, pressure, k, and omega are observed. According to this figure, the modelling is considered stable, since all values are less than 10^{-3}.

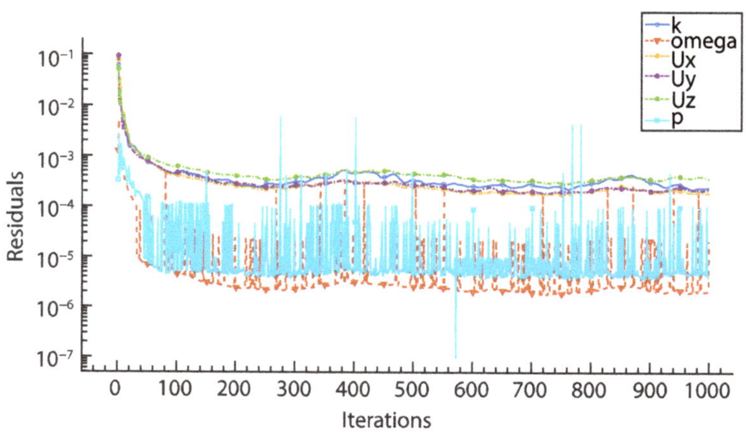

Figure 9. Stability and convergence of the model.

As can be seen, the simulation performed with OpenFOAM presents satisfactory results, and therefore, the model is considered validated. The errors comparing the proposed expressions and the simulations are quite acceptable near the BEP for these sorts of numerical models.

3.2. Analytical Expressions Validation

3.2.1. Analytical Expressions—New Expressions to Predict PATs Behaviour

Considering that, in the case of PATs, the information to select the suitable machines is not known because it is not provided by the manufacturers [54], polynomial expressions have been proposed as a function of semiempirical methods to estimate the characteristic curves in PATs when the rotational speed is constant [20,54–56]. However, considering that flow rates in water systems are variable due to user demand, an optimal energy analysis for PATs cannot be performed if the rotational speed is considered constant. Therefore, strategies have been proposed to maximise energy when the machine works at different rotational speeds, called the variable operation strategy (VOS) [19]. Plua et al. [21] proposed new empirical expressions applying the VOS strategy in water systems for different rotational speeds of 15 different machines and analysing 87 different curves with 56,450 operating points.

Through a mathematical analysis of 10 general expressions (6 polynomials and 4 potentials) considering specific variables as the ratio of rotational speed α and the ratio Q/QBEP, it was possible to adjust a polynomial function for experimental values of head and efficiency and a potential function for power. These expressions are observed in Equations (3)–(7) and present the lowest errors (30 to 50% compared to other models) in the respective analyses performed where the RMSE, MAD, MRD, and BIAS indices were calculated. Equations (3)–(7) correspond to the expressions proposed in [21] to calculate the flow number (q), the head number (h), the efficiency number (e), and the torque number (p), which are dimensionless parameters and correspond to the relationship between the current conditions of the PATs and the best efficient point (BEP) of the machine, to predict the characteristics curves of the PATs when the pump is used in turbine mode. Figure 6 in [21] shows a head and efficiency curve comparison between the proposed model, experimental data, and other models.

$$q = -0.1525\left(\alpha\frac{Q}{Q_{BEP}}\right) + 0.1958\left(\frac{Q}{Q_{BEP}}\right)^2 - 0.0118\left(\frac{Q}{Q_{BEP}}\right) - 0.6429\alpha^2 + 1.8489\alpha - 0.2241 \tag{3}$$

$$h = -0.31070\left(\alpha\frac{Q}{Q_{BEP}}\right) + 0.3172\left(\frac{Q}{Q_{BEP}}\right)^2 - 0.0546\left(\frac{Q}{Q_{BEP}}\right) + 0.242\alpha^2 + 1.1708\alpha - 0.3426 \tag{4}$$

$$e = 0.8271\left(\alpha\frac{Q}{Q_{BEP}}\right) - 0.3187\left(\frac{Q}{Q_{BEP}}\right)^2 - 0.1758\left(\frac{Q}{Q_{BEP}}\right) - 1.035\alpha^2 + 1.1815\alpha + 0.5019 \tag{5}$$

$$p = \alpha^{2.4762} \tag{6}$$

$$q = \alpha^{0.7439} \tag{7}$$

where

$$q = \frac{Q_i}{Q_{BEP}}; h = \frac{H_i}{H_{BEP}}; e = \frac{\eta_i}{\eta_{BEP}}; p = \frac{P_i}{P_{BEP}} = qhe$$

3.2.2. Analytical Expressions Validation

The head value H for different rotational speeds at points close to the BEP was compared with the experimental head obtained in [51], the expressions proposed in [21], and the mathematical model (see Figure 10). As can be seen, the relationship between Q and H increased in all cases. Figure 11d,e present the best results. It is also observed that the numerical modelling is always closer to reality; the growth slopes are very similar, unlike the other methodology, where this slope is lower. The predictions made in the numerical simulation with OpenFOAM and with the new expressions present values close to the experimental ones when the operation of the machine approaches the BEP.

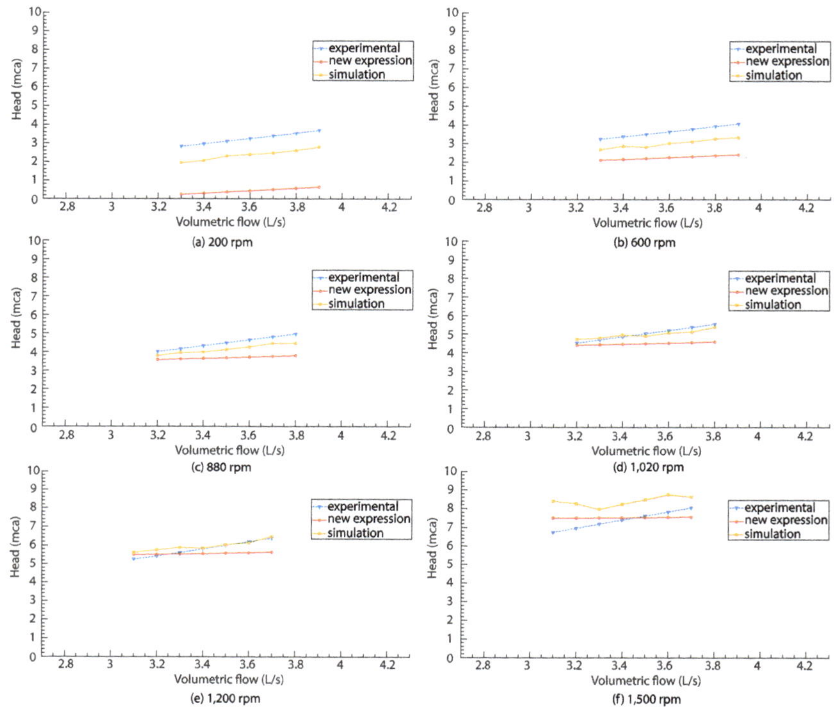

Figure 10. Experimental head [51] vs. head obtained with new expressions in [21] and CFD OpenFOAM simulation.

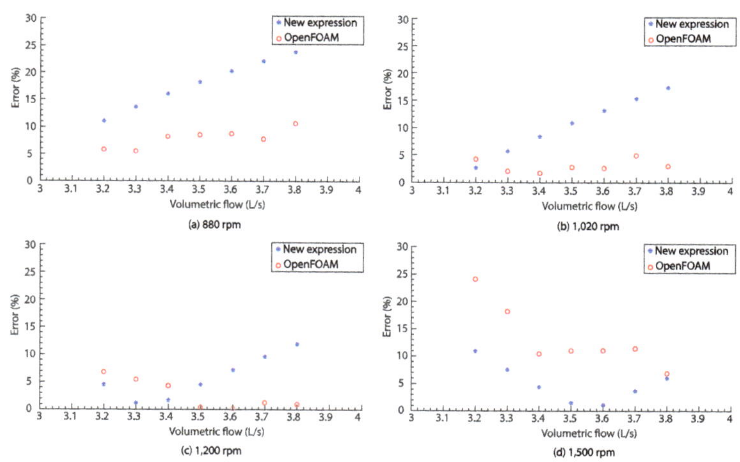

Figure 11. Error correlations.

3.2.3. Error Analysis

The error indices obtained in the predictions made as a function of the rotational speed are presented in Figure 12. As can be seen, as the conditions approach those of the BEP, the predictions reflect values closer to reality. As in other cases, the best results occur when the speed is between 1020 rpm and 1200 rpm. The results of the calculations of the absolute errors are shown in Table 7. In the case of the CFD methodology, the range is from 0 to 11%, while, with the new expressions, this range varies between 1 and 24%.

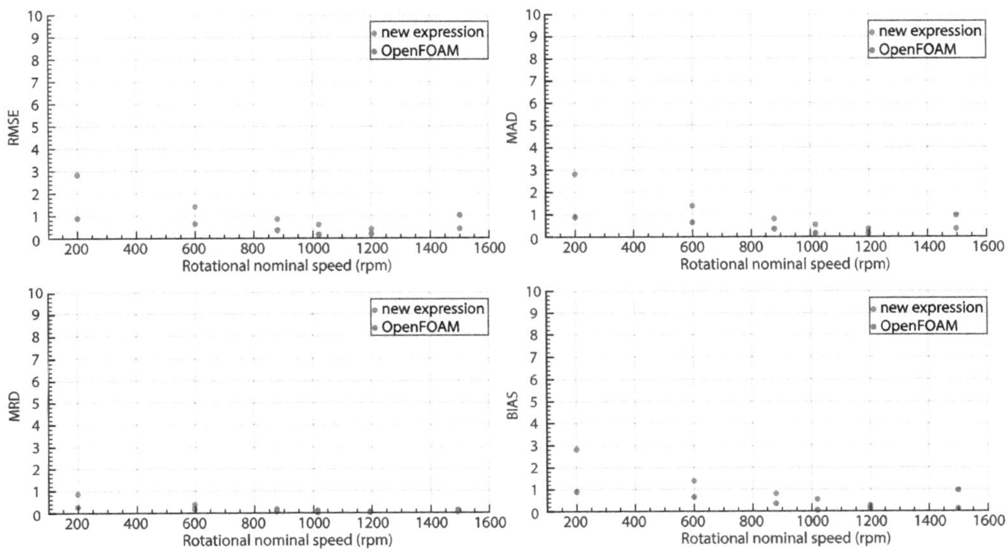

Figure 12. Error indices condidered as RMSE, MAD, MRD and BIAS.

Table 7. Calculated absolute errors.

	Range of Absolute Error			
	OpenFOAM		New Expressions [21]	
n (rpm)	Min (%)	Max (%)	Min (%)	Max (%)
880	5	11	11	24
1020	2	5	3	17
1200	0	7	1	12
1500	1	11	7	24

4. Conclusions

This research proposes to validate as a prediction methodology of flow (Q) vs. head (H) curves of variable speed PATs, the numerical simulation with OpenFOAM 3D Free Code Package, depending on its configuration and working conditions. It also proposes to validate the new expressions submitted by Plua [21]. It was demonstrated that the simulation presents adequate results once the mathematical model and the nominal curve of Pérez-Sánchez's [51] research were calibrated. Furthermore, based on experimental data from a PAT, the Q vs. H curves were calculated through the new expressions [21], as well as with the numerical simulation performed in OpenFOAM, presenting satisfactory results as the operation point of the work approached the BEP, since the trend of the generated curves, the slope thereof, and the error indices demonstrated acceptable values. However, when moving away from the BEP conditions, the error increased.

In summary, it was possible to validate the prediction methodologies of the Q vs. H characteristic curves of the PATs and verify the range in which they present the best results. This study can be extended through the tuning of the coefficients of the proposed analytical expressions based on the feedback with new experimental data.

Author Contributions: Conceptualisation, M.P.-S. and P.A.L.-J.; methodology, F.A.P., F.-J.S.-R., and M.P.-S. software, writing—original draft preparation, V.H., F.A.P., and M.P.-S.; writing—review and editing, P.A.L.-J. and M.P.-S.; visualisation M.P.-S. and P.A.L.-J.; and supervision, M.P.-S. and P.A.L.-J. All authors have read and agreed to the published version of the manuscript.

Funding: Grant PID2020-114781RA-I00 funded by MCIN/AEI/10.13039/501100011033.

Institutional Review Board Statement: Not applicable.

Informed Consent Statement: Not applicable.

Data Availability Statement: Not applicable.

Conflicts of Interest: The authors declare no conflict of interest.

References

1. Farley, M. *Leakage Management and Control*; WHO: Geneva, Switzerland, 2001; pp. 1–98. Available online: http://whqlibdoc.who.int/hq/2001/WHO_SDE_WSH_01.1_pp1-98.pdf (accessed on 1 May 2023).
2. Niu, W.-J.; Feng, Z.-K. Evaluating the performances of several artificial intelligence methods in forecasting daily streamflow time series for sustainable water resources management. *Sustain. Cities Soc.* **2020**, *64*, 102562. [CrossRef]
3. Ebrahimi, S.; Riasi, A.; Kandi, A. Selection optimization of variable speed pump as turbine (PAT) for energy recovery and pressure management. *Energy Convers. Manag.* **2020**, *227*, 113586. [CrossRef]
4. Orouji, H.; Haddad, O.B.; Fallah-Mehdipour, E.; Mariño, M.A. Modeling of Water Quality Parameters Using Data-Driven Models. *J. Environ. Eng.* **2013**, *139*, 947–957. [CrossRef]
5. Vakilifard, N.; Anda, M.; Bahri, P.A.; Ho, G. The role of water-energy nexus in optimising water supply systems—Review of techniques and approaches. *Renew. Sustain. Energy Rev.* **2018**, *82*, 1424–1432. [CrossRef]
6. Ferrarese, G.; Malavasi, S. Perspectives of Water Distribution Networks with the GreenValve System. *Water* **2020**, *12*, 1579. [CrossRef]
7. Islam, M.S.; Babel, M.S. Economic Analysis of Leakage in the Bangkok Water Distribution System. *J. Water Resour. Plan. Manag.* **2013**, *139*, 209–216. [CrossRef]
8. Bibri, S.E.; Krogstie, J. Smart sustainable cities of the future: An extensive interdisciplinary literature review. *Sustain. Cities Soc.* **2017**, *31*, 183–212. [CrossRef]
9. Bach, P.M.; Rauch, W.; Mikkelsen, P.S.; McCarthy, D.T.; Deletic, A. A critical review of integrated urban water modelling—Urban drainage and beyond. *Environ. Model. Softw.* **2014**, *54*, 88–107. [CrossRef]
10. Raboaca, M.S.; Bizon, N.; Trufin, C.; Enescu, F.M. Efficient and Secure Strategy for Energy Systems of Interconnected Farmers' Associations to Meet Variable Energy Demand. *Mathematics* **2020**, *8*, 2182. [CrossRef]
11. Morani, M.C.; Carravetta, A.; D'ambrosio, C.; Fecarotta, O. A New Preliminary Model to Optimize PATs Location in a Water Distribution Network. *Environ. Sci. Proc.* **2020**, *2*, 57. [CrossRef]
12. Gupta, A.; Bokde, N.; Marathe, D.; Kulat, K. Leakage Reduction in Water Distribution Systems with Efficient Placement and Control of Pressure Reducing Valves Using Soft Computing Techniques. *Eng. Technol. Appl. Sci. Res.* **2017**, *7*, 1528–1534. [CrossRef]
13. Ávila, C.A.M.; Sánchez-Romero, F.J.; López-Jiménez, P.A.; Pérez-Sánchez, M. Improve leakage management to reach sustainable water supply networks through by green energy systems. Optimized case study. *Sustain. Cities Soc.* **2022**, *83*, 103994. [CrossRef]
14. Ramos, H.M.; Borga, A. Pumps as turbines: An unconventional solution to energy production. *Urban Water* **1999**, *1*, 261–263. [CrossRef]
15. Gallagher, J.; Harris, I.M.; Packwood, A.J.; McNabola, A.; Williams, A.P. A strategic assessment of micro-hydropower in the UK and Irish water industry: Identifying technical and economic constraints. *Renew. Energy* **2015**, *81*, 808–815. [CrossRef]
16. Ávila, C.A.M.; Sánchez-Romero, F.-J.; López-Jiménez, P.A.; Pérez-Sánchez, M. Leakage Management and Pipe System Efficiency. Its Influence in the Improvement of the Efficiency Indexes. *Water* **2021**, *13*, 1909. [CrossRef]
17. Ávila, C.A.M.; Sánchez-Romero, F.-J.; López-Jiménez, P.A.; Pérez-Sánchez, M. Definition of the Operational Curves by Modification of the Affinity Laws to Improve the Simulation of PATs. *Water* **2021**, *13*, 1880. [CrossRef]
18. Ávila, C.A.M.; Sánchez-Romero, F.-J.; López-Jiménez, P.A.; Pérez-Sánchez, M. Optimization tool to improve the management of the leakages and recovered energy in irrigation water systems. *Agric. Water Manag.* **2021**, *258*, 107223. [CrossRef]
19. García, I.F.; Novara, D.; Mc Nabola, A. A Model for Selecting the Most Cost-Effective Pressure Control Device for More Sustainable Water Supply Networks. *Water* **2019**, *11*, 1297. [CrossRef]

20. Pérez-Sánchez, M.; Sánchez-Romero, F.J.; Ramos, H.M.; López-Jiménez, P.A. Improved Planning of Energy Recovery in Water Systems Using a New Analytic Approach to PAT Performance Curves. *Water* **2020**, *12*, 468. [CrossRef]
21. Plua, A.F.; Sánchez-Romero, F.-J.; Hidalgo, V.; López-Jiménez, P.A.; Pérez-Sánchez, M. New Expressions to Apply the Variation Operation Strategy in Engineering Tools Using Pumps Working as Turbines. *Mathematics* **2021**, *9*, 860. [CrossRef]
22. Lee, S.; Pomeroy, C.; Burian, S. Setting Future Water Rates for Sustainability of a Water Distribution System. *J. Water Resour. Plan. Manag.* **2021**, *147*, 04020108. [CrossRef]
23. Ahmadi, S.; Saboohi, Y.; Vakili, A. Frameworks, quantitative indicators, characters, and modeling approaches to analysis of energy system resilience: A review. *Renew. Sustain. Energy Rev.* **2021**, *144*, 110988. [CrossRef]
24. Stefanizzi, M.; Capurso, T.; Balacco, G.; Binetti, M.; Camporeale, S.M.; Torresi, M. Selection, control and techno-economic feasibility of Pumps as Turbines in Water Distribution Networks. *Renew. Energy* **2020**, *162*, 1292–1306. [CrossRef]
25. Renzi, M.; Rudolf, P.; Štefan, D.; Nigro, A.; Rossi, M. Installation of an axial Pump-as-Turbine (PaT) in a wastewater sewer of an oil refinery: A case study. *Appl. Energy* **2019**, *250*, 665–676. [CrossRef]
26. Sánchez, M.P.; Sánchez-Romero, F.J.; Ramos, M.H.; López-Jiménez, P.A. *Bombas Operando Como Turbinas (PAT): Principios de Funcionamiento y Selección*; Universidad Politécnica de Valencia: Valencia, Spain, 2020.
27. Binama, M.; Su, W.-T.; Li, X.-B.; Li, F.-C.; Wei, X.-Z.; An, S. Investigation on pump as turbine (PAT) technical aspects for micro hydropower schemes: A state-of-the-art review. *Renew. Sustain. Energy Rev.* **2017**, *79*, 148–179. [CrossRef]
28. Novara, D.; Carravetta, A.; McNabola, A.; Ramos, H.M. Cost Model for Pumps as Turbines in Run-of-River and In-Pipe Microhydropower Applications. *J. Water Resour. Plan. Manag.* **2019**, *145*, 04019012. [CrossRef]
29. Thoma, D.; Kittredge, C.P. Centrifugal pumps operated under abnormal conditions. *J. Power Sources* **1931**, *73*, 881–884.
30. Stepanoff, A.J. *Centrifugal and Axial Flow Pumps: Theory, Design, And Application*; Krieger Publishing Company: Malabar, FL, USA, 1957.
31. Childs, S. Convert pumps to turbines and recover HP. *Hydrocarb. Process. Pet. Refin.* **1962**, *41*, 173–174.
32. Jain, S.V.; Patel, R.N. Investigations on pump running in turbine mode: A review of the state-of-the-art. *Renew. Sustain. Energy Rev.* **2014**, *30*, 841–868. [CrossRef]
33. Fecarotta, O.; McNabola, A. Optimal Location of Pump as Turbines (PATs) in Water Distribution Networks to Recover Energy and Reduce Leakage. *Water Resour. Manag.* **2017**, *31*, 5043–5059. [CrossRef]
34. Moazeni, F.; Khazaei, J. Optimal energy management of water-energy networks via optimal placement of pumps-as-turbines and demand response through water storage tanks. *Appl. Energy* **2021**, *283*, 116335. [CrossRef]
35. Rossi, M.; Renzi, M. A general methodology for performance prediction of pumps-as-turbines using Artificial Neural Networks. *Renew. Energy* **2018**, *128*, 265–274. [CrossRef]
36. Prasad, V.; Shukla, S.N.; Joshi, S.G. Performance characteristics of pump as turbine. *Indian Pumps* **2006**, *38*, 5–9.
37. Williams, A. Pumps as turbines for low cost micro hydro power. *Renew. Energy* **1996**, *9*, 1227–1234. [CrossRef]
38. Páscoa, J.C.; Silva, F.J.; Pinheiro, J.S.; Martins, D.J. A new approach for predicting PAT-pumps operating point from direct pumping mode characteristics. *J. Sci. Ind. Res.* **2012**, *71*, 144–148.
39. Rossi, M.; Righetti, M.; Renzi, M. Pump-as-turbine for Energy Recovery Applications: The Case Study of An Aqueduct. *Energy Procedia* **2016**, *101*, 1207–1214. [CrossRef]
40. Plua, F.; Hidalgo, V.; Cando, E.; Pérez-Sánchez, M.; López-Jiménez, P. Pumps as Turbines (PATs) by Analysis with CFD Models. *Int. J. Adv. Sci. Eng. Inf. Technol.* **2022**, *12*, 1098–1104. [CrossRef]
41. Rawal, S.; Kshirsagar, J.T. Numerical Simulation on a Pump Operating in a turbine mode. In Proceedings of the 23rd International Pump Users Symposium, Houston, TX, USA, 5–8 March 2007.
42. Nautiyal, H.; Kumar, A. Reverse running pumps analytical, experimental and computational study: A review. *Renew. Sustain. Energy Rev.* **2010**, *14*, 2059–2067. [CrossRef]
43. Páscoa, J.; Silva, F.J.; Pinheiro, J.S.; Martins, D.J. Accuracy details in realistic CFD modeling of an industrial centrifugal pump in direct and reverse modes. *J. Therm. Sci.* **2010**, *19*, 491–499. [CrossRef]
44. Carravetta, A.; Fecarotta, O.; Ramos, H. Numerical simulation on Pump As Turbine: Mesh reliability and performance concerns. In Proceedings of the 2011 International Conference on Clean Electrical Power (ICCEP), Ischia, Italy, 14–16 June 2011; pp. 169–174.
45. Carravetta, A.; Fecarotta, O.; Del Giudice, G.; Ramos, H. Energy Recovery in Water Systems by PATs: A Comparisons among the Different Installation Schemes. *Procedia Eng.* **2014**, *70*, 275–284. [CrossRef]
46. Pugliese, F.; De Paola, F.; Fontana, N.; Giugni, M.; Marini, G.; Francos, J.F. Experimental and numerical investigation of centrifugal Pumps As Turbines. In Proceedings of the 10th International Conference on Energy Efficiency in Motor Driven System, Rome, Italy, 6–8 September 2017; pp. 6–7.
47. Plua, F.; Hidalgo, V.; López-Jiménez, P.A.; Pérez-Sánchez, M. Analysis of Applicability of CFD Numerical Studies Applied to Problem When Pump Working as Turbine. *Water* **2021**, *13*, 2134. [CrossRef]
48. Nautiyal, H.; Kumar, V.; Thakur, A. CFD Analysis on Pumps Working as Turbines. *Hydro Nepal J. Water Energy Environ.* **2011**, *6*, 35–37. [CrossRef]
49. Hidalgo, V.; Velasco, M.; Cando, E.; Valencia, E.; Simbaña, S.; Puga, D.; Mora, C.; Escaler, X. Rotatory 3D structured mesh study using openFOAM to simulate the flow in francis turbine. *Mater. Today Proc.* **2021**, *49*, 142–148. [CrossRef]
50. Hidalgo, V.; Escaler, X.; Valencia, E.; Peng, X.; Erazo, J.; Puga, D.; Luo, X. Scale-Adaptive Simulation of Unsteady Cavitation Around a Naca66 Hydrofoil. *Appl. Sci.* **2019**, *9*, 3696. [CrossRef]

51. Pérez-Sánchez, M.; Simão, M.; López-Jiménez, P.A.; Ramos, H.M. CFD Analyses and Experiments in a PAT Modeling: Pressure Variation and System Efficiency. *Fluids* **2017**, *2*, 51. [CrossRef]
52. Hidalgo, V.; Díaz, C.; Erazo, J.; Simbaña, S.; Márquez, D.; Puga, D.; Velasco, R.; Mafla, C.; Barragán, G.; Parra, C.; et al. Simplified simulation of a small Pelton turbine using OpenFOAM. *IOP Conf. Ser. Earth Environ. Sci.* **2021**, *774*, 012075. [CrossRef]
53. Huang, S.; Wei, Y.; Guo, C.; Kang, W. Numerical Simulation and Performance Prediction of Centrifugal Pump's Full Flow Field Based on OpenFOAM. *Processes* **2019**, *7*, 605. [CrossRef]
54. Fecarotta, O.; Aricò, C.; Carravetta, A.; Martino, R.; Ramos, H.M. Hydropower Potential in Water Distribution Networks: Pressure Control by PATs. *Water Resour. Manag.* **2014**, *29*, 699–714. [CrossRef]
55. Novara, D.; McNabola, A. A model for the extrapolation of the characteristic curves of Pumps as Turbines from a datum Best Efficiency Point. *Energy Convers. Manag.* **2018**, *174*, 1–7. [CrossRef]
56. Shah, S.R.; Jain, S.V.; Patel, R.N.; Lakhera, V.J. CFD for centrifugal pumps: A review of the state-of-the-art. *Procedia Eng.* **2013**, *51*, 715–720. [CrossRef]

water

MDPI

Article

Investigation of the Internal Flow in a Francis Turbine for Comparing the Flow Noise of Different Operation Conditions

Tao Zhang [1], Gensheng He [2], Weilong Guang [3], Jiahao Lu [3], Xijie Song [4], Di Zhu [3] and Zhengwei Wang [4,*]

1 Technical Management Department, Y. R. Wanjiazhai Water Multi-Purpose Dam Project Co., Ltd., Taiyuan 030000, China; 15047748589@163.com
2 Power Station Management Bureau, Y. R. Wanjiazhai Water Multi-Purpose Dam Project Co., Ltd., Taiyuan 030000, China; hgs5567@126.com
3 College of Engineering, China Agricultural University, Beijing 100083, China; g_weilong@163.com (W.G.); lujiahao2209@cau.edu.cn (J.L.); zhu_di@cau.edu.cn (D.Z.)
4 State Key Laboratory of Hydroscience and Engineering & Department of Energy and Power Engineering, Tsinghua University, Beijing 100084, China; songxijie@mail.tsinghua.edu.cn
* Correspondence: wzw@mail.tsinghua.edu.cn

Abstract: Francis turbines are commonly used for water energy utilization in medium- to high-head sections. The high head may cause strong flow-induced noise problems and adverse effects during operation. In order to explore the causes and specific locations of flow-induced noise, this study evaluated the flow-induced noise of a Francis turbine under different loads. By using computational fluid dynamics simulation methods and sound power level evaluation methods, flow-induced noise analysis could be performed based on turbulent flow simulations. By comparing experimental and simulated values, three different load conditions were selected located in the allowed, restricted, and prohibited regions. The results indicated that the flow state of the operating points in the allowed region was good and the flow-induced noise was low, except near the guide vane. The swirling flow in restricted and prohibited regions was strong; in particular, the flow-induced noise in the draft tube was relatively high. Overall, the local flow-induced noise of the runner blade inlet edge was the strongest, and the relationship between the flow-induced noise of the draft tube and the rotating flow was the strongest. The flow-induced noise near the guide vane was the most related to the opening angle, and the larger the opening angle, the greater the noise was. This study can assist in the diagnosis of Francis turbine noise problems and related low-noise design in engineering.

Keywords: Francis turbine; flow-induced noise; swirling flow; computational fluid dynamics; sound power level

Citation: Zhang, T.; He, G.; Guang, W.; Lu, J.; Song, X.; Zhu, D.; Wang, Z. Investigation of the Internal Flow in a Francis Turbine for Comparing the Flow Noise of Different Operation Conditions. *Water* **2023**, *15*, 3461. https://doi.org/10.3390/w15193461

Academic Editor: Helena M. Ramos

Received: 4 September 2023
Revised: 21 September 2023
Accepted: 26 September 2023
Published: 30 September 2023

1. Introduction

In recent years, in order to alleviate the negative effects of excessive use of fossil energy, the development and utilization of clean energy have become particularly important around the world, and water utilization is the most important. In order to efficiently utilize the water energy of rivers and oceans, many scholars have created and improved mechanical devices related to the utilization of water energy, among which the Francis turbine is the most widely used. This is a typical impact turbine with a compact structure and high operation efficiency under rated working conditions, and it can operate stably under a wide head range [1].

With the continuous improvement of the single capacity of the Francis turbine, the runner diameter is becoming larger, and more and more attention is being paid to the flow stability of the internal flow field by more and more scholars [2–4]. In particular, when a Francis turbine is operating under conditions that deviate from the design conditions, the flow mechanism of the fluid in the flow passage becomes more complex, as seen in situations such as deflow at the inlet edge of the blade, low-frequency vortex zones in the draft tube, cavitation in various parts of the flow passage [5,6], etc., which have a direct impact on the vibration of the unit and workshop, cavitation erosion and cracks in the runner blade, and the rupture of the cone section of the draft tube. This can affect the safe and stable operation of hydropower stations and the efficient utilization of water energy [7].

In general, high- and low-frequency noise and abnormal noise generated by structural resonance are likely to occur when mechanical vibration, electromagnetic vibration and hydraulic excitation occur in a unit [8,9]. In accordance with the actual situation, relevant means can be adopted to avoid and eliminate vibration so as to solve the problem of noise in the power station. Wang et al. [10] studied the influence of the guide vane opening angle on the induced noise of fluid flow in a Francis turbine. Adjusting the guide vane opening angle and improving the elbow section of the draft tube could reduce the noise pressure generated by fluid flow. Pang et al. [11] found that hydraulic excitation would cause resonance and abnormal noise in the pressure steel pipe in the unit. By improving the flow passage environment, the diversion plate installed at the inlet of the volute could significantly improve the situation. Liu et al. [12] studied the dynamic and static interference in the bladeless region and the hydraulic excitation in the volute of a Francis turbine and confirmed that the dynamic and static interference in the bladeless region was related to the abnormal vibration and noise of the unit. The improved blading–sand combination can effectively achieve vibration and noise reduction in power stations. Kubo et al. [13] improved the operation stability of the unit and reduced noise by repairing and improving the guide vane and runner of the turbine. Luo et al. [14] found that, when the Francis turbine was running under load conditions, the pressure shock and noise in the turbine could be effectively inhibited by proper air flow into the unit.

In general, with the continuous progress of modern manufacturing technology, mechanical noise can be greatly reduced, but hydraulic vibration noise still exists in Francis turbines [15]. As noise is easy to detect and monitor during the operation and maintenance of power stations, the noise characteristics generated by the internal flow field of Francis turbines under different working conditions can be compared and analyzed. By analyzing the causes and characteristics of different noises, the unit structure and workshop can be repaired and improved, therefore reducing the noise itself and the sound source and preventing the unit from being in a state of unsafe operation for a long time. To analyze and predict a unit's unstable operation, Wang et al. [16] studied the operation of the Francis turbine under off-design conditions and established an evaluation method for pressure pulsation in the diversion tube that can be used to evaluate the pressure pulsation in the unit and the noise generated by the pulsation. Favrel et al. [17] carried out an excitation resonance test on a model Francis turbine in a seal hydraulic system to improve the prediction of an underwater acoustic model.

In this study, CFD software was used to model a Francis turbine, analyze its internal flow conditions and noise characteristics under different working conditions, and deeply explore the mechanism of the easily appearing noise of the Francis turbine and the commonality of related noise so as to provide a relevant technical reference for subsequent research on the noise of Francis turbines [18–21].

2. Research Object

The object of this study was a Francis turbine for a medium head section, and its relevant parameters are shown in Table 1. This Francis turbine had five key flow-passing components: the spiral case, stay vane, guide vane, runner, and draft tube. The runner was a 13-blade Francis runner. The stay vane had 24 blades and the guide vane also had 24 blades. The fluid domain was modeled as shown in Figure 1 for the computational fluid dynamics (CFD) simulation in the following sections. A set of x-y-z coordinates were defined, where z was the runner axis direction, x was the spiral case inflow direction, and y was perpendicular to the x–z plane.

Table 1. Parameters of the Francis turbine.

Parameter	Value
Rotor diameter D_{rn}	5.80 m
Rated output power P_r	183.7 MW
Rated flow rate Q_r	301 m^3/s
Rated rotation speed n_r	100 r/min
Rated head H_r	68.0 m
Maximum output power P_{max}	204.1 MW
Turbine installation height H_{ins}	895.00 m

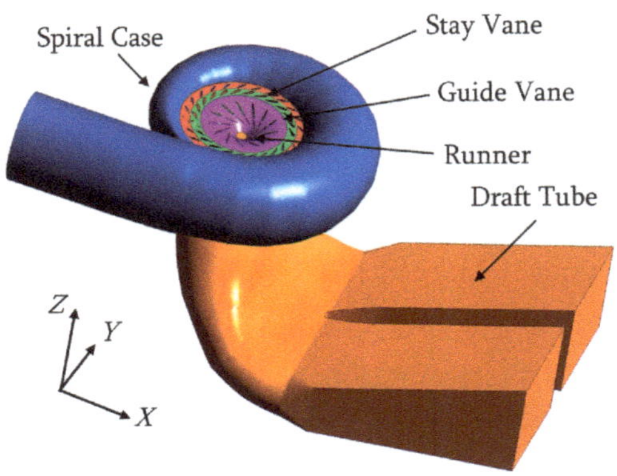

Figure 1. The main components of the Francis turbine.

3. Mathematical Methods

3.1. Basic Equations

In this case, time-averaged and Reynolds-averaged Navier–Stokes equations were used to decompose the quantities into a time-averaged component (for example, $\overline{u}$) and fluctuating component (for example, u') [22]. Thus, the continuity equation and momentum equation were:

$$\frac{\partial \overline{u_i}}{\partial x_i} = 0 \tag{1}$$

$$\rho \frac{\partial \overline{u_i}}{\partial t} + \rho \overline{u_j} \frac{\partial \overline{u_i}}{\partial x_j} = \frac{\partial}{\partial x_j}\left(-\overline{p}\delta_{ij} + 2\mu \overline{S_{ij}} - \rho \overline{u_i' u_j'}\right) \tag{2}$$

where u is the velocity, t is the time, ρ is the density, x is the coordinate component, δ_{ij} is the Kroneker delta, and μ is the dynamic viscosity. The term $\rho u_i' u_j'$ is called the Reynolds stress. S_{ij} is the mean rate of the strain tensor:

$$\overline{S_{ij}} = \frac{1}{2}\left(\frac{\partial \overline{u_i}}{\partial x_j} + \frac{\partial \overline{u_j}}{\partial x_i}\right) \tag{3}$$

Here, the total energy equation is:

$$\frac{\partial}{\partial t}(\rho h_{tot}) - \frac{\partial p}{\partial t} + \frac{\partial}{\partial x_j}(\rho u_j h_{tot}) = \frac{\partial}{\partial x_j}\left(\lambda_t \frac{\partial T}{\partial x_j} - \overline{u_j h_{sta}}\right) + \frac{\partial}{\partial x_j}\left[u_j\left(2\mu \overline{S_{ij}} - \rho \overline{u_i' u_j'}\right)\right] \tag{4}$$

where T is temperature, h_{sta} is the static enthalpy, and h_{tot} is the total enthalpy:

$$h_{tot} = h_{sta} + \frac{1}{2}u^2 \tag{5}$$

λ_t is the thermal conductivity.

3.2. Turbulence Model

As the time-averaged equation is not closed, the relationship between the Reynolds stress and eddy viscosity μ_t can be built as follows [23]:

$$-\rho\overline{u_i'u_j'} = 2\mu_t\overline{S_{ij}} - \frac{2}{3}k\delta ij \tag{6}$$

where k is the turbulence kinetic energy. In this study, the SST model [24] was used, which can be written as:

$$\frac{\partial(\rho k)}{\partial t} + \frac{\partial(\rho u_i k)}{\partial x_i} = P_k - \frac{\rho k^{3/2}}{l_{k-w}} + \frac{\partial}{\partial x_i}\left[(\mu + \sigma_k\mu_t)\frac{\partial k}{\partial x_i}\right] \tag{7}$$

$$\frac{\partial(\rho\omega)}{\partial t} + \frac{\partial(\rho u_i\omega)}{\partial x_i} = C_\omega P_\omega - \beta\rho\omega^2 + \frac{\partial}{\partial x_i}\left[(\mu_l + \sigma_\omega\mu_t)\frac{\partial\omega}{\partial x_i}\right] + 2(1 - F_1)\frac{\rho\sigma_{\omega 2}}{\omega}\frac{\partial k}{\partial x_i}\frac{\partial\omega}{\partial x_i} \tag{8}$$

where $l_{k\text{-}\omega}$ is the turbulence scale, P_k and P_ω are the production term, F_1 is the coefficient of the production term, σ_k is the blending function, and σ_ω and β_k are model constants.

3.3. Flow-Induced Noise Analysis

The Lighthill acoustic analogy method was applied based on the flow simulation [25]. It can be used to determine the near-field noise according to the sound power level:

$$W_A = \alpha_\varepsilon\rho\varepsilon M_t^5 \tag{9}$$

where α_ε is a constant equal to 0.1, ε is the eddy dissipation rate, and M_t is the specific turbulence kinetic energy:

$$M_t = \frac{\sqrt{2k}}{V_c} \tag{10}$$

where V_c is the speed of sound, which is 340 m/s. The flow-induced sound power level L_{sp} is calculated as follows:

$$L_{sp} = 10\log_{10}\left(\frac{W_A}{W_{ref}}\right) \tag{11}$$

where W_{ref} is the reference sound power of 1×10^{-12} W/m^3.

4. Setup of CFD

In this study, the commercial software ANSYS CFX was used for CFD simulation. When simulating, the fluid medium was set as water at 20 °C. The pressure reference value was 0 Pa, which means that the simulated pressures were all absolute values. On the basis of the spiral case, stay vane, guide vane, runner, and draft tube, the computing domain connected and transmitted data to each domain through interfaces. The inlet of the spiral case was set as a total pressure-type inlet boundary, with the direction of the inflow perpendicular to the boundary and the pressure in the zero-gradient form. The outlet of the draft tube was set as a static pressure-type outlet boundary, and based on the given pressure value, the velocity was in zero-gradient form. The grid in the computational domain adopted unstructured tetrahedral elements as the main body, and the blade surface was encrypted with prism-type grids to meet the requirements of the wall function. Figure 2a shows the grid convergence check based on the convergence index (GCI) [26]. Three grid schemes following an increasing rate of mesh spacing of 1.35 were checked based on the evaluation of efficiency η. The GCI value of the coarse medium grid was 3.91% and the GCI value of the fine medium grid was 0.03%. After the checking of each component grid, the grid element number was determined, as shown in Table 2 and Figure 2a. The total number of elements was 2.75 million. The schematic map of the grid is shown in Figure 2b.

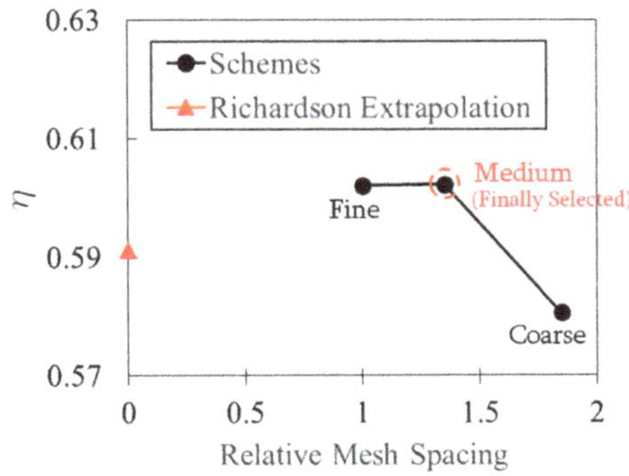

(**a**) Grid convergence check details

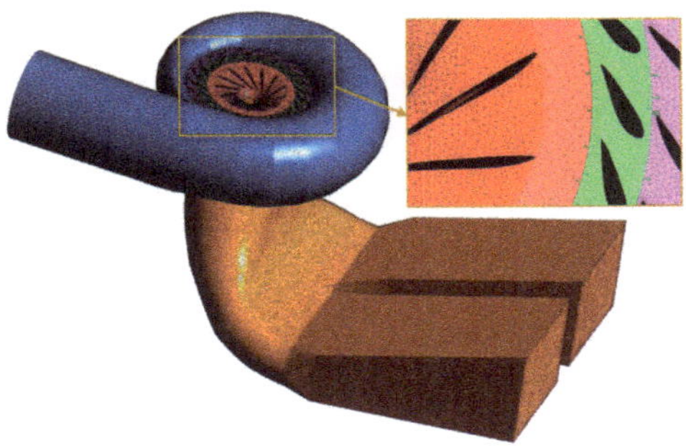

(**b**) The finally selected grid

Figure 2. Grid for Francis turbine's fluid domain.

Table 2. Grid number details.

Component	Grid Element Number
Spiral case	209,260
Stay vane	558,230
Guide vane	623,824
Runner	897,680
Draft tube	464,692
Total	2,753,686

When running CFD, we first performed a 600-timestep steady simulation and monitored the residual between the continuity equation and the momentum equation. If it was less than 0.0001, convergence was considered to take place. Then, based on this, we carried out transient calculations for three runner revolutions, with 360 timesteps iterated for each cycle and 10 iterations for each step. The residual between the continuity equation and the momentum equation was still monitored, and convergence was considered to be less than 0.0001.

5. Comparison of Experimental and Numerical Results

Figure 3 shows the operation characteristic of the Francis turbine tested on site. The best efficiency region (BER) had an efficiency of 95%. The operation region was divided into three subregions: the allowed region, restricted region, and prohibited region. The division was based on the operation stability. The BER was in the allowed region. We found that the flow-induced noise was high in the restricted region and prohibited region. Therefore, three points were compared, as indicated in Figure 2. Point one was in the allowed region, and the guide vane opening angle was 24 degrees. Point two was in the restricted region, and the guide vane opening angle was 20 degrees. Point three was in the prohibited region, and the guide vane opening angle was 12 degrees. Table 3 shows a comparison of experimental and numerical results. The unit speed n_{11} and unit flow rate Q_{11} were defined as:

$$n_{11} = \frac{nD_{rn}}{\sqrt{H}} \tag{12}$$

$$Q_{11} = \frac{Q}{D_{rn}^2\sqrt{H}} \tag{13}$$

where n is the rotational speed, H is the head, and Q is the flow rate.

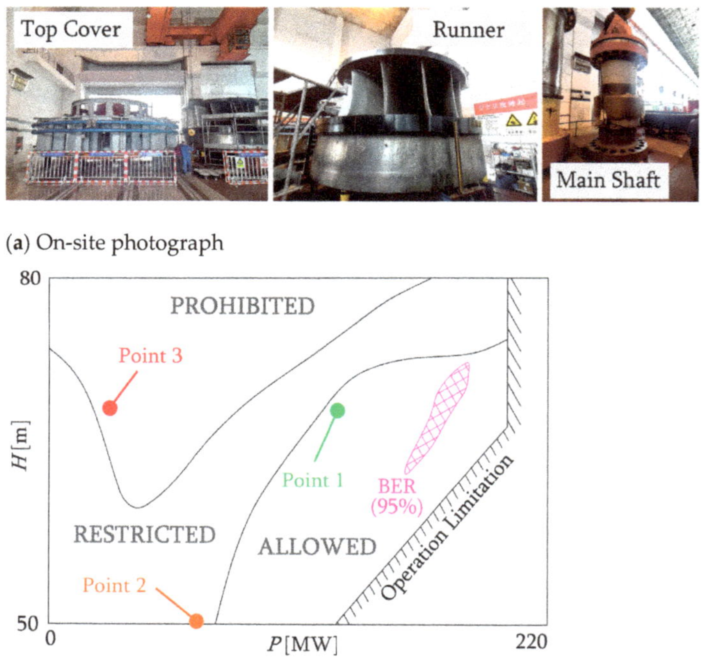

Figure 3. The operation characteristics of the Francis turbine.

Table 3. Comparison of experimental and numerical results.

Point	CFD Efficiency	Experimental Efficiency	n_{11}	Q_{11}
Point One	91.05%	90.51%	77.85	0.7864
Point Two	71.72%	70.32%	78.55	0.3678
Point Three	55.25%	52.88%	91.21	0.3495

6. Results of CFD

6.1. Streamlines

Figure 4 shows the distribution of streamlines in the Francis turbine's fluid domain representing the turbulence of the flow regime, with colors used to represent the flow velocity. From Figure 4a, it can be seen that, at point one, the flow was smoother due to the operating conditions being near the BER and in the allowed operation region. The flow velocity in the spiral case was also medium-high, with an average of around 10 m/s. The flow regime in the draft tube was very smooth. From Figure 4b, it can be seen that, at point two, due to the condition being in the restricted region, the flow pattern became bad. The flow velocity in the volute decreased to approximately 5~6 m/s. The flow velocity between the guide vane and the runner was extremely high, reaching up to 25~28 m/s. There was a significant twisting flow in the draft tube, which was related to the formation of vortex rope. From Figure 4c, it can be seen that, at point three, the flow pattern became worse as the operating conditions entered the prohibited region. The flow velocity in the spiral case was also relatively low, only about 5~6 m/s. The flow velocity in the area between the guide vane and runner slightly decreased to around 16~18 m/s. There was still a significant swirling flow in the draft tube, which was related to the formation of vortex rope. In general, the flow pattern was strongly related to the operation region. The relationship between flow-induced noise and condition should be analyzed in detail.

6.2. Flow-Induced Noise

6.2.1. Reference Planes for Plotting

To analyze the specific situation of the flow-induced noise, we created a series of reference planes, as shown in Figure 5. Among them, plane one was the middle section of the spiral case and vane and was an x–y plane. Planes two and three were also x–y planes located in the straight part of the draft tube. Planes four and five were y–z planes in the diffusion part of the draft tube, which was downstream of the elbow part.

6.2.2. Contours of Flow-Induced Noise

Figure 6a shows the distribution of the sound power level L_{sp} at the point-one condition. On plane one of the point-one condition, it can be seen that there was an area with a very high sound power level near the inlet of the spiral case. This area was concentrated in the middle of the flow channel and closely related to the compression between fluids. As the cross-section of the spiral case continued to decrease, the sound power level gradually decreased. When the fluid entered the stay vane and guide vane, the sound power level increased. As the fluid entered the runner blade channel and began to interact with the blade, the sound power level reached its maximum value at the inlet edge of the blade, approximately 78.0 dB. At the position near the cone (plane two) of the draft tube, the sound power level gradually rose from the center to the wall up to 36.3 dB. In the middle of the straight section of the draft tube (plane three), the sound power level also showed a gradual rise from the center to the wall up to 25.0 dB. After the fluid flowed through the elbow section, near its position (plane four), the sound power level was higher at the upper wall and lower at the lower wall, reaching a maximum of 21.6 dB. At the diffusion section (plane five) downstream of the draft tube, the distribution of the sound power level showed high levels on the left and low levels on the right without obvious regularity, and the maximum value was 12.7 dB.

Figure 6b shows the distribution of the sound power level L_{sp} at the point-two condition. On plane one of the point-two condition, there was still a high-sound-power-level area near the inlet of the spiral case, with a larger range and high intensity. The sound power level became very low near the stay vane and guide vane, which was related to the extremely small opening angle of the guide vane, which greatly hindered the flow. In the runner, especially at the blade inlet edge, the sound power level increased. The maximum sound power level on plane one was approximately 89.9 dB, which was higher than the point-one operating condition. At the position of the draft tube near the cone (plane two), the sound power level from the center to the wall showed a trend of rising first and then falling, reaching a value of up to 63.1 dB, far higher than the point-one working condition. In the middle of the straight section of the draft tube (plane three), the sound power level also showed a trend of rising first and then falling from the center to the wall, reaching a value of up to 80.1 dB, far higher than the point-one working condition. Near the elbow (plane four), the sound power level was still higher at the upper wall and lower at the lower wall, reaching a maximum of 73.9 dB, higher than point one. At the diffusion section (plane five) downstream of the draft tube, the distribution of the sound power level showed low levels on the left and high levels on the right with no obvious regularity. The maximum value was 55.0 dB, higher than point one.

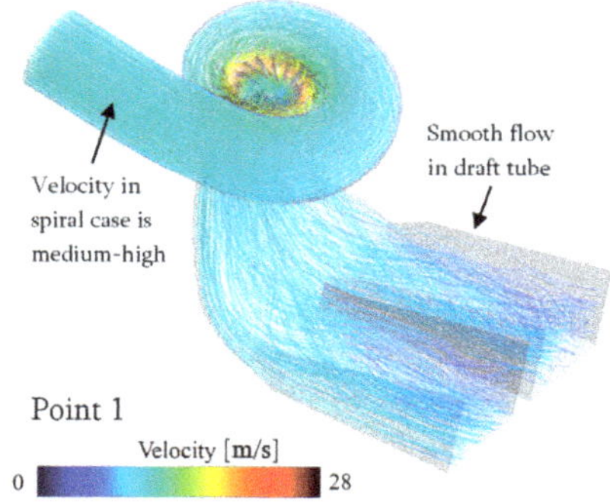

(**a**) Streamlines in turbine at point one

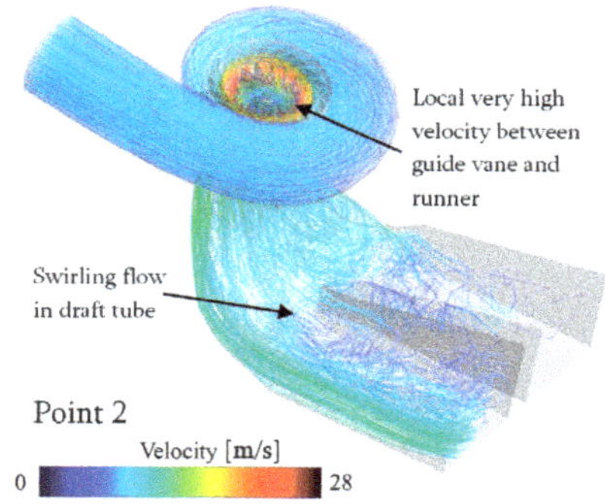

(**b**) Streamlines in turbine at point two

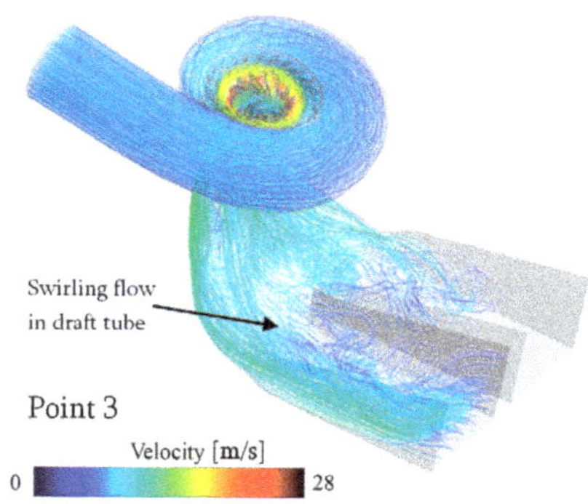

(**c**) Streamlines in turbine at point three

Figure 4. The streamlines in the Francis turbine fluid domain.

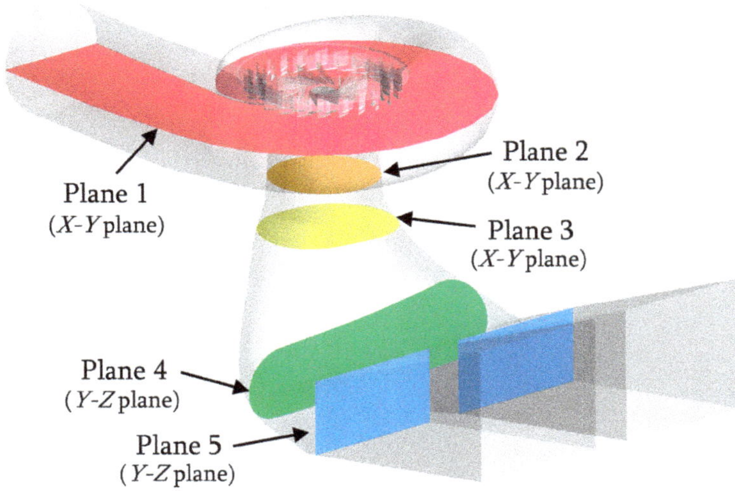

Figure 5. The reference planes in the Francis turbine.

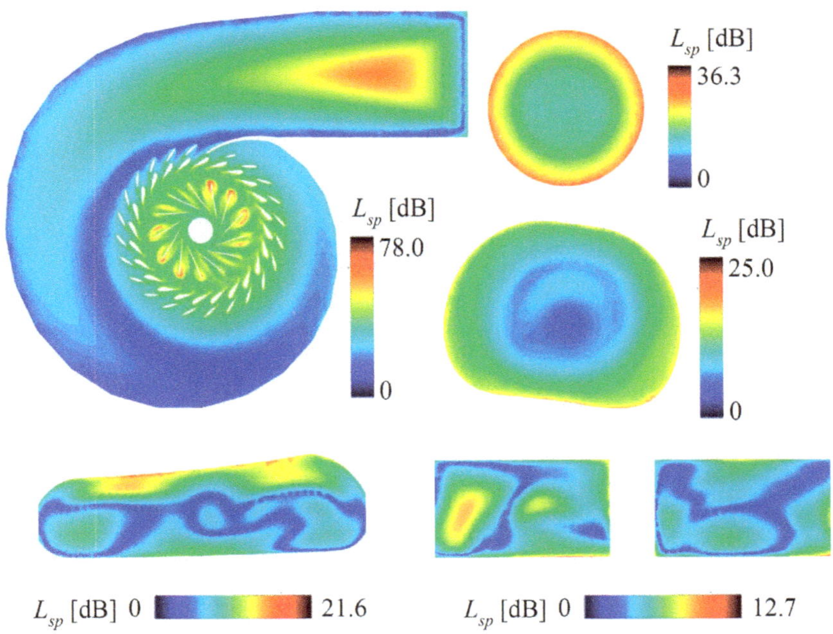

(**a**) Point one

Figure 6. *Cont.*

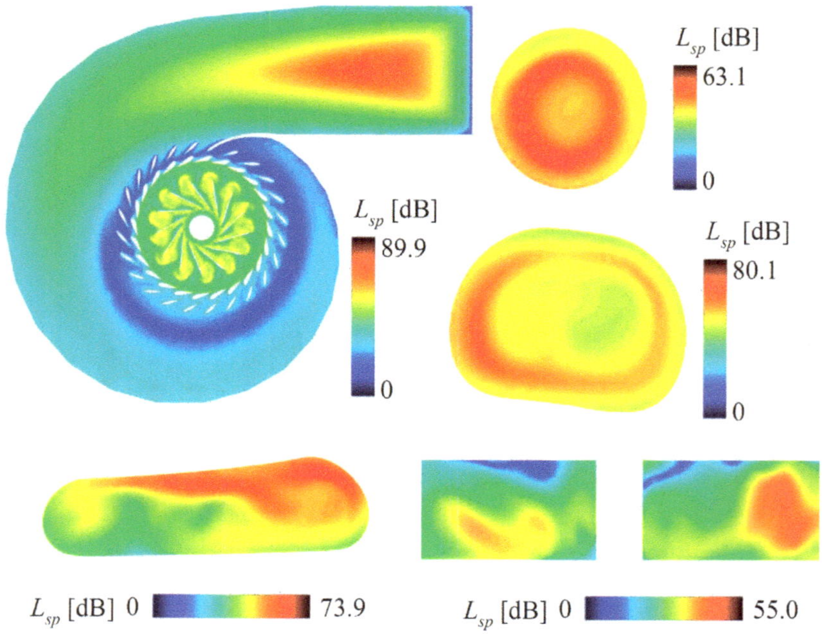

(**b**) Point two

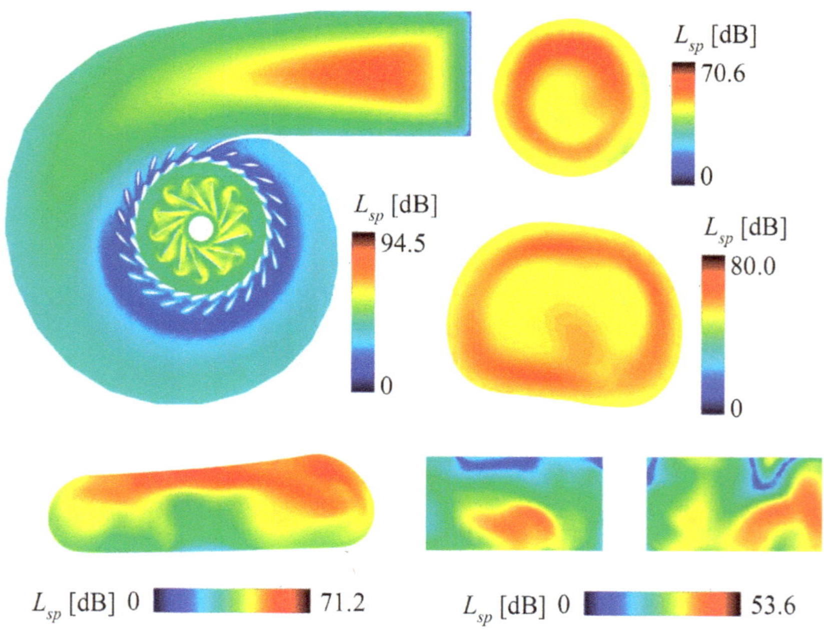

(**c**) Point three

Figure 6. The contours of flow-induced noise L_{sp} in the Francis turbine.

Figure 6c shows the distribution of the sound power level L_{sp} at the point-three condition. On plane one of the point-three condition, there was still a high-sound-power-level area near the inlet of the spiral case, and the range and intensity were larger than those of point two. The sound power level became very low near the stay vane and guide vane, similarly to point two. This was related to the flow retardation caused by the extremely small guide vane opening angle. In the runner, especially at the blade inlet edge, the sound power level increased. The maximum sound power level on plane one was approximately 94.5 dB, which was higher than both the point-one and point-two operating conditions. At the position of the draft tube near the cone (plane two), the sound power level rose first and then decreased from the center to the wall, reaching 70.6 dB at most, which was similar to point two. In the middle of the straight section of the draft tube (plane three), the sound power level also showed the trend of rising first and then falling from the center to the wall, reaching a maximum of 80.0 dB, which was similar to point two. Near the elbow (plane four), the sound power level was still higher at the upper wall and lower at the lower wall, reaching a maximum of 71.2 dB, similarly to point two. At the diffusion section (plane five) downstream of the draft tube, the distribution of the sound power level showed no obvious trend, and the maximum value was 53.6 dB, which was similar to point two.

6.2.3. The Variation in Flow-Induced Noise with Flow Direction

In order to analyze the variation in the maximum flow-induced noise with the flow direction under three working conditions, a total of seven positions labeled A~G were defined, as shown in Table 4. A was at the spiral case inlet. B was between the stay vane and guide vane. C was in the runner blade channels. D was near the runner cone in the draft tube. E was in the middle of the draft tube's straight section. F was near the draft tube elbow section. G was in the draft tube diffusion section.

Table 4. Positions for comparing the maximum flow-induced noise.

Symbol of Position	Position Description
A	At spiral case inlet
B	Between stay vane and guide vane
C	In runner blade channels
D	Near runner cone in draft tube
E	Middle of draft tube straight section
F	Near draft tube elbow section
G	In draft tube diffusion section

Figure 7 shows a comparison of the maximum sound power level at different positions. At position A, point one had the lowest sound power level and point three had the highest sound power level. Overall, the differences between the three operating conditions were not too significant. At position B, point one had the highest sound power level, while point two and point three had lower sound power levels. At position C, point one had the lowest sound power level and point three had the highest sound power level. The flow-induced noise generated by the interaction between the high-speed rotating blades and fluid was the strongest within the entire hydraulic turbine. At positions D, E, F, and G, the situation was quite similar. Both instances of point one had lower sound power levels, while point two and point three had higher sound power levels. In the draft tube, the gap between point one and point two/three was obvious. Along the direction of flow, flow-induced noise tended to decrease. Overall, the sound power level in the vanes was related to the opening angle of the guide vane, and flow-induced noise was lower when the opening of the guide vane was small. The sound power level in the runner was the highest, mainly at the inlet edge of the blade, and not closely related to the operating conditions. The sound power level in the draft tube had an obvious relationship with the condition: the lower the flow rate and the more serious the flow diffusion, the stronger the flow-induced noise in the draft tube.

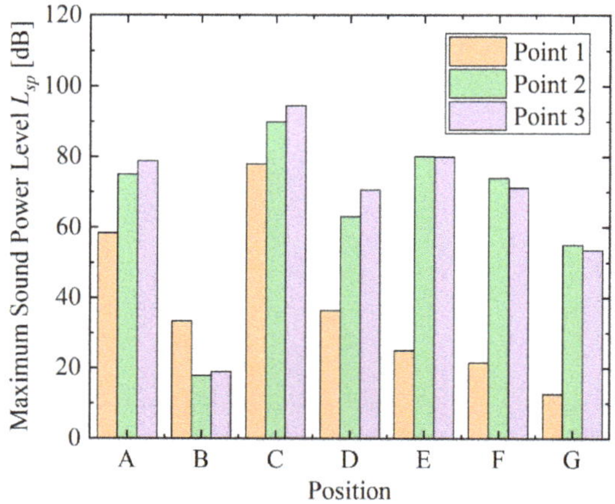

Figure 7. Comparison of the maximum sound power levels at different positions.

7. Discussion

From the results of this study, it can be seen that the Francis turbine produces four main areas of flow-induced noise under different loads (Figure 8). Firstly, this occurs at the point of flow merging; in this case, mainly in the import of the spiral case. Intensification of the reduction in the flow channel area or the channel curvature can cause different flows to squeeze each other, resulting in strong flow-induced noise. Strong flow-induced noise then appears at the leading edge of the vane or runner blade. This is due to the flow and wall impact. In addition, circulating flow is one of the possible causes of flow-induced noise, as significant flow-induced noise was observed between the vane and runner. Finally, the strongly rotating vortex belt of the draft tube may interact with the wall or fluid under low loads, resulting in strong flow-induced noise.

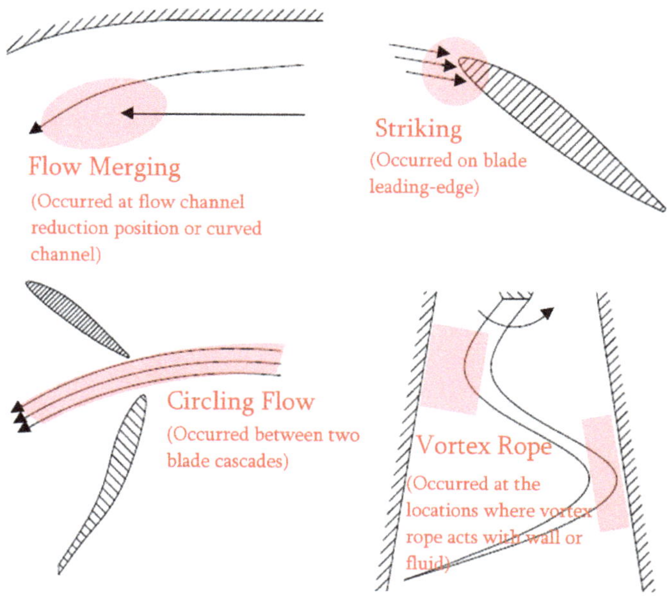

Figure 8. The schematic map of the locations where high flow-induced noise occurs.

8. Conclusions

In general, this study found some significant patterns for flow-induced noise in Francis turbines. The following three conclusions can be drawn:

(a) The main locations where flow-induced noise occurred included spiral case inlets, runner blade inlet edges, draft tube walls, and the middle of the draft tube. The flow-induced noise was closely related to the flow convergence and extrusion, fluid–blade interaction, draft tube rotating flow, etc. The noise caused by the interaction between the fluid and high-speed rotating runner was often the strongest;

(b) Flow-induced noise was related to the operation condition. When the turbine was operating in the allowed region (point one), the noise was lower. When the turbine was operating in the restricted region (point two) and the prohibited region (point three), the noise was higher. An exception occurred in the region near the vanes, as it was mainly affected by the opening of the guide vanes;

(c) The difference in sound power level was related to the flow regime. When swirling flow was strong, it could cause high noise. When flow was smooth, the induced noise was low. In addition, flow-induced noise was not significant when flow was obstructed. In contrast, when the flow passed through the guide vanes at high speed, the flow-induced noise was strong.

How to reduce flow-induced noise through design methods is crucial for hydraulic turbines. This study can provide a reference basis for this issue and scientific references for engineering cases.

Author Contributions: Conceptualization, T.Z. and X.S.; methodology, Z.W.; software, D.Z.; validation, D.Z., W.G., and J.L.; formal analysis, J.L.; investigation, W.G. and G.H.; resources, Z.W.; data curation, T.Z.; writing—original draft preparation, G.H., T.Z., and J.L.; writing—review and editing, X.S. and W.G.; supervision, Z.W.; project administration, Z.W.; funding acquisition, Z.W. All authors have read and agreed to the published version of the manuscript.

Funding: This research was funded by the Open Research Fund Program of the State Key Laboratory of Hydroscience and Engineering (no. sklhse-2022-E-01).

Data Availability Statement: Data sharing is not applicable to this article.

Acknowledgments: We acknowledge the support provided for the research by the Open Research Fund Program of the State Key Laboratory of Hydroscience and Engineering.

Conflicts of Interest: The authors declare no conflict of interest.

References

1. Zheng, Y.; Chen, D.X. *Hydraulic Turbine*, 3rd ed.; China Water Resources and Hydropower Press: Beijing, China, 2008; pp. 219–247.
2. Huang, J.F.; Wang, C.Y.; Wen, J.; Rao, B.Y.; Long, L.Y. Research on Vortex of a Francis Hydro-turbine Based on CFD Technology. *J. Yunnan Agric. Univ.* **2010**, *25*, 147–151.
3. Wan, Y.; Tan, Z.G. Vibration characteristics test-based analysis and evaluation on stability of large francis turbine-generator unit. *Water Conserv. Hydropower Technol.* **2014**, *45*, 15–19.
4. Chen, T.; Zhang, Y.N.; Li, S.C. Instability of large-scale prototype Francis turbines of Three Gorges power station at part load. *Proc. Inst. Mech. Eng.* **2016**, *230*, 619–632. [CrossRef]
5. Liu, D.M.; Liu, X.B.; Zhao, Y.Z. Experimental Investigation of Inter-Blade Vortices in a Model Francis Turbine. *Chin. J. Mech. Eng.* **2017**, *30*, 854–865. [CrossRef]
6. Liu, D.M.; Liu, X.B.; Zhang, H.K. Introduction the Original Reason of Channel Vortex Based on the Cavitation Model. In Proceedings of the 35th World Congress of the International-Association-for-Hydro-Environment-Engineering-and-Research (IAHR), Chengdu, China, 8–13 September 2013.
7. Kamal, M.; Saini, G.; Abbas, A.; Prasad, V. Prediction and analysis of the cavitating performance of a Francis turbine under different loads. *Energy Sources Part A-Recovery Util. Environ. Eff.* **2021**, 1–25. [CrossRef]
8. Guo, T.; Xu, L.H.; Luo, Z.M. Influence of Upstream Disturbance on Vortex Rope Evolution and Pressure Fluctuation in Draft Tube. *Trans. Chin. Soc. Agric. Mach.* **2022**, *53*, 192–201.
9. Frunzăverde, D.; Muntean, S.; Mărginean, G.; Campian, V.; Marşavina, L.; Terzi, R.; Şerban, V. Failure analysis of a Francis turbine runner. In Proceedings of the 25th IAHR Symposium on Hydraulic Machinery and Systems, Timisoara, Romania, 20–24 September 2010.

10. Wang, H.L.; Zheng, Y.; Sun, A.R.; Zhang, F.L.; Gao, C.H.; Zhou, Y. The influence of guide vane opening on pressure pulsation characteristics and flow-induced noise in Francis turbine. *South-North Water Transf. Water Sci. Technol.* **2019**, *16*, 186–192.
11. Pang, L.J.; Zhong, S.; Bu, L.F.; Hu, J.W. Analysis of abnormal noise of francis turbine by scene test. *J. Vib. Shock* **2012**, *31*, 39–42.
12. Liu, P.; Chen, X.L.; Wang, Q.; Li, D. Analysis of rotor-stator interaction and vibration in high-head Francis turbines. *J. Hydroelectr. Eng.* **2016**, *35*, 91–98.
13. Kubo, K.; Nakashima, T.; Shima, R.; Enomoto, Y. CFD-based rehabilitation for the high specific speed Francis turbine. In Proceedings of the 29th IAHR Symposium on Hydraulic Machinery and Systems (IAHR), Kyoto, Japan, 16–21 September 2018.
14. Luo, X.; Yu, A.; Yu, W.; Wang, L.; Xu, H. Pressure oscillation suppression by air admission in a Francis turbine draft tube. In Proceedings of the 29th IAHR Symposium on Hydraulic Machinery and Systems (IAHR), Kyoto, Japan, 16–21 September 2018.
15. Wang, H.Z.; Zhou, D.Q.; Zhang, L.G. Numerical simulation of flow noise of low specific speed Francis turbine. *Hydropower Energy Sci.* **2014**, *32*, 153–156.
16. Wang, W.; Chen, Q.; Yan, D.; Geng, D. A novel comprehensive evaluation method of the draft tube pressure pulsation of Francis turbine based on EEMD and information entropy. *Mech. Syst. Signal Process.* **2019**, *116*, 772–786. [CrossRef]
17. Favrel, A.; Landry, C.; Müller, A.; Avellan, F. Experimental identification and study of hydraulic resonance test rig with Francis turbine operating at partial load. In *IOP Conference Series: Earth and Environmental Science*; IOP Publishing: Bristol, UK, 2012.
18. Chen, D.Y.; Zhao, W.L.; Sprott, J.C.; Ma, X.Y. Application of Takagi-Sugeno fuzzy model to a class of chaotic synchronization and anti-synchronization. *Nonlinear Dyn.* **2013**, *73*, 1495–1505. [CrossRef]
19. Xu, B.B.; Chen, D.Y.; Zhang, H.; Zhou, R. Dynamic analysis and modeling of a novel fractional-order hydro-turbine-generator unit. *Nonlinear Dyn.* **2015**, *81*, 1263–1274. [CrossRef]
20. Xu, B.B.; Chen, D.Y.; Zhang, H.; Wang, F.F. Modeling and stability analysis of a fractional-order francis hydro-turbine governing system. *Chaos Solitons Fractals* **2015**, *75*, 50–61. [CrossRef]
21. Xu, B.B.; Wang, F.F.; Chen, D.y.; Zhang, H. Hamiltonian modeling of multi-hydro-turbine governing systems with sharing common penstock and nonlinear dynamic analyses under shock load. *Energy Convers. Manag.* **2016**, *108*, 478–487. [CrossRef]
22. Chen, M. *Fundamentals of Viscous Fluid Dynamics*; Higher Education Press: Beijing, China, 1993.
23. Terentiev, L. *The Turbulence Closure Model Based on Linear Anisotropy Invariant Analysis*; VDM Verlag: Saarbrucken, Germany, 2008.
24. Menter, F.R. Two-Equation Eddy-Viscosity Turbulence Models for Engineering Applications. *AIAA J.* **1994**, *32*, 1598–1605. [CrossRef]
25. Proudman, I. The generation of noise by isotropic turbulence. *Proc. R. Soc. A* **1952**, *214*, 119–132.
26. Celik, I.B.; Ghia, U.; Roache, P.J.; Freitas, C.J.; Coleman, H.; Raad, P.E. Procedure for Estimation and Reporting of Uncertainty due to Discretization in CFD Applications. *J. Fluids Eng. Trans. ASME* **2008**, *130*, 078001.

water

Article

Comparison of Pressure Pulsation Characteristics of Francis Turbine with Different Draft Tube Arrangement Direction

Tao Zhang [1], Zilong Hu [2], Xinjun Liu [3], Jiahao Lu [2], Xijie Song [4], Di Zhu [5] and Zhengwei Wang [4,*]

1 Technical Management Department, Y. R. Wanjiazhai Water Multi-Purpose Dam Project Co., Ltd., Taiyuan 030000, China; 15047748589@163.com
2 College of Water Resources and Civil Engineering, China Agricultural University, Beijing 100083, China; huzilong1999@cau.edu.cn (Z.H.); lujiahao2209@cau.edu.cn (J.L.)
3 Power Station Management Bureau, Y. R. Wanjiazhai Water Multi-Purpose Dam Project Co., Ltd., Taiyuan 030000, China; hwgslxj_wjz@163.com
4 State Key Laboratory of Hydroscience and Engineering, Department of Energy and Power Engineering, Tsinghua University, Beijing 100084, China; songxijie@mail.tsinghua.edu.cn
5 College of Engineering, China Agricultural University, Beijing 100083, China; zhu_di@cau.edu.cn
* Correspondence: wzw@mail.tsinghua.edu.cn

Abstract: Hydroelectric power generation is an important clean energy source, and the stability of water turbine operation determines the quality of hydro energy utilization. For hydro turbines, the layout direction of the draft tube is often only considered from a structural perspective, ignoring the hydrodynamic characteristics. This study adopts the computational fluid dynamics method and verifies the effectiveness of numerical simulation with experimental results, analyzing the influence of asymmetric draft tube layout direction on pressure pulsation of mixed flow turbine. The results show that under different working conditions, there is basically no difference in efficiency corresponding to different inclined directions of the draft tube, and the relative difference in performance values is less than 1%. From the perspective of internal flow, the working condition has a greater impact on the flow, and the draft tube tilt has a smaller impact. Under strong swirling flow conditions in the draft tube, the variational mode decomposition of pressure fluctuation is carried out. Research has found 7 characteristic frequency bands including 140 Hz, 80–90 Hz, 40–46 Hz, 5.5–6.5 Hz, 2.5–3 Hz, 1.67 Hz, and <1 Hz. The frequency of the dominant mode corresponding to the left tilt is higher, while the amplitude of the runner frequency is slightly lower. In general, left tilt is not only more suitable for unit layout but also has a better effect on draft tube pressure fluctuation, which is worthy of application in engineering.

Keywords: Francis turbine; pressure pulsation; draft tube design; hydrodynamic analysis; variational mode decomposition

Citation: Zhang, T.; Hu, Z.; Liu, X.; Lu, J.; Song, X.; Zhu, D.; Wang, Z. Comparison of Pressure Pulsation Characteristics of Francis Turbine with Different Draft Tube Arrangement Direction. *Water* **2023**, *15*, 4028. https://doi.org/10.3390/w15224028

Academic Editor: Vittorio Di Federico

Received: 1 October 2023
Revised: 3 November 2023
Accepted: 7 November 2023
Published: 20 November 2023

1. Introduction

The excessive use of fossil energy has led to the emission of a large number of greenhouse gases, the rise of global temperature, the melting of glaciers, and the rise of sea level. In order to alleviate this dilemma and reduce the need for fossil energy, we have been looking for a new type of sustainable, renewable, and efficient energy. Studies have shown that with the development of renewable energy technology, the use of fossil energy as the main energy source has seen a huge decline in the past decade, and new renewable energy, such as hydro energy [1], wind energy [2], marine energy [3], and solar energy [4], has gradually occupied a dominant position in the energy structure. The carbon reduction of conventional energy-consuming machinery has also become a key research object [5–7]. Among all the energies, hydropower is a type of clean energy and occupies the largest proportion of renewable energy [8]. Francis turbine plays a vital role as an important component of hydropower generation [9]. It converts the gravitational potential energy of the water flow into fluid kinetic energy, which drives the rotor to generate electricity. In recent years, due to the development of other energy technologies, such as wind energy and photovoltaics, the randomness of these energy sources makes the turbine operate under off-design conditions to meet the requirements of a stable grid [10]. However, due to the inherent performance of the Francis turbine, when the runner runs under off-design conditions, the increase or decrease in the runner's flow will exert a rotational component on the flow in the runner, resulting in the formation of flow gaps in draft tube, and flow gaps cause reflux in draft tube [11,12]. The reflux further develops into the vortex and the subsequent pressure pulsation. The generation of pressure pulsation will reduce the operation efficiency of the unit, causing vibration and noise in the operation of the unit [13,14]. Excessive pressure pulsations will reduce the operation life of the unit and even induce serious engineering accidents [15,16]. So for the Francis turbine draft tube, a pressure pulsation study is very necessary.

The draft tube plays a very important role as one of the components of the Francis turbine. The draft tube is located below the runner, and the fluid in the draft tube undergoes energy conversion from fluid kinetic energy to static pressure. Approximately 80% of energy conversion occurs in the draft tube cone, which is approximately 10% of the total length of the draft tube cone. The flow field of the draft tube can vary with different operating conditions [17,18]. For the Francis turbine, we can change the unit operating flow by controlling the opening of the movable guide vane. Due to changes in flow conditions, a vortex rope of the draft tube is formed. A large number of researchers have conducted a lot of studies on the formation of vortex ropes of the draft tube. Liu et al. [19] conducted a comprehensive and in-depth analysis of the unsteady turbulence and pressure pulsation of the Three Gorges hydroelectric unit, demonstrating the low-frequency characteristics of vortex rope pulsation. Liao et al. [20], based on unsteady calculation, determined typical locations where low-frequency pulsations are concentrated, such as the elbow section of the draft tube. Zhang et al. [21] revealed the spatial asymmetry of pressure pulsation distribution in draft tube vortex belts through numerical simulation. The numerical study of Pasche et al. [22] shows that the vortex rope and the draft tube wall are coupled, resulting in the pressure pulsation being transformed into the violent vibration of the unit structure. Existing research has shown that the phenomenon of pressure pulsation is extremely complex and may exhibit significant differences when slightly affected. The issue of the layout direction of the draft tube of the hydraulic turbine was originally a slight geometric difference, but it may cause a change in pressure pulsation. However, there is still a lack of research and clear conclusions.

By analyzing the draft tube pressure pulsation signal, we can find out the cause of draft tube pressure pulsation. A large number of scholars have used fast Fourier transform (FFT) to study the decomposition of pressure pulsation signals, and FFT has certain advantages in the convenience of signal decomposition. Jin et al. [23] conducted a fast Fourier transform analysis of the leakage flow in the Kaplan turbine, focusing on the impact of the flow of the tip gap on turbine performance and hydraulic efficiency. Wu et al. [24] compared the DMD method with the FFT method and retrograde method for the dynamic pattern structure analysis of the Savonius turbine near the wake region, providing a new idea for flow state analysis. Tian et al. [25] improved the traditional FFT and proposed that ZOOM-FFT can effectively diagnose and analyze the fault of submersible sewage pumps, aiming at the operation of mixed-flow pumps as turbines. Wang et al. [26] studied the pressure pulsation at the inlet of the volute, the second section of the volute, and the draft tube under out-of-control conditions of turbine runaway and analyzed the frequency domain characteristics of the pressure pulsation based on FFT. Tang [27] used adaptive noise complete geometric empirical mode decomposition (CEEMDAN) to extract time domain features and carried out two fast Fourier changes to extract deep frequency domain features. The results show that this method is effective and applicable in the actual fault diagnosis of wind turbine rolling bearings. The pulsation tracking network (PTN) [28,29] combined with the fast Fourier transform (FFT) can provide better visualization of draft tube pressure pulsations and pulse signal decomposition, revealing the pulse frequency, amplitude, and phase. The method support is provided to reveal the influence law of the draft tube pressure pulsation. From the above-related research, it can be seen that although conventional analysis methods for pressure pulsation are effective, they lack intuitiveness. To determine the difference between two or more pressure pulsation signals, mode decomposition becomes crucial.

In engineering cases, the draft tube is generally arranged according to the civil structure, sometimes asymmetrically. This method does not fully consider the impact of fluid dynamics, which becomes a hidden danger in the operation process. On the basis of the above research, this paper carries out an analysis of the asymmetric layout of the draft tube and compares the left side tilt and the right side tilt. We hope to find out which direction is better for the arrangement of the draft tube. How to determine the specific placement direction of the draft tube is an unresolved issue. Based on computational fluid dynamics and Fourier's transformation, it can more clearly and directly reveal the law of pressure pulsation under the two situations. It is significant for the operation of the Francis turbine and the design of the draft tube for hydro turbines. This article conducts research based on practical engineering problems, and the innovative results provided can provide assistance and reference for other projects.

2. Francis Turbine Unit

2.1. Basic Parameters

Figure 1 shows the on-site Francis turbine unit with a three-dimensional model of the flow-passing components. This turbine unit is a prototype from the "Wanjiazhai Water Multi-purpose Dam Project" in Shanxi province, China. The top cover, main shaft, and runner are shown in the on-site pictures. The flow-passing components include the volute and stay vane for flow assembly, the guide vane for flow rate regulation, the runner for energy conversion, and the draft tube for flow guidance and pressure energy recovery. This turbine includes three cascades: the runner with 13 blades, the stay vane with 24 blades, and the guide vane also with 24 blades.

Figure 1. *Cont.*

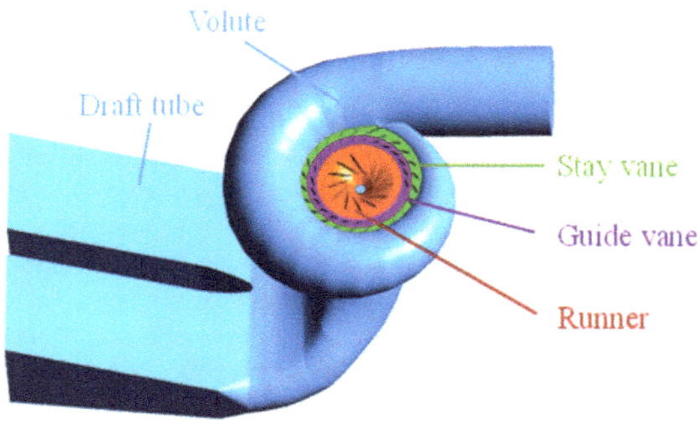

Figure 1. The on-site pictures and the flow-passing components of the Francis turbine.

The basic parameters of this turbine are listed as follows. The runner diameter D_{rn} is 5.80 m. The rated output power P_r is 183.7 MW. The rated flow rate Q_r is 301 m^3/s. The rated rotation speed n_r is 100 rpm. The rated water head H_r is 68.0 m. The maximum output power P_{max} is 204.1 MW. The turbine installation height H_{ins} is 895 m.

2.2. Asymmetric Arrangement of Draft Tube

In this study, the asymmetric layout of two types of draft tubes is considered, as shown in Figure 2. According to the rotation direction of the runner shown in Figure 2, the inlet of the volute is on the left side of the figure, and the water gradually rotates to the right. When the draft tube is asymmetrically arranged, the split buttress in the middle is not on the centerline of the rotation axis but has a certain offset. Left tilt (LT) is defined as the left deviation of the buttress and the draft tube flow passage. On the contrary, right tilt (RT) is defined as the deviation of the buttress and draft tube flow channel to the right as shown in the figure. From a hydrodynamic perspective, the two should not differ significantly due to the constant cross-sectional area of the flow. From the structural point of view, the draft tube of the left tilt occupies less additional space due to the large left area of the volute. On the contrary, the draft tube of the right tilt will occupy some additional space on the right side. In fact, because the rotating direction of the runner is fixed and the rotating flow is excited in the draft tube, the RT and LT arrangement of the draft tube will be different in pressure pulsation. This is the focused issue of this study.

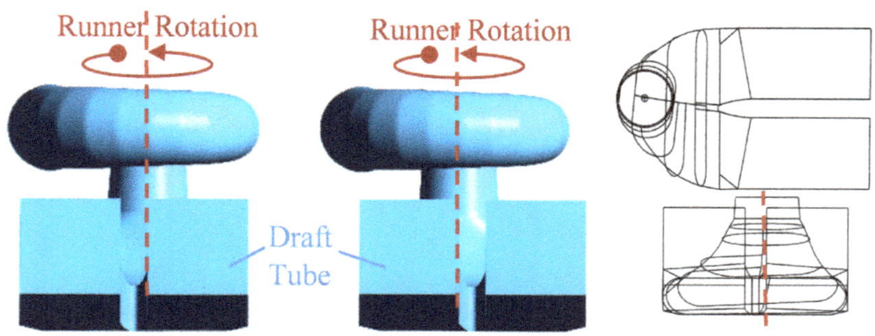

(**a**) Left Tilt(LT) (**b**) Right Tilt (RT) (**c**) Left Tilt Draft

Figure 2. The left tile (LT) and right tilt (RT) of a draft tube of the Francis turbine.

3. Numerical Setup

3.1. Governing Equations

This study uses the Reynolds averaged Navier–Stokes equation to solve flow problems. Quantities are decomposed into averaged terms and pulsating terms [30]. The continuity equation and momentum equation can be written as follows:

$$\frac{\partial \overline{u_i}}{\partial x_i} = 0 \tag{1}$$

$$\rho \frac{\partial \overline{u_i}}{\partial t} + \rho \overline{u_j} \frac{\partial \overline{u_i}}{\partial x_j} = \frac{\partial}{\partial x_j}\left(-\overline{p}\delta_{ij} + 2\mu \overline{S_{ij}} - \rho \overline{u_i' u_j'}\right) \tag{2}$$

where t is time, ρ is density, u is velocity, μ is dynamic viscosity, x is coordinate component, and δ_{ij} is the Kroneker delta. The term named Reynold stress, represented as $\rho u_i' u_j'$, is not closed, so Boussinesq introduced the turbulence isotropy assumption, and it is developed to build the relationship between Reynolds stress and eddy viscosity μ_t [31]:

$$-\rho \overline{u_i' u_j'} = 2\mu_t \overline{S_{ij}} - \frac{2}{3} k \delta i j \tag{3}$$

where k is the turbulence kinetic energy, and S_{ij} is the mean rate of strain tensor:

$$\overline{S_{ij}} = \frac{1}{2}\left(\frac{\partial \overline{u_i}}{\partial x_j} + \frac{\partial \overline{u_j}}{\partial x_i}\right) \tag{4}$$

The eddy viscosity μ_t is connected to turbulence kinetic energy k and turbulence eddy frequency ω based on the experiment, and the shear stress transport (SST) model [32] is established by:

$$\frac{\partial(\rho k)}{\partial t} + \frac{\partial(\rho u_i k)}{\partial x_i} = P_k - \frac{\rho k^{3/2}}{l_{k-\omega}} + \frac{\partial}{\partial x_i}\left[(\mu + \sigma_k \mu_t)\frac{\partial k}{\partial x_i}\right] \tag{5}$$

$$\frac{\partial(\rho \omega)}{\partial t} + \frac{\partial(\rho u_i \omega)}{\partial x_i} = C_\omega P_\omega - \beta \rho \omega^2 + \frac{\partial}{\partial x_i}\left[(\mu_l + \sigma_\omega \mu_t)\frac{\partial \omega}{\partial x_i}\right] + 2(1 - F_1)\frac{\rho \sigma_{\omega 2}}{\omega}\frac{\partial k}{\partial x_i}\frac{\partial \omega}{\partial x_i} \tag{6}$$

where P_k and P_ω are the production terms, F_1 is the coefficient of the production term, σ_k is the blending function, and σ_ω and β_k are model constants. $l_{k-\omega}$ is the turbulence scale, formulated as follows:

$$l_{k-\omega} = k^{1/2} \beta_k \omega \tag{7}$$

This model is good at both the strong shear flow and large pressure gradient. Applying the $k-\omega$ mode in the near-wall region overcomes the disadvantages of the $k-\varepsilon$ model series in capturing wall flow. With the mode of $k-\varepsilon$, the physical field gradient in the flow channel is described with great precision. Therefore, the SST model combines the advantages of both modes. This model provides a good solution for predicting flow problems in engineering cases.

3.2. CFD Setup with Monitoring Points

In this study, computational fluid dynamics (CFD) simulation is based on the commercial software ANSYS CFX 18.0. In total five components, i.e., volute, stay vane, runner, guide vane, and draft tube, are considered. Because the runner is rotating and other components are fixed, the multiple reference frame is used. Therefore, the fluid domain is also divided into five parts. General grid interfaces are given between each two parts. The volute inlet is set as the total pressure inlet boundary. The draft tube outlet is set as the static pressure outlet boundary. All walls are set as non-slip walls.

To evaluate the transient flow and pressure pulsation, a transient state simulation is conducted in this study. In total 10 runner revolutions are considered. In each revolution, 720 timesteps are considered. In each timestep, the iteration number is set as 5~20 and the convergence criterion is 1×10^{-5} for the residuals of both continuity and momentum equations. Figure 3 shows the monitoring points set in the draft tube. To investigate the pressure pulsation caused by vortex rope, P1~P4 are set as indicated. P1 and P2 are on the same higher plane and located on the left side and right side near the wall. P3 and P4 are on the same lower plane and located on the left side and right side also near the wall. In this CFD simulation work, we used a high-performance parallel computing workstation. The AMD EPYV 7813 64-core processor is used as the CPU with 128 threads. The frequency is 2.25 GHz. The memory of the workstation is 128 GB. For each transient simulation, the CPU time is approximately 82,800~90,000 s.

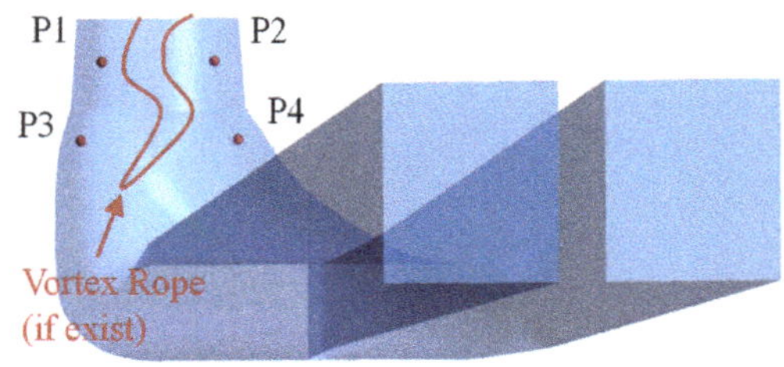

Figure 3. The CFD monitoring points in the draft tube of the Francis turbine.

3.3. Determination of Mesh

A reasonable mesh scheme is helpful to balance the computing cost and accuracy. ICEMCFD platform is used to generate three mesh schemes to conduct a check named "grid convergence index" (GCI) [33]. These three schemes are increased in spacing by 1.35. The efficiency η evaluated by CFD at rated conditions is used as the index. Figure 4 shows the check detail with a coarse mesh, a medium mesh, a fine mesh, and the Richardson extrapolated value. The GCI value of the coarse and medium grid is 3.91% and the GCI value of the medium and fine grid is 0.03%. The smaller the GCI value, the more accurate the simulation results. When the GCI value is less than 5%, the simulation result is acceptable. In order to balance the computational workload and accuracy, the medium scheme with 2.75 million mesh is chosen as the final mesh scheme as shown in Figure 5 and Table 1.

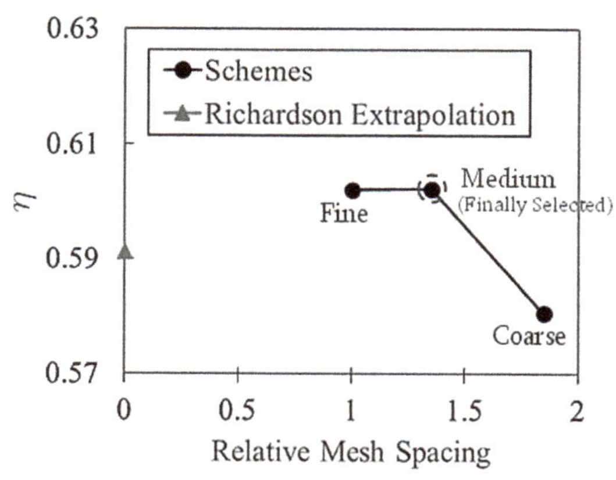

Figure 4. The mesh check based on GCI.

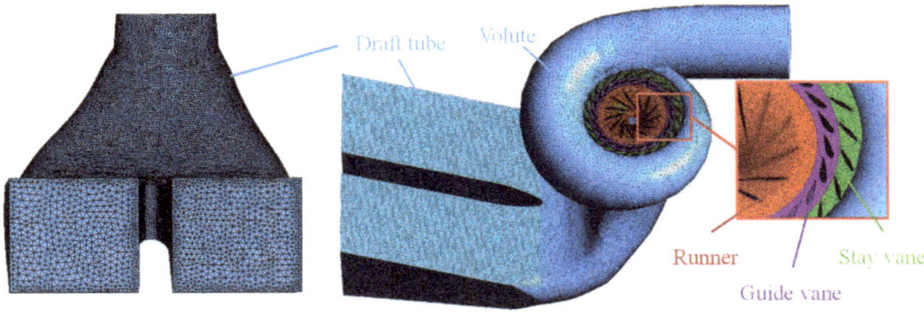

Figure 5. The schematic map of the finally chosen mesh.

Table 1. Mesh number details.

Component	Mesh Number	Range of y+	Average of y+
Volute	209,260	11–230	53
Stay vane	558,230	2–165	38
Guide vane	623,824	2–165	45
Runner	897,680	1.5–211	31
Draft tube	464,692	16–253	88
Total	2,753,686	—	—

4. Results of Performance Analysis

4.1. Efficiency Comparison

Figure 6 shows the operating curve of the water turbine. Its horizontal axis represents turbine power P. Its ordinate represents the water head H. The right side of the figure shows the power limit line, and the left side shows the area where the unit may operate. The optimal efficiency zone on the graph is close to the right. From the bottom right corner to the top left corner, instability gradually increases. Figure 6 shows three conditions for validation and verification. They are illustrated on the P-H map. The prototype Francis turbine's draft tube is left tilt. On the left side of the operation limitation line, with the decrease in power P and increase in head H, there are three regions defined as the "allowed region", "restricted region", and "prohibited region". The three chosen conditions are tagged by C_{alw}, C_{rst}, and C_{phb} with the unit rotation speed n_{11} of 77.85, 78.55, and 91.21 and the unit flow rate Q_{11} of 0.7864, 0.3678, and 0.3495. The definitions for n_{11} and Q_{11} are as follows:

$$n_{11} = \frac{nD_{rn}}{\sqrt{H}} \tag{8}$$

$$Q_{11} = \frac{Q}{D_{rn}^2\sqrt{H}} \tag{9}$$

where n is the rotational speed and Q is the flow rate. Figure 7 shows the comparison of efficiency η between CFD and experimental data. At C_{alw}, the CFD efficiency is 91.05% for LT and 91.08% for RT, and the experimental efficiency is 90.51%. At C_{rst}, the CFD efficiency is 71.72% for LT and 71.69% for RT, and the experimental efficiency is 70.32%. At C_{phb}, the CFD efficiency is 55.25% for LT and 55.23% for RT, and the experimental efficiency is 52.88%. RT and LT show a good match on the performance. The CFD value shows a good prediction of unit performance and can be used for subsequent analysis.

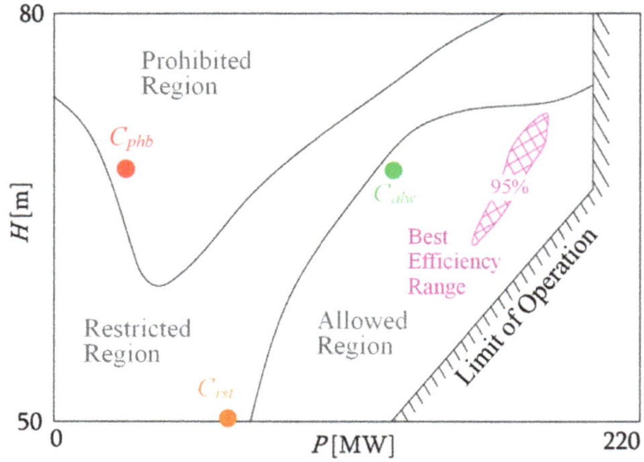

Figure 6. The operation characteristic *P-H* map.

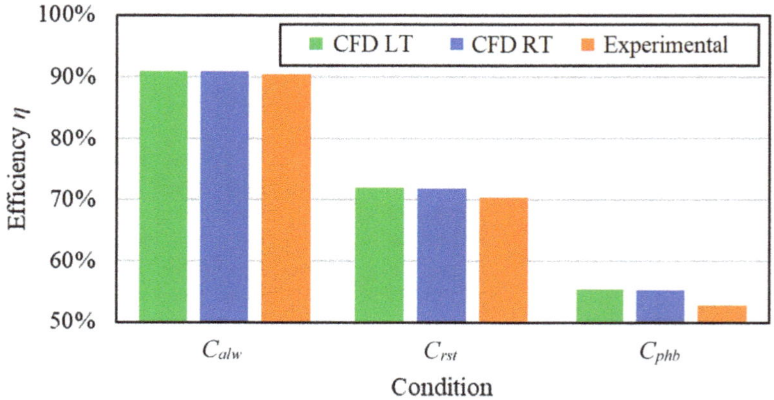

Figure 7. The comparison of efficiency for CFD-experiment verification and validation.

4.2. The Internal Flow Pattern

Figure 8 shows the flow inside the Francis turbine, using colored streamlines to represent the velocity and direction of the flow. It can be seen that the performance of LT and RT in internal flow is very similar, and there is almost no difference. This is also why the performance of the two is so close. At the C_{alw} point in the allowed region, the flow is relatively stable and the streamline is smooth. Velocity in runner is uniform and up to about 20 m/s. At the C_{rst} point in the restricted region, the velocity in the runner increases to a maximum of about 38 m/s and the distribution of velocity becomes non-uniform. The flow in the draft tube develops many vortexes. The vortical flow dominates the component downstream to the runner and seems out of control. At the C_{phb} point in the prohibited region, the velocity in the runner is still relatively high (maximum value is about 35 m/s). The vortical flow in the draft tube further intensifies. In comparison, the flow status is relatively worse in prohibited and restricted operating regions.

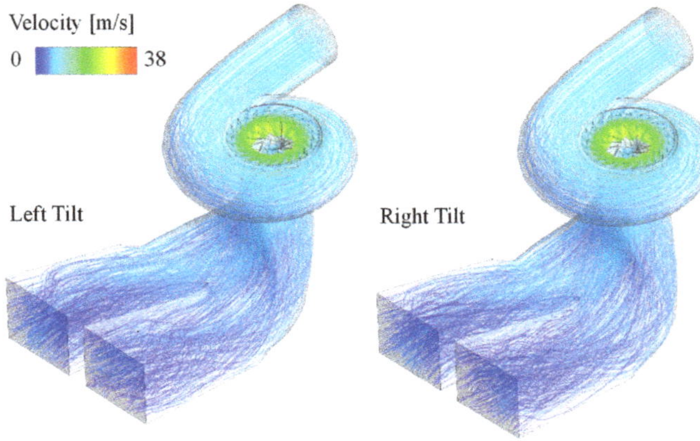

(**a**) Condition C_{alw}

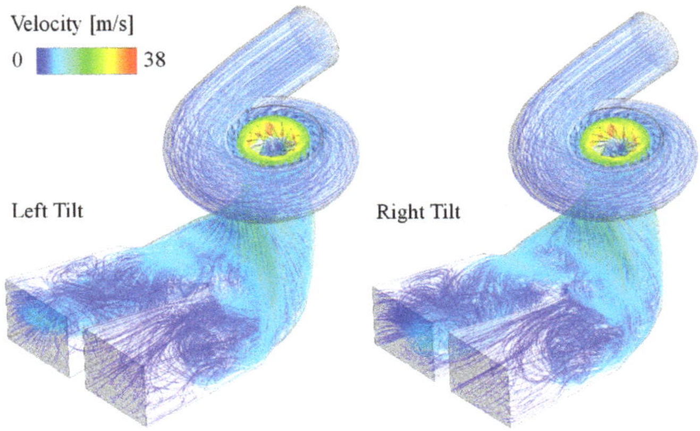

(**b**) Condition C_{rst}

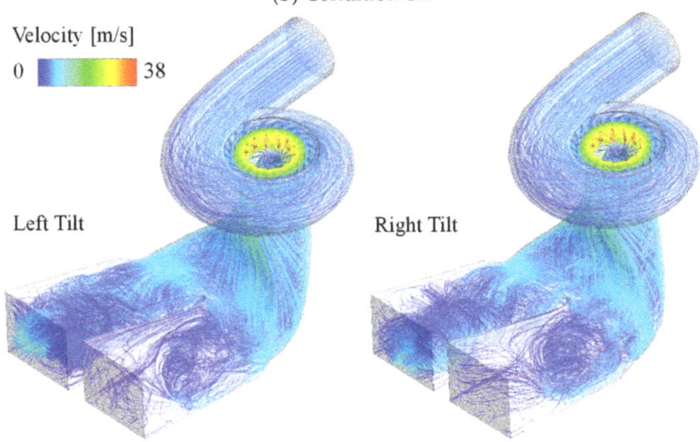

(**c**) Condition C_{phb}

Figure 8. Internal flow pattern in Francis turbine.

5. Results of Pressure Pulsation Analysis

Variational modal decomposition (VMD) is an adaptive, completely non-recursive method for modal decomposition and signal processing, which can decompose pressure pulsation signals into a set of IMF components with different frequencies. Based on the points P1~P4 shown in Figure 3, analyses are conducted using the variational mode decomposition (VMD) method, which is used to decompose the pressure pulsation of four monitoring points under the conditions of the left and right tilts of the draft tube of the turbine. This is the first time that the VMD method has been introduced into the study of pressure pulsation in the Francis turbine, which provides assistance in clearly extracting signal features and obtaining clearer results. The time domain diagram of each component and its corresponding frequency domain diagram obtained from the decomposition are shown in Figures 9–12. Several intrinsic mode functions (IMFs) are indicated for a better analysis.

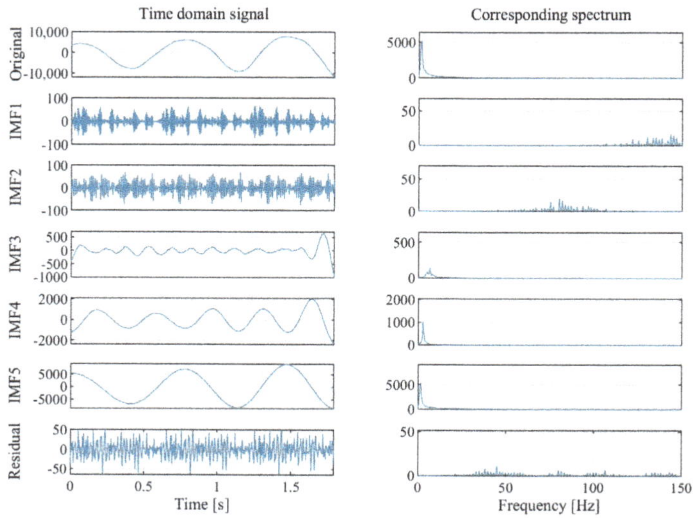

(**a**) Left Tilt

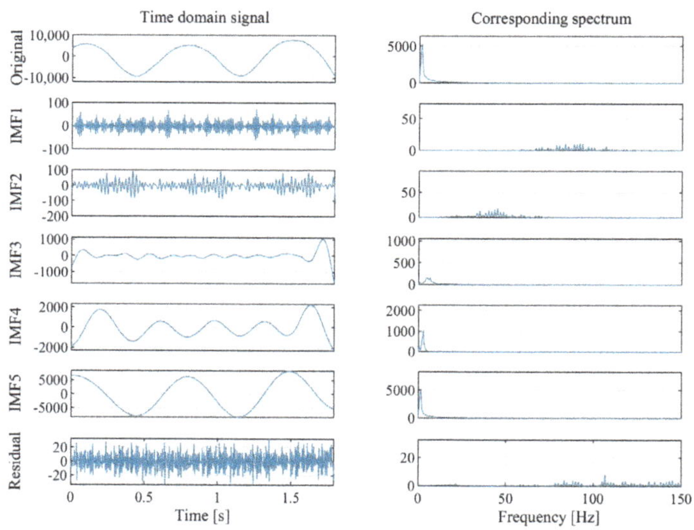

(**b**) Right Tilt

Figure 9. Time domain diagram and corresponding spectrum of P1 pressure pulsation signal based on VMD decomposition under different operating conditions.

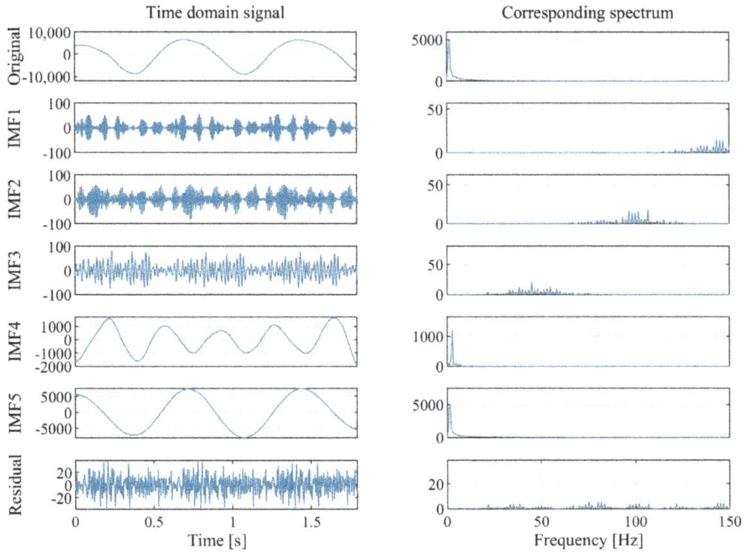

(**a**) Left Tilt

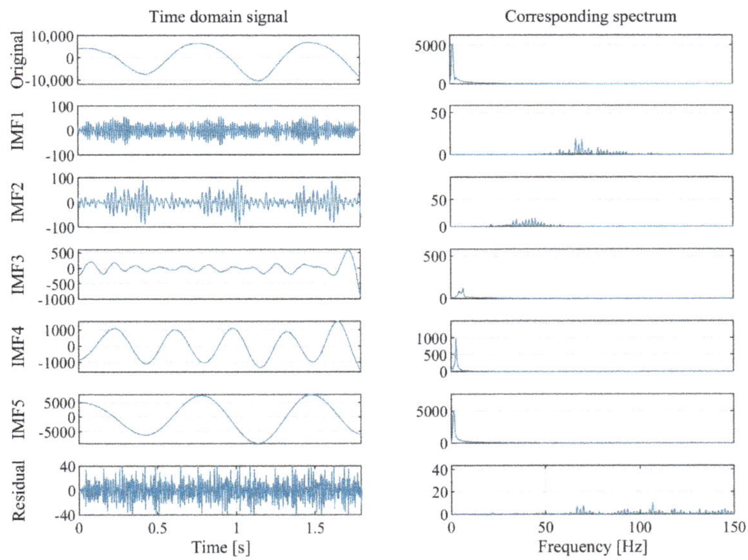

(**b**) Right Tilt

Figure 10. Time domain diagram and corresponding spectrum of P2 pressure pulsation signal based on VMD decomposition under different operating conditions.

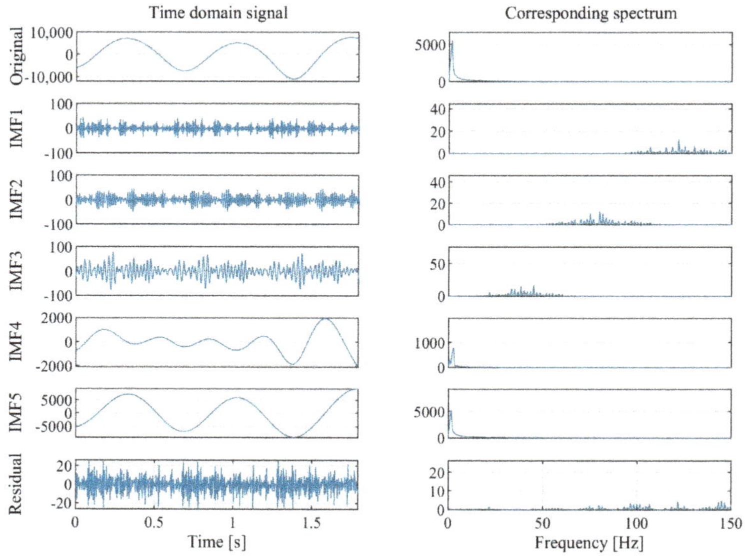

(a) Left Tilt

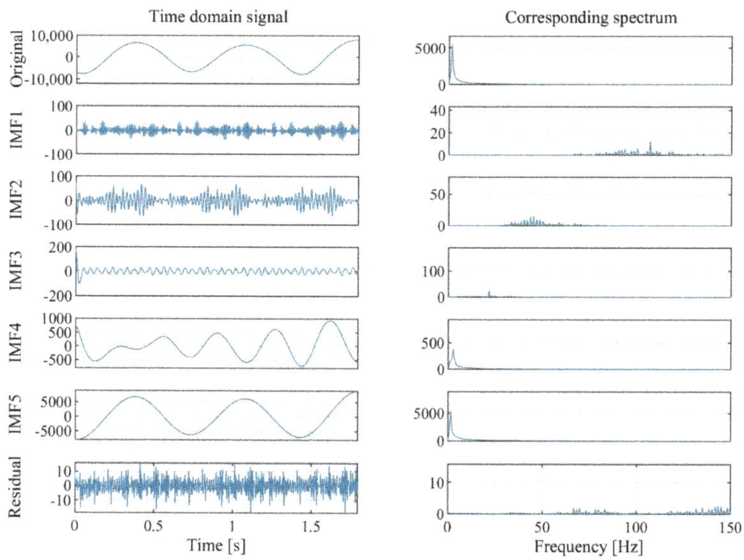

(b) Right Tilt

Figure 11. Time domain diagram and corresponding spectrum of P3 pressure pulsation signal based on VMD decomposition under different operating conditions.

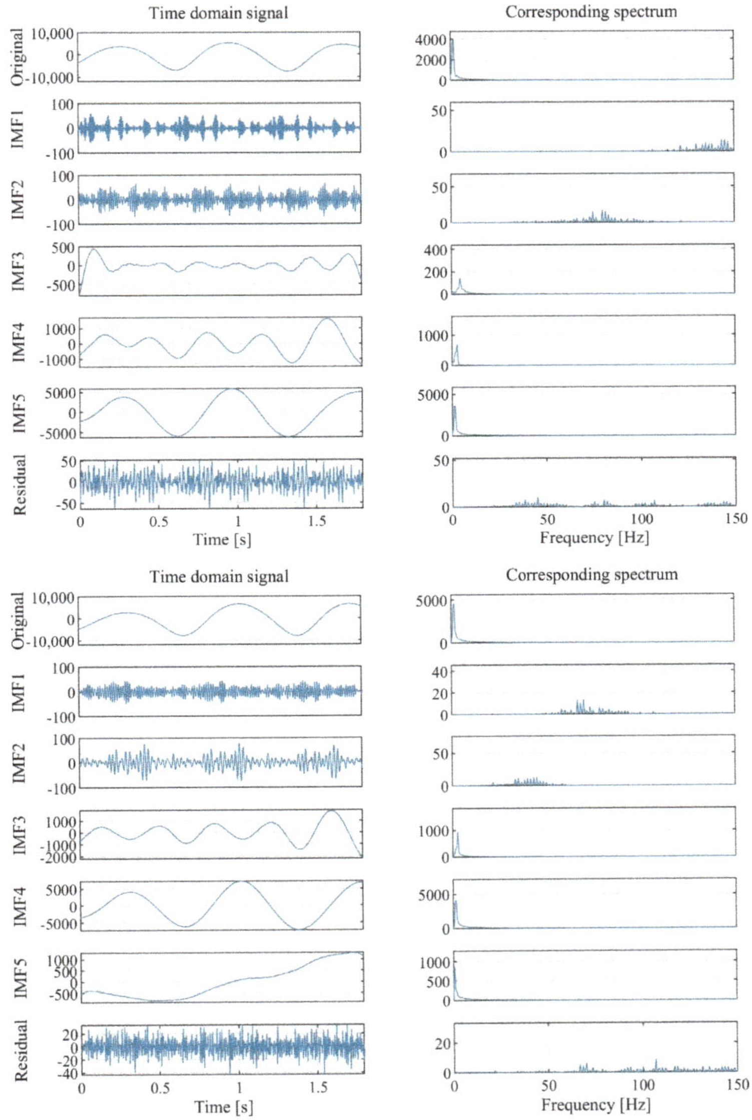

(**b**) Right Tilt

Figure 12. Time domain diagram and corresponding spectrum of P4 pressure pulsation signal based on VMD decomposition under different operating conditions.

5.1. Analysis of Time Domain and Frequency Domain

Figure 9 shows the time domain diagrams and corresponding frequency spectra of each component obtained by decomposing the pressure pulsation signal of the P1 monitoring point using VMD under different working conditions. It can be seen from the figure that the frequency domain diagram and time domain diagram of each component obtained by VMD decomposition are basically similar under the conditions of the left and right tilts of the draft tube. There is only one wave peak in the frequency domain diagram of IMF4 and IMF5 components. When the draft tube tilts to the left, the peak value of the frequency domain diagram of IMF5 is 5314 Pa, the peak frequency is the rotating frequency of the runner f_r (1.67 Hz), the peak value of the frequency domain diagram of IMF4 is 1032 Pa, and the peak frequency is 2.78 Hz. When the draft tube tilts to the right, the peak value of the frequency domain diagram of IMF5 is 5126 Pa, the peak frequency is 1.67 Hz, the peak value of the frequency domain diagram of IMF4 is 1041 Pa, and the peak frequency is 2.78 Hz. It can be seen that the pressure pulsation of the draft tube is mainly affected by the rotating frequency of runner f_r, while the left and right tilts of the draft tube have little influence on the IMF1 and IMF2 components. The frequency domain diagram of IMF1–IMF3 is composed of multiple high-frequency signals with low amplitude. When the draft tube tilts to the left, the center frequencies of IMF1–IMF3 components are 137.1 Hz, 81.36 Hz, and 6.24 Hz, respectively. When the draft tube tilts to the right, the center frequencies of IMF1–IMF3 components are 87.7 Hz, 42.3 Hz, and 5.51 Hz, respectively. Different draft tube tilt conditions have a greater impact on the center frequency of high-frequency components. When the draft tube tilts to the left, the center frequency of the IMF1–IMF3 component of pressure pulsation at the P1 monitoring point is significantly greater than that of the draft tube tilts to the right.

Figure 10 shows the time domain diagrams and corresponding frequency spectra of each component obtained by decomposing the pressure pulsation signal of the P2 monitoring point using VMD under different working conditions. When the draft tube tilts to the left, the amplitude of the IMF5 component is 4851 Pa, and the amplitude of the IMF4 component is 1196 Pa. When the draft tube tilts to the right, the amplitude of the IMF5 component is 5126 Pa, and the amplitude of the IMF4 component is 991 Pa. Compared with the right tilt of the draft tube, the left tilt of the draft tube can reduce the pressure pulsation at monitoring point P2 to a certain extent, which is of great significance to the stable operation of the unit. The P1 and P2 monitoring points are located on both symmetrical sides, and the difference in pressure pulsation between them may be related to the asymmetric volute. At the P2 monitoring point, the center frequencies of IMF1–IMF3 components are 141.41 Hz, 99.89 Hz, and 46.85 Hz, respectively, when the draft tube tilts to the left. When the draft tube tilts to the right, the center frequencies of IMF1–IMF3 components are 78.28 Hz, 42.06 Hz, and 5.78 Hz, respectively. When the draft tube tilts to the left, the center frequency of IMF1–IMF3 components at the P2 monitoring point is significantly different from that at the P1 monitoring point. When the draft tube tilts to the right, the center frequencies of the P1 and P2 monitoring points are basically the same, which indicates that the left tilt of the draft tube has a greater impact on the internal pressure pulsation of the draft tube.

Figure 11 shows the time domain diagrams and corresponding frequency spectra of each component obtained by decomposing the pressure pulsation signals of P3 monitoring points using VMD under different operating conditions. When the draft tube tilts to the left, the amplitude of the IMF5 component is 5193 Pa, and the amplitude of the IMF4 component is 782 Pa. When the draft tube tilts to the right, the amplitude of the IMF5 component is 5370 Pa, and the amplitude of the IMF4 component is 379 Pa. Similar to the P1 monitoring point, different draft tube tilt conditions have little impact on the amplitude of pressure pulsation at the P3 monitoring point. Under the two conditions, the difference in pressure pulsation amplitude is only 3.3%. At the P3 monitoring point, when the draft tube tilts to the left, the center frequencies of IMF1–IMF3 components are 139.33 Hz, 93.67 Hz, and 41.29 Hz, respectively. When the draft tube tilts to the right, the center frequencies of IMF1–IMF3 components are 102.43 Hz, 45.16 Hz, and 20.28 Hz, respectively. When the draft tube tilts to the left, the center frequency of the IMF1–IMF3 component is significantly higher than that of the draft tube tilts to the right, and the center frequencies of the IMF4 and IMF5 components are basically the same, which is basically the same as the impact of draft tube tilt on the center frequency of the P1 monitoring point.

Figure 12 shows the time domain diagrams and corresponding frequency spectra of each component obtained by decomposing the pressure pulsation signals of P4 monitoring points using VMD under different operating conditions. When the draft tube tilts to the left, the amplitude of the IMF5 component is 3626 Pa, and the amplitude of the IMF4 component is 695 Pa. When the draft tube tilts to the right, the amplitude of the IMF5 component is 871 Pa, and the amplitude of the IMF4 component is 4119 Pa. Different from other monitoring points, the amplitude–frequency of the IMF5 component at the P4 monitoring point is 0.56 Hz when draft tube tilts to the right, and the amplitude–frequency of IMF4 is runner frequency f_r. Compared with the draft tube tilting to the right, the pressure pulsation amplitude of the draft tube tilting to the left can be reduced by 11.9%. When the draft tube is tilted to the left, the center frequencies of IMF1–IMF3 components are 141.41 Hz, 99.89 Hz, and 46.85 Hz, respectively. When the draft tube tilts to the right, the center frequencies of IMF1–IMF3 components are 78.28 Hz, 42.06 Hz, and 5.78 Hz, respectively. At this point, there is a significant difference in the center frequency of the pressure pulsation component between the P4 monitoring point and the P3 monitoring point under both operating conditions. This shows that as the monitoring point moves downstream of the draft tube, the larger the draft tube tilt, the greater the impact of the tilt on the draft tube pressure pulsation, and the more significant the impact of the left tilt of the draft tube on reducing the amplitude of pressure pulsation.

5.2. The Law of Pressure Pulsation

Table 2 summarizes the law of pressure pulsation of all four monitoring points with the comparison between left tilt and right tilt. Overall, there are seven characteristic frequency bands, including approximately 140 Hz, 80–90 Hz, 40–46 Hz, 5.5–6.5 Hz, 2.5–3 Hz, f_r (1.67 Hz), and <1 Hz. The frequency value of the same mode of left tilt is higher than that of the right tilt. In addition, the intensity of the runner frequency f_r is very important, as analyzed in Figure 13. On P1, the amplitudes of LT and RT are similar. On P2, RT has a higher amplitude of f_r than that of LT. On P3, RT has a slightly higher amplitude of f_r than that of LT. On P4, RT has a higher amplitude of f_r than that of LT and the difference becomes bigger. Generally, the right tilt draft tube may trigger higher pulsation of the pressure field.

Table 2. IMFs of all four monitoring points.

	Left Tilt				Right Tilt			
	P1	P2	P3	P4	P1	P2	P3	P4
IMF1 (Hz)	137.10	141.41	139.33	138.67	87.70	78.28	102.43	71.15
IMF2 (Hz)	81.36	99.89	93.67	78.72	42.30	42.06	45.16	40.25
IMF3 (Hz)	6.24	46.85	41.29	4.67	5.51	5.78	20.28	2.58
IMF4 (Hz)	2.89	2.77	2.27	2.35	2.60	2.80	2.75	1.34
IMF5 (Hz)	1.41	1.38	1.33	1.33	1.35	1.40	1.37	0.34

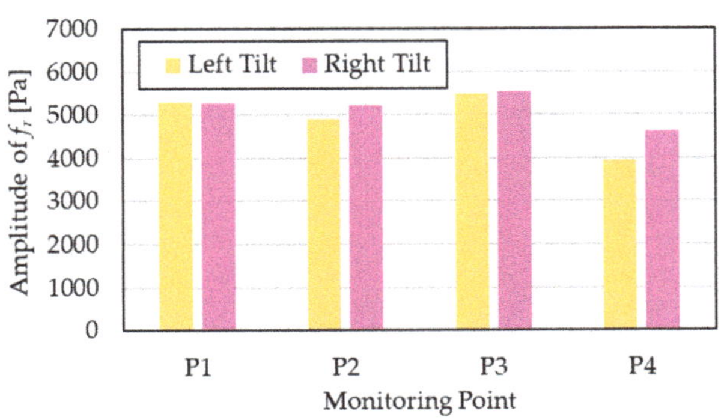

Figure 13. Pulsation amplitude of runner frequency f_r on all four monitoring points of the left tilt and right tilt of the draft tube.

6. Conclusions

This study focuses on the influence of the asymmetric arrangement (tilt) direction of a draft tube on the pressure pulsation of a Francis turbine, and conclusions can be drawn as follows:

(1) Left-tilt arrangement and right-tilt arrangement of the draft tube have little impact on the unit performance of the Francis turbine. By comparing the CFD calculation and analysis results with the prototype measurement results, it can be found that the direction of the asymmetric arrangement of the draft tube has little effect on the efficiency. As long as the cross-sectional area remains unchanged, the variation in total pressure remains unchanged, and the impact of left and right tilts on different working conditions is relatively small, typically less than 1% on efficiency. Therefore, from the perspective of hydraulic performance, the tilt direction of the draft tube can be left or right.

(2) From the internal flow, it can be seen that there are significant differences in the flow inside the unit under different operating conditions. In the allowed region with higher efficiency, the flow smoothness is relatively high, and there are almost no obvious vortices or other forms of adverse flow patterns visible. In restricted and prohibited regions, the velocity distribution will be uneven, and the local velocity may be very high. Flow in the draft tube has also become chaotic, mainly with large-scale strong rotating flow. Judging from the characteristics of the flow direction, both the left-tilt and right-tilt draft tubes are acceptable. The impact of the left-tilt arrangement and right-tilt arrangement is still not significant, and the difference is almost invisible.

(3) Based on the mode decomposition of the pressure fluctuation signal by VMD, it can be seen that different arrangement directions of the draft tube will have some effects on the flow. Overall, the frequency characteristics of the two are relatively similar with no huge differences. However, the dominant mode of the left tilt corresponds to a higher frequency, while the right tilt corresponds to a lower frequency. In addition, in terms of the amplitude of the important runner frequency f_r, the value corresponding to the right tilt is slightly higher than that of the left tilt. From a hydrodynamic perspective, the geometric features of flow-passing components that are more adaptable to flow are more acceptable. If the left tilt is used, the fluctuation in the flow field may be more stable due to the same arrangement direction of the draft tube and volute.

In general, this study gives a hydrodynamic reference of the design of draft tube of Francis turbines and can be applied to other cases of hydro turbines. The innovation of this article lies in relying on real engineering cases to conduct in-depth and detailed analyses of small problems, providing great assistance for the promotion of the results. In the future, we will carry out construction and renovation of the tailwater pipes of the power plant and compare the impact of different inclined tailwater pipes on the stability of the unit during actual operation. In addition, through numerical simulations, we will also compare the effects of other units and different degrees of inclination of the draft tube, deepen our research, and provide more support for the design of hydropower stations.

Author Contributions: Conceptualization, T.Z. and X.S.; methodology, T.Z.; software, D.Z.; validation, D.Z., Z.H. and J.L.; formal analysis, J.L.; investigation, Z.H. and X.L.; resources, Z.W.; data curation, T.Z.; writing—original draft preparation, X.L., T.Z. and J.L.; writing—review and editing, X.S. and Z.H.; supervision, Z.W.; project administration, Z.W.; funding acquisition, Z.W. All authors have read and agreed to the published version of the manuscript.

Funding: This research was funded by the Open Research Fund Program of the State Key Laboratory of Hydroscience and Engineering (No. sklhse-2022-E-01).

Data Availability Statement: Data are contained within the article.

Acknowledgments: We acknowledge the Open Research Fund Program of the State Key Laboratory of Hydroscience and Engineering to support the research. We also acknowledge the Wanjiazhai Water Multi-purpose Dam Project for providing the chance to study this Francis turbine, and some of the authors (T. Zhang and X. Liu) spent a significant amount of time on this study based on Wanjiazhai Project.

Conflicts of Interest: Authors Tao Zhang and Xinjun Liu were employed by the company Wanjiazhai Water Multi-Purpose Dam Project Co., Ltd. The remaining authors declare that the research was conducted in the absence of any commercial or financial relationships that could be construed as a potential conflict of interest.

References

1. Kose, F.; Kaya, M.N. Analysis on meeting the electric energy demand of an active plant with a wind-hydro hybrid power station in Konya, Turkey: Konya water treatment plant. *Renew. Energy* **2013**, *55*, 196–201. [CrossRef]
2. Wu, Y.; Zhang, J.; Yuan, J.; Geng, S.; Zhang, H. Study of decision framework of offshore wind power station site selection based on ELECTRE-III under intuitionistic fuzzy environment: A case of China. *Energy Convers. Manag.* **2016**, *113*, 66–81. [CrossRef]
3. Tao, R.; Lu, J.; Jin, F.; Hu, Z.; Zhu, D.; Luo, Y. Evaluation of the rotor eccentricity added radial force of oscillating water column. *Ocean Eng.* **2023**, *276*, 114222. [CrossRef]
4. Dong, J.; Feng, T.-T.; Yang, Y.-S.; Ma, Y. Macro-site selection of wind/solar hybrid power station based on ELECTRE-II. *Renew. Sustain. Energy Rev.* **2014**, *35*, 194–204.
5. Jafari, B.; Seddiq, M.; Mirsalim, S.M. Impacts of diesel injection timing and syngas fuel composition in a heavy-duty RCCI engine. *Energy Convers. Manag.* **2021**, *247*, 114759. [CrossRef]
6. Ping, X.; Yang, F.; Zhang, H.; Zhang, J.; Xing, C.; Yan, Y.; Yang, A.; Wang, Y. Information theory-based dynamic feature capture and global multi-objective optimization approach for organic Rankine cycle (ORC) considering road environment. *Appl. Energy* **2023**, *348*, 121569. [CrossRef]
7. Jafari, B.; Seddiq, M.; Mirsalim, S.M. Assessment of the impacts of combustion chamber bowl geometry and injection timing on a reactivity controlled compression ignition engine at low and high load conditions. *Int. J. Engine Res.* **2021**, *22*, 2852–2868. [CrossRef]
8. Zhou, M.; Ye, Q.; Liu, Z. Climate impacts on hydro-power development in China. *Bot. Mar.* **2005**, *19*, 1–4.
9. Pereira, J.G.; Vagnoni, E.; Favrel, A.; Landry, C.; Alligne, S.; Nicolet, C.; Avellan, F. Prediction of unstable full load conditions in a Francis turbine prototype. *Mech. Syst. Signal Process.* **2022**, *169*, 108666. [CrossRef]
10. Kalantar, M.; Mousavi, S.M.G. Dynamic behavior of a stand-alone hybrid power generation system of wind turbine, microturbine, solar array and battery storage. *Appl. Energy* **2010**, *87*, 3051–3064. [CrossRef]
11. Zhu, D.; Tao, R.; Xiao, R.; Pan, L. Solving the runner blade crack problem for a Francis hydro-turbine operating under condition-complexity. *Renew. Energy* **2020**, *149*, 298–320. [CrossRef]
12. Li, P.; Xiao, R.; Tao, R. Study of vortex rope based on flow energy dissipation and vortex identification. *Renew. Energy* **2022**, *198*, 1065–1081. [CrossRef]
13. Zhu, D.; Yan, W.; Guang, W.; Wang, Z.; Tao, R. Influence of guide vane opening on the runaway stability of a pump-turbine used for hydropower and ocean power. *J. Mar. Sci. Eng.* **2023**, *11*, 1218. [CrossRef]
14. Tamura, Y.; Tani, K.; Okamoto, N. Experimental and numerical investigation of unsteady behavior of cavitating vortices in draft tube of low specific speed Francis turbine. *IOP Conf. Ser. Earth Environ. Sci.* **2014**, *22*, 032011. [CrossRef]
15. Arthur, F.; Muller, A.; Landry, C.; Yamamoto, K.; Avellan, F. LDV survey of cavitation and resonance effect on the precessing vortex rope dynamics in the draft tube of Francis turbines. *Exp. Fluids* **2016**, *57*, 168.
16. Zhang, J.; Appiah, D.; Zhang, F.; Yuan, S.; Gu, Y.; Asomani, S.N. Experimental and numerical investigations on pressure pulsation in a pump mode operation of a pump as turbine. *Energy Sci. Eng.* **2019**, *7*, 1264–1279. [CrossRef]
17. Bosioc, A.I.; Susan-Resiga, R.; Muntean, S.; Tanasa, C. Unsteady pressure analysis of a swirling flow with vortex rope and axial water injection in a discharge cone. *J. Fluids Eng.* **2012**, *134*, 669–679. [CrossRef]
18. Geng, C.; Li, Y.; Tsujimoto, Y.; Nishi, M.; Luo, X. Pressure oscillations with ultra-low frequency induced by vortical flow inside Francis turbine draft tubes. *Sustain. Energy Technol. Assess.* **2022**, *51*, 101908. [CrossRef]
19. Liu, S.; Shao, Q.; Yang, J.; Wu, Y.; Dai, J. Unsteady turbulent simulation of Three Gorges hydraulic turbine and analysis of pressure in the whole passage. *J. Hydroelectr. Eng.* **2004**, *23*, 97–101.
20. Liao, W.; Ji, J.; Lu, P.; Luo, X. Unsteady flow analysis of francis turbine. *Chin. J. Mech. Eng.* **2009**, *45*, 134–140. [CrossRef]
21. Zhang, Y.; Liu, S.; Wu, Y. Detailed simulation and analysis of pressure pulsation in Francis turbine. *J. Hydroelectr. Eng.* **2009**, *28*, 183–186.
22. Pasche, S.F.; Gallaire, F.; Avellan, F. Origin of the synchronous pressure pulsations in the draft tube of Francis turbines operating at part load conditions. *J. Fluids Struct.* **2019**, *86*, 13–33. [CrossRef]
23. Kim, H.H.; Rakibuzzaman, M.; Kim, K.; Suh, S.-H. Flow and fast fourier transform analyses for tip clearance effect in an operating kaplan turbine. *Energies* **2019**, *12*, 264. [CrossRef]

24. Wu, Y.; Guang, W.; Tao, R.; Liu, J. Dynamic mode structure analysis of the near-wake region of a Savonius-type hydrokinetic turbine. *Ocean Eng.* **2023**, *282*, 114965. [CrossRef]
25. Tian, L.; Wu, J. Fault diagnosis and analysis of submersible sewage pump based on zoom-fft. *Mach. Des. Res.* **2018**, *34*, 171–174.
26. Wang, W.; Tai, G.; Shen, J. Experimental investigation on pressure pulsation characteristics of a mixed-flow pump as turbine at turbine and runaway conditions. *J. Energy Storage* **2022**, *55 Pt C*, 105562. [CrossRef]
27. Tang, Z.; Wang, M.; Ouyang, T.; Fei, C. A wind turbine bearing fault diagnosis method based on fused depth features in time–frequency domain. *Energy Rep.* **2022**, *8*, 12727–12739. [CrossRef]
28. Jin, F.; Tao, R.; Lu, Z.; Xiao, R. A spatially distributed network for tracking the pulsation signal of flow field based on CFD simulation: Method and a case study. *Fractal Fract.* **2021**, *5*, 181. [CrossRef]
29. Lu, Z.; Tao, R.; Jin, F.; Li, P.; Xiao, R.; Liu, W. The Temporal-Spatial Features of Pressure Pulsation in the Diffusers of a Large-Scale Vaned-Voluted Centrifugal Pump. *Machines* **2021**, *9*, 266. [CrossRef]
30. Pope, S.B. *Turbulent Flows*; Cambridge University Press: Cambridge, UK, 2000.
31. Terentiev, L. *The Turbulence Closure Model Based on Linear Anisotropy Invariant Analysis*; VDM Verlag: Saarbrucken, Germany, 2008.
32. Menter, F.R. Two-Equation Eddy-Viscosity Turbulence Models for Engineering Applications. *AIAA J.* **1994**, *32*, 1598–1605. [CrossRef]
33. Celik, I.B.; Ghia, U.; Roache, P.J.; Freitas, C.J.; Coloman, H.; Raad, P.E. Procedure for estimation and reporting of uncertainty due to discretization in CFD applications. *J. Fluids Eng.* **2008**, *130*, 78001.

water

MDPI

Article

Study on the Cavitation Characteristics of Shroud Clearance in Prototype and Model of a Kaplan Turbine

Yali Zhang [1], Wendong Luo [2], Tao Chen [2], Lingjiu Zhou [1,3] and Zhengwei Wang [4,*]

1 College of Water Resources and Civil Engineering, China Agricultural University, Beijing 100083, China; 15838868943@163.com (Y.Z.); zlj@cau.edu.cn (L.Z.)
2 Guangxi Datengxia Gorge Water Conservancy Development Co., Ltd., Guiping 537226, China; lwd420@163.com (W.L.); total_00@163.com (T.C.)
3 Beijing Engineering Research Center of Safety and Energy Saving Technology for Water Supply Network System, Beijing 100083, China
4 State Key Laboratory of Hydroscience and Engineering, Department of Energy and Power Engineering, Tsinghua University, Beijing 100084, China
* Correspondence: wzw@mail.tsinghua.edu.cn

Abstract: Model tests and model calculations are the most basic means currently available to study the characteristics of the axial-flow pumps and Kaplan turbines in a systematic manner. Large and medium-sized turbine units and axial-flow pumps must rely on model tests and model calculations to ensure the performances of prototype units before designing. The conversions between models and prototypes are mainly carried out through similarity criteria. However, it is difficult to meet all the similarity criteria in the model tests and the similarity conversions, and the hydraulic and cavitation performances of the model and the prototype are often different. In this paper, numerical calculations of shroud clearance cavitation are performed on both the prototype and model using different cavitation coefficients. The results indicate that the prototype and model have a similar clearance cavitation flow regularity when the cavitation coefficient changes, but they have different energy characteristics and cavitation characteristics. In cavitation conditions, the prototype has higher energy characteristics than the model and the critical cavitation coefficient is similar to the model. When the cavitation coefficient is higher than the critical cavitation coefficient, compared to the model, the blade cavitation performance of the prototype is worse, and the clearance cavitation and runner chamber cavitation are more serious. If the cavitation coefficient decreases to the device cavitation coefficient, the runner chamber of the prototype will cavitate, even though the model has not cavitated yet. The comparison of shroud clearance cavitation between the prototype and the model can be used as a reference for the accuracy of similarity conversion results between the model and the prototype. It also has a positive impact on the design and operation of the prototype.

Keywords: prototype Kaplan turbine; model Kaplan turbine; shroud clearance flow; cavitation; clearance leakage vortex

Citation: Zhang, Y.; Luo, W.; Chen, T.; Zhou, L.; Wang, Z. Study on the Cavitation Characteristics of Shroud Clearance in Prototype and Model of a Kaplan Turbine. *Water* **2023**, *15*, 3960. https://doi.org/10.3390/w15223960

Academic Editor: Giuseppe Pezzinga

Received: 9 October 2023
Revised: 9 November 2023
Accepted: 10 November 2023
Published: 14 November 2023

1. Introduction

Cavitation performance is one of the key indicators to evaluate the overall performance of turbine units. Problems caused by cavitation, such as degradation of hydraulic performance, cavitation wear of overcurrent components, and vibration intensification, have been becoming major problems in unit operation [1]. In the Kaplan turbine, the leakage flow and vortex flow in the runner clearance region are very complicated because of the influence of the double coordination of the guide vane and blade, and the clearance cavitation forms are various [2]. Clearance cavitation will not only reduce the hydraulic performance and efficiency of the unit [3] and induce low-pressure pulsation [4] but also cause cavitation damage at the runner chamber, runner hub, and blade tip [5]. Although the damage scope is generally small, the degree of damage is serious. Therefore, it is important to focus on the flow and cavitation characteristics in the shroud clearance region before designing the prototype unit for production.

Model tests and calculations are the most basic means of studying the flow of shroud clearance. Xiao et al. [6] found that the clearance leakage vortex is the main reason for the highest total pressure drop and the highest total pressure loss in an axial turbine runner by the experiment, and the lowest pressure occurs in the region of clearance leakage vortex. Amiri et al. [7] tested and found that the clearance leakage flow enhances the velocity pulsation from the leading edge to the middle on the suction surface of the blade. Ma et al. [8] simulated the shroud clearance flow of a Kaplan turbine, and the results show that increasing the shroud clearance can reduce the cavitation performance in the region of the blade tip, near the blade tip on the suction surface, and the middle on the suction of the blade. In the tubular turbine, increasing the clearance not only enhances the pressure pulsation in the region of the blade tip but also the bladeless region between the runner and the guide vane [9]. Additionally, some scholars have conducted numerical simulations of the shroud clearance flow, and the simulation results are in good agreement with the experimental results [10–13].

The model tests and the model numerical simulations have improved the understanding of the shroud clearance flow and provided a reference for the prototype. However, some prototype units still have problems such as clearance cavitation in the actual operation, which is because the characteristics of the prototype are predicted by the similarity conversion. The model and prototype should strictly meet all the similarity criteria to fully reflect the actual working conditions of the prototype unit. However, in engineering practice, it is impossible to meet all the similarity criteria simultaneously, and the model test can only approximate the flow of the prototype [14]. To improve the accuracy of the characteristics of the prototype, many scholars have revised the similarity conversion formulas. Some scholars such as Osterwalder [15], Zheng [16], Hutton [17], and Ida et al. [18–20] have revised the conversion formula for hydraulic efficiency. Chen et al. [21] analyzed a variety of hydraulic efficiency conversion formulas and found significant differences among them, which had certain applicability and limitations. The correction of the cavitation scale effect has been studied by some scholars. McCormick [22] found that the critical cavitation coefficient is influenced by the Reynolds number and angle of attack when the wing shape is consistent. Based on the assumption that the initial cavitation coefficient of the vortex is proportional to the density of the boundary layer on the compression surface of the hydrofoil, the scale correction model of the cavitation coefficient is established. Ren et al. [23] proposed that, in addition to the diameter of the runner, the head of the unit, Reynolds number, air content in water, water quality factors such as the radius of air nuclei in water, and the tensile strength of water would also affect the scale effect of cavitation, and provided a corresponding correction formula for the scale effect of cavitation. Ni [24] proposed the relationship of the incipient cavitation coefficient between the prototype and the model based on the cavitation dynamics equation, but also emphasized the deficiencies in the process of deduction. Despite the many similar conversion correction formulas between the prototype and model, the results obtained from different correction formulas differ significantly. The same correction formula is not applicable to all hydraulic machines. To establish the conversion relationship between the prototype and the model as accurately as possible, it is necessary to fully understand the differences between the prototype and the model.

In recent years, many scholars have started to improve their understanding of the differences between prototypes and models by comparing both results. Angulo et al. [25] compared the prototype and model cavitation tests of Kaplan turbines and proposed that the model test was consistent with the results of the prototype test in the time and frequency domains, and analyzed the reasons for the differences between test results from the aspects of scale effect and test techniques. In addition, Angulo et al. [26] also conducted air injection tests on the prototype and model of the Kaplan turbine, aiming to evaluate the effect of air injection on reducing pressure pulsation. The beneficial effect of air injection was predicted in both model and prototype tests. However, the prediction effect of the model is relatively weak, and the efficiency loss was overestimated in model tests. The hydraulic performance of the prototype and model in the bucket turbine was predicted using the numerical simulation method by Zeng et al. [27]. The results showed that the hydraulic performance differences between the prototype and the model mainly occurred during the high torque and torque decline stages. Li et al. [28] performed a pump-turbine simulation and discovered that the pressure pulsation of the prototype and the model in the rotor–stator interaction region exhibited a high resemblance. However, the similarity between the two in the downstream of the rotor–stator interaction region and the tailpipe region would be greatly weakened. The results of the model were not sufficient to support the operation of the prototype at this time.

To conclude, it is still challenging to accurately predict the clearance cavitation flow characteristics of the prototype. A complete understanding of the clearance cavitation flow characteristics of the prototype and the model is required. At present, numerical simulation of the prototype and the model is an effective means of understanding the differences between them. Therefore, this paper studies the clearance cavitation flow characteristics of the prototype and the model for the Kaplan turbine by numerical simulation. The differences in the clearance cavitation flow obtained in this paper confirm that model testing or numerical simulation alone cannot accurately predict the cavitation characteristics of the prototype. And the cavitation margin of the prototype still needs to be taken into account. Furthermore, the results of this paper can be used as a reference to correct the similarity conversion formulas between the prototype and the model in the Kaplan turbine.

2. Research Object and Methodology

2.1. Numerical Techniques

The cavitation flow in the Kaplan turbine is simulated by the commercial Computational Fluid Dynamics (CFD) software ANSYS CFX 19.2 (Ansys, Inc., Canonsburg, PA, USA). The fluid in the cavitation flow field is considered a mixture of water and water vapor. The mixture is assumed to be homogeneous in this paper. The continuity equation, the momentum equation, and the mass-transport equation for the homogeneous flow are as follows:

$$\frac{\partial \rho_{\mathrm{m}}}{\partial t}+\frac{\partial(\rho_{\mathrm{m}} u_j)}{\partial x_j}=0 \tag{1}$$

$$\frac{\partial(\rho_{\mathrm{m}} u_i)}{\partial t}+\frac{\partial(\rho_{\mathrm{m}} u_i u_j)}{\partial x_j}=-\frac{\partial p}{\partial x_i}+\frac{\partial}{\partial x_j}\left[(\mu_{\mathrm{m}}+\mu_{\mathrm{t}})\left(\frac{\partial u_i}{\partial x_j}+\frac{\partial u_j}{\partial x_i}-\frac{2}{3}\frac{\partial u_k}{\partial x_k}\delta ij\right)\right] \tag{2}$$

$$\frac{\partial}{\partial t}(\alpha_{\mathrm{l}}\rho_{\mathrm{l}})+\frac{\partial}{\partial x_j}(\alpha_{\mathrm{l}}\rho_{\mathrm{l}} u_j)=m \tag{3}$$

where ρ_{m} and μ_{m} can be calculated respectively by Formulas (4) and (5):

$$\rho_{\mathrm{m}}=\alpha_{\mathrm{l}}\rho_{\mathrm{l}}+\rho_{\mathrm{v}}(1-\alpha_{\mathrm{l}}) \tag{4}$$

$$\mu_{\mathrm{m}}=\alpha_{\mathrm{l}}\mu_{\mathrm{l}}+\mu_{\mathrm{v}}(1-\alpha_{\mathrm{l}}) \tag{5}$$

In Formulas (1) to (5), ρ_{m} represents the density of the mixture; ρ_{l} and ρ_{v} represent the component densities of the water and water vapor, respectively; μ_{m} represents the dynamic viscosity of the mixture; μ_{l} and μ_{v} represent the component dynamic viscosity of the water and water vapor, respectively; α_{l} represents the component volume fraction of the water; u_{a} (a = i, j, k; $i, j, k = 1, 2, 3$) represent the three direction components of the velocity in the Cartesian coordinate system; x_{a} represent the coordinates; p represents the static pressure; μ_{t} represents the turbulent viscosity; δ_{ij} represents the Kronecker delta (if $i = j$, $\delta_{ij}=1$; if $i \neq j$, $\delta_{ij}=0$); and m represents the source terms of cavitation, which is used to control the mass transfer rate between the water vapor phase and the water phase.

In the calculation of the cavitation flow field, the turbulence and cavitation models are established to close the equations. The various turbulence models and their advantages and disadvantages in simulating clearance flow are introduced in detail in reference [29]. The SST-CC turbulence model can be used to accurately predict the rotation and streamline curvature characteristics of the clearance leakage vortex flow and the leakage vortex cavitation flow [29]. Therefore, the SST-CC turbulence model is used in this paper. The Zwart–Gerber–Belamri (ZGB) cavitation model is also used for simulation. The ZGB cavitation model has good compatibility with various turbulence models and can be selected directly in ANSYS CFX software 19.2.

2.2. Object of Study

The study is focused on the prototype and model of a Kaplan turbine, with the model being created by scaling the prototype turbine to equal proportions. The basic parameters of the Kaplan turbine are shown in Table 1. Figure 1 shows the three-dimensional modeling of the whole passage, including five parts: spiral casing, stay vanes, guide vanes, runner, and draft tube.

Table 1. Basic parameters of Kaplan turbine.

Geometry	Prototype	Model
Runner diameter D_1	29.71 D_m	D_m
Stay vanes Z_s	25	25
Guide vanes Z_g	28	28
Runner blades Z	6	6
shroud clearance width ω	4% D_m~4.57% D_m	1.34‰ D_m~1.54‰ D_m

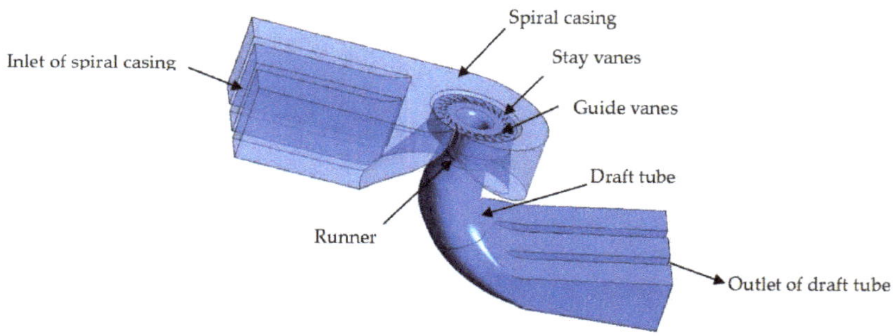

Figure 1. The whole passage of the Kaplan turbine.

2.3. Computation Domain

Considering the large geometric size of the prototype turbine, the finer meshes mean a large quantity, which can obtain more accurate calculation results. Also, the clearance size is small, and the clearance leakage vortex demands more mesh cells, resulting in an even larger quantity of mesh cells. Huge computing resources and high computing costs will inevitably be incurred due to the increase in the number of mesh cells. Therefore, the single-blade channel is intended to be used as the computing domain in this calculation, as shown in Figure 2.

To verify the rationality of the calculation of the single-blade computing domain, Table 2 shows the results of the runner region of the whole passage and the single-blade computing domain of the model turbine. The operating parameters are as follows: the head is 8 m, the blade angle is 0°, and the guide vane angle is 28°. Since the whole passage includes six blades, for the convenience of comparison, the calculation results of the single-blade computing domain in the table are also converted into the results under six blades. The parameters in Table 2 include the energy characteristics of the runner region and the minimum pressure coefficient (C_P) of the clearance leakage vortex cores. The energy characteristics mainly include flow coefficient (Q^*), head coefficient (H^*), power coefficient (P^*), and efficiency (η). As shown by the energy characteristics, the results obtained from the single-blade channel and the whole passage are very close to each other, indicating that the single-blade channel calculation method can predict the energy characteristics of the runner domain very well. As shown by the minimum pressure coefficient of the clearance leakage vortex cores, the pressure coefficient obtained from the single-blade channel calculation is lower, indicating that it can better capture the cavitation phenomenon of the clearance leakage vortex. Therefore, it is valid to use a single-blade computing domain for shroud clearance cavitation flow calculations in this paper.

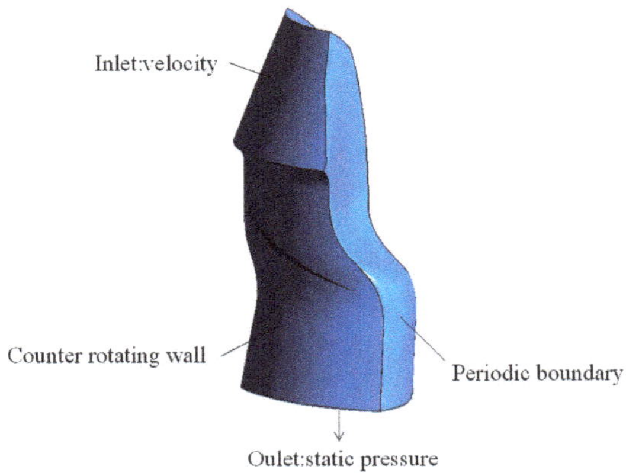

Figure 2. The single-blade computation domain and boundary conditions.

Table 2. The calculation results of the whole passage and the single-blade computing domain.

Computational Domain	Q^*	H^*	P^*	η	C_P
Whole passage	0.545	0.266	1349	95.13%	−0.929
Single-blade	0.541	0.263	1324	95.19%	−1.074

The parameters in Table 2 are defined as follows:

$$Q^* = \frac{Q_v}{D^3 n} \tag{6}$$

$$H^* = \frac{H_r}{D^2 n^2} \tag{7}$$

$$P^* = \frac{2\pi n T}{D^5 n^3} \tag{8}$$

$$\eta = \frac{P^*}{\rho g Q^* H^*} \tag{9}$$

$$C_P = \frac{p - p_{in}}{\frac{1}{2}\rho V_{tip}^2} \tag{10}$$

In the formulas, D represents the runner diameter, n represents the spindle speed, Q_v represents the volume flow rate, H_r represents the head of the runner domain, T represents the output torque of the spindle, ρ represents the density of water, g represents the gravity acceleration, p represents the minimum pressure at vortex cores, p_{in} represents the average pressure at the inlet of the runner, and V_{tip} is the circumferential velocity of the runner tip, calculated using the product of the tip radius and the rotational angular velocity of the runner.

2.4. Operating Conditions and Boundary Conditions

In this paper, the differences in cavitation characteristics of the shroud clearance are studied with the variation of the cavitation coefficient between the prototype and the model. The operating parameters of the prototype and the model are as follows: the blade angle (β) is 0°, the guide vane angle (α) is 28°, the unit flow rate (Q_{11}) is 1.05 m^3/s, and the unit speed (n_{11}) is 115.38 r/min. The unit flow rate (Q_{11}), the unit speed (n_{11}), and the cavitation coefficient (N^*) are the same for the prototype and the model under similar operating conditions. The three parameters are defined as

$$Q_{11} = \frac{Q_v}{D^2\sqrt{H}} \tag{11}$$

$$n_{11} = \frac{nD}{\sqrt{H}} \tag{12}$$

$$N^* = \frac{H_a - H_{va} - H_s}{H} \tag{13}$$

In the formula, H represents the head of the unit, and H_a, H_{va}, and H_s are the atmospheric pressure, the liquid vapor pressure, and the suction height of the turbine, respectively, which are expressed in terms of the height of the liquid column.

Valid boundary conditions for the inlet and outlet are necessary for the calculation of the single-blade calculation domain. To obtain a more accurate understanding of the inlet speed of the runner, the whole passage calculation domain is first calculated in this paper. Three velocity components (axial, circumferential, and radial) are set as boundary conditions for the inlet of the single-blade calculation domain. The specific values are obtained by averaging circumferentially the three velocity components on the inlet face of the runner in the whole passage calculation. The outlet of the single-blade calculation domain is set as a static pressure outlet, the value of which is the average static pressure value on the outlet of the runner in the whole passage calculation. The calculation domain is set to the rotating domain, and the runner chamber is set to the counter-rotating wall, indicating that it is stationary regarding the stationary domain. Both sides of the blade are set to periodic boundary conditions, and the wall surface is used as a smooth no-slip wall surface. The boundary conditions for the main parts are depicted in Figure 2. Gravity is considered in the calculation and is used as the external force source term. The initial value for the cavitation calculation is determined by performing a no-cavitation calculation before the cavitation calculation. The simulation results are obtained after the mass and momentum equations converge and the efficiency of the runner domain becomes stable.

2.5. Mesh Independence Verification

The calculation focuses on the energy characteristics of the runner domain and the flow characteristics of the shroud clearance. The efficiency of the runner domain and the minimum pressure coefficient of the clearance leakage vortex cores are used as evaluation criteria for verifying mesh independence. Using the same calculation settings, six sets of meshes were computed for the prototype and five sets of meshes for the model. The results of different numbers of mesh cells are shown in Figure 3.

Compared to the results of the 4.2 million mesh cells and the 5.2 million mesh cells, the relative error of the model efficiency is about 0.028%, and the relative error of the minimum pressure coefficient of the clearance leakage vortex cores is about 0.37%. Compared to the results of the 5.2 million mesh cells and the 6.2 million mesh cells, the relative error of the prototype efficiency is about 0.014%, and the relative error of the minimum pressure coefficient of the clearance leakage vortex cores is about 0.75%. The evaluation criteria for mesh independence verification in the prototype and the model are not more than 1%. It is evident that the mesh independence validation for the 4.2 million mesh cells in the model and 5.2 million mesh cells in the prototype has been met. In addition, the significant differences between the prototype and the model can be observed in the Section 3. The differences were obtained by comparing the energy characteristics of the runner domain, clearance leakage characteristics, and clearance leakage vortex cavitation characteristics. The calculation results in the Section 3 are obtained by using 5.2 million mesh cells in the prototype and 4.2 million mesh cells in the model. Therefore, the final number of mesh cells in the prototype is 5.2 million, while the number of mesh cells in the model is 4.2 million. The number of mesh layers of shroud clearance is 20. The average Y+ value of the wall of the prototype runner blade is about 100 and that of the model runner blade is about 30. The meshes of the calculation domain, blade, and shroud surface are shown in Figure 4.

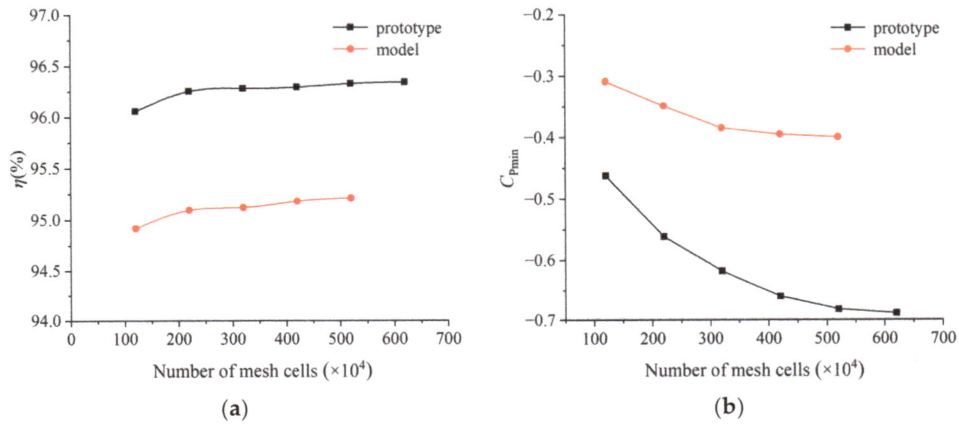

Figure 3. The relation curves between the validation indexes and the mesh number. (**a**) Efficiency; (**b**) the minimum pressure coefficient of the clearance leakage vortex cores.

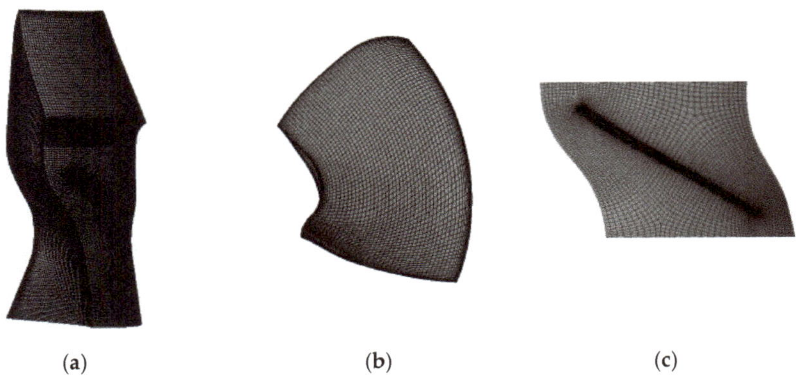

Figure 4. Meshes of the calculation domain. (**a**) Mesh in the single-blade calculation domain; (**b**) mesh on the blade surface; (**c**) mesh on the shroud surface.

3. Calculation Results

3.1. Analysis of Calculation Results

3.1.1. Energy Characteristics

As one of the main parameters of turbine performance, the cavitation coefficient has a significant impact on both the cavitation characteristics and energy characteristics of the unit. Figure 5 shows the relationship curves between the energy characteristics and the cavitation coefficients (N^*). The head coefficient (H^*), power coefficient (P^*), and efficiency (η) of the runner domain in the prototype and the model have a similar variation pattern when the cavitation coefficient decreases.

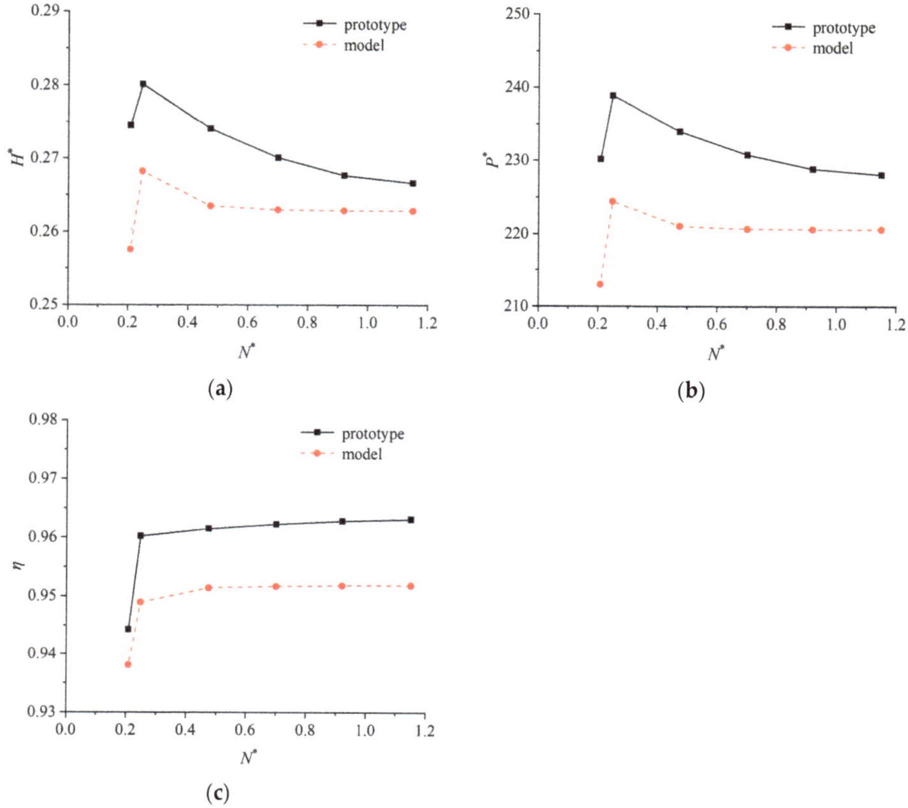

Figure 5. The relationship curves between the energy characteristics and the cavitation coefficients. (**a**) Head coefficient; (**b**) power coefficient; (**c**) efficiency.

When the cavitation coefficient (N^*) decreases from 1.15 to 0.248, the head coefficient (H^*) of the prototype and the model increases by 5.07% and 2.04%, respectively, as illustrated in Figure 5a. This shows that the energy characteristics of the runner domain, especially the prototype runner domain, will be slightly improved with the reduction in the cavitation coefficient in this range. When the cavitation coefficient (N^*) continued to decline to 0.208, the water head coefficient (H^*) began to decline rapidly. Compared to the cavitation coefficient $N^* = 0.248$, the head coefficient of the prototype and the model decreased by 2.02% and 3.97%, respectively. This indicates that when the cavitation coefficient (N^*) is below 0.248, the cavitation phenomenon will be detrimental to the working ability of the runner, especially the model runner. The variant rule of power coefficient (P^*) with cavitation coefficient (N^*) is similar to that of the head coefficient (H^*), as shown in Figure 5b. With the decrease in the cavitation coefficient, the power coefficients (P^*) of the prototype and the model first increase by 4.75% and 1.73%, respectively, and then decrease by 3.65% and 5.03%, respectively, relative to their peak points. The relationship curves between the efficiency and the cavitation coefficients can be seen in Figure 5c. Different from the changes in the head coefficient (H^*) and the power coefficient (P^*), the efficiencies (η) of the prototype and the model remain constant when the cavitation coefficient (N^*) is greater than 0.474, and then start to decline with the decrease in the cavitation coefficient (N^*). The efficiency drops sharply when the cavitation coefficient (N^*) is below 0.248. The cavitation coefficient when the efficiency of the runner domain decreases by 1% is defined as the critical cavitation coefficient. The critical cavitation coefficient of the prototype and the model was estimated by interpolation to be 0.230 and 0.222, respectively. Although the critical cavitation coefficient of the prototype is higher than that of the model, they are still comparable. The critical cavitation coefficient calculated by the model can be considered as a reference to the prototype.

Comparing the prototype and the model, the energy characteristics of the prototype are higher than those of the model at the same cavitation coefficient. As the cavitation coefficient increases, so does the difference. At the point where the energy characteristics drop, the energy characteristics of the prototype are about 4.43%, 6.42%, and 1.19% higher than those of the model, respectively.

3.1.2. Clearance Leakage Characteristics

Figure 6 shows the leakage flow and the leakage vortex in the region of the shroud clearance, as well as the relationship curves between the clearance leakage flow rate and cavitation coefficients (N^*). The value of the clearance leakage flow rate (Q_{leak}) is normalized by the flow rate of the inlet (Q_{in}). Both the prototype and the model exhibit a gradual decrease in clearance leakage flow rate as the cavitation coefficient (N^*) decreases. The clearance leakage flow rate changes are small in both the prototype and the model when the cavitation coefficient (N^*) is in the range of 1.15 to 0.7. When the cavitation coefficient (N^*) decreases to 0.474, the clearance leakage flow rate of the prototype decreases by 1.66% relative to when the cavitation coefficient (N^*) is 0.7, while that of the model decreases by 0.33%. The decrease rate of the clearance leakage flow rate of the prototype is obviously higher than that of the model, which is related to the cavitation state of the clearance region. When the flow channel near the clearance is blocked by the clearance cavitation, the clearance leakage flow rate will be reduced. The clearance leakage flow rate decreases as the flow path becomes more blocked. When the cavitation coefficient (N^*) decreases further, the degree of cavitation near the clearance region becomes more severe, and the clearance leakage flow rate also continues to decrease. The clearance leakage flow rates of the prototype and model decrease quickly, with the cavitation coefficient (N^*) decreasing from 0.248 to 0.208, resulting in a decrease of 1.29% and 2.53%, respectively. During this process, the energy characteristics of the prototype and the model also decrease rapidly.

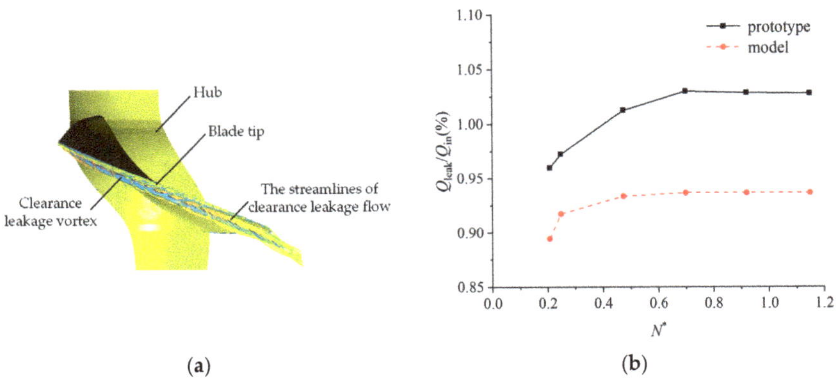

Figure 6. Clearance leakage characteristics. (**a**) The streamlines of clearance leakage flow and the clearance leakage vortex. (**b**) The relationship curves between the clearance leakage flow rate and cavitation coefficients.

3.1.3. Cavitation Distribution near the Blade Tip

The tip clearance cavitation and clearance leakage vortex cavitation in the shroud clearance region are shown in Figure 7. Reference [30] describes the formation reasons for them.

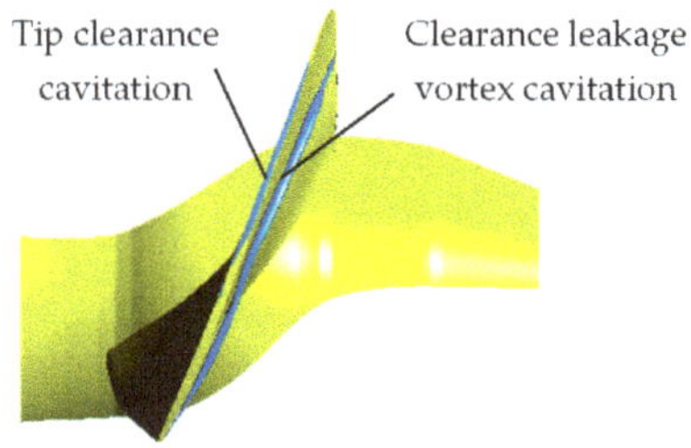

Figure 7. The tip clearance cavitation and clearance leakage vortex cavitation in the shroud clearance region.

The cavitation distribution near the blade tip can be seen in Figure 8 by utilizing the isosurface of vapor volume fraction ($\alpha_V = 0.1$). The degree of cavitations in the blade tip region of both the prototype and the model becomes more intense with the decrease in the cavitation coefficient (N^*). Specifically, when cavitation occurs, the clearance leakage vortex cavitation starting from the head of the blade occurs earlier than the tip clearance cavitation, which is in the middle and the tail of the blade. With a decreased cavitation coefficient (N^*), the clearance leakage vortex cavitation diameter thickens and continues to extend to the outlet side of the blade, while the tip clearance cavitation develops bidirectionally towards the head and the tail of the blade. The interaction between tip clearance cavitation and clearance leakage vortex cavitation occurs when the cavitation coefficient (N^*) is 0.248. The tail of the blade was completely covered by the cavitation region. As the cavitation coefficient (N^*) continues to decrease, the cavitation degree near the tip intensifies further, resulting in a stronger blocking effect on the flow passage.

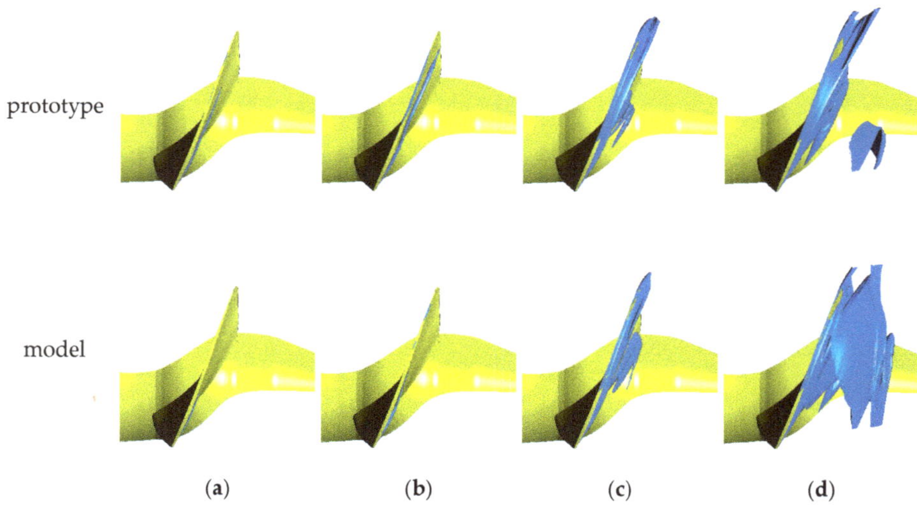

Figure 8. The isosurfaces of the vapor volume fraction $\alpha_V = 0.1$ under different cavitation coefficients. (**a**) $N^* = 0.7$; (**b**) $N^* = 0.474$; (**c**) $N^* = 0.248$; (**d**) $N^* = 0.208$.

Comparing the cavitation region near the blade tip of the prototype and the model, it can be seen that when the cavitation coefficient (N^*) is bigger than 0.248, the degree of tip clearance cavitation and clearance leakage vortex cavitation of the prototype is more serious than that of the model under the same cavitation coefficient. The clearance leakage vortex cavitation of the prototype is further away from the suction surface of the blade. When the cavitation coefficient (N^*) is below 0.248, the degree of cavitation near the blade tip of the model is more obvious and more serious than that of the prototype.

3.1.4. Volume Change of the Runner Cavitation

As can be seen from Figure 8, the tip clearance cavitation and the clearance leakage vortex cavitation can be seen more clearly than other cavitations in Figure 8a,b. As the cavitation coefficient decreases, all the cavitations increase in intensity. Other cavitations can also be seen in Figure 8c,d. In Figure 8d, it should be noted that other cavitations have been more serious than tip clearance cavitation and clearance leakage vortex cavitation. To quantitatively analyze the variation in clearance cavitation near the blade tip with the cavitation coefficient, the flow field was divided into the flow field near the blade tip and the other flow fields. According to the cavitation morphology near the blade tip when the cavitation coefficient (N^*) is 0.208, the position of the blade-spreading length with the shroud facing inwards by 20% is divided into the flow field near the blade tip. The position remains constant under different cavitation coefficients. The cavitation in the flow field near the blade tip is called tip cavitation, and the cavitation in other flow fields is called other cavitation. For ease of differentiation, the volume of the entire flow passage was denoted as V_{RV}, the total cavitation volume in the flow passage was denoted as $V_{\text{cav-total}}$, the tip cavitation volume was denoted as $V_{\text{cav-tip}}$, and the other cavitation volumes were denoted as $V_{\text{cav-other}}$.

Figure 9a,b show the variations of the ratios of the tip cavitation volume and other cavitation volumes to the total cavitation volume in the runner domain under the different cavitation coefficients, respectively. Figure 9c,d are the variations of the ratios of the tip cavitation volume and other cavitation volumes to the total volume of the flow field under the different cavitation coefficients, respectively. Combined with Figure 9a,b, it can be seen that when the cavitation coefficient (N^*) is bigger than 0.248, the maximum value of the ratio of the tip cavitation volume to the total cavitation volume in the computational domain is about 99% in the prototype, while it is about 98% in the model. With the decrease in the cavitation coefficient (N^*), the ratio of the tip cavitation volume to the total cavitation volume is continuously decreasing, and the ratio of the other cavitation volumes to the total cavitation volume is continuously increasing. When the cavitation coefficient (N^*) is 0.208, the ratio of the tip cavitation volume to the total cavitation volume in the computational domain is about 49% in the prototype, while it is about 37% in the model. The indication is that other cavitation in the runner domain is more serious than tip clearance cavitation at the moment. Combining Figure 9c,d, it can be seen that the tip cavitation volume and other cavitation volumes of the prototype and the model increase exponentially in the flow passage with the decrease in the cavitation coefficient (N^*). The ratio of the tip cavitation volume to the flow field volume of the prototype is higher than that of the model when the cavitation coefficient (N^*) is greater than 0.248. The ratio of the tip cavitation volume to the flow field volume in the prototype is lower than that in the model when the cavitation factor (N^*) is 0.208. The conclusions of cavitation morphology in the runner domain are in agreement with this. In addition, it can be seen that when the cavitation coefficient (N^*) is below 0.248, the volumes of tip cavitation and other cavitation in the flow passage increase sharply. The energy characteristics of the runner domain show a rapid decline trend at this time, indicating that the change in cavitation volume has a great influence on the energy characteristics of the runner domain.

3.1.5. Distribution of Pressure Coefficients on the Blade Surfaces

When the cavitation coefficient (N^*) is 0.474, the cloud plots of the distribution of the pressure coefficients (C_{P}) on the suction surfaces of the prototype and model are shown in Figure 10. The pressure coefficient is determined by using Formula (10), with the p here representing the static pressure on the surface of the blade. It can be seen that the distribution of pressure coefficients on the suction surface of the blade in the prototype and model is similar. Due to the influence of the clearance leakage vortex, a strip of low-pressure region similar to the trajectory of the clearance leakage vortex appears at the tip of the blade.

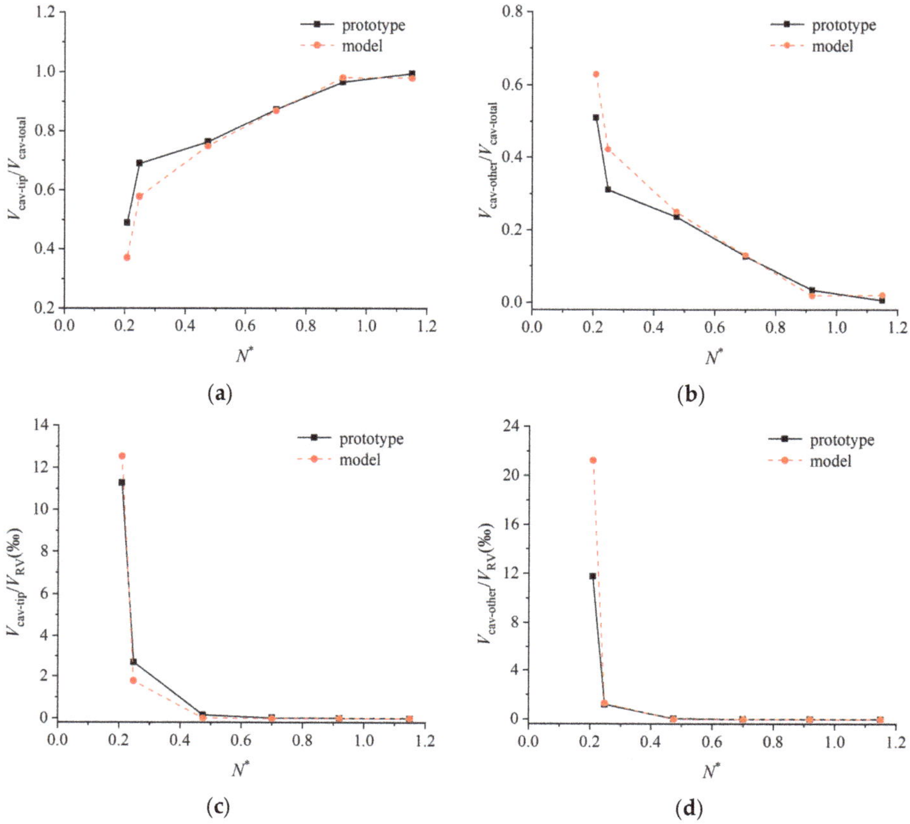

Figure 9. The relationship curves between the ratios of different cavitation volumes and cavitation coefficients. (**a**) $V_{cav\text{-}tip}/V_{cav\text{-}total}$; (**b**) $V_{cav\text{-}other}/V_{cav\text{-}total}$; (**c**) $V_{cav\text{-}tip}/V_{RV}$; (**d**) $V_{cav\text{-}other}/V_{RV}$.

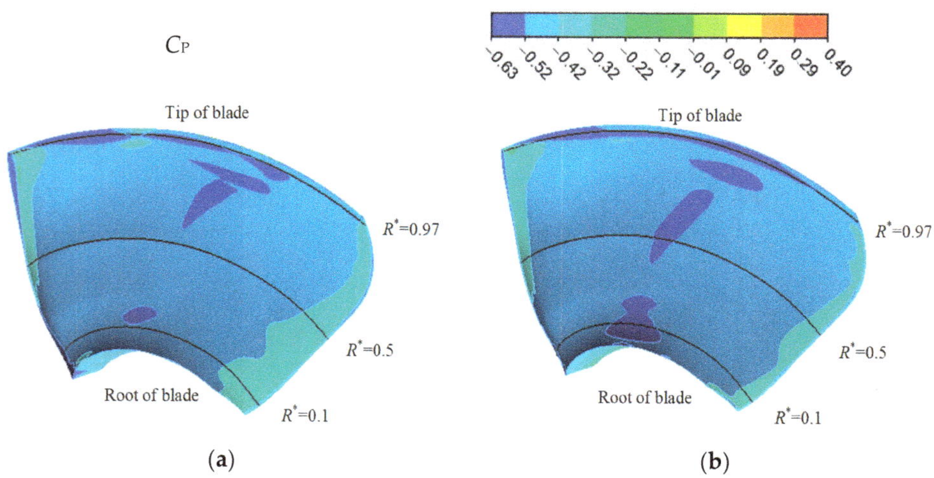

Figure 10. Distribution of pressure coefficients on the blade surfaces at $N^* = 0.474$. (**a**) Prototype; (**b**) model.

The positions $R^* = 0.97$, $R^* = 0.5$, and $R^* = 0.1$ of the blade span were selected to explore the distribution of pressure coefficients on the blade surfaces under different cavitation coefficients, as shown in Figure 11. Taking the device cavitation coefficient as an example, it can be seen that the prototype and model have similar changes in pressure coefficient on their blade surfaces under the same cavitation coefficient. The fluctuations in the pressure coefficients at the positions $R^* = 0.97$ and $R^* = 0.1$ are stronger than those at the position $R^* = 0.5$ along the flow direction, which is mainly influenced by the shroud clearance flow and hub clearance flow. At the position $R^* = 0.97$, it can be seen that the pressure drop occurs at the same chord length position for the prototype and the model. This indicates that the initial point of the clearance leakage vortex cavitation along the chord length is close to each other. However, the extent of the low-pressure region of the prototype is longer than that of the model along the chord length, which means that the clearance leakage vortex of the prototype has a greater influence on the suction surface of the blade. The distribution of pressure coefficients on the pressure surface remains constant regardless of the cavitation coefficient. The suction surface of the blade has a greater variation in pressure coefficients, particularly at the tip and root positions. This is because the clearance flows of the shroud and hub are greatly affected by the cavitation coefficient. At position $R^* = 0.97$, the local pressure on the suction surface of the blade increases as the cavitation coefficient decreases. The local pressure increases will reduce the pressure difference between the pressure surface and suction surface of the blade to a certain extent and thus reduce the pressure load of the blade. Although the position of the pressure drop point on the blade surface does not change much with the decrease in the cavitation coefficient, the range of low-pressure regions across the chord length expands.

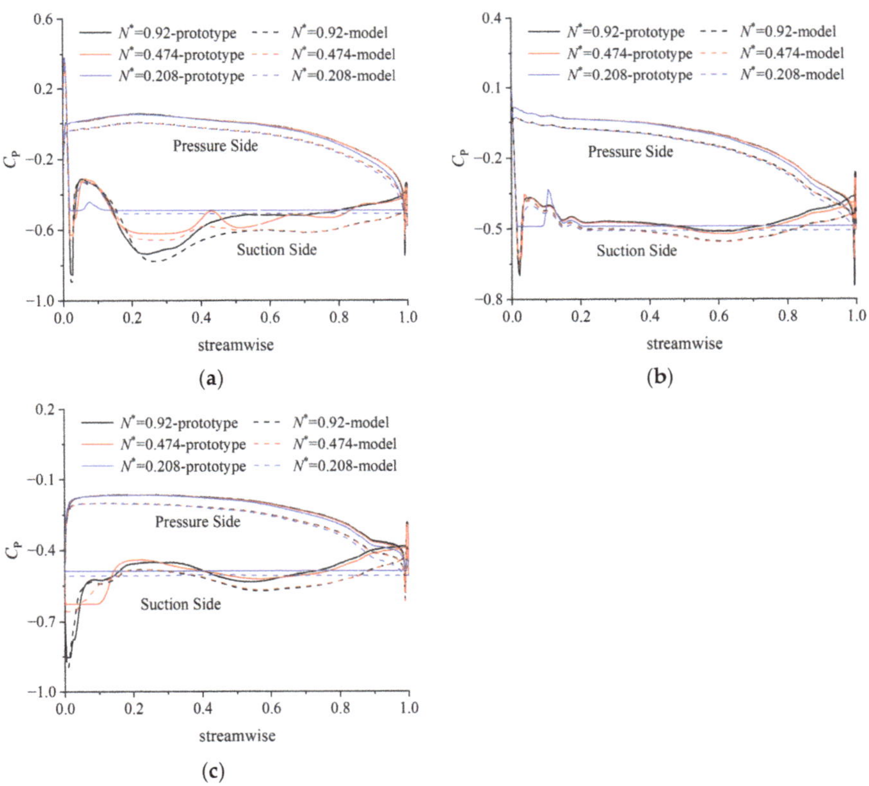

Figure 11. Distribution of pressure coefficients on the blade surfaces under different cavitation coefficients. (**a**) $R^* = 0.97$; (**b**) $R^* = 0.5$; (**c**) $R^* = 0.1$.

3.1.6. Cavitation Characteristics of the Runner Chamber

The distribution cloud map of the vapor volume fraction of the shroud surface can be seen in Figure 12. The cavitation characteristics of the runner chamber can be analyzed using these cloud maps. The cavitation of the runner chamber in the prototype and the model mainly occurs in the region from the middle of the blade to the exit edge of the blade. With the decrease in the cavitation coefficient, the cavitation of the runner chamber in the prototype occurs earlier than that in the model. When the cavitation coefficient is bigger than the device cavitation coefficient, the maximum vapor volume fraction on the shroud surface of the prototype and model is approximately 0. When the cavitation coefficient is 0.474, the maximum air volume fraction of the shroud surface of the prototype is about 0.32, while that of the model is about 0. It can be thought that the cavitation may occur in the runner chamber of the prototype under the device cavitation coefficient, but not in the runner chamber of the model. When the cavitation coefficient continued to decrease to close to the critical cavitation coefficient, the runner chamber of the model began to cavitate. when the cavitation coefficient is 0.208, the cavitation region on the runner chamber of the model is larger than that of the prototype, which is caused by the more serious cavitation degree at the tip of the model.

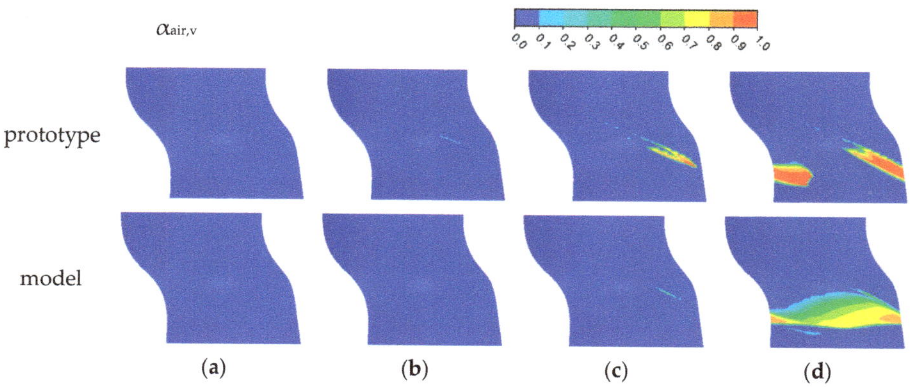

Figure 12. The vapor volume fraction on the shroud surface under different cavitation coefficients. (**a**) $N^* = 0.7$; (**b**) $N^* = 0.474$; (**c**) $N^* = 0.248$; (**d**) $N^* = 0.208$.

4. Conclusions

As the change in cavitation coefficients, the rules of the clearance cavitation flow between the prototype and the model are similar, but there are large differences in the energy characteristics and cavitation characteristics between the two. The specific characteristics are as follows.

1. Under the cavitation conditions, the energy characteristics (head coefficient, power coefficient, and efficiency) of the prototype are higher than those of the model. The operating conditions where the energy characteristics drop rapidly are the same. In addition, the critical cavitation coefficient of the prototype is close to that of the model. To some extent, the critical cavitation coefficient calculated by the model has reference significance for the prototype.
2. When the cavitation coefficient is larger than the critical cavitation coefficient, the cavitation characteristics of the blade and the runner chamber in the prototype are more serious than those in the model. When the cavitation coefficient is less than the critical cavitation coefficient, the degree of cavitation in the runner domain of the model is intensified. The cavitation characteristic of the runner chamber of the model is more serious than that of the prototype.
3. With the decrease in the cavitation coefficient, the cavitation of the runner chamber in the prototype occurs earlier than that in the model. The runner chamber in the

prototype experiences cavitation at the device cavitation coefficient, while the runner chamber in the model does not. When the cavitation coefficient is reduced to close to the critical cavitation coefficient, the energy characteristics of the runner do not change much, but the cavitation significantly intensifies in the prototype and model.

Author Contributions: Conceptualization, Y.Z.; methodology, W.L. and L.Z.; software, Y.Z.; validation, T.C.; investigation, Y.Z. and W.L.; writing—original draft preparation, Y.Z.; writing—review and editing, Z.W. and W.L; visualization, Y.Z. and T.C.; supervision, L.Z. and W.L.; project administration, Z.W. and T.C. All authors have read and agreed to the published version of the manuscript.

Funding: This research was funded by the National Natural Science Foundation of China (No. 52079141).

Institutional Review Board Statement: Not applicable.

Informed Consent Statement: Not applicable.

Data Availability Statement: All data generated or analyzed during this study are included in this article.

Conflicts of Interest: Authors Wendong Luo and Tao Chen were employed by the company Guangxi Datengxia Gorge Water Conservancy Development Co. Ltd. The remaining authors declare that the research was conducted in the absence of any commercial or financial relationships that could be construed as a potential conflict of interest.

References

1. Zhang, A.M.; Huang, W.C.; Li, D.F. Problems in the operation of Kaplan turbine. *China Urban Econ.* **2011**, *30*, 260.
2. Liao, W.L.; Liu, S.Z.; Zhang, L.F. Experimental investigation of the tip clearance cavitations on blades flange of Kaplan turbine. *J. Hydroelectr. Eng.* **2005**, *4*, 67–72.
3. Chen, Q.F.; Luo, Y.Y.; Li, X.G.; Ahn, S.H.; Zhang, J.; Wang, Z.W.; Xiao, Y.X. Influence of runner clearance on efficiency and cavitation in Kaplan turbine. *IOP Conf. Ser. Earth Environ. Sci.* **2018**, *163*, 012068. [CrossRef]
4. You, D.; Wang, M.; Moin, P.; Mittal, R. Vortex dynamics and low-pressure fluctuations in the vortex dynamics and tip-clearance flow. *J. Fluids Eng.* **2007**, *129*, 1002–1014. [CrossRef]
5. Tan, D.Q.; Han, X.L.; Liu, B.; Liu, W.J.; Wei, X.Z. Research on differences between clearance cavitation and airfoil cavitation in axial flow turbine. *Large Electr. Mach. Hydraul. Turbine* **2012**, *2*, 34–37.
6. Xiao, X.W.; McCarter, A.A.; Lakshminarayana, B. Tip clearance effects in a turbine rotor: Part I—Pressure field and loss. *J. Turbomach.* **2001**, *123*, 296–304. [CrossRef]
7. Amiri, K.; Mulu, B.; Cervantes, M.J. Experimental investigation of the interblade flow in a Kaplan runner at several operating points using Laser Doppler Anemometry. *J. Fluids Eng.* **2016**, *138*, 021106. [CrossRef]
8. Ma, Y.; Qian, B.; Feng, Z.G.; Wang, X.; Shi, G.T.; Liu, Z.K.; Liu, X.B. Flow behaviors in a Kaplan turbine runner with different tip clearances. *Adv. Mech. Eng.* **2021**, *13*, 1069494471. [CrossRef]
9. Wu, Y.Z.; Wang, X.H.; Yang, X.L.; Ding, J.F.; Zhu, D.; Tao, R.; Wang, H.M.; Xiao, R.F. Prediction of the influence of runner tip clearance on the performance of tubular turbine. *J. Mar. Sci. Eng.* **2022**, *10*, 136. [CrossRef]
10. Liu, J.-S.; Bozzola, R. Three-dimensional Navier-Stokes analysis of tip clearance flow in linear turbine cascades. *AIAA J.* **1993**, *31*, 2068–2074. [CrossRef]
11. Nikiforova, K.; Semenov, G.; Kuznetsov, I.; Spiridonov, E. Numerical investigation of tip clearance cavitation in Kaplan runners. *IOP Conf. Ser. Earth Environ. Sci.* **2016**, *49*, 092008. [CrossRef]
12. Kim, H.-H.; Rakibuzzaman, M.; Kim, K.; Suh, S.-H. Flow and fast fourier transform analyses for tip clearance effect in an operating Kaplan turbine. *Energies* **2019**, *12*, 264. [CrossRef]
13. Wang, L.K.; Lu, J.L.; Liao, W.L.; Zhao, Y.P.; Wang, W. Numerical simulation of the tip leakage vortex characteristics in a semi-open centrifugal pump. *Appl. Sci.* **2019**, *9*, 5244. [CrossRef]
14. Zhou, J.L. Selection problems for pump station equipment. *Drain. Irrig. Mach.* **2001**, *19*, 3–12.
15. Osterwalder, J. Efficiency scale-up for hydraulic turbo-machines with due consideration of surface roughness. *J. Hydraul. Res.* **2010**, *16*, 55–76. [CrossRef]
16. Zheng, Y.C.; He, Z.R.; Chen, J.; Liu, D.X. Discussion on conversion method of performance parameters of the large pump unit. *J. Hydraul. Eng.* **1995**, *10*, 47–52.
17. Hutton, S.P. Component losses in Kaplan turbines and the prediction of efficiency from model tests. *Proc. Inst. Mech. Eng.* **1954**, *168*, 743–762. [CrossRef]
18. Ida, T. Analysis of scale effects on performance characteristics of hydraulic turbines—Part I. *J. Hydraul. Res.* **1989**, *27*, 809–831. [CrossRef]
19. Ida, T. Analysis of scale effects on performance characteristics of hydraulic turbines—Part II. *J. Hydraul. Res.* **1990**, *28*, 93–104. [CrossRef]

20. Ida, T. New formulae for scaling-up hydraulic efficiency of hydraulic turbines. *J. Hydraul. Res.* **1995**, *33*, 147–162. [CrossRef]
21. Chen, S.S.; Ma, X.Z.; Chen, J.Q.; Zhou, Z.F.; He, Z.N. The conversion method of characteristic parameters of pump and pump set between prototype and model. *J. Yangzhou Univ. (Nat. Sci. Ed.)* **2015**, *18*, 45–48.
22. McCormick, B.W. On cavitation produced by a vortex trailing from a lifting surface. *J. Basic Eng.* **1962**, *84*, 369–378. [CrossRef]
23. Ren, J.; Chang, J.S. The scale effect of cavitation in the large hydraulic unit. *Water Resour. Hydropower Eng.* **1998**, *29*, 22–25.
24. Ni, H.G. Correction of incipient cavitation coefficient for scale effect. *J. Hydraul. Eng.* **1999**, *30*, 28–32.
25. Angulo, M.; Luciano, C.; Botero, F.; Rivetti, A.; Liscia, S. Dynamic measurements on a Kaplan turbine: Model—Prototype comparison. *IOP Conf. Ser. Earth Environ. Sci.* **2019**, *240*, 022006. [CrossRef]
26. Angulo, M.; Rivetti, A.; Díaz, L.; Liscia, S. Air injection test on a Kaplan turbine: Prototype—Model comparison. *IOP Conf. Ser. Earth Environ. Sci.* **2016**, *49*, 022006. [CrossRef]
27. Zeng, C.J.; Xiao, Y.X.; Zhang, J.; Gui, Z.H.; Wang, S.H.; Luo, Y.Y.; Fan, H.G.; Wang, Z.W. Numerical prediction of hydraulic performance in model and homologous prototype Pelton turbine. *IOP Conf. Ser. Earth Environ. Sci.* **2018**, *163*, 012016. [CrossRef]
28. Li, Z.J.; Wang, Z.W.; Bi, H.L. Numerical study of similarity in prototype and model pumped turbines. *IOP Conf. Ser. Earth Environ. Sci.* **2014**, *22*, 032049. [CrossRef]
29. Guo, Q. Study on the Characteristics of the Blade Tip Leakage Vortex Flow and the Cavitating Flow Field. Ph.D. Thesis, Agricultural University, Beijing, China, 2017.
30. Zhang, Y.L.; Wu, Y.B.; Wei, J.W.; Wang, Z.W.; Zhou, L.J. Clearance flow field characteristics of Kaplan turbine under different flange clearance. *IOP Conf. Ser. Earth Environ. Sci.* **2022**, *1037*, 012023. [CrossRef]

water

Article

Fractional-PID and Its Parameter Optimization for Pumped Storage Units Considering the Complicated Conduit System

Xuan Zhou [1], Yang Zheng [1,*], Bo Xu [2], Wushuang Liu [1], Yidong Zou [1] and Jinbao Chen [1]

1 School of Power and Mechanical Engineering, Wuhan University, Wuhan 430072, China; 2022202080048@whu.edu.cn (X.Z.); 2017302650012@whu.edu.cn (W.L.); yidongzou@whu.edu.cn (Y.Z.); 2020102080020@whu.edu.cn (J.C.)
2 Technology and Research Center, China Yangtze Power Co., Ltd., Yichang 443002, China; xu_bo1@cypc.com.cn
* Correspondence: zhengyang@whu.edu.cn

Abstract: Speed governing control is significant in ensuring the stable operation of pumped storage units. In this study, a state-space equation mathematical model of the pumped storage governing system considering the complex hydraulic pipeline structure of the pumped storage plant is proposed to describe the system's dynamic behaviors under small disturbance conditions. Considering the frequent operating condition transitions and the complicated nonlinear dynamic characteristics of the pumped storage units, the fractional-order PID (FOPID) scheme that possesses a higher degree of control freedom than the traditional PID scheme is discussed in detail. To optimize the control parameters of the unit governor, an improved gravitational search algorithm (IGSA) that combines the basic searching mechanisms of the gravitational search algorithm and chaotic search, elastic sphere boundary treatment, and elite guidance strategy is developed. Comparative studies have been carried out under frequency and load disturbance conditions. Simulation results indicate that the control performance of FOPID is better than that of PID under diverse operating conditions and the proposed IGSA has satisfactory parameter optimization capability.

Keywords: pumped storage unit; governing system; fractional-order PID; improved gravitational search algorithm

Citation: Zhou, X.; Zheng, Y.; Xu, B.; Liu, W.; Zou, Y.; Chen, J. Fractional-PID and Its Parameter Optimization for Pumped Storage Units Considering the Complicated Conduit System. *Water* **2023**, *15*, 3851. https://doi.org/10.3390/w15213851

Academic Editor: Giuseppe Pezzinga

Received: 29 September 2023
Revised: 25 October 2023
Accepted: 2 November 2023
Published: 4 November 2023

1. Introduction

With the gradual increase in the proportion of new energy sources such as wind power and solar energy in the electrical network [1,2], pumped storage is becoming more and more important as an effective regulation power source [3]. As the core control system of pumped storage units, the pumped storage governing system (PSGS) undertakes the important tasks of manipulating the start-up and shutdown of the unit, the working condition conversions, and the peak regulation and frequency regulation [4]. PSGS is a nonlinear complex system, which may present complicated dynamic behavior during operation. Therefore, the establishment of an accurate and effective simulation model [5] as well as the optimization of governor control parameters [6] can achieve a better control effect on the unit and ensure the stable operation of the unit.

With the development of pumped storage power plants, extensive studies have been reported on the modeling and control of PSGS. A typical PSGS consists of a hydraulic system, pump-turbine, generator, governor and electro-hydraulic servomechanism [7]. In system modeling, the PSGS is similar to the conventional hydraulic turbine governing system (HTGS) in turbine mode. Usually, time-domain models are utilized to simulate the evolution of state variables during the transition process. The state-of-the-art methods include the method of characteristics (MOC) [8,9], the finite difference method [8,10], and the differential equation model [10], etc. Under small disturbance conditions, the system model ignores the system nonlinearities and selects a rigid water hammer model for the pipes and six-parameter model for the pump-turbine [11]; hence, the PSGS model can be described by linearized state space equations [12]. For the speed governor, PID controller is the most commonly used for PSGS due to its structural simplicity and applicability. With the in-depth study of PID control laws, PID-like control laws such as nonlinear PID [13] and FOPID [14] are increasingly used. FOPID is widely used in various engineering fields [15], e.g., the chemical industry [16], nuclear power [17], and aerospace [18]. Compared with traditional PID control, FOPID control possesses two more adjustable parameters, i.e., the differential order and the integral order. It allows the FOPID controller to have better robustness and control performance than the PID controller, but at the same time increases the difficulty of parameter optimization [19].

In pumped storage regulation systems, the governor is a very important component. The system's control performances are mainly decided by the tunable control parameters, and the optimization of these parameters can prominently enhance the system stability level. For the controller parameter optimization problem, the traditional rectification methods include the orthogonal test approach [20] and the simplex approach [21]. Although these methods are simple to operate, they are not accurate enough to obtain the optimal parameters under complex operating circumstances. The meta-heuristic algorithms [22] (MA) have achieved better results in optimizing the controller parameters of hydropower units. MA include the Genetic Algorithm [23] (GA) based on Darwinian evolutionary theory, the Particle Swarm Optimization algorithm [24] (PSO) based on bird predation, the Differential Evolution Algorithm [25] (DEA) that uses the differences between random vectors to generate new vectors, the Gradient Descent Algorithm [26] (GDA) that uses the gradient search algorithm to keep approaching the optimal result, the Pattern Search Algorithm [27] (PSA) with axial exploration in the direction of feasible descent, and the Simulated Annealing Algorithm [28] (SA) based on the annealing process of solids. The Sine Cosine Algorithm (SCA) achieves close to the optimal result by the exact solution of the sine–cosine function [29], and the Gravity Search Algorithm [30] (GSA) based on gravity, etc. Unlike GA and PSO, which are based on biological phenomena, GSA is an optimization method based on the laws of gravity and mass interaction in physics. GSA has been verified to be more efficient than other optimization methods in optimizing the controller parameters [31], but still suffers from localized and premature convergence problems. However, almost all of the above research works have used the conventional PID controller which ultimately leads to a sub-optimal response. The superiority of the FOPID controllers are demonstrated in the results of the research work in this paper.

The accuracy and efficiency of optimization algorithms are crucial in solving engineering problems. In recent years, scholars have performed various research works in related fields and proposed many effective improvement ideas, including boundary processing, strategy improvement, algorithm fusion and parameter improvement. To improve the capability of GSA, Li et al. [32] proposed an improved gravitational search algorithm combining PSO and GSA. Sarafrazi et al. [33] proposed fusing GSA with a physics-inspired Kepler algorithm to speed up the algorithm's search and improve accuracy. Zhang et al. [34] designed and proposed a hybrid strategy based on Cauchy and Gaussian mutation to improve the exploration of GSA. Yin et al. [35] incorporate a cross-search in the GSA to improve the algorithm's development capability. He et al. [36] introduced the concept of repulsive force and proposed an improved gravitational search algorithm under the joint effect of repulsive and gravitational forces. Tian et al. [37] combined the Water Cycle Algorithm (WCA) with the GSA and use the concepts of watersheds and evapotranspiration to enhance the search capability. Based on previous research, this paper proposes an improved GSA (IGSA) in which the following four improvement strategies are incorporated into the GSA. Firstly, the chaotic operator is added to increase the diversity of the population and the randomness of the search; secondly, the adaptive gravitational decay factor is introduced to improve the change rule of the gravitational constant; subsequently, the elastic sphere strategy is carried out in the searching boundary treatment process; finally, the population of elite particles are applied to accelerate the convergence rate of the algorithm.

When using MA to optimize controller parameters, defining a proper objective function is crucial for the optimization of results. Usually, the input of the objective function is the parameter to be optimized, and the output value is the fitness value. In this paper, we select an objective function that comprehensively takes account of the rotational speed error and the water pressure fluctuation during the transition process.

Most of the traditional models do not take into account the hydraulic structures of the pumped storage power plant, such as the surge tanks, and cannot reflect the transient processes of the crucial components. In order to improve the dynamic description ability of the model, this paper establishes an accurate state-space model of the PSGS which fully considers the hydraulic characteristics of the pipes, surge tanks and pump-turbine. In order to obtain better control performance, the FOPID controller with higher control degrees of freedom is chosen and compared to the conventional PID controller. In addition, the standard GSA algorithm was improved in this study to better optimize the control parameters.

The rest of this paper is organized as follows: In Section 2, the mathematical equations of the PSGS derived from the state-space equations are established. In Section 3, the fractional order calculus and the structure of the FOPID controller are introduced. Subsequently, Section 4 introduces the GSA algorithm and the proposed improvements. Then, the simulation results are analyzed in Section 5. Finally, the conclusions of this study are condensed in Section 6.

2. Mathematical Model

The hydroelectric power plant is an important component in the modern power system that converts water energy into electricity. Pumped storage power plant is a special kind of hydroelectric power plant, which uses water as an energy storage medium to store and manage electrical energy through the mutual conversion of electrical energy and potential energy.

The PSGS is composed of upstream and downstream reservoirs, penstock, surge tanks, pump-turbine, generator and speed governor. Water flows from the upstream reservoir through the diversion pipeline, upstream surge tank and penstock to the pump-turbine inlet, the kinetic energy of the running water drives the runner of the pump-turbine to rotate and generates mechanical energy, the shaft of the pump-turbine is connected to the generator and drives the generator rotor to rotate synchronously. The speed governor guarantees the system's frequency stability by regulating the guide vane opening.

This paper divides the PSGS into five parts: hydraulic system, pump-turbine, generator and load, speed governor, and hydraulic servomechanism, with combined modeling of the PSGS based on mathematical models of each component.

2.1. Hydraulic Systems' Modeling

A typical layout of a pumped storage plant with two surge tanks is shown in Figure 1. In order to study the control strategy for the speed governor, this work divides the PSGS into five parts: the hydraulic system, pump-turbine, generator and load, speed governor, and hydraulic servomechanism.

As shown in Figure 1, the water level of the upstream reservoir in the pumped storage power plant is H_u (in meters), the cross-sectional areas of the diversion tunnel, penstock and downstream tailrace tunnel are A_1,A_2,A_3,A_4 (in square meters), the lengths are L_1,L_2,L_3,L_4 (in meters), the cross-sectional areas of the upstream and downstream surge tanks are A_{s1},A_{s2}, and the water levels are H_{s1},H_{s2}. Finally, a downstream reservoir with a water level of H_d can be seen on the far right of the Figure 1, where the reservoir levels H_u and H_d are considered as constants.

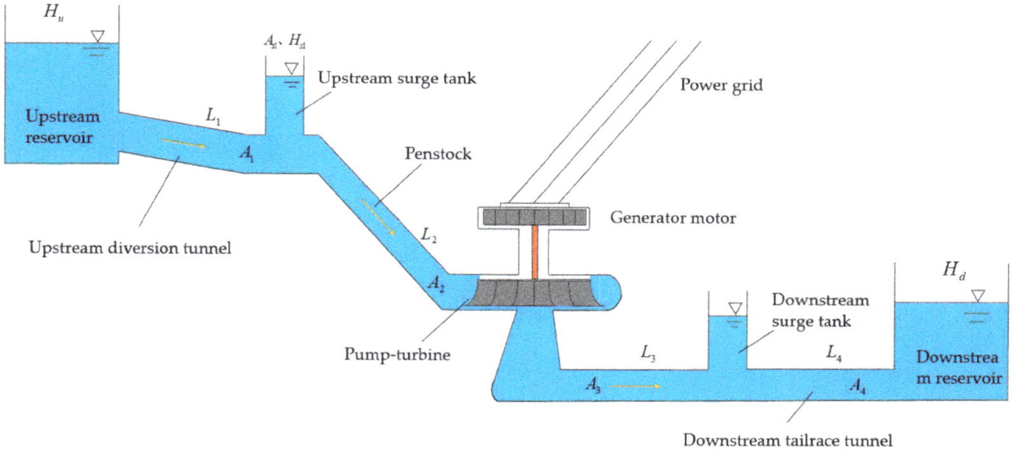

Figure 1. Typical layout of a pumped storage plant with two surge tanks.

2.1.1. Overall Pipeline Modeling

In applied hydraulic transients [38], if the pipe length is less than 600–800 m, or the governing system operates in small fluctuations conditions, the error of the rigid water hammer model can meet the engineering requirements.

If head loss is considered, the rigid water strike transfer function can be obtained as follows:

$$G(s) = \frac{H(s)}{Q(s)} = -\left(T_w S + h_f\right) \tag{1}$$

where T_w is water inertia time constant; h_f is the head loss of the tunnel.

As depicted in Figure 1, the overall pipeline is divided into three parts: the upstream diversion tunnel, the penstock and the downstream tailwater tunnel. Using the rigid water strike transfer function, the mathematical model of the dynamic relationship between these three sections of the tunnel is given below.

1. Upstream diversion tunnel:

As displayed in Figure 1, the upstream diversion tunnel connects the upstream reservoir and the upstream surge tank. Since the upstream reservoir H_u is constant, the water pressure deviation remains constant. The end of the diversion tunnel is connected to the upstream surge tank, so the pressure deviation of the surge tank can be considered the same as the pressure deviation of the diversion tunnel. The mathematical model of the upstream diversion tunnel is derived from the rigid water strike Equation (1):

$$h_1 = -\left(T_{w1}s + h_{f1}\right)q_1 \qquad h_1 = h_{s1} \tag{2}$$

where:

T_{w1}—water inertia time constant of the diversion tunnel;

h_{f1}—relative head loss of the diversion tunnel;

h_1,h_{s1}—relative water pressure deviation of the diversion tunnel and the upstream surge tank;

q_1—the relative flow deviation of the diversion tunnel.

2. Penstock:

As presented in Figure 1, the penstock connects the upstream surge tank and pump-turbine inlet, and the tailrace tunnel connects the pump-turbine outlet and downstream surge tank. The inertia of water flow is also considered in the penstock, and the mathematical model of the penstock is deduced as follows:

$$\begin{cases} h_{s1} - h_2 = \left(T_{w2}s + h_{f2}\right)q_t \\ h_3 - h_{s2} = \left(T_{w3}s + h_{f3}\right)q_t \end{cases} \quad (3)$$

where:

T_{w2},T_{w3}—water inertia time constant of the penstock and the tailrace tunnel;
h_{f2},h_{f3}—relative head loss of the penstock and the tailrace tunnel;
h_{s2},h_2,h_3—relative water pressure deviation of the upstream surge tank, the penstock and the tailrace tunnel;
q_t—relative flow deviation of the pump-turbine.

3. Downstream tailwater tunnel

As displayed in Figure 1, the tailrace tunnel connects the downstream surge tank and the downstream reservoir, the downstream reservoir H_d is constant and the water pressure deviation is kept constant, the mathematical model of the downstream tailrace tunnel is deduced as follows:

$$h_{s2} = \left(T_{w4}s + h_{f4}\right)q_4 \quad (4)$$

where

T_{w4}—water inertia time constant of the downstream tailrace tunnel;
h_{f4}—relative head loss of the downstream tailrace tunnel;
h_{s2}—relative water pressure deviation of the downstream tailrace tunnel;
q_4—relative flow deviation of the downstream tailrace tunnel.

2.1.2. Surge Tank Modeling

As shown in Figure 1, the upstream and downstream surge tanks are straight cylinder surge tanks. The surge tank plays the role of reflecting water hammer waves and reducing water hammer pressure. In omitting the surge tank inlet damping, the straight cylinder surge tank can be described as:

$$\Delta H = \int \frac{\Delta Q dt}{As} \quad (5)$$

where:

ΔH—water level in the surge tank changes, and water level rises as positive;
ΔQ—flow in and out of the surge tank, inflow is positive;
A_s—cross-sectional area of the surge tank.
Taking the relative value of Equation (5):

$$\frac{h_s(s)}{q_s(s)} = \frac{1}{T_j s} \quad (6)$$

where:

T_j—time constant of the surge tank;
h_s—relative water pressure deviation;
q_s—relative flow deviation.
Considering the flow continuity, the relationship between the flows in the upstream and downstream surge tanks can be deduced as follows [38]:

$$\begin{cases} q_1 = q_{s1} + q_2 \\ q_3 = q_{s2} + q_4 \\ q_2 = q_t = q_3 \end{cases} \quad (7)$$

where q_2,q_3,q_{s1},q_{s2}, respectively, are the relative flow deviations of the penstock, the tailrace tunnel, the upstream surge tank, and the downstream surge tank.

2.2. Pump-Turbine Modeling

For the study of the mathematical model of the pump-turbine, a six-parameter linear model can be used to describe the dynamic characteristics of the hydraulic turbine for small fluctuation problems, and the parameters of the pump-turbine are shown in Appendix A, and near a stable operating point, the linear model of the hydraulic turbine can be expressed as [11]:

$$\begin{cases} m_t = e_x x + e_y y + e_h h_t \\ q_t = e_{qx} x + e_{qy} y + e_{qh} h_t \end{cases} \tag{8}$$

where:

m_t—relative value of hydraulic turbine torque deviation;
h_t—relative value of hydraulic turbine head deviation;
x—relative value of rotation speed deviation;
y—relative value of guide vane opening deviation;
The six transfer functions are $e_x = \frac{\partial M_t}{\partial x}$, $e_y = \frac{\partial M_t}{\partial y}$, $e_h = \frac{\partial M_t}{\partial h_t}$, $e_{qx} = \frac{\partial Q}{\partial x}$, $e_{qy} = \frac{\partial Q}{\partial y}$, $e_{qh} = \frac{\partial Q}{\partial h_t}$.

2.3. Generator Modeling

In the modeling of the governing system of a pumped storage power plant, the most frequently used model for the synchronous generator is the first-order model. That is, considering only the dynamic response process of rotor motion, and treating the generating motor as a rotating rigid shaft, which is coaxial with the pump-turbine and connected by a coupling, and the parameters of the generator are shown in Appendix B. The dynamic model considering the load is expressed as [11]:

$$T_a \frac{dx}{dt} + (e_g - e_x)x = m_t - m_{g0} \tag{9}$$

where:

m_{g0}—load torque;
$e_g = \frac{\partial m_g}{\partial x}$—load self-regulation factor;
T_a—time constant of mechanical inertia of the generator, expressed as:

$$T_a = \frac{GD^2 n_r^2}{3580 P_r}$$

where GD^2 is the flywheel torque of the rotating part of the unit; n_r is the rated speed of the unit; P_r is the rated output of the unit.

The transition process of the pumped storage power plant governing system, the change in m_{g0} is usually regarded as a load disturbance to the PSGS.

2.4. Modeling of Electro-Hydraulic Servomechanism

The electro–hydraulic servomechanism is the actuator of the governor. It receives output signals from the controller and performs electro–hydraulic conversion and hydraulic amplification of the electrical control signals. Eventually, the electrical signal is converted into the displacement signal of the receiver. The electro–hydraulic servo mechanism is usually simplified to a first-order inertia link, as stated in Equation (10) [39],

$$\frac{y}{u} = \frac{1}{T_y S + 1} \tag{10}$$

where:

u—control signal output by the controller;
y—stroke of the servo motor;
T_y—servomotor response time.

2.5. PSGS Model

Based on the separate mathematical models of each link of the governing system of the pumped storage power plant established earlier, a block diagram of the transfer function of the pumped storage power plant governing system can be obtained, as presented in Figure 2:

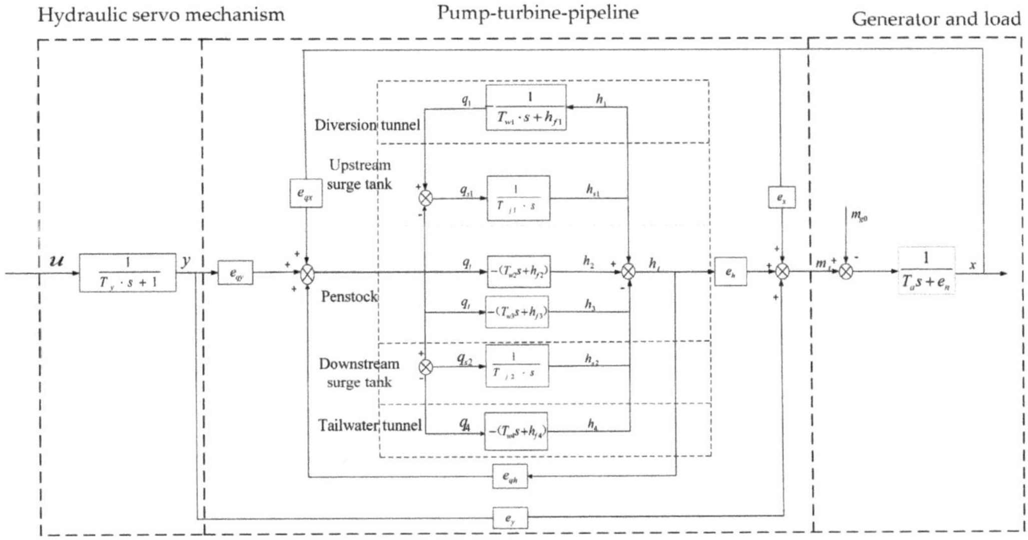

Figure 2. Block diagram of the transfer function of the pumped storage power plant governing system.

As shown in Figure 2, the physical state variables x, y, q_1, h_{s1}, q_t, h_{s2}, q_4, h_t that can be measured, are chosen to describe the overall PSGS model in the form of the state space equations, as displayed in Equation (11),

$$\begin{cases} \dot{x} = -\frac{e_n}{T_a}x + \frac{e_y}{T_a}y + \frac{e_h}{T_a}h_t - \frac{1}{T_a}m_{g0} \\ \dot{y} = -\frac{1}{T_y}y + \frac{1}{T_y}u \\ \dot{q_1} = -\frac{h_{f1}}{T_{w1}}q_1 - \frac{1}{T_{w1}}h_{s1} \\ \dot{h_{s1}} = \frac{1}{T_{j1}}q_1 - \frac{1}{T_{j1}}q_t \\ \dot{q_t} = \frac{1}{T_{w2}+T_{w3}}h_{s1} - \frac{h_{f2}+h_{f3}}{T_{w2}+T_{w3}}q_t - \frac{1}{T_{w2}+T_{w3}}h_{s2} - \frac{1}{T_{w2}+T_{w3}}h_t \\ \dot{h_{s2}} = \frac{1}{T_{j2}}q_t - \frac{1}{T_{j2}}q_4 \\ \dot{q_4} = \frac{1}{T_{w4}}h_{s2} - \frac{h_{f4}}{T_{w4}}q_4 \\ \dot{h_t} = \frac{e_{qx}e_n}{T_a e_{qh}}x + \left(\frac{e_{qy}}{T_y e_{qh}} - \frac{e_{qx}e_y}{T_a e_{qh}}\right)y + \frac{1}{(T_{w2}+T_{w3})e_{qh}}h_{s1} - \frac{h_{f2}+h_{f3}}{(T_{w2}+T_{w3})e_{qh}}q_t \\ \quad -\frac{1}{(T_{w2}+T_{w3})e_{qh}}h_{s2} - \left(\frac{1}{(T_{w2}+T_{w3})e_{qh}} + \frac{e_h e_{qx}}{T_a e_{qh}}\right)h_t - \frac{e_{qy}}{T_y e_{qh}}u + \frac{e_{qx}}{T_a e_{qh}}m_{g0} \end{cases} \quad (11)$$

Based on the mathematical model of the penstock, the equation is discretized, and the flow values obtained using adjacent discrete sampling times are used to calculate Δq, then, h_2,h_3 can be expressed as:

$$\begin{cases} h_3 = T_{w3}\frac{q_t - q_{t-1}}{T_s} + h_{s2} + h_{f3} \\ h_2 = h_3 + h_t \end{cases} \quad (12)$$

3. Fractional PID Algorithm of the Speed Governor

3.1. Definition of Fractional Calculus

Fractional calculus [40] is a mathematical theory that studies the properties of calculus and integral operators of arbitrary order, and it extends the order of calculus to the field of fractions and even plurals. Fractional calculus is a generalization of differentiation and integration of non-integer order, and its operation basic operation operator is ${}_aD_t^{\alpha}$ (a, t is the upper and lower bounds of the operation operator, α is the order of the calculus).

The calculus operation operator unifies the differentiation and integration, and a variety of definitions of fractional calculus emerge from the theory of fractional calculus, such as the *Cauchy* definition, the $Grünwald-Letnikov(G-L)$ definition, the $Riemann-Liouville(R-L)$ and the *Caputo* definition etc.

In this paper, using the $G-L$ definition, the calculus of order α for a certain continuously derivable function $f(t)$ is:

$$ {}_aD_t^{\alpha}f(t) = \lim_{h\to 0}\frac{1}{h^{\alpha}}\sum_{j=0}^{\left[\frac{t-a}{h}\right]}(-1)^j\binom{\alpha}{j}f(t-jh) \quad (13) $$

where $\binom{\alpha}{j} = \frac{\alpha(\alpha-1)(\alpha-2)\cdots(\alpha-j+1)}{j!} = \frac{\alpha!}{j!(\alpha-j)!}\cdot\left[\frac{t-\alpha}{h}\right]$ is the largest integer smaller than the real number $\frac{t-\alpha}{h}$, and h is the calculation step.

3.2. FOPID Controller Structure

Compared with the traditional integer-order PID, FOPID controller introduces differential order μ and integral order λ, which has two more degrees of freedom and can achieve better control effects. The PID and FOPID structure diagrams are given in Figure 3. The transfer function of the FOPID controller can be expressed as:

$$ C(s) = \frac{u(s)}{e(s)} = K_p + K_i s^{-\lambda} + K_d s^{\mu} \quad (14) $$

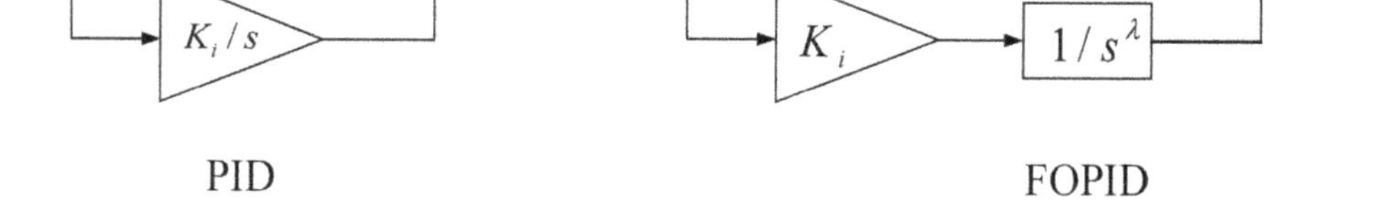

Figure 3. Comparison of the PID and FOPID structure diagram.

Its time domain expression is:

$$ u(t) = K_p e(t) + K_i D^{-\lambda}e(t) + K_d D^{\mu}e(t) \quad (15) $$

where $e(t)$ is the error signal, $u(t)$ is the controller output signal, K_p,K_i,K_d are the controller gain parameters, λ, μ are the fractional calculus orders, and $\lambda, \mu \in [0,2]$. When $\lambda = \mu = 1$, the FOPID controller is transformed into the integer-order PID controller.

4. Improved Chaotic Gravitational Search Algorithm

4.1. Standard Gravitational Search Algorithm

Inspired by the law of universal gravity, Rashedi et al. proposed a new swarm intelligence optimization algorithm—the gravitational search algorithm (GSA) in 2009 [30]. The law of gravity is a law that explains the relationship between objects interacting with each other. In the gravitational search algorithm, gravity is equivalent to an information transfer tool that enables information sharing among individuals and optimal search by the group under the effect of gravity.

Assuming that there are N particles, the position and velocity of the ith individual in the D-dimensional space are denoted as $X_i = \left(x_i^1, \cdots, x_i^d, \cdots, x_i^D\right)$ and $V_i = \left(v_i^1, \cdots, v_i^d, \cdots, v_i^D\right)$, where x_i^d and v_i^d, respectively, denote the position and velocity components of individual i in the d-dimensions, and the position denotes the solution of the problem. The position and velocity are first initialized in the solution space and velocity space, the objective function values of each individual are calculated and evaluated, the mass, gravitational force and acceleration of each individual are calculated, and finally the velocity and position of the individual are updated [41].

According to Newton's gravitation theory, in the d dimensions, the gravitational force of individual j on i is expressed as:

$$F_{ij}^d(t) = G(t)\frac{M_i(t) \times M_j(t)}{R_{ij}(t) + \varepsilon}\left(x_j^d(t) - x_i^d(t)\right) \tag{16}$$

where $G(t)$, $R_{ij}(t)$, respectively, are the universal gravitational constants at the tth iteration and the Euclidean distances of individuals i and j, with the following expressions:

$$G(t) = G_0 \times exp(-\alpha \times \frac{t}{T})\ R_{ij}(t) = ||X_i(t), X_j(t)||_2 \tag{17}$$

where G_0, α and ε are constants, t is the current number of iterations and T is the maximum number of iterations, $i,j\epsilon\{1, 2, \cdots, N\}$ and $i \neq j$.

In the d dimensions, the combined force on individual i is:

$$F_i^d(t) = \sum\nolimits_{j\epsilon Kbest,\ j\neq i}^{N} rand_j \times F_{ij}^d(t) \tag{18}$$

where $rand_j$ is a random variable obeying a uniform distribution between $[0, 1]$, and $Kbest$ is the collection of the top k individuals with the optimal fitness value and maximum quality.

The updated velocity and position can be obtained as:

$$\begin{cases} v_i^d(t+1) = rand_i \times V_i^d(t) + a_i^d(t) \\ x_i^d(t+1) = x_i^d(t) + v_i^d(t+1) \end{cases} \tag{19}$$

4.2. Improved Gravitational Search Algorithm (IGSA)

In order to improve the global search capability of GSA and avoid falling into a local optimum, the following four improvement strategies are incorporated in this paper:

Chaos operator.

Chaos [42] is a seemingly random movement that cannot be repeated precisely that occurs in nature. Chaotic motion is a complex state of motion unique to deterministic nonlinear dynamical systems, it has characteristics such as class randomness, initial value sensitivity, and ergodicity.

Chaos mapping is a mapping (evolutionary function) that exhibits some chaotic behavior, some common chaos mappings include Logistic map, Sinusoidal map, Tent map, Sine map and Bernoulli map.

In this paper, we introduce the chaos operator to improve the GSA by choosing the ergodic logistic mapping, namely:

$$r_{t+1} = ur_t \times (1 - r_t) \tag{20}$$

where control parameters $u \in [0,4]$, $r \in (0,1)$ and $r_0 \notin [0.25, 0.5, 0.75, 1]$, $r_t \in (0,1)$ is the number of chaos generated in the tth iteration.

In the position update phase, chaos sequences are introduced to improve the convergence of the algorithm, while chaos perturbations can help individuals escape from their current position when they fall into a local optimum. The steps are as follows:

(1) A d-dimensional random vector $c_1^d \in [0,1]$ is generated, and the control parameter $u = 4$, that is a fully chaos state is reached by Equation (20): $c_i^d = 4c_{i-1}^d \times (1 - c_{i-1}^d)$, $(i = 1, 2, \cdots, k)$, generating a chaos vector with k denoting the number of individuals;

(2) The generated chaos vector is added to the search space, at which point the speed update equation in the improved algorithm is:

$$v_i^d(t+1)\prime = \left[rand_i \times v_i^d(t) + \xi\left(c_i^d - 0.5\right)\right] + a_i^d(t) \tag{21}$$

where ξ is the factor that controls the range of chaos;

(3) Using the improved velocity update equation to calculate the position at the next moment:

$$x_i^d(t+1)\prime = x_i^d(t) + v_i^d(t+1)\prime \tag{22}$$

Adaptive universal gravitational constant decay factor.

From the gravitational force calculation Equation (16), we can see that the universal gravitational constant $G(t)$ is positively related to the gravitational force value, and its value plays an important role in the calculation of the gravitational force. When $G(t)$ takes a larger value, the search range of the algorithm is wider to avoid falling into the local optimum, while when $G(t)$ takes a smaller value, the algorithm can better converge to the global optimum. From Equation (17), it can be seen that the universal gravitational constant decay factor α in the standard GSA is taken as a constant, which limits the performance of the algorithm. Therefore, an adaptive universal gravitational constant decay factor is proposed:

$$\alpha(t) = \alpha_0 + \overline{w} \times \sinh(\delta \times (\frac{t}{T} - \theta)) \tag{23}$$

where α_0 is the initial value of the gravitational decay factor, $\overline{w}$ and δ are scaling factors, $\overline{w} \in [0,1]$, $\delta > 1$, θ is the shift factor, which is taken in this paper as $\alpha_0 = 20$, $\overline{w} = 1$, $\delta = 100$, $\theta = 0.35$.

As the number of iterations increases, the universal gravitational constant shows a nonlinear decrease, then the improved algorithm has a large gravitational constant at the beginning to enhance the search ability, and the algorithm reduces the gravitational constant at the later stage to accelerate the convergence of the algorithm.

Elastic ball boundary treatment.

Boundary constraints are added when initializing individuals, but as the iteration proceeds, there may be individuals that cross the boundary after the position update, and for the individuals that cross the boundary, the elastic ball boundary is handled by:

$$\begin{cases} x_i^d(t) = Ub(d) - up, up = x_i^d(t) - Ub(d), if x_i^d(t) > Ub(d) \\ x_i^d(t) = Lb(d) + down, down = Lb(d) - x_i^d(t), if x_i^d(t) < Lb(d) \end{cases} \tag{24}$$

It can be seen from Equation (24) that the elastic ball boundary treatment increases the individual positions, enhances the search capability of the algorithm, and reduces the risk of the algorithm falling into a local optimum.

Elite Guidance.

The velocity update formula of GSA only considers the effect of acceleration, while the velocity update in the particle swarm seeking algorithm takes into account the individual memory and group information exchange, it is more convergent for particle seeking. Therefore, the velocity update of IGSA combined with the chaos arithmetic and PSO velocity update mechanism is obtained:

$$\begin{aligned} v_i^d(t+1) = & \left[rand_i \times v_i^d(t) + \xi\left(c_i^d - 0.5\right)\right] + a_i^d(t) \\ & + c_1 \times rand \times \left(F_{best}^d - x_i^d(t)\right) + c_2 \times rand \times \left(X_{best}^d - x_i^d(t)\right) \end{aligned} \tag{25}$$

where c_1, c_2 are the adaptive learning factors that vary with the number of iterations, F_{best}^d and X_{best}^d denote the individual optimal position and the global optimal position in the d-dimensions, respectively.

Because the addition of the optimal position in the velocity update equation will reduce the search capability of the algorithm, the adaptive learning factor is added to balance the global search capability and local optimization capability of the algorithm. c_1, c_2 update equations [43] are:

$$c_1 = 1 - \exp(-30 \times (\frac{t}{T})^2) \qquad c_2 = \exp(-30 \times (\frac{t}{T})^2) \tag{26}$$

Based on the above improvement strategies, a new improved gravitational search algorithm (IGSA) is proposed in this paper with the following steps:

Step 1: Random initialization of the population and setting parameters. The setup parameters include the population size N, the maximum number of iterations T, the boundary of the position $[Lb, Ub]$, the initial value of the universal gravitational force G_0 and the initial value of the gravitational decay factor α_0. The population individual positions are randomly initialized.

Step 2: It is determined whether the updated individual position is out of bounds. The transgression uses the elastic ball boundary strategy of Equation (24) to assign a new position.

Step 3: Calculate the fitness value of all individuals in the population, and determine whether the fitness value is NaN (Not A Number) or not, if it is, then randomly initialize the position of the individual and calculate the fitness value again, and repeat this step until the fitness value is an output table value.

Step 4: Update global optimal position X_{best}^d and individual optimal position F_{best}^d, if $fzbest(t) < fzbest$, then $X_{best}^d = zbest(t)$, $fzbest = fzbest(t)$.

Step 5: Calculating the individual mass M_i, the gravitational parameter G is calculated according to Equations (17) and (23), the gravitational force F_i^d is calculated according to Equations (16) and (18), finally calculating the acceleration a_i^d.

Step 6: The adaptive learning factor is calculated according to Equations (26), and the speed and position are updated according to Equations (25) and (19).

Step 7: The number of iterations $t = t + 1$, if $t < T$, go to step 2, otherwise end the loop and obtain the optimal objective function value $fzbest$ and the optimal position vector X_{best}^d.

The algorithm flow chart of IGSA is indicated in Figure 4:

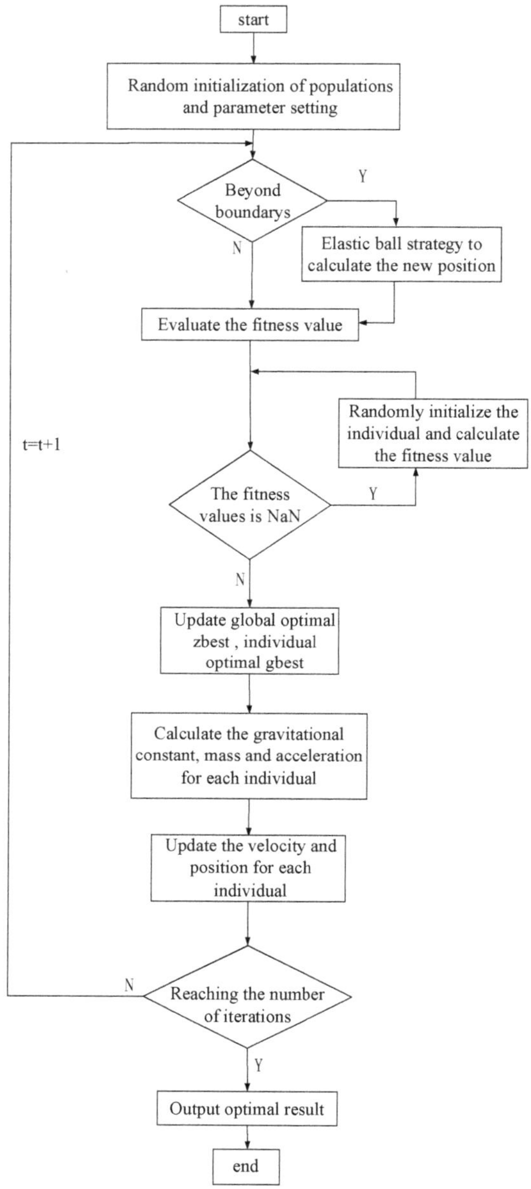

Figure 4. Algorithm flow chart of IGSA.

4.3. Objective Function

Time multiplied by the integral of the absolute value of the error (ITAE) is often used as an evaluation index for the performance of the speed regulation system of hydropower units. The objective function proposed in this paper takes into account the rotational speed error value and the water pressure deviation. The calculated ratio of ITAE to the number of samples is taken as the first output, and the second output is the absolute value of water pressure deviation minimum value during the transition process. By adjusting the weights of both during the parameter optimization process to ensure that the two output values are approximately equal. The objective function can be presented as:

$$f = w_1 \times \frac{\int t|e(t)|dt}{N} + w_2 \times |h_{tmin}| \quad (27)$$

Its discrete expression is given by:

$$f = w_1 \times \frac{\sum_{k=1}^{N} T(k)|x_{ref} - x(k)|}{N} + w_2 \times |h_{tmin}| \quad (28)$$

where w_1, w_2 are weighting factors, N is the number of samples, T indicates the time sequence and h_{tmin} indicates the minimum value of water pressure deviation.

5. Simulation Results Analysis

5.1. Simulation Parameter Setting

In this section, the HTGS model proposed in Section 2.5 is simulated in the MATLAB R2020b software environment, the simulation environment is under a 2.4 GHz inteli7 CPU and 16G RAM. Tables 1 and 2 show the simulation parameters of HTGS and the transfer coefficients of the pump-turbine, respectively. The values of the boundary on the FOPID and PID tuning parameters are given in Table 3.

Table 1. HTGS simulation parameters [44].

Parameters	Values
Water inertia time constant $T_{w1}, T_{w2}, T_{w3}, T_{w4}$	0.6, 0.5, 0.5, 0.6
Head loss of the tunnel $h_{f1}, h_{f2}, h_{f3}, h_{f4}$	0.0026, 0.003, 0.003, 0.0015
Time constant of the surge tan k T_{j1}, T_{j2}	10, 10
Time constant of mechanical inertia of the generator T_a	3.4
Load self − regulation factor e_g	0.05
Servomotor response time T_y	0.2

Table 2. Transfer coefficients of water pump-turbine.

e_x	e_y	e_h	e_{qx}	e_{qy}	e_{qh}
−0.45	1.0	1.5	0.0	1.0	0.5

Table 3. The values of the boundary on the FOPID and PID tuning parameters.

	K_p	K_i	K_d	λ	μ
FOPID	[0, 5]	[0, 5]	[0, 5]	[0, 2]	[0, 2]
PID	[0, 5]	[0, 5]	[0, 5]	-	-

Based on the above unit parameters, the frequency disturbance and load disturbance simulation tests are performed in this study.

5.2. Load Disturbance Scenario

According to the model established in this paper, under a 10% load disturbance, the optimization performances of four parameter optimization algorithms (i.e., SA, GA, PSO and IGSA) and the control effects of the PID and FOPID are compared, respectively.

5.2.1. Comparison of SA, GA, PSO and IGSA

The parameters of SA are set as follows: initial temperature $T_s = 100$, termination temperature $T_e = 0.01$, temperature decay factor $\alpha = 0.95$, maximum number of iterations $T = 100$, maximum number of iterations at temperature T_k Lk = 30.

The parameters of GA are set as follows: population size $N = 30$, maximum number of iterations $T = 100$, crossover probability $P_c = 0.7$, variance probability $P_m = 0.01$.

The parameters of PSO are set as follows: population size $N = 30$, maximum number of iterations $T = 100$, learning factor $c_1 = c_2 = 2$, velocity weight $w = 0.6$.

The parameters of IGSA are set as follows: population size $N = 30$, maximum number of iterations $T = 100$, $G_0 = 20$, the gravitational constant decay factor parameter is set the same as revealed in Section 4.2.

When the weight factor $w_1 = 0.05$, $w_2 = 0.95$, after 100 iterations, the convergence curves of the fitness functions of SA, GA, PSO and IGSA are shown in Figure 5:

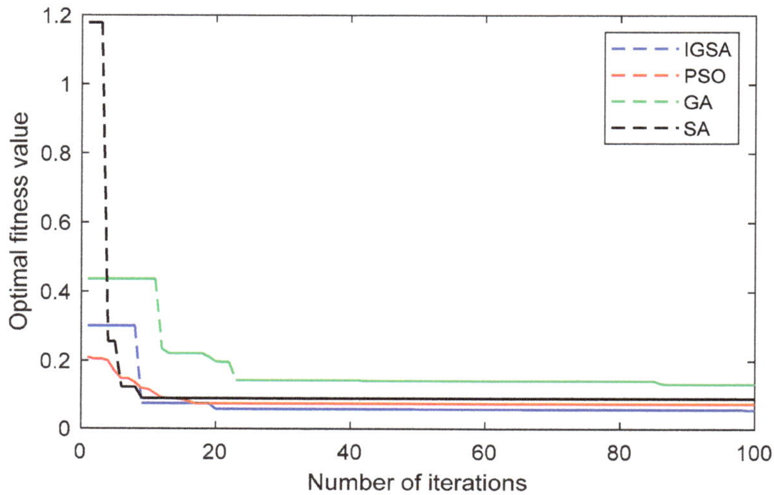

Figure 5. Convergence curves of the fitness functions under load disturbance.

According to the fitness convergence trend displayed in Figure 5, it can be seen that among the four algorithms, the final fitness value of GA is the largest and that of IGSA is the smallest. From the convergence curve, SA starts with the largest fitness value and decreases the fastest, PSO decreases the fitness more slowly during the descent process, and GA converges the slowest. Comparing the final fitness values, it can be seen that SA, GA and PSO easily fall into the local optimum and have difficulty in jumping out of it, while the final fitness value of IGSA is smaller. This shows that as the number of iterations increases, the fitness value of IGSA has been decreasing. The chaos perturbations added in the position update phase enables IGSA to jump out of the local optimum continuously, which indicates that IGSA has better global exploration ability.

Table 4 shows the dynamic indicators for SA, GA, PSO and IGSA with optimal control parameters:

Table 4. Dynamic indicators of SA, GA, PSO and IGSA optimal control parameters under load disturbance.

	f	Overshoot	Stabilization Time	$\lvert h_{min} \rvert$
SA	0.0509	--	45.0	0.034
GA	0.1230	--	60.6	0.036
PSO	0.0482	--	40.1	0.033
IGSA	0.0452	--	21.2	0.029

Figure 6 compares the relative values of rotation speed deviation x, guide vane opening deviation y and hydraulic turbine head deviation h_t variation curves under the optimal control parameters of SA, GA, PSO and IGSA during load disturbance:

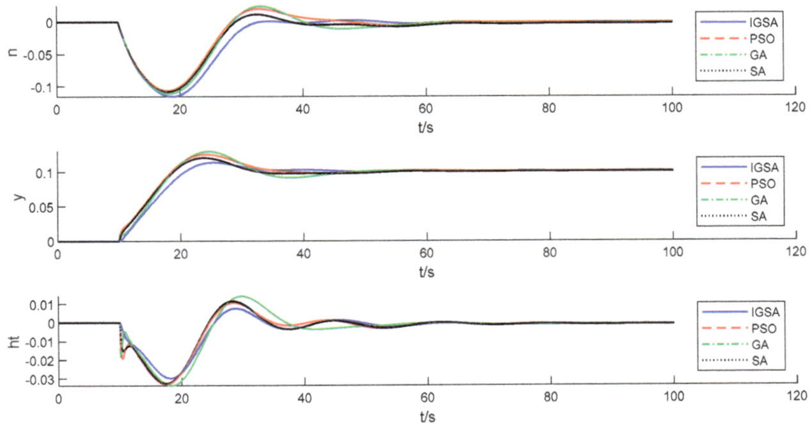

Figure 6. Variation curves of n, y, h_t of four algorithms under load disturbance.

Combining Table 4 and Figure 6, the dynamic indexes under the optimal control parameters of IGSA are significantly better than the other three algorithms. In Table 4, IGSA has the smallest objective function, stabilization time and absolute value of the minimum value of water pressure deviation. It can also be seen in Figure 6 that the unit speed deviation transition process optimized by the IGSA algorithm is significantly improved compared to the other three algorithms. The speed increase process is smoother, the stabilization time is shorter, and the absolute value of the minimum value of the water pressure deviation is also the smallest, which achieves better control parameter optimization. Overall, the transition process of x, y, h_t under the optimal control parameters of IGSA is obviously improved.

5.2.2. Comparison of PID and FOPID

In order to prove that FOPID control has a better control effect than PID control in the parameter optimization of pumped storage speed regulation systems, tests select the optimal control parameters for both controls. Figure 7 compares x, y and h_t variation curves of the model under the load disturbance of the two controllers:

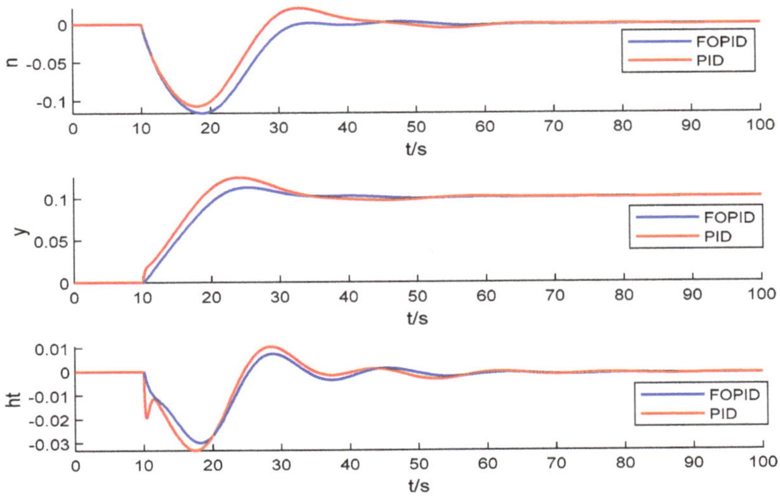

Figure 7. Variation curves of n, y, h_t of two controllers under load disturbance.

Table 5 shows the dynamic indicators for PID and FOPID with optimal control parameters:

Table 5. Dynamic indicators of PID and FOPID optimal control parameters under load disturbance.

	f	Overshoot	Stabilization Time	$\vert h_{min} \vert$
PID	0.0505	--	40.2	0.0332
FOPID	0.0452	--	21.2	0.0290

In Figure 7, it is obvious that the dynamic performance of the FOPID controller is significantly better than that of the PID controller during load disturbance. From the data presented in Table 5, it can be concluded that the values of the objective function, stabilization time and absolute value of the minimum value of water pressure deviation under the FOPID control are reduced by 0.0053, 19 and 0.0042, respectively, which verifies the superiority of the FOPID control.

5.3. Frequency Disturbance Scenario

Under a 5% frequency disturbance, this section compares four different parameter optimization algorithms of SA, GA, PSO and IGSA and the control effects of the two controllers PID and FOPID.

5.3.1. Comparison of SA, GA, PSO and IGSA

The SA, GA, PSO and IGSA parameters are set the same as in Section 5.2.1. When the weight factor $w_1 = 0.115$, $w_2 = 0.885$, after 100 iterations, the convergence curves of the fitness functions of SA, GA, PSO and IGSA are displayed in Figure 8:

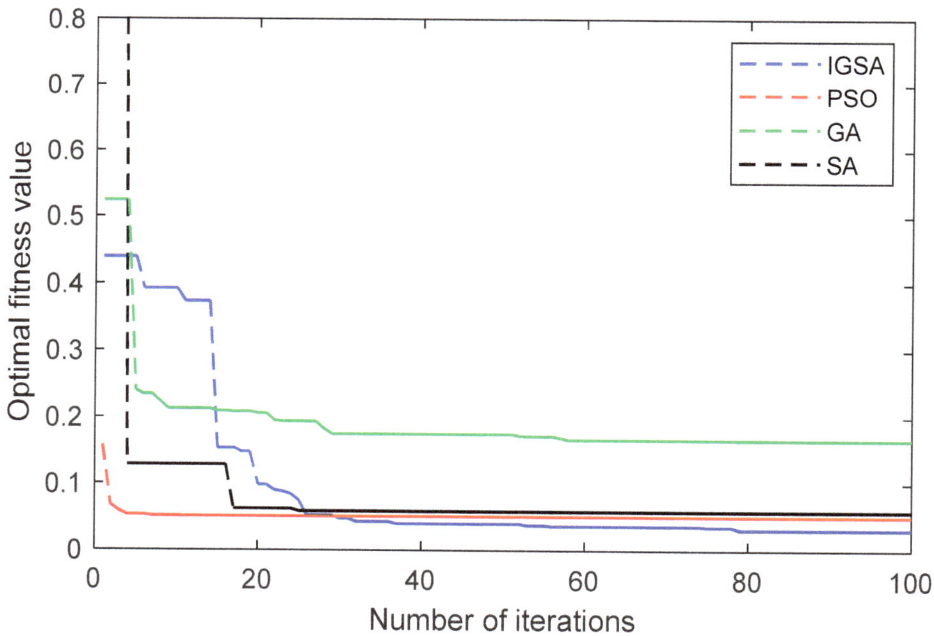

Figure 8. Convergence curves of the fitness functions under frequency disturbance.

According to the fitness convergence trend displayed in Figure 8, in the 5% frequency disturbance, SA starts with the largest fitness value and the fitness value falls into the local optimum after two decreases, PSO falls into the local optimum after about seven iterations, and GA has the largest fitness value after the continuous optimization search. Comparing the final fitness values of the four algorithms, it is observed that SA, GA and PSO easily fall into the local optimum and have difficulty jumping out of it. The proposed IGSA has a wider search range and better search capability, and the optimal fitness value decreases continuously during the iterative process. It proves that the IGSA has better applicability and explorability.

Table 6 shows the dynamic indicators for SA, GA, PSO and IGSA with optimal control parameters:

Table 6. Dynamic indicators of SA, GA, PSO and IGSA optimal control parameters under frequency disturbance.

	f	Overshoot	Stabilization Time	$\lvert h_{min} \rvert$
SA	0.0587	23.26%	44.85	0.0180
GA	0.1667	42.22%	67.70	0.0213
PSO	0.0319	10.14%	32.10	0.0140
IGSA	0.0254	2.28%	17.05	0.0084

Figure 9 compares x, y and h_t variation curves under the optimal control parameters of SA, GA, PSO and IGSA during frequency disturbance:

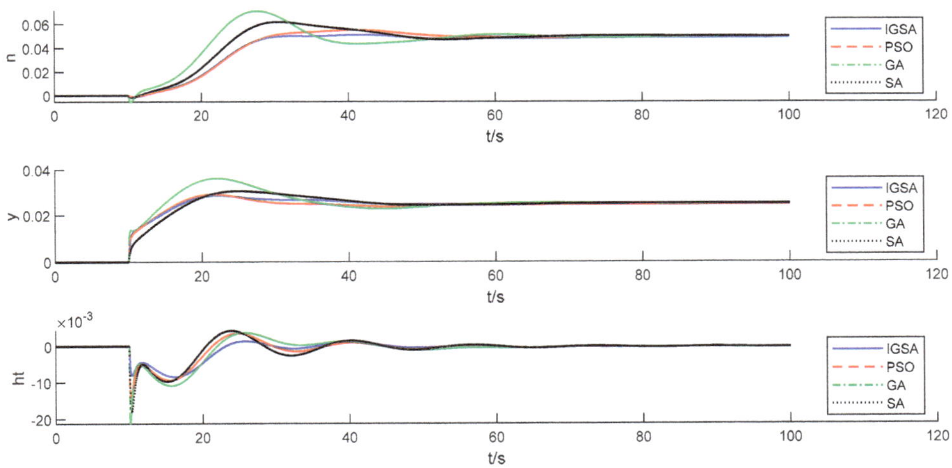

Figure 9. Variation curves of n, y, h_t of four algorithms under frequency disturbance.

Combining Table 6 and Figure 9, the dynamic indexes under the optimal control parameters of IGSA are significantly better than the other three algorithms. In Table 6, the values of the objective function, overshoot, stabilization time and absolute value of the minimum value of water pressure deviation of IGSA are lower than the other three algorithms. From Figure 9, the speed deviation transition process of the unit optimized by the IGSA algorithm is significantly improved compared to the other three algorithms. The speed increase process is smoother and reaches stability faster, the speed overshoot is almost nothing, and the absolute value of the minimum value of the hydraulic pressure deviation is also the smallest, which achieves a better optimization of the control parameters. In conclusion, the transition process of x, y, h_t under the optimal control parameters of IGSA is obviously improved.

5.3.2. Comparison of PID and FOPID

Figure 10 compares x, y and h_t variation curves of the model under the frequency disturbance of the two controllers:

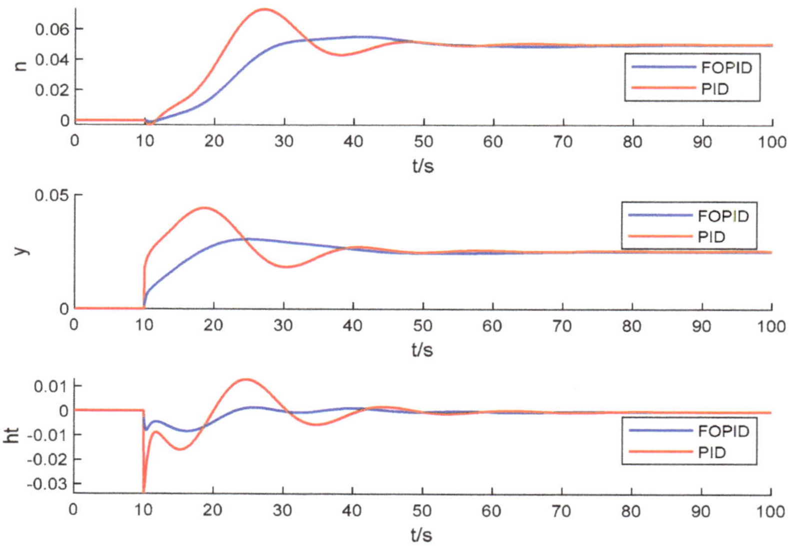

Figure 10. Variation curves of n, y, h_t of two controllers under frequency disturbance.

Table 7 shows the dynamic indicators for PID and FOPID with optimal control parameters:

Table 7. Dynamic indicators of PID and FOPID optimal control parameters under frequency disturbance.

	f	Overshoot	Stabilization Time	$\lvert h_{min} \rvert$
PID	0.0452	45.98%	30.75	0.0339
FOPID	0.0254	2.28%	17.05	0.0084

It is evident from Figure 10 that the FOPID controller achieves better results in PSGS control at frequency disturbance. By analyzing the results of each parameter in Table 6, it is found that the values of the objective function, overshoot, stabilization time and absolute value of the minimum value of water pressure deviation are reduced by 0.0198, 43.7%, 13.7 and 0.0255, respectively. Combined with Table 7 and Figure 10, it is found that under frequency disturbance, the adjustment quality improvement under FOPID control is more obvious.

5.4. Results Analysis

Through the frequency disturbance and load disturbance tests on the model, comparing the four algorithms of SA, GA, PSO and IGSA, it is proved that the optimization results of IGSA are better than those of SA, GA and PSO, and IGSA has better search ability, while the FOPID parameters optimized by IGSA can make the pumped storage regulation system have better dynamic performance; comparing the control effects of both FOPID and PID controllers, it can be seen that the dynamic performance obtained through the action of the FOPID controller is better than that of the PID controller.

6. Conclusions

This paper establishes a simulation model of a pumped storage speed regulation system with double surge tanks and compares the control effects of two controllers PID and FOPID. The IGSA is proposed to optimize the governor parameters. By comparing the convergence curves of the fitness function under load disturbance and frequency disturbance, it is verified that the IGSA proposed in this paper has better search capability than SA, GA and PSO, and it can address the algorithm precociousness and the local optimum trapping problem. In addition, under load disturbance and frequency disturbance, comparing the control effects under two different controllers PID and FOPID, it is found that FOPID has better comprehensive control performances.

In summary, the main contribution of this paper lies in the following two aspects:

(1). A new PSGS state-space equation model which fully considers the hydraulic characteristics of the pipes, surge tanks and pump-turbine and the eletromechanical behaviors of the generator and hydraulic servomechanism is proposed.

(2). An improved GSA optimizer combining the basic searching mechanisms of the gravitational search algorithm and chaotic search, elastic sphere boundary treatment, and elite guidance strategy is developed for the control parameter optimization of the FOPID scheme. Through comparative case studies under load disturbance and frequency disturbance, it is proved that the proposed IGSA shows superiority over parameter optimizers on governor parameter optimization.

Future work will also focus on comparing simulation results with experimental results.

Author Contributions: Conceptualization, Y.Z. (Yang Zheng) and X.Z.; writing—original draft preparation, X.Z.; methodology, B.X.; writing—review and editing, W.L. and Y.Z. (Yidong Zou); visualization, J.C. All authors have read and agreed to the published version of the manuscript.

Funding: This work was supported by Hubei Provincial Nature Science Foundation of China [Grant No. 2022CFD165] and the Fundamental Research Funds for the Central Universities (Grant No. 2042022kf1022).

Data Availability Statement: For the sake of information security, the original data used in this paper will not be disclosed.

Conflicts of Interest: B.X. was employed by the company Technology and Research Center, China Yangtze Power Co., Ltd. The remaining authors declare that the research was conducted in the absence of any commercial or financial relationships that could be construed as a potential conflict of interest.

Nomenclature

PSGS:	
T_w	water inertia time constant
h_f	head loss of the tunnel
$T_{w1}, T_{w2}, T_{w3}, T_{w4}$	water inertia time constant of diversion tunnel, penstock, tailrace tunnel and downstream tailrace tunnel
h_s	relative water pressure deviation
h_{s1}, h_{s2}	relative water pressure deviation of upstream surge tank and downstream tailrace tunnel
h_1, h_2, h_3	relative water pressure deviation of the diversion tunnel, penstock, tailrace tunnel
$h_{f1}, h_{f2}, h_{f3}, h_{f4}$	relative head loss of diversion tunnel, penstock, tailrace tunnel and downstream tailrace tunnel
q_s	relative flow deviation
$q_1, q_2, q_3, q_4, q_{s1}, q_{s2}$	relative flow deviation of diversion tunnel, penstock, tailrace tunnel, downstream tailrace tunnel, upstream surge tank and downstream surge tank
ΔH	water level in the surge tank changes
ΔQ	flow in and out of surge tank
A_s	cross-sectional area of surge tank
T_j	time constant of surge tank
T_{j1}, T_{j2}	time constant of upstream surge tank and downstream surge tank
m_t	relative value of hydraulic turbine torque deviation
h_t	relative value of hydraulic turbine head deviation
q_t, q_{t-1}	relative flow deviation of the pump-turbine at moments t and $t-1$
x, y	relative value of rotate speed deviation and guide vane opening deviation

y (servo mechanism)	stroke of the servo motor
e_x, e_y, e_h	partial derivatives of the torque with respect to head, guide vane and turbine speed
e_{qx}, e_{qy}, e_{qh}	partial derivatives of the flow with respect to head, guide vane and turbine speed
T_a	time constant of mechanical inertia of the generator
m_{g0}	load torque
e_g	load self-regulation factor
e_n	synthetic self-regulation coefficient
GD^2	flywheel torque of rotating part of the unit
n_r	rated speed of the unit
P_r	rated output of the unit
u	control signal output by the controller
T_y	servomotor response time
FOPID:	
${}_aD_t^{\alpha}$	operation basic operation operator of fractional calculus
$a,\ t$	upper and lower bounds of the operation operator
α	order of the calculus
h	calculation step
$\left[\frac{t-\alpha}{h}\right]$	the largest integer smaller than the real number $\frac{t-\alpha}{h}$
K_p	proportional adjustment coefficient
K_i	integral adjustment coefficient
K_d	differential adjustment coefficient
$e(t)$	error signal
$u(t)$	controller output signal
$\lambda,\ \mu$	integral order and differential order
s	the Laplace operator
IGSA:	
x_i^d, v_i^d	the position and velocity components of individual i in the d-dimensions
$F_{ij}^d(t)$	gravitational force of individual j on i
$G(t),\ R_{ij}(t)$	the universal gravitational constants at the tth iteration and the Euclidean distances of individuals i and j
$G_0,\ \alpha,\ \varepsilon$	gravitational constant
t, T	current number of iterations and the maximum number of iterations
$rand_j$	a random variable obeying a uniform distribution between $[0,1]$
$Kbest$	the collection of the top k individuals with the optimal fitness value and maximum quality
$a_i^d(t)$	the acceleration of individual i in the d-dimensions
r_t	the number of chaos generated in the tth iteration
u	chaos control parameters
c_1^d	a d-dimensional random vector
ξ	factor that controls the range of chaos
α_0	initial value of the gravitational decay factor
$\overline{w},\ \delta$	scaling factors
θ	shift factor
$Ub(d),\ Lb(d)$	the upper and lower boundaries of the positions of the individuals in d-dimension
c_1, c_2	adaptive learning factors
F_{best}^d, X_{best}^d	individual optimal position and the global optimal position in the d-dimensions
f	fitness value
w_1, w_2	weighting factors
N, T	the number of samples and time sequence
h_{tmin}	the minimum value of water pressure deviation

Appendix A

Table A1. The parameters of the pump-turbine.

Parameters	Values
Rated rotation speed	500 r/min
Power frequency	50 Hz
Maximum head	565 m
Rated head	540 m
Minimum head	526 m
Rated discharge	62.09 m^3/s
100% GVO	20.47°
Rated Power output	300 MW

Appendix B

Table A2. The parameters of the generator.

Parameters	Values
Rated capacity (generator)	334 MVA
Rated power (generator)	300 MW
Rated capacity (electric motor)	338 MVA
Rated power (electric motor)	325 MW
Rated voltage	15.75 kV
Power factor (generator)	0.9
Power factor (electric motor)	0.975
Rated rotation speed	500 rpm
Number of rotor magnetic poles	12 (Pole logarithm 6)
Rated frequency	50 Hz

References

1. Liang, Y.Y.; Yu, B.Y.; Wang, L. Costs and benefits of renewable energy development in China's power industry. *Renew. Energy* **2019**, *131*, 700–712. [CrossRef]
2. Zhu, L.; Xu, Y.; Pan, Y.J. Enabled comparative advantage strategy in China's solar PV development. *Energy Policy* **2019**, *133*, 9. [CrossRef]
3. Barbour, E.; Wilson, I.A.G.; Radcliffe, J.; Ding, Y.L.; Li, Y.L. A review of pumped hydro energy storage development in significant international electricity markets. *Renew. Sust. Energ. Rev.* **2016**, *61*, 421–432. [CrossRef]
4. Kong, Y.G.; Kong, Z.G.; Liu, Z.Q.; Wei, C.M.; Zhang, J.F.; An, G.C. Pumped storage power stations in China: The past, the present, and the future. *Renew. Sust. Energ. Rev.* **2017**, *71*, 720–731. [CrossRef]
5. Jiang, X.P.; Chen, X.F.; Wang, Z.W.; IOP. Research on Modeling and Control Strategy of Hydraulic Turbine Governing System Based on Improved Genetic Algorithm. In Proceedings of the 30th IAHR Symposium on Hydraulic Machinery and Systems (IAHR), Electr Network, Lausanne, Switzerland, 21–26 March 2021.
6. Jiang, C.W.; Ma, Y.C.; Wang, C.M. PID controller parameters optimization of hydro-turbine governing systems using deterministic-chaotic-mutation evolutionary programming (DCMEP). *Energy Convers. Manag.* **2006**, *47*, 1222–1230. [CrossRef]
7. Yu, X.D.; Yang, X.W.; Zhang, J. Stability analysis of hydro-turbine governing system including surge tanks under interconnected operation during small load disturbance. *Renew. Energy* **2019**, *133*, 1426–1435. [CrossRef]
8. Chaudhry, H.M. *Applied Hydraulic Transients*; Springer: London, UK, 2014.
9. Ramos, H.M.; Coronado-Hernandez, O.E.; Morgado, P.A.; Simao, M. Mathematic Modelling of a Reversible Hydropower System: Dynamic Effects in Turbine Mode. *Water* **2023**, *15*, 2034. [CrossRef]
10. Yang, J.D.; Wang, M.J.; Wang, C.; Guo, W.C. Linear Modeling and Regulation Quality Analysis for Hydro-Turbine Governing System with an Open Tailrace Channel. *Energies* **2015**, *8*, 11702–11717. [CrossRef]
11. Zhang, H.; Chen, D.Y.; Xu, B.B.; Wang, F.F. Nonlinear modeling and dynamic analysis of hydro-turbine governing system in the process of load rejection transient. *Energy Convers. Manag.* **2015**, *90*, 128–137. [CrossRef]
12. Zheng, Y.; Chen, Q.J.; Yan, D.L.; Liu, W.Y. A two-stage numerical simulation framework for pumped-storage energy system. *Energy Convers. Manag.* **2020**, *210*, 18. [CrossRef]
13. Ibrahim, A.A.S.; IEEE. Nonlinear PID controller design using fuzzy logic. In Proceedings of the 11th IEEE Mediterranean Electrotechnical Conference (IEEE MELECON 2002), Cairo, Egypt, 7–9 May 2002; pp. 595–599.

14. Li, J.Y.; Chen, Q.J. Fractional Order Controller Designing with Firefly Algorithm and Parameter Optimization for Hydroturbine Governing System. *Math. Probl. Eng.* **2015**, *2015*, 11. [CrossRef]
15. Padula, F.; Visioli, A. Tuning rules for optimal PID and fractional-order PID controllers. *J. Process. Control.* **2011**, *21*, 69–81. [CrossRef]
16. Singh, A.; Sharma, V.; IEEE. Concentration Control of CSTR Through Fractional Order PID Controller by Using Soft Techniques. In Proceedings of the 4th International Conference on Computing, Communications and Networking Technologies (ICCCNT), Tiruchengode, India, 4–6 July 2013.
17. Das, S.; Pan, I.; Das, S. Fractional order fuzzy control of nuclear reactor power with thermal-hydraulic effects in the presence of random network induced delay and sensor noise having long range dependence. *Energy Convers. Manag.* **2013**, *68*, 200–218. [CrossRef]
18. Saraji, A.M.; Ghanbari, M. Fractional order PID controller for improvement of PMSM speed control in Aerospace Applications. In Proceedings of the 10th International Conference on Mathematical Problems in Engineering, Aerospace and Sciences (ICNPAA), Narvik Univ Coll, Narvik, Norway, 15–18 July 2014; pp. 959–967.
19. Chen, Z.H.; Yuan, X.H.; Ji, B.; Wang, P.T.; Tian, H. Design of a fractional order PID controller for hydraulic turbine regulating system using chaotic non-dominated sorting genetic algorithm II. *Energy Convers. Manag.* **2014**, *84*, 390–404. [CrossRef]
20. Li, Z.H.; Malik, O.P. An orthogonal test approach based control parameter optimization and its application to a hydro-turbine governor. *IEEE Trans. Energy Convers.* **1997**, *12*, 388–393. [CrossRef]
21. Hedlund, P.; Gustavsson, A. Design and evaluation of an improved simplex method. *Anal. Chim. Acta* **1998**, *371*, 9–21. [CrossRef]
22. Yang, X.S. Nature-inspired optimization algorithms: Challenges and open problems. *J. Comput. Sci.* **2020**, *46*, 8. [CrossRef]
23. Rasheed, K.; Hirsh, H. Learning to be selective in genetic-algorithm-based design optimization. *AI EDAM-Artif. Intell. Eng. Des. Anal. Manuf.* **1999**, *13*, 157–169. [CrossRef]
24. Kennedy, J.; Eberhart, R. Particle swarm optimization. In Proceedings of the IEEE International Conference on Neural Networks—Conference Proceedings, Perth, WA, Australia, 27 November–1 December 1995; pp. 1942–1948.
25. Ramezani, F.; Lotfi, S. The Modified Differential Evolution Algorithm (MDEA). In Proceedings of the 4th International Scientific Asian Conference on Intelligent Information and Database Systems (ACIIDS), Kaohsiung, Taiwan, 19–21 March 2012; pp. 109–118.
26. Ahmadianfar, I.; Bozorg-Haddad, O.; Chu, X.F. Gradient-based optimizer: A new metaheuristic optimization algorithm. *Inf. Sci.* **2020**, *540*, 131–159. [CrossRef]
27. Chun, J.; Wu, Y.L.; Dai, Y.F.; Li, S.Y. Fiber optic active alignment method based on a pattern search algorithm. *Opt. Eng.* **2006**, *45*, 045005. [CrossRef]
28. Lv, P.; Yuan, L.; Zhang, J.F. Cloud theory-based simulated annealing algorithm and application. *Eng. Appl. Artif. Intell.* **2009**, *22*, 742–749. [CrossRef]
29. Bhookya, J.; Jatoth, R.K. Optimal FOPID/PID controller parameters tuning for the AVR system based on sine-cosine-algorithm. *Evol. Intell.* **2019**, *12*, 725–733. [CrossRef]
30. Rashedi, E.; Nezamabadi-Pour, H.; Saryazdi, S. GSA: A Gravitational Search Algorithm. *Inf. Sci.* **2009**, *179*, 2232–2248. [CrossRef]
31. Norlina, M.S.; Mazidah, P.; Sin, N.D.M.; Rusop, M.; IEEE. Application of Metaheuristic Algorithms in Nano-process Parameter Optimization. In Proceedings of the IEEE Congress on Evolutionary Computation (CEC), Sendai, Japan, 25–28 May 2015; pp. 2625–2630.
32. Li, C.S.; Zhou, J.Z. Parameters identification of hydraulic turbine governing system using improved gravitational search algorithm. *Energy Convers. Manag.* **2011**, *52*, 374–381. [CrossRef]
33. Sarafrazi, S.; Nezamabadi-Pour, H.; Seydnejad, S.R. A novel hybrid algorithm of GSA with Kepler algorithm for numerical optimization. *J. King Saud. Univ.—Comput. Inform. Sci.* **2015**, *27*, 288–296. [CrossRef]
34. Zhang, N.; Li, C.S.; Li, R.H.; Lai, X.J.; Zhang, Y.C. A mixed-strategy based gravitational search algorithm for parameter identification of hydraulic turbine governing system. *Knowl.-Based Syst.* **2016**, *109*, 218–237. [CrossRef]
35. Yin, B.Y.; Guo, Z.L.; Liang, Z.P.; Yue, X.Z. Improved gravitational search algorithm with crossover. *Comput. Electr. Eng.* **2018**, *66*, 505–516. [CrossRef]
36. He, S.S.; Zhu, L.; Wang, L.; Yu, L.; Yao, C.H. A Modified Gravitational Search Algorithm for Function Optimization. *IEEE Access* **2019**, *7*, 5984–5993. [CrossRef]
37. Tian, M.N.; Liu, J.H.; Yue, W.; Zhou, J. A Novel Integrated Heuristic Optimizer Using a Water Cycle Algorithm and Gravitational Search Algorithm for Optimization Problems. *Mathematics* **2023**, *11*, 1880. [CrossRef]
38. Fang, H.Q.; Chen, L.; Dlakavu, N.; Shen, Z.Y. Basic Modeling and simulation tool for analysis of hydraulic transients in hydroelectric power plants. *IEEE Trans. Energy Convers.* **2008**, *23*, 834–841. [CrossRef]
39. Zeng, Y.; Guo, Y.K.; Zhang, L.X.; Xu, T.M.; Dong, H.K. Nonlinear hydro turbine model having a surge tank. *Math. Comput. Model. Dyn. Syst.* **2013**, *19*, 12–28. [CrossRef]
40. Shah, P.; Agashe, S. Review of fractional PID controller. *Mechatronics* **2016**, *38*, 29–41. [CrossRef]
41. Jiang, S.H.; Wang, Y.; Ji, Z.C. Convergence analysis and performance of an improved gravitational search algorithm. *Appl. Soft. Comput.* **2014**, *24*, 363–384. [CrossRef]
42. Lanza, M.L. Nonlinear dynamics: Chaos and catastrophe theory. *J. Nurs. Care Qual.* **2000**, *15*, 55–65. [CrossRef] [PubMed]

43. Li, M.A.; Li-tao, L.I.U. Analysis and Improvement of Gravitational Search Algorithm. *Microelectron. Comput.* **2015**, *32*, 76–80. [CrossRef]
44. Wei, S.P. *Simulation of Hydraulic Turbine Regulation System*; Huazhong University of Science and Technology Press: Wuhan, China, 2011; p. 457.

water

MDPI

Article

Analysis of the Energy Loss Mechanism of Pump-Turbines with Splitter Blades under Different Characteristic Heads

Zhonghua Gui [1], Zhe Xu [2], Dongkuo Li [1], Fei Zhang [1], Yifeng Zhao [1], Lianchen Xu [2], Yuan Zheng [2,3] and Kan Kan [2,3,*]

[1] Pumped Storage Technological and Economic Research Institute, State Grid Xinyuan Co., Ltd., Beijing 100761, China; guizh@163.com (Z.G.); lidongkuo1992@163.com (D.L.); spiritgiant@126.com (F.Z.); thpzyf@sina.com (Y.Z.)
[2] College of Water Conservancy and Hydropower Engineering, Hohai University, Nanjing 210098, China; xuzhe@hhu.edu.cn (Z.X.); xulianchen@hhu.edu.cn (L.X.); zhengyuan@hhu.edu.cn (Y.Z.)
[3] College of Energy and Electrical Engineering, Hohai University, Nanjing 211100, China
* Correspondence: kankan@hhu.edu.cn

Abstract: The operating efficiency of high-head pump turbines is closely related to the internal hydraulic losses within the system. Conventional methods for calculating hydraulic losses based on pressure differences often lack detailed information on their distribution and specific sources. Additionally, the presence of splitter blades further complicates the hydraulic loss characteristics, necessitating further study. In this study, Reynolds-averaged Navier–Stokes (RANS) simulations were employed to analyze the performance of a pump turbine with splitter blades at three different head conditions and a guide vane opening (GVO) of 10°. The numerical simulations were validated by experimental tests using laser doppler velocimetry (LDV). Quantitative analysis of flow components and hydraulic losses was conducted using entropy production theory in combination with an examination of flow field distributions to identify the origins and features of hydraulic losses. The results indicate that higher heads are associated with lower growth rates of total hydraulic losses. In particular, the significant velocity gradients at the trailing edge of the splitter blades contribute to higher hydraulic losses. Furthermore, the hydraulic losses in the runner (RN) region are predominantly influenced by velocity gradients and not by vortices, with the flow conditions in the RN region impacting the hydraulic losses in the draft tube (DT).

Keywords: energy loss; pump turbine with splitter blades; entropy production; laser doppler velocimetry; helicity

Citation: Gui, Z.; Xu, Z.; Li, D.; Zhang, F.; Zhao, Y.; Xu, L.; Zheng, Y.; Kan, K. Analysis of the Energy Loss Mechanism of Pump-Turbines with Splitter Blades under Different Characteristic Heads. *Water* **2023**, *15*, 2776. https://doi.org/10.3390/w15152776

Academic Editors: Helena M. Ramos and Wencheng Guo

Received: 12 June 2023
Revised: 26 July 2023
Accepted: 28 July 2023
Published: 31 July 2023

1. Introduction

As the global demand for electricity continues to rise, the share of renewable energy power plants in power grids is growing rapidly. To cope with fluctuations in the power system caused by the disruption of renewable energy sources such as photovoltaics, hydroelectric power plants of different capacities are playing a crucial role in power grids [1]. Among these, pumped storage power plants, with their significant regulation capabilities, have become an important part of the grid regulation system [2]. As one of the central elements of pumped storage power plants, the pump turbine is of considerable research importance.

The performance of pump turbines in turbine mode is somewhat limited due to the need to operate in both directions. To optimize the performance of pump turbines, extensive hydraulic design studies have been conducted. Researchers typically employ numerical simulation methods [3–7] and model tests [8–12] as approaches for research on hydraulic machines. The hydraulic characteristics of pump turbines in turbine mode have been extensively investigated. Deng et al. [13] analyzed the vortex motion patterns and pressure pulsation characteristics of pump turbines in vaneless space in turbine mode using numerical simulations and experiments. Lai et al. [14,15] conducted experimental studies on the flow characteristics in the DT of a pump turbine in turbine mode and determined the connection between velocity distribution and vortex rope motion in the DT. While most researchers have focused on flow characteristics, hydraulic loss remains the key factor affecting turbine efficiency.

When studying hydraulic losses, the pressure difference method is typically used first. However, this method does not provide a specific distribution of hydraulic losses. Therefore, researchers have proposed various alternative methods to study hydraulic losses, such as the entropy production method and the local hydraulic loss method. The entropy production method, derived from thermodynamics, has been improved and widely employed in hydraulic loss studies.

In the initial decade of this century, scholars primarily focused on fundamental theoretical research. They engaged in theoretical derivations and modeling analyses to address core issues, utilizing entropy production theory to explore topics such as laminar and turbulent flow, the viscous layer of turbulent wall flows, and more [16]. Subsequently, an increasing number of researchers started employing entropy production theory as a crucial framework for investigating process mechanisms and fluid flow states in specific types of hydraulic machinery. This application aimed at optimizing performance, understanding energy dissipation in flows, and studying flow characteristics [16]. Yan et al. [17] utilized the entropy production method to investigate the hydraulic losses of a pump turbine in the S characteristic zone and found that the vortex flow in the guide vane (GV) zone and the DT contributed significantly to the hydraulic losses. Li et al. [18] applied the entropy production method to examine the hydraulic losses of a pump turbine in the hump region and concluded that rotating stall in this region was a major source of hydraulic losses. Therefore, the entropy production method is well suited for studying hydraulic losses in pump turbines. The local hydraulic loss method, established by Qin et al. [19], was specifically applied to the study of hydraulic losses in pump turbines. Currently, the entropy production method is more widely used due to its superior accuracy. Although entropy production methods have been widely used in the study of energy losses in pump turbines, research on pump turbines with splitter blades has not been conducted.

As a means of validating and supplementing numerical simulations, non-contact measurement methods are widely used in hydromechanics. Two main techniques commonly used in hydraulic machines are particle image velocimetry (PIV) and LDV. PIV is well suited for studying the evolution of flow structures [20,21], while LDV offers higher accuracy in local velocity measurements [22,23]. When the velocity of a point needs to be measured in pump turbines, the LDV technique has become the preferred experimental approach due to its superior accuracy.

This paper investigates the mechanism of hydraulic losses in pump turbines operating in turbine mode at different heads. Numerical simulations are performed and validated by LDV tests. The study utilizes the entropy production method to analyze hydraulic losses and explores the relationship between losses and heads in conjunction with flow field analysis. The findings contribute to a better understanding of the hydraulic performance of pump turbines and provide insights into improving efficiency and operation.

2. Theory of the Simulation

2.1. Turbulence Model and Entropy Production Theory

In the numerical simulations, Navier–Stokes equations based on Reynolds time averaging were used, and the SST k-ω turbulence model was used [24–28].

The entropy production rate (EPR) of the time-averaged Reynolds flow includes the direct entropy rate caused by the time-averaged velocity and the indirect entropy rate caused by the pulsation velocity; the EPR is shown in Equation (1). The EPR caused by direct dissipation (EPDD) is shown in Equation (2), and the EPR caused by turbulence dissipation (EPTD) is calculated using Equation (3).

$$\dot{S}_D''' = \dot{S}_{\overline{D}}''' + \dot{S}_{D'}''' \tag{1}$$

$$\begin{aligned}\dot{S}_{\overline{D}}''' = \frac{2\mu_{eff}}{T}\left[\left(\frac{\partial\overline{u}_1}{\partial x_1}\right)^2 + \left(\frac{\partial\overline{u}_2}{\partial x_2}\right)^2 + \left(\frac{\partial\overline{u}_3}{\partial x_3}\right)^2\right] \\ + \frac{\mu_{eff}}{T}\left[\left(\frac{\partial\overline{u}_2}{\partial x_1} + \frac{\partial\overline{u}_1}{\partial x_2}\right)^2 + \left(\frac{\partial\overline{u}_3}{\partial x_1} + \frac{\partial\overline{u}_1}{\partial x_3}\right)^2 + \left(\frac{\partial\overline{u}_2}{\partial x_3} + \frac{\partial\overline{u}_3}{\partial x_2}\right)^2\right]\end{aligned} \tag{2}$$

$$\dot{S}_{D'}''' = \beta\frac{\rho\omega k}{T} \tag{3}$$

where $\dot{S}_D'''$, $\dot{S}_{\overline{D}}'''$, and $\dot{S}_{D'}'''$ denote EPR, EPDD, and EPTD, respectively, in m^{-3}·K^{-1}; $\overline{u}_1$, $\overline{u}_2$, and $\overline{u}_3$ denote the time-averaged velocity components in m/s; u_1', u_2', and u_3' denote the pulsating velocity components in m/s; T indicates temperature in Kelvin; μ_{eff} is the effective dynamic viscosity of fluid in Pa·s; β is an empirical constant, approximated as 0.09; k denotes the turbulent kinetic energy in m^2/s^2; and ω denotes the turbulent vortex frequency in s^{-1}.

The EPR caused by wall shear stress (EPWS) is calculated using Equation (4):

$$\dot{S}_W'' = \frac{\vec{\tau} \cdot \vec{v}}{T} \tag{4}$$

where $\dot{S}_W''$ denotes EPWS, W·m^{-2}·K^{-1}; $\vec{\tau}_w$ and $\vec{v}_w$ indicate the shear stress and velocity near the wall, respectively.

The TEP (sum of the integrals for each entropy production rate) can be obtained by integrating $\dot{S}_{\overline{D}}'''$, $\dot{S}_{D'}'''$ and $\dot{S}''_W$, respectively, over the computational domain with subsequent summation.

$$S_{pro,\overline{D}} = \int_V \dot{S}_{\overline{D}}''' dV \tag{5}$$

$$S_{pro,D'} = \int_V \dot{S}_{D'}''' dV \tag{6}$$

$$S_{pro,W} = \int_A \frac{\vec{\tau}_w \cdot \vec{v}_w}{T} dA \tag{7}$$

$$S_{pro} = S_{pro,\overline{D}} + S_{pro,D'} + S_{pro,W} \tag{8}$$

where S_{pro} denotes the TEP.

2.2. Calculation Domain and Mesh Generation

The pump turbine model is derived from real units in high-head pumped storage power plants and is created using Unigraphics NX software to achieve high-precision modeling across the entire computational domain based on design blueprints. The pump turbine model used in the numerical simulation is shown in Figure 1, and the geometric parameters of the pump turbine are shown in Table 1.

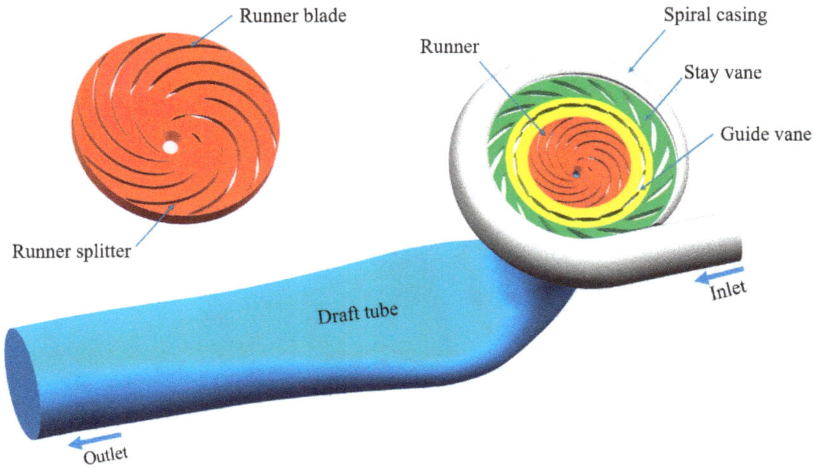

Figure 1. Calculation domain of pump-turbine.

Table 1. Geometric parameters of pump-turbine.

Domain	Parameter	Variable/Unit	Value
RN	Blades	Z_b	5
	Splitters	Z_s	5
GV	Guide vanes	Z_G	16
Spiral casing (SC)	Wrap angle	$\Phi/(°)$	360
Stay Vane (SV)	Stay vanes	Z_S	16

The full flow channel of the pump turbine was meshed using ICEM-CFD. The final grid information is shown in Table 2. The full computational domain grid is shown in Figure 2. The cell growth ratio was set to 1.2 at all the wall boundaries.

Table 2. Pump turbine grid information.

Domain	Grid Type	Number of Grid Cells	y+
RN	Hexahedral	3,011,474	<15
GV	Tetrahedral	2,988,878	<20
SV	Hexahedral	1,699,875	<30
DT	Hexahedral	1,665,874	<30
SC	Hexahedral	1,856,253	<50
Total	/	11,222,354	/

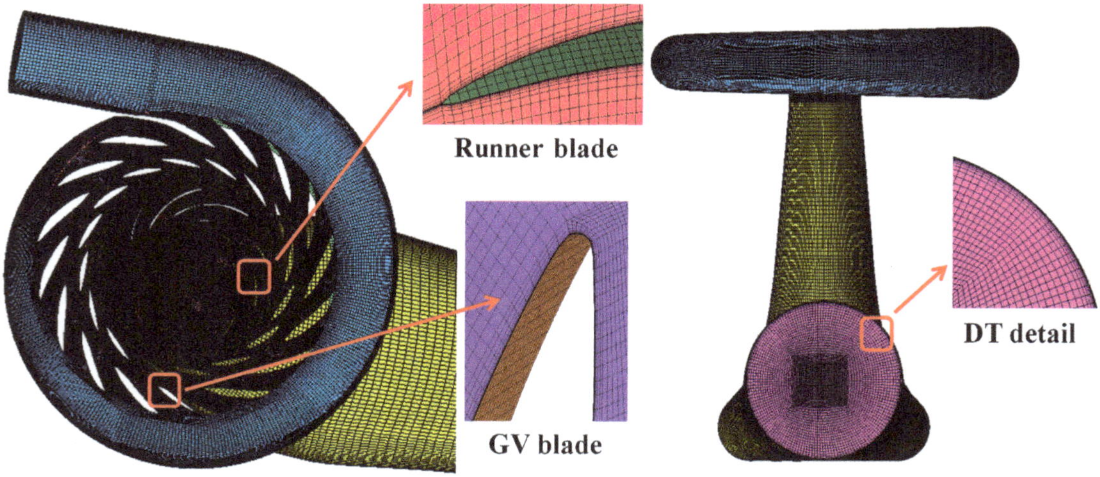

Figure 2. Grid scheme of the pump turbine.

2.3. Grid Independence Verification

Seven grids were formed, and the numbers of grid cells were about 2,130,000, 3,150,000, 4,210,000, 6,380,000, 8,390,000, 9,560,000, and 11,220,000, respectively. A sensitivity analysis of the key parameters of the external characteristics of the unit, namely, efficiency and torque, was carried out, as shown in Figure 3. It was found that an increase in the number of grid cells could significantly improve the accuracy of the key parameters. The efficiency and torque values remain stable after a grid number greater than 8 million. To further determine the influence of the grid on the calculations, an independence analysis of the grids was performed.

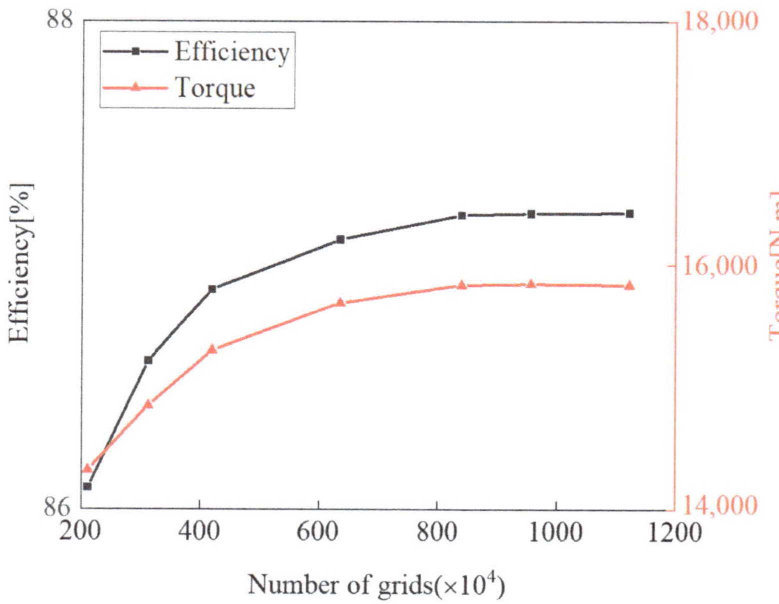

Figure 3. Grid sensitivity analysis.

To confirm grid independence in this study, the Richardson extrapolation method was utilized [29–32]. Moreover, the grid convergence index (GCI) was employed to quantitatively assess the convergence of the computational results. The GCI can be calculated using the following formula:

$$GCI^{21} = \frac{F_S e_a^{21}}{r_{21}^{\zeta} - 1} \tag{9}$$

$$e_a^{21} = \left| \frac{\varphi_1 - \varphi_2}{\varphi_1} \right| \tag{10}$$

$$r_{21} = \sqrt[3]{G_1/G_2} \tag{11}$$

$$\zeta = \frac{1}{\ln(r_{21})} |\ln|\varepsilon_{32}/\varepsilon_{21}| + q(\zeta)| \tag{12}$$

$$q(\zeta) = \ln(\frac{r_{21}^{\zeta} - s}{r_{32}^{\zeta} - s}) \tag{13}$$

$$s = 1 \cdot \mathrm{sgn}(\varepsilon_{32}/\varepsilon_{21}) \tag{14}$$

where F_S is the safety factor, typically 1.25; e_a is the relative error of the two sets of grid numerical value φ_1 and φ_2; G_1 is the grid number of G1; r is grid refinement factor; ζ is the convergence accuracy; ε_{32} and ε_{21} are the difference between the two sets of numerical values of the grid.

Three groups of grid plans were selected, namely, G1: 11,220,000, G2: 8,390,000, and G3: 3,150,000, which were involved in the verification of grid accuracy with flow rate and efficiency. The GCIs of flow and efficiency involved in the grid independence verification were less than 3.0% when the grid number was 11,220,000, indicating that the grid number of G1 met the requirements of computational accuracy. Grid independence verification information is shown in Table 3.

Table 3. Grid independence verification.

Parameters	φ_1	φ_2	φ_3	F_S	ζ	GCI
Q	381.75	381.25	378.35	1.2	10.78	1.55%
Efficiency	87.3	87.2	86.9	1.2	3.89	1.35%

2.4. Boundary Conditions and Case Information

ANSYS-Fluent software (Ansys, Inc., Canonsburg, PA, USA) was used for the numerical simulation, and the pressure boundary was used for both the inlet and outlet of the computational domain. The inlet boundary pressure is specified using a head value, while the outlet boundary pressure is set to 0. The rotating component moves using the Moving reference frame, and the rotational speed is set to a test value. The SIMPLEC algorithm is used to solve the flow fields, the second order upwind format is used to discretize the convective terms and diffusive terms, the wall condition is set to no-slip wall, and the root mean square of the residuals is set to 10^{-5} [28]. During the simulation, the interface between the rotating domain and the static domain is dealt with by means of the frozen rotor. This method is usually used for steady calculations, which reduce computational requirements by assuming that the flow field does not change significantly during one rotor revolution. The characteristic operating points during the actual operation of the power plant were selected for the study. The case information is shown in Table 4.

Table 4. Simulation case selection (all data in the table is from test data are used as boundary conditions for simulation).

Case	H (m)	GVO (°)	n (rev/min)	Q (m^3/s)	N (MW)	η (%)	Δh (m)
Case 1	580	10	500	36.6	180.2	86.69	77.20
Case 2	600	10	500	38.4	198.9	88.10	71.40
Case 3	640	10	500	41.3	233.9	90.30	62.08

3. Experimental Test and Validation

3.1. Test Rig

The model pump turbine was tested on a general hydraulic–mechanical test rig, as shown in Figure 4, and the external characteristics of the unit and the velocity distribution in the DT were obtained. The test bench parameters are shown in Table 5.

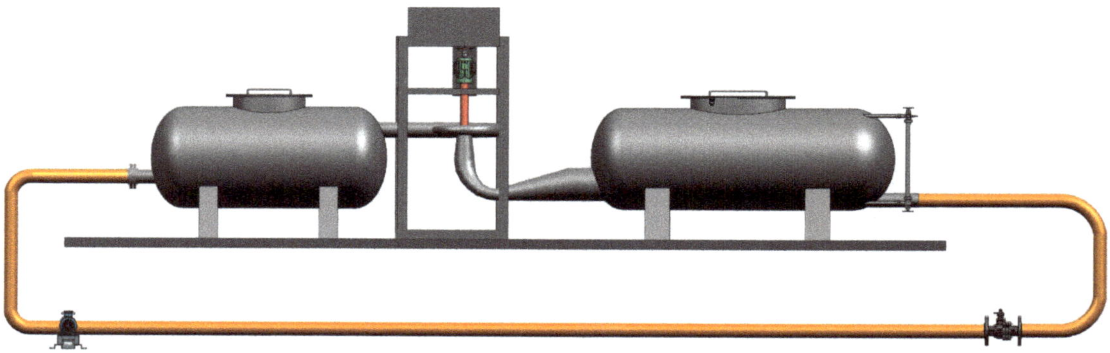

Figure 4. General test rig of reaction turbine.

Table 5. Test rig parameters.

Parameters	Value
Maximum test flow rate (m^3/s)	1.5
Maximum test head (m)	150
Maximum test speed (r/min)	2500
Model RN diameter (mm)	250–500
Dynamometer maximum power (kW)	500
Rated power of pump motor (kW)	2 × 850
DT pressure (kPa)	−85 to +250
Uncertainty of efficiency measurements	$\leq \pm 0.25\%$
Model type	Reaction turbine

The structure of the LDV system is shown in Figure 5a, the parameters are shown in Table 6, and the test site is shown in Figure 5b.

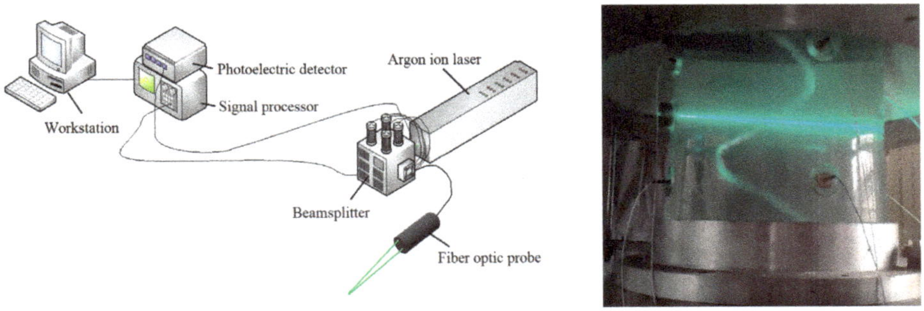

(a) (b)

Figure 5. LDV test system. (**a**) Structure of LDV test system. (**b**) Test site.

Table 6. Parameters of LDV test system.

Parameters	Value
Speed range	–150–1000 m/s
Measurement error	0.1%
Sample frequency	400–800 MHZ
Maximum processing frequency	175 MHz
Minimum processing frequency	300 Hz
Bits	8

3.2. Computational Validation

In order to facilitate the comparison of the velocity distribution in the DT obtained from the numerical simulation and the experiment, the velocity in the DT must be dimensionless.

$$C_u = \frac{V_u}{nD_1} \tag{15}$$

$$C_m = \frac{V_m}{nD_1} \tag{16}$$

where C_u denotes the tangential velocity coefficient; C_m denotes the axial velocity coefficient; V_u and V_m denote the velocity obtained by the test or numerical simulation, respectively, in m/s; n is the rotational speed in rev/min; and D_1 is the RN inlet diameter of the pump turbine in m.

The head H_M of the model test is determined from Equation (17):

$$H_M = \left(\frac{n_M}{n}\right)^2 \left(\frac{D_M}{D_1}\right)^2 H \tag{17}$$

where H is the head; H_M is the model test head, m; n_M is the rotational speed of model pump-turbine, rev/min; D_M is the RN inlet diameter of model pump-turbine, m.

The external characteristics and velocity distribution of the DT of Case 1 were compared, as shown in Figure 6.

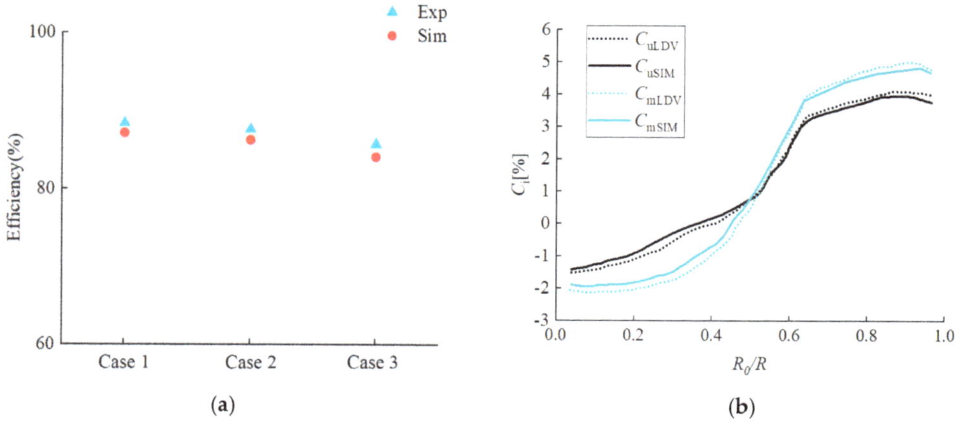

Figure 6. Comparison between numerical simulation and experimental results: (**a**) efficiency; (**b**) velocity distribution of DT (C_{uSim} and C_{mSim} denote the velocity coefficient obtained in CFD, and C_{uLDV} and C_{mLDV} denote the velocity coefficient obtained in LDV).

As seen in Figure 6a, the efficiency values obtained from the three cases were very close to the efficiency values obtained from experimental testing, indicating that the simulated external characteristics were accurate. As seen in Figure 6b, the distribution of the simulated axial velocity coefficient and tangential velocity coefficient in the draft tube exhibited the same trend and almost identical values as the experimental results, indicating that the simulated internal flow within the unit was accurate.

4. Results and Analysis

4.1. Total Energy Loss

In order to accurately calculate the energy loss in the pump turbine under characteristic heads, the entropy production method was utilized; it mainly consists of three terms, namely, EPDD, EPTD, and EPWS. In addition, to describe the TEP increment ratio quantitatively compared to Case 1, the growth ratio (γ) is defined as

$$\gamma = \frac{\gamma_n - \gamma_1}{\gamma_1}(n = 2,3) \tag{18}$$

where γ_1, γ_2, and γ_3 represent the TEP values in Cases 1, 2, and 3, respectively.

Figure 7 illustrates the entropy production values and growth rates for different entropy terms under three operating conditions. The TEP gradually decreased from Case 1 to Case 3, indicating lower total energy losses at higher heads, consistent with the data in Table 4. Among the three entropy terms, EPTD was the dominant factor, followed by EPWS and EPDD. Specifically, the EPTD accounted for over 98% in all three cases (Figure 7a). Under the turbine condition with the high head, the flow rate of the unit was large, resulting in a large velocity gradient inside the unit, and the EPDD term was closely related to the velocity gradient, which led to a significant increase in the EPDD term in Cases 2 and 3. The efficiency of the unit was higher under the condition of a high water head, indicating that the flow regime of a high head is better. Since the EPTD of the simplified calculation was controlled by the turbulent kinetic energy, the EPTD terms of Cases 2 and 3 were significantly reduced. The large flow in the unit under the high head led to the increase of the velocity in the near-wall region. Since the EPWS item was affected by the near-wall velocity, the EPWS growth rate of Case 2 and Case 3 was significant. Additionally, the negative growth rates of −3.03% and −9.42% for Cases 2 and 3, respectively, compared to the EPTD in Case 1, indicated a decrease in intra-unit turbulence intensity with increasing flow rates (Figure 7b).

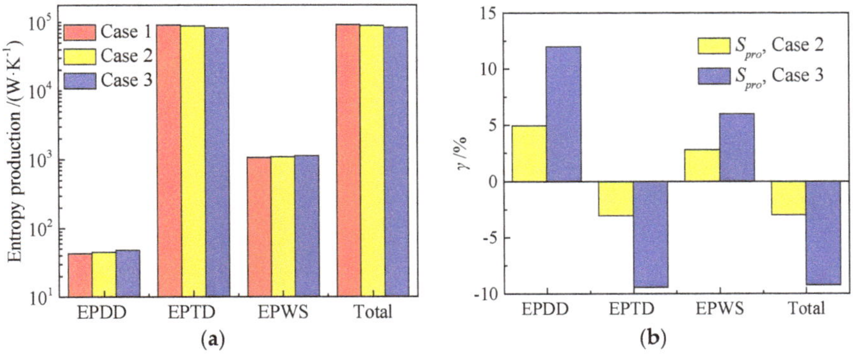

Figure 7. Entropy production of different entropy terms in three cases and the growth rate relative to case 1: (**a**) Entropy production; (**b**) growth rate.

Figure 8 depicts the TEP of the five components for the three cases and their growth rates compared to Case 1. Among the five components, the DT contributed the most to the TEP, followed by the RN, GV, SV, and SC. It was observed that the energy losses increased significantly along the flow direction (Figure 8a). The SC, SV, and GV components were mainly affected by inlet boundary conditions since they are inlet components of the unit. Consequently, the growth rates of the SC, SV, and GV components were more pronounced in Case 2 and Case 3 than the TEP growth rate in Case 1. Furthermore, the TEP growth rate of the impeller in Case 3 was significantly lower than that in Case 2 (−15.74% vs. −6.46%). This indicates that the RN's efficiency zone is at a higher head (Figure 8b).

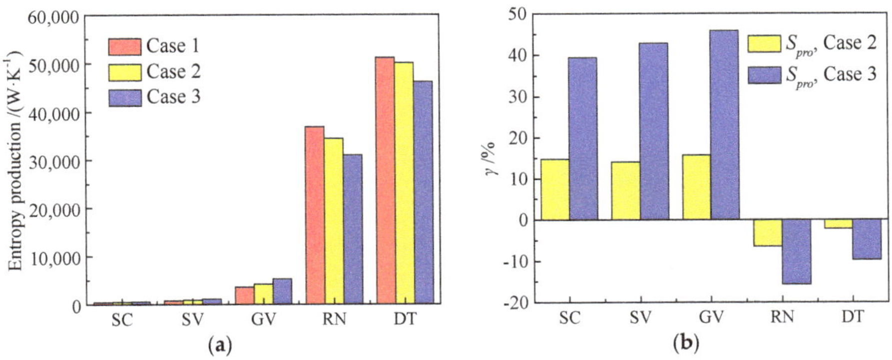

Figure 8. TEP of different unit components for three investigated cases, and the growth rate relative to Case 1: (**a**) TEP; (**b**) growth rate.

4.2. Analysis of Flow Characteristics in Inlet Components

In turbine mode, the pressure difference between the inlet and outlet of the pump turbine served as the primary driving force for the water flow, with energy losses occurring during energy conversion. The improved flow patterns in the inlet components (SC, SV, GV) contributed to lower energy losses in these components, accounting for 5.2%, 6.2%, and 8.3% in the three cases, respectively (Figure 8a).

To facilitate the comparison of the pressure distribution of the inlet part of the pump turbine in three cases (Figure 9), the dimensionless pressure coefficient C_p can be written as

$$C_p = \frac{p - p_0}{\rho g H} \tag{19}$$

where p_0 means the average static pressure at the inlet of the SC, pa.

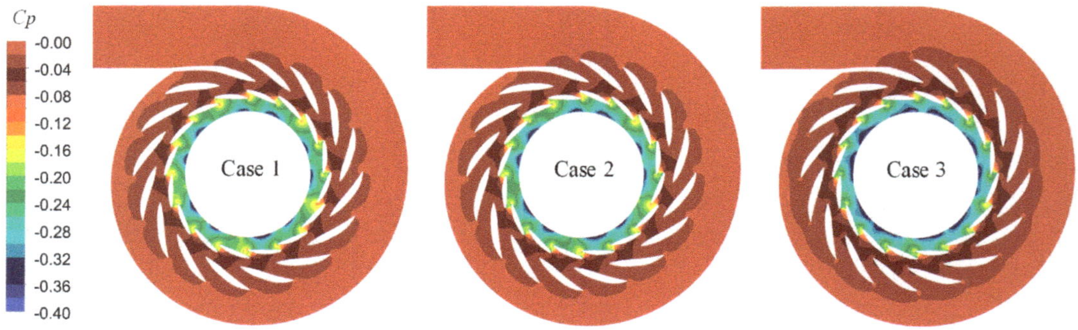

Figure 9. C_p value distributions on the horizontal plane of the inflow part of the pump turbine in three cases.

As shown in Figure 9, from the inlet of the worm shell to the outlet of the GV, the pressure gradually decreased, resulting in a more evenly distributed C_p value. This indicates that the number of SV and GV was set correctly, effectively balancing the water pressure created by high heads. The presence of a distinct high C_p region at the leading edge of SV and GV indicated a higher static pressure in this region. Compared to Case 1, the region with low C_p values (dark blue) near the interface between GV and RN was larger in Case 3, indicating a faster pressure drop in the inlet component under high head conditions.

Figure 10 illustrates the streamline distribution on the horizontal surface of the inlet components. In all three cases, the streamlines in the inflow part were smooth, without significant vortex formation, which was related to the proper GV angle and proper boundary conditions. Additionally, the velocity gradually increased along the flow direction, with the most pronounced increase occurring after the GV, as the water flow section through the GV area decreased. Since the GVO remained the same at 10° in all three cases, there was no significant difference in the streamline patterns. However, comparing Case 1 to Case 3, the velocity at the outlet of GV was higher in Case 3 due to the elevated flow rate under high head conditions.

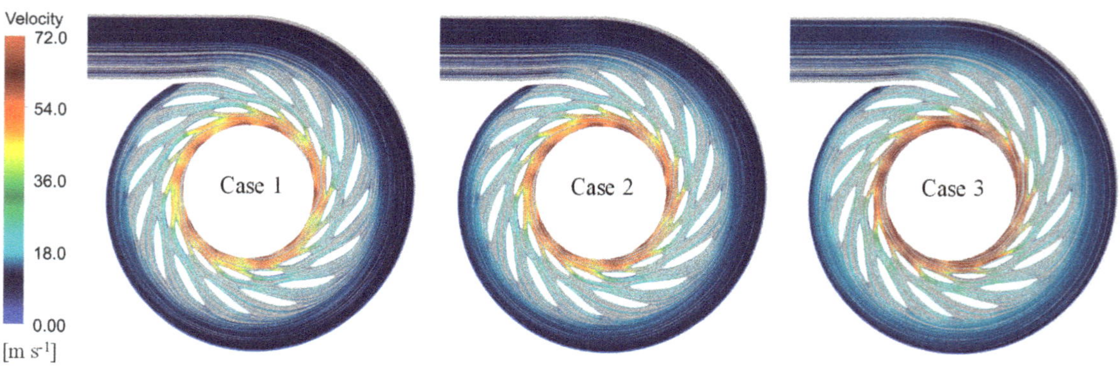

Figure 10. Velocity streamlines on the horizontal plane of the inflow part of the pump turbine in three cases.

Figure 11 shows the EPR distribution on the horizontal plane of the inlet components, specifically EPDD and EPTD. In all three cases, a distinct region of high EPR was observed at the trailing edge of the GV, corresponding to a significant velocity gradient within this area. Additionally, the energy loss at the SV trailing edge was smaller than that at the GV trailing edge, while the EPR value was lowest in the SC region, which agreed with the TEP distribution of the three inlet components (Figure 8a). It is noteworthy that the increase in EPR values was most pronounced in Case 3 when compared to Case 1, indicating that the higher flow rate led to larger energy losses within the inlet components and consequently to a higher TEP growth rate for the inlet components in Case 3 (Figure 8b).

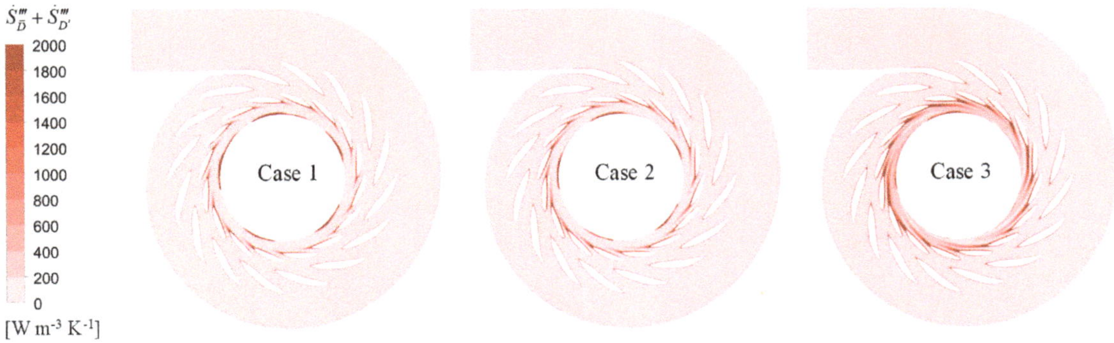

Figure 11. EPR distributions on the horizontal plane of the inflow part of the pump turbine in three cases.

4.3. Analysis of the Flow Characteristics in RN

The RN is the key component responsible for energy conversion in a water pump turbine, and its internal flow characteristics have a direct impact on energy conversion efficiency. The pressure difference on both sides of the blade is the main driving force for RN rotation, which will greatly affect the flow state inside the unit. Figure 12 displays the C_p value distribution at different span-wise surfaces of RN in the three cases. Here, span represents the dimensionless distance from hub to shroud, indicating the position of the blade-to-blade surface. For example, span = 0 indicates the blade-to-blade surface at the hub, and span = 1 indicates the blade-to-blade surface at the shroud. It can be clearly seen that the pressure inside the RN gradually decreased from the inlet to the outlet. At span = 0.05, the Cp value of the pressure side (PS) of the rotor blade was much higher than that of the suction side (SS), indicating that the pressure difference between the PS side and the SS side of the rotor blade drives the impeller to rotate. Moreover, the pressure difference on both sides of the blades was greater than that of the splitters, indicating that the ability of the splitter to drive the rotation of the impeller is weaker than that of the blades. At span = 0.5 and 0.95, the Cp distribution in the RN was consistent with that at span = 0.05, indicating that the pressure distribution of different spans is basically the same. Under the three different cases, the distribution of Cp inside the RN was basically the same, indicating that the design of the impeller of the unit is reasonable, and the pressure distribution inside the RN is basically unchanged within the operating range of the characteristic head.

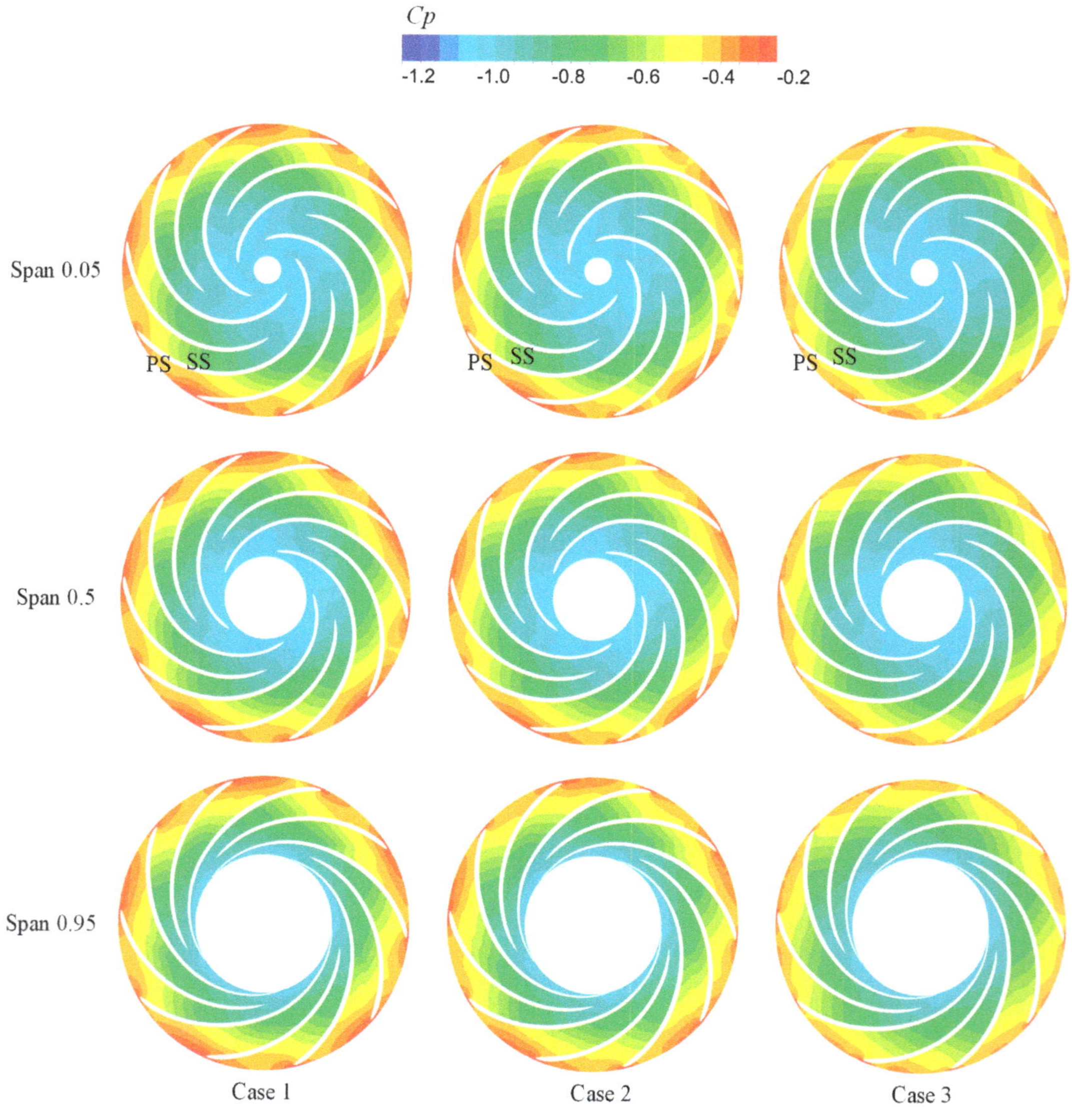

Figure 12. *Cp* value distribution at different span-wise surfaces of RN in the three cases.

In order to allow for a clearer comparison of the impeller flow characteristics in the three cases, Figure 13 intentionally shows the relative velocity streamlines at various deployed surfaces within the RN domain. In Case 1, where span = 0.05, the speed decreased continuously during water flow from the leading edge to the trailing edge of the blade. At this point, the energy of the water was gradually converted into the rotational mechanical energy of the blade. At this point, the velocity on the PS of the rotor blade was much higher than on the SS, suggesting that the water flow on the PS side was the main driving force for the rotation of the RN. Simultaneously, the alternating distribution of blades and splitters also influenced the flow pattern, making the flow pattern inside the impeller asymmetric. Compared to the velocity on the PS side of the blades, the velocity on the PS side of the splitters was higher, while the velocity on the SS side of the splitter was lower. Starting from the energy conversion of rotating machinery, the distribution of pressure and flow velocity at the inlet and outlet sections of the impeller was relatively uniform. However, considering that the geometric length of the splitter along the flow direction was shorter than that of the blades, the force bearing area of the splitter was smaller, and the velocity gradient nearby was larger. Compared to the case with span = 0.05 in Case 1, the velocity on the PS of the splitters at span = 0.5 was slightly lower, and the velocity at the RN outlet was also slightly lower. At the same time, the relative velocity streamlines between the blades were smoother at span = 0.5, indicating a better flow pattern at the center line of the blade profile. Furthermore, at span = 0.95, the relative velocity flow direction on the SS side of the leading edge of the splitters deviated from the center line of the blade profile, and there was noticeable low-speed recirculation on the SS side of the trailing edge of the long blades. Comparing the three different cases, it was noticeable that in Case 3, the velocity difference between the two sides of the RN was smaller, with a lower velocity on the PS side and a higher velocity on the SS side. Additionally, the relative velocity streamlines between the blades in Case 3 were smoother and aligned better with the center line of the blade profile. This indicated that Case 3 had the best flow pattern, followed by Case 2, and Case 1 had the worst performance.

To investigate the relationship between vortices and energy loss within the RN, absolute helicity, defined as the absolute value of the dot product between the velocity vector and vorticity vector, was used to characterize the degree of vortex spiralization. Energy loss is represented by EPR. Figures 14 and 15 show the distribution of absolute helicity and EPR at different unfolded surfaces of the RN domain. For span = 0.05 in Case 1, the leading edge of the splitter on the SS exhibited high helicity and high EPR, indicating significant vortex spiraling and energy loss in this area. At span = 0.5 in Case 1, there was high EPR and high-velocity flow at the trailing edge of the splitter, but no appreciable helicity, suggesting that the energy loss in this region was mainly due to the large velocity gradient and not the presence of vortices. In Case 1, the helicity distribution inside the impeller at span = 0.95 presented obvious asymmetry, and the leading edge of the splitter had obvious high helicity and energy loss, indicating that the high-speed flow off the centerline existed at the leading edge of the splitter. Significant helicity leads to significant energy loss. At the same time, compared with span = 0.5, the flow direction of the relative velocity near the leading edge of the SS side splitter at span = 0.95 deviated more from the centerline of the blade profile, resulting in higher helicity and EPR, thus exacerbating the impeller internal flow state asymmetry. Comparing the three cases, it was observed that the RN domain in Case 3 had the smoothest streamlines, the lowest helicity, and the lowest EPR. This contributed to the lowest TEP within the RN in Case 3. Consequently, as the head increased, the flow pattern within the RN domain improved, resulting in a reduction in overall helicity and EPR and ultimately a reduction in TEP within the RN (Figure 8).

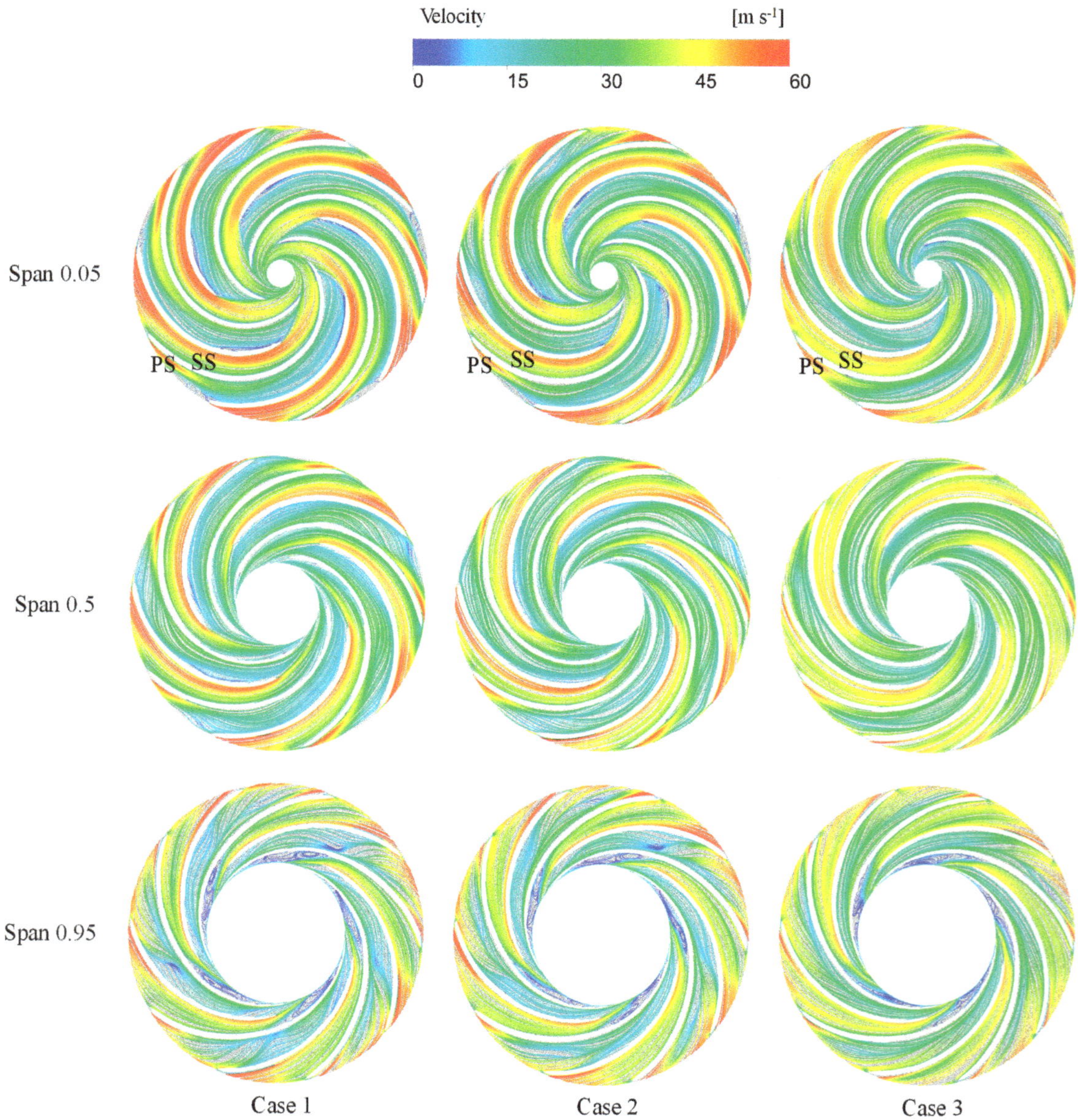

Figure 13. Relative velocity streamlines at different span-wise surface of RN in three cases.

Figure 14. Helicity distributions at different span-wise surface of RN in three cases.

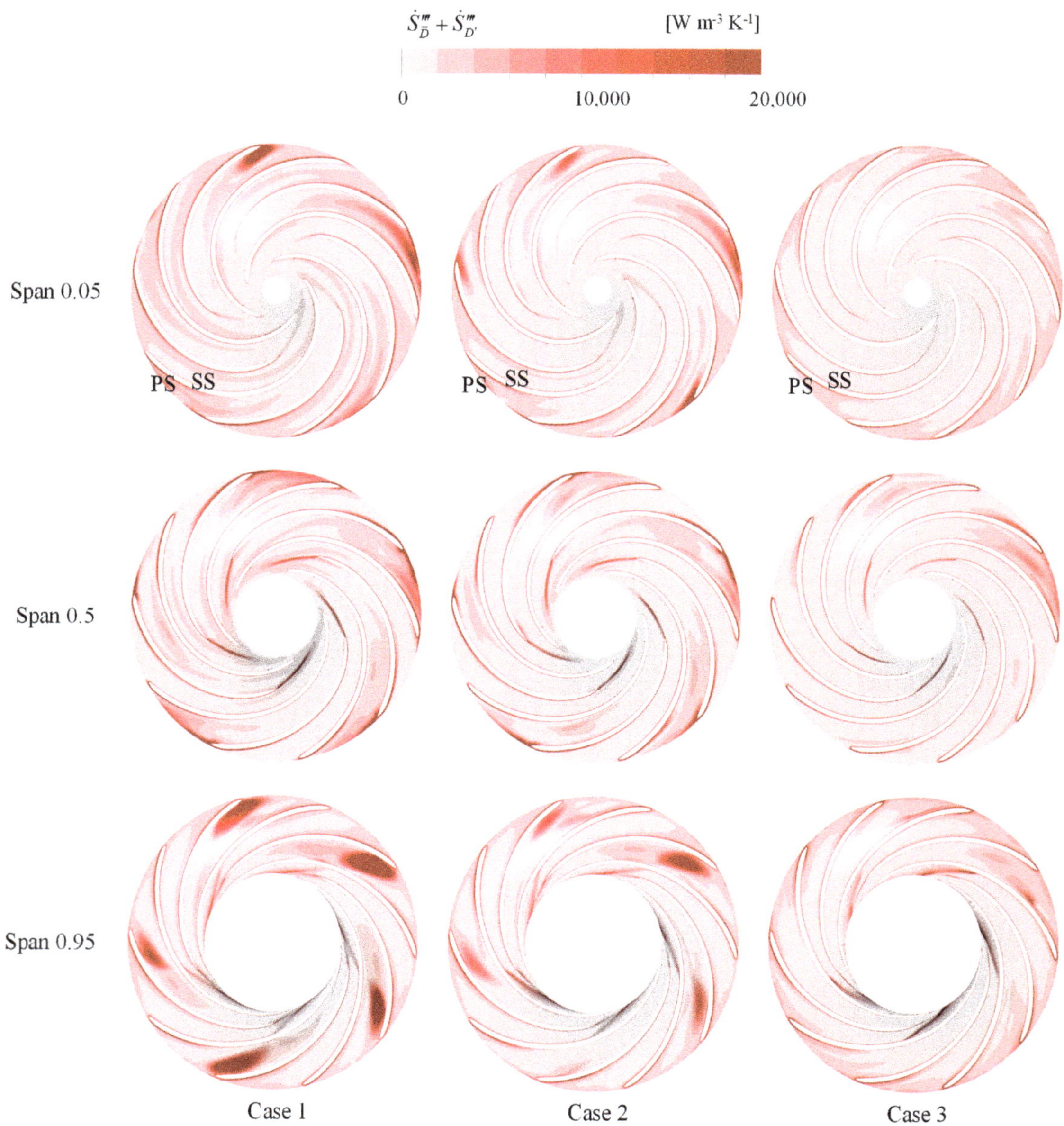

Figure 15. EPR distributions at different span-wise surface of RN in three cases.

In the process of flowing from the leading edge of the RN to the trailing edge, the water flow gradually transitioned from the state of rotating around the axis to the state of axial outflow. The swirling degree is closely related to the operating efficiency of the unit and the flow state in the RN. In order to quantitatively describe the degree of swirl in the RN under three characteristic water heads, the swirl number (S_w) is calculated as follows [33]:

$$S_w = \frac{\int_0^R U_a U_t r^2 \mathrm{d}r}{R\int_0^R U_a^2 r \mathrm{d}r} \tag{20}$$

where U_a is the axial velocity, U_t is tangential velocity, and R is the hydraulic radius, representing the impeller radius.

Figure 16 represents the S_w values on different horizontal planes of RN in the three cases. According to the physical structure of the RN, it could be divided into three regions: R1 is the region where the water flows in tangentially, R2 is the transition region, and R3 is the region where the water flows out axially. It can be clearly seen that the S_w value in the R1 domain was much greater than 1.0, indicating that the tangential velocity of the water flow was much greater than the axial velocity. Moreover, the S_w value dropped sharply along the axial direction, which showed that the kinetic energy of the water flow was quickly converted into the rotational mechanical energy of RN, which was also related to the greater pressure difference between the two sides of the blade in the R1 domain. In the R2 domain, the S_w value was less than 1.0, indicating that the main flow direction of the water flow was axial. At the same time, the S_w value decreased slowly along the axial direction, and it could be seen that the energy transferred by the water flow to RN was reduced. In the R3 domain, the S_w value was maintained at a low level, far below 1.0, and the velocity circulation still existed in the outlet domain. Comparing the three cases, it could be seen that on the same level, the S_w value of case 1 was the highest, followed by case 2, and that of case 3 was the least. Especially in the R3 domain, the average S_w values of the three cases were 0.23, 0.22, and 0.19, respectively. It could be seen that the velocity circulation of the water flow in Case 3 was the smallest, which was also the performance of Case 3 with less energy loss.

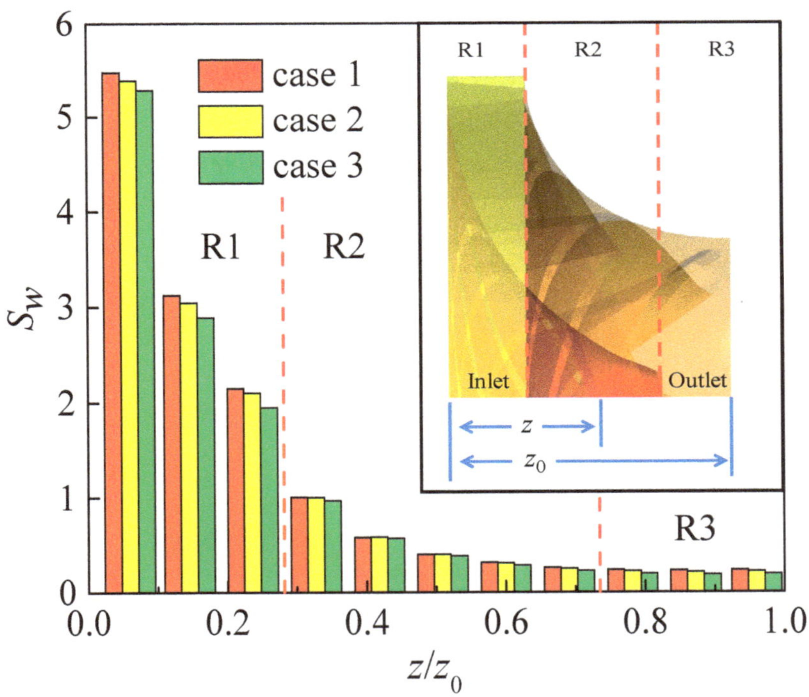

Figure 16. Swirl number on different horizontal planes of RN in three cases.

4.4. Analysis of the Flow Characteristics in DT

Based on the calculation results of the entropy production in Figure 7, it was observed that the TEP in the DT was larger than that in the RN domain. This was partly due to the presence of significant turbulence at the RN outlet and the larger integration volume of EPR in the DT. Since EPTD was the main contributor to the TEP in the system and was calculated indirectly from the turbulent kinetic energy, analyzing the level of turbulence in the DT could help identify the source of the energy loss. By examining the streamlines in the DT domain, as shown in Figure 17, it could be seen that there was significant tangential velocity at the inlet of the DT and that the velocity near the wall was much higher than that in the central region of the DT. Additionally, the overall flow velocity decreased from the inlet to the outlet of the DT. Figures 18–20 represent the distribution of TKE, spiralization, and EPR on the vertical plane of the DT, respectively. In Case 1, there was significant spiralization near the wall in the inlet region of the DT, consistent with the tangential flow distribution. Moreover, the inlet region of the DT exhibited high TKE and EPR values, and their distributions closely matched, indicating that the high level of turbulence in the inlet region resulted in significant energy loss. Moreover, regions of high EPR in the DT exhibited relatively low spiralization, suggesting that the increase in TKE was caused by the velocity gradient rather than the presence of vortices. Comparing the three different cases, it is clear that Case 3 had significantly lower TKE and EPR values compared to Case 1 and Case 2, indicating that the improved flow properties in the RN domain reduced the turbulence level and energy loss in the DT.

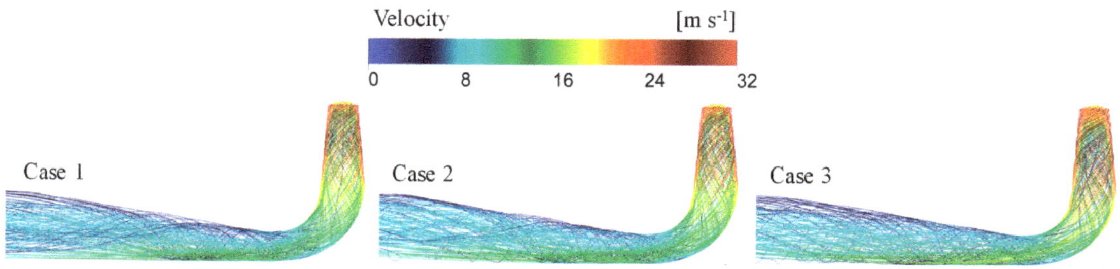

Figure 17. Velocity streamlines in DT of the pump turbine in three cases.

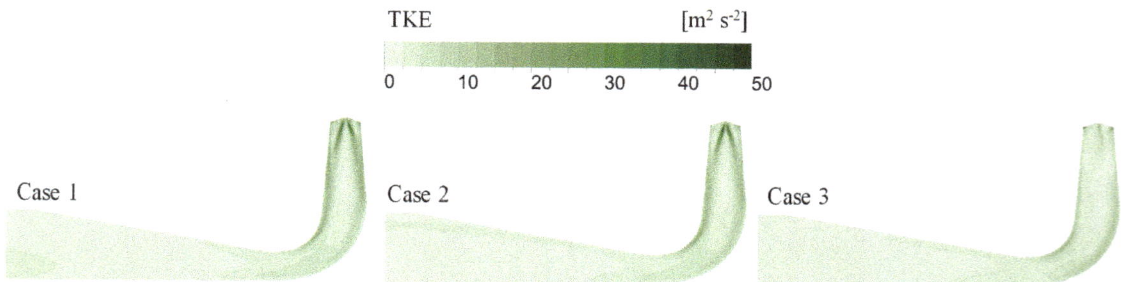

Figure 18. Turbulence Kinetic Energy distributions on the vertical plane of DT in three cases.

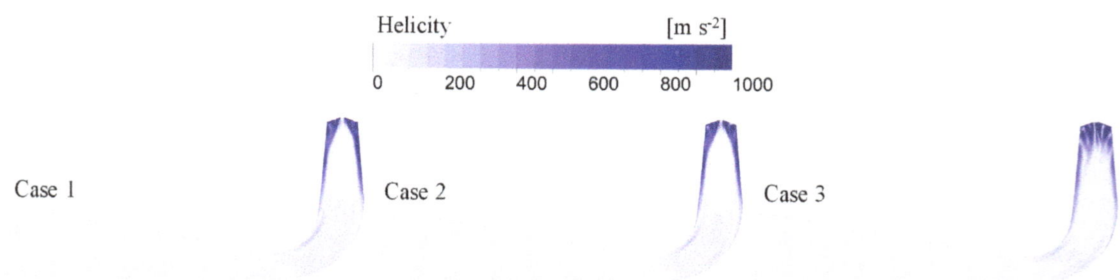

Figure 19. Helicity distributions on the vertical plane of DT in three cases.

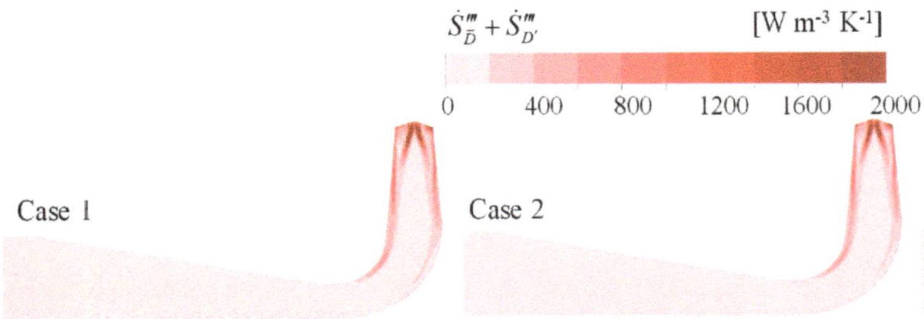

Figure 20. EPR distributions on the vertical plane of DT in three cases.

5. Conclusions

In this paper, the hydraulic loss mechanism of a high-head pump turbine with splitter blades was studied using three-dimensional visualization techniques. The numerical simulation was verified by an experimental LDV test, and the velocity coefficients at the section of the draft tube were in good agreement. Using entropy generation theory combined with the flow field distribution, the hydraulic loss and internal flow state changes of the unit under three characteristic heads, such as velocity, pressure, helicity, and EPR, were compared to explore the root causes and characteristics of the hydraulic loss. The following main conclusions were drawn:

(1) Among the three terms of entropy production, EPTD dominates, accounting for over 98% of the TEP. Within the five flow components, RN and DT play a dominant role, and TEP increases significantly along the flow direction. In all three cases, the growth rate of TEP decreases with increasing head, suggesting that the high-efficiency region of the turbine is at high-head operating conditions.

(2) In the inlet components, the presence of a large velocity gradient at the trailing edge of the GV leads to a significant EPR. Furthermore, under high head conditions, the higher flow rate increases the velocity in the GV area, increasing the velocity gradient at the trailing edge and causing more energy loss. The average energy loss growth rate of inlet components in case 2 and case 3 are 14.87% and 42.69%, respectively.

(3) In the RN domain, at spans of 0.05 and 0.95, the high spiralization and high EPR are observed at the leading edge of the splitter on SS where the flow direction deviates from the center line of the blade profile. This deviation, together with the high flow velocity, leads to significant spiraling and significant energy losses. At span = 0.5, the high EPR and high flow velocity are observed at the trailing edge of the splitter, with no appreciable spiralization occurring. This indicates that the energy loss in this region is mainly due to the large velocity gradient and not to the presence of vortices. In the RN domain, the S_w value continues to decrease along the axial direction, indicating that the water kinetic energy is continuously converted into the rotational mechanical energy of the RN. Especially in the R3 domain, the average S_w values of the three cases are 0.23, 0.22, and 0.19, respectively.

(4) In Case 1, the region close to the wall in the inflow section of the DT exhibits a high tangential velocity, which leads to significant spiralization. Additionally, the central region of the inflow section with large velocity gradients generates higher TKE and EPR. Case 3 shows significantly lower TKE and EPR values compared to Case 1, indicating that the improved flow properties in the RN domain reduce turbulence and energy losses in the DT.

Author Contributions: Conceptualization, Z.G. and D.L.; methodology, Z.X.; software, L.X.; validation, Z.X., F.Z. and Y.Z. (Yifeng Zhao); formal analysis, Z.X.; investigation, Z.X.; resources, Z.G.; data curation, L.X.; writing—original draft preparation, Z.G.; writing—review and editing, K.K.; visualization, Y.Z. (Yuan Zheng); supervision, K.K.; project administration, Y.Z. (Yuan Zheng); funding acquisition, Z.G. All authors have read and agreed to the published version of the manuscript.

Funding: This work was supported by the Science and Technology Project of State Grid Xinyuan Co., LTD. (Grant No. SGXYKJ-2022-044).

Data Availability Statement: Not applicable.

Acknowledgments: The authors are also grateful for the support of Hohai University.

Conflicts of Interest: The authors declare no conflict of interest.

References

1. Kan, K.; Binama, M.; Chen, H.; Zheng, Y.; Zhou, D.; Su, W.; Muhirwa, A. Pump as turbine cavitation performance for both conventional and reverse operating modes: A review. *Renew. Sustain. Energy Rev.* **2022**, *168*, 112786. [CrossRef]
2. Mahfoud, R.J.; Alkayem, N.F.; Zhang, Y.; Zheng, Y.; Sun, Y.; Alhelou, H.H. Optimal Operation of Pumped Hydro Storage-Based Energy Systems: A Compendium of Current Challenges and Future Perspectives. *Renew. Sustain. Energy Rev.* **2023**, *178*, 113267. [CrossRef]
3. Xu, L.; Liu, D.; Li, Z.; Zhao, X.; Liu, X. Experimental and Numerical Simulation Research on Flow Characteristics of Model Pump-Turbine in Four-Quadrant Operating Quadrants. *J. Energy Storage* **2022**, *54*, 105083. [CrossRef]
4. Li, W.; Li, Z.; Han, W.; Li, Y.; Yan, S.; Zhao, Q.; Gu, Z. Pumping-Velocity Variation Mechanisms of a Ferrofluid Micropump and Structural Optimization for Reflow Inhibition. *Phys. Fluids* **2023**, *35*, 052005. [CrossRef]
5. Kan, K.; Chen, H.; Zheng, Y.; Zhou, D.; Binama, M.; Dai, J. Transient Characteristics during Power-off Process in a Shaft Extension Tubular Pump by Using a Suitable Numerical Model. *Renew. Energy* **2021**, *164*, 109–121. [CrossRef]
6. Li, W.; Li, Z.; Han, W.; Li, Y.; Yan, S.; Zhao, Q.; Chen, F. Measured Viscosity Characteristics of Fe3O4 Ferrofluid in Magnetic and Thermal Fields. *Phys. Fluids* **2023**, *35*, 012002. [CrossRef]
7. Kan, K.; Zhang, Q.; Xu, Z.; Zheng, Y.; Gao, Q.; Shen, L. Energy Loss Mechanism Due to Tip Leakage Flow of Axial Flow Pump as Turbine under Various Operating Conditions. *Energy* **2022**, *255*, 124532. [CrossRef]
8. Jia, X.; Lv, H.; Zhu, Z. Research on The Influence of Impeller Tip Clearance on The Internal Flow Loss of Axial Circulating Pump under Unpowered Driven Condition. *ASME J. Fluids Eng.* **2022**, *145*, 021202. [CrossRef]
9. Xu, Z.; Liu, X.; Xu, X.; Li, Y.; Zheng, Y.; Zhong, Z.; Wan, J.; Dai, Q. Similarity of pressure pulsation between prototype and model pumps. *J. Drain. Irrig. Mach. Eng.* **2023**, *41*, 224–230. (In Chinese)
10. Li, W.; Li, Z.; Qin, Z.; Yan, S.; Wang, Z.; Peng, S. Influence of the Solution PH on the Design of a Hydro-Mechanical Magneto-Hydraulic Sealing Device. *Eng. Fail. Anal.* **2022**, *135*, 106091. [CrossRef]
11. Jia, X.; Lv, H.; Zhu, Z. Unsteady Flow Characteristics of an Axial Flow Pump Based on Energy Loss and Vortex Identification under Rotating Stall Condition. *J. Process Mech. Eng.* **2022**. [CrossRef]
12. Li, D.; Qin, Y.; Wang, J.; Zhu, Y.; Wang, H.; Wei, X. Optimization of blade high-pressure edge to reduce pressure fluctuations in pump-turbine hump region. *Renew. Energy* **2022**, *181*, 24–38. [CrossRef]
13. Deng, W.; Xu, L.; Li, Z.; Tang, W.; Wang, X.; Shang, L.; Liu, D.; Liu, X. Stability Analysis of Vaneless Space in High-Head Pump-Turbine under Turbine Mode: Computational Fluid Dynamics Simulation and Particle Imaging Velocimetry Measurement. *Machines* **2022**, *10*, 143. [CrossRef]
14. Lai, X.-D.; Liang, Q.-W.; Ye, D.-X.; Chen, X.-M.; Xia, M.-M. Experimental Investigation of Flows inside Draft Tube of a High-Head Pump-Turbine. *Renew. Energy* **2019**, *133*, 731–742. [CrossRef]
15. Lai, X.; Chen, X.; Liang, Q.; Ye, D.; Gou, Q.; Wang, R.; Yan, Y. Experimental and Numerical Investigation of Vortex Flows and Pressure Fluctuations in a High-Head Pump-Turbine. *Renew. Energy* **2023**, *211*, 236–247. [CrossRef]
16. Zhou, L.; Hang, J.; Bai, L.; Krzemianowski, Z.; El-Emam, M.A.; Yasser, E.; Agarwal, R. Application of Entropy Production Theory for Energy Losses and Other Investigation in Pumps and Turbines: A Review. *Appl. Energy* **2022**, *318*, 119211. [CrossRef]
17. Qin, Y.; Li, D.; Wang, H.; Liu, Z.; Wei, X.; Wang, X. Investigation on Hydraulic Loss Component and Distribution in Hydraulic Machinery: A Case Study of Pump-Turbine in Pump Mode. *J. Energy Storage* **2022**, *52*, 104932. [CrossRef]
18. Yan, X.; Kan, K.; Zheng, Y.; Chen, H.; Binama, M. Entropy Production Evaluation within a Prototype Pump-Turbine Operated in Pump Mode for a Wide Range of Flow Conditions. *Processes* **2022**, *10*, 2058. [CrossRef]
19. Li, D.; Gong, R.; Wang, H.; Xiang, G.; Wei, X.; Qin, D. Entropy Production Analysis for Hump Characteristics of a Pump Turbine Model. *Chin. J. Mech. Eng.* **2016**, *29*, 803–812. [CrossRef]
20. Liu, D.; Xu, W.; Zhao, Y. Experimental Study of the Flow Field of a High Head Model Pump Turbine Based on PIV Technique. *J. Hydrodyn.* **2021**, *33*, 1045–1055. [CrossRef]
21. Xu, L.; Jin, X.; Li, Z.; Deng, W.; Liu, D.; Liu, X. Particle Image Velocimetry Test for the Inter-Blade Vortex in a Francis Turbine. *Processes* **2021**, *9*, 1968. [CrossRef]
22. Deng, W.; Li, Z.; Ji, L.; Shang, L.; Liu, D.; Liu, X. Laser Doppler Velocimetry Test of Flow Characteristics in Draft Tube of Model Pump Turbine. *Processes* **2022**, *10*, 1323. [CrossRef]
23. Ji, L.; Xu, L.; Peng, Y.; Zhao, X.; Li, Z.; Tang, W.; Liu, D.; Liu, X. Experimental and Numerical Simulation Study on the Flow Characteristics of the Draft Tube in Francis Turbine. *Machines* **2022**, *10*, 230. [CrossRef]
24. Han, Y.; Tan, L. Experimental investigation on spatial-temporal evolution of tip leakage cavitation in a mixed flow pump with tip clearance. *Int. J. Multiph. Flow* **2023**, *164*, 104445. [CrossRef]

25. Kan, K.; Xu, Z.; Chen, H.; Xu, H.; Zheng, Y.; Zhou, D.; Muhirwa, A.; Maxime, B. Energy Loss Mechanisms of Transition from Pump Mode to Turbine Mode of an Axial-Flow Pump under Bidirectional Conditions. *Energy* **2022**, *257*, 124630. [CrossRef]
26. Menter, F.R. Two-Equation Eddy-Viscosity Transport Turbulence Model for Engineering Applications. *AIAA J.* **1994**, *32*, 1598–1605. [CrossRef]
27. Chen, G.; Gu, C.; Hajaiej, H.; Morris, P.J.; Paterson, E.G.; Sergeev, A. OpenFOAM computation of interacting wind turbine flows and control (I): Free rotating case. *Int. J. Hydromechatron.* **2021**, *4*, 1–26. [CrossRef]
28. Ciappi, L.; Stebel, M.; Smolka, J.; Cappietti, L.; Manfrida, G. Analytical and Computational Fluid Dynamics Models of Wells Turbines for Oscillating Water Column Systems. *J. Energy Resour. Technol.* **2021**, *144*, 050903. [CrossRef]
29. Xu, Z.; Zheng, Y.; Kan, K.; Chen, H. Flow Instability and Energy Performance of a Coastal Axial-Flow Pump as Turbine under the Influence of Upstream Waves. *Energy* **2023**, *272*, 127121. [CrossRef]
30. Ji, L.; Li, W.; Shi, W.; Tian, F.; Agarwal, R. Effect of blade thickness on rotating stall of mixed-flow pump using entropy generation analysis. *Energy* **2021**, *236*, 121381. [CrossRef]
31. Kan, K.; Yang, Z.; Lyu, P.; Zheng, Y.; Shen, L. Numerical Study of Turbulent Flow Past a Rotating Axial-Flow Pump Based on a Level-Set Immersed Boundary Method. *Renew. Energy* **2021**, *168*, 960–971. [CrossRef]
32. Li, J.; Sui, T.; Dong, X.; Gu, F.; Su, N.; Liu, J.; Xu, C. Large eddy simulation studies of two-phase flow characteristics in the abrasive flow machining of complex flow ways with a cross-section of cycloidal lobes. *Int. J. Hydromechatron.* **2022**, *5*, 136–166. [CrossRef]
33. Morris, C.E.; O'Doherty, D.M.; Mason-Jones, A.; O'Doherty, T. Evaluation of the swirl characteristics of a tidal stream turbine wake. *Int. J. Mar. Energy* **2016**, *14*, 198–214. [CrossRef]

water

MDPI

Article

Numerical Investigation of Inner Flow Characteristics of a Prototype Pump Turbine with a Single Pier in Draft Tube at Part Load Conditions

Haiping Hu [1], Ming Xia [2], Xianghui Song [1], Zhengwei Wang [2,*] and Mu Qiao [1]

[1] Baishan Storage Power Station, State Grid Xinyuan Co., Ltd., Huadian 132400, China; sea3219@163.com (H.H.); sxh17@163.com (X.S.); qiaomu_51@163.com (M.Q.)
[2] Department of Energy and Power Engineering, Tsinghua University, Beijing 100084, China; xiam17@mails.tsinghua.edu.cn
* Correspondence: wzw@mail.tsinghua.edu.cn

Abstract: Pump turbines operate under various off-design conditions, resulting in complex internal flow patterns. This study employs Reynolds-averaged Navier–Stokes (RANS) numerical methods to investigate the flow characteristics of a prototype pump turbine with a single draft tube pier in turbine mode, and then, the flow characteristics inside the draft tube are discussed with emphasis. Asymmetry between the pier-divided draft tube passage flows is inevitable due to the elbow section's curvature. Most of the fluid flows out of one passage, while vortex motion dominates the interior of the other one, resulting in completely different pressure fluctuation characteristics for the two flow passages. The large-flow passage is mainly characterized by the wide band in the frequency domain, corresponding to the recirculation zone, while some of the measured points in the low-discharge passage exhibit frequency splitting under kinematic progression. Further analysis demonstrates a low-frequency peak corresponding to the complementary shape between the vortex rope and the recirculation zone. This work elucidates the effects of the pier on the flow behavior and pressure fluctuation characteristics inside the draft tube and fills the research gap on piers in the field of pump turbines.

Keywords: pump turbine; single pier; vortex rope; pressure fluctuation; numerical simulation

Citation: Hu, H.; Xia, M.; Song, X.; Wang, Z.; Qiao, M. Numerical Investigation of Inner Flow Characteristics of a Prototype Pump Turbine with a Single Pier in Draft Tube at Part Load Conditions. *Water* **2024**, *16*, 13. https://doi.org/10.3390/w16010013

Academic Editor: Helena M. Ramos

Received: 12 November 2023
Revised: 15 December 2023
Accepted: 16 December 2023
Published: 20 December 2023

1. Introduction

With the development of new energy, pumped storage power stations play an important role in peak shaving and valley filling of the power grid, which can make up for the instability of wind and solar power stations [1]. In some offshore pumped storage power stations, it can also solve important problems such as seawater desalination and power supply on isolated islands [2,3].

Compared with typical Francis or Kaplan turbines, pump turbines need to operate under a wider range of off-design conditions in turbine mode, while also being able to switch to pump mode frequently. For this purpose, the internal flow characteristics of pump turbines under different conditions have been extensively studied by means of experiments or numerical simulations. Many studies have confirmed that the region in which strong pressure fluctuations are likely to occur in pump turbines is the vaneless space [4]. When the operating condition of a pump turbine changes, it may enter the "S-region", in which the pressure fluctuations in the vaneless space rise sharply, which can affect structural components, and this poses a threat to the safe operation of the unit when serious [5]. A detailed PIV model test study elaborated that the S-region characteristic of pump turbines is caused by the blocking of a sub-synchronous stall cell in the guide vane flow passage. The numerical study by Zhu et al. [6] indicated that the intense pressure fluctuation is related to the non-uniform distribution of vortex structures in the vaneless space. The simulation results by Hu et al. [7] found triangular or quadrilateral flow patterns in the upper crown cavity, which is an important reason for the low-frequency components in this region. Asomani et al. [8] reviewed the influence of design parameters on the internal flow characteristics of pump turbines, and the main influence of geometric parameters is the RSI characteristics of different units. Fu et al. [9] considered the clearance flow between the hub and the shroud, and studied the pressure fluctuation and shafting mechanical characteristics during startup. It was found that the influence of the clearance dominated the axial thrust much more than the influence of the runner blades. Yang et al. investigated the transient characteristics of the internal flow of pump turbines during start-up and shutdown by numerical simulation [10], and further studied the influence of flow-induced vibration on the structural stress characteristics of non-rotating components [11].

The flow pattern in the draft tube is also an important research focus on pump turbines. Compared with Francis and Kaplan turbines, pump turbines are influenced by the S-region characteristic, and the pressure fluctuation characteristics are affected by both the rotor–stator interaction and the draft tube vortex rope under some conditions [12]. Kim et al. [13] studied the internal flow in the draft tube of a pump turbine under partial flow rate conditions. By comparing the swirling number and the shape of the vortex rope, it is indicated that the vortex rope phenomenon is more obvious at medium flow rates, but at lower flow rates the swirling intensity is stronger. Tridon et al. [14] obtained an analytical expression for the radial velocity based on the formula of a conical diffuser and the superposition of three Batchelor vortices. The study by Lin et al. [15] showed that the draft tube vortex rope is also significantly affected by leakage flow and the blade vortex. Pang et al. studied the influence of the cavitation number on the draft tube vortex rope [16]. Numerical simulation revealed that, under different cavitation numbers, the vortex rope presents a spiral shape or a torch-like shape, and the torch-like vortex rope under a lower cavitation number has a higher swirling intensity and a lower dominant frequency.

In the above studies, most of the draft tubes are single-channel ones without a pier. The pier can increase the pressure recovery coefficient of the diffuser section [17], thereby improving the overall efficiency of the hydraulic turbine. Since the design of piers aims at improving the power generation efficiency, the current research on piers mainly focuses on Francis [18,19], Kaplan [20,21], and a small number of bulb turbines [22,23], with little attention paid to pump turbines. A well-known study was carried out by Tridon et al. [24,25]. Through LDV tests, they observed that, at a flow rate coefficient of 0.38, secondary flow caused a main vortex and a vertical vortex in the left flow passage of the draft tube, and the flow imbalance between the two passages was an important reason for the unexpected efficiency decrease at low flow rates after the replacement of the impeller with a high-efficiency one. DUPRAT [26] performed large eddy simulation (LES) and also found similar back-flow. Li et al. numerically studied a Francis turbine with a pier in the draft tube [27]. Their results showed that, at low flow rates, the discharge in the draft tube was concentrated in the flow passage on the inside of the pier, while the outer flow passage was dominated by the vortex. As the flow rate increased, the flow in the outer passage also gradually increased. Wang et al. [28] performed numerical simulation on a Francis turbine unit and believed that the draft tube pier can restrain the development of a vortex and reduce pressure fluctuations. In an early study, Paik et al. [29] simulated a draft tube with two piers, in which the vortex rope broke up and transported to three flow passages. Pasche et al. [30] carried out an asymptotic analysis on a Francis turbine with a pier in the draft tube and proposed that the synchronous pressure wave inside the draft tube results from the combined effects of the vortex rope and wall disturbances.

In this study, we performed unsteady numerical simulation on a pump turbine with a single pier in turbine mode. After examining the flow in the guide vane and runner passages, we focused on analyzing the flow patterns and spectral characteristics inside the draft tube under two typical loads. In simple terms, the draft tube pier has some influence on the evolution process of the vortex rope, which has not been considered in previous studies in the field of pump turbines. The influence of the rotor–stator interaction the and draft tube vortex rope on the pressure fluctuation characteristics was preliminarily investigated, filling the research gap on the stability of pump turbines in the presence of a pier.

2. Research Object and Numerical Method

The physical model constructed for this paper used a prototype pump turbine with the specific parameters shown in Table 1. According to the definition of the specific speed in Equation (1), the specific speed of the pump turbine was n_s = 228.

$$n_s = \frac{n\sqrt{P}}{H^{5/4}} \quad (1)$$

where n is the rated speed, P is the rated power, and H is the rated head.

Table 1. Basic parameters of the unit.

Parameter	Unit	Value
Runner inlet diameter (D)	m	5.22
Rotational speed (n)	r/min	200
Rated power (P)	MW	139
Rated flow (Q)	m^3/s	148.8
Number of blades (Z_b)	/	7
Number of stay vanes (Z_s)	/	20
Number of guide vanes (Z_g)	/	20
Rated head (H)	m	105.8

The overall 3D model is shown in Figure 1. In order to highlight this research's focus, we separately adopted structured grids in the runner and draft tube parts, as shown in Figures 2 and 3. The Y+ values at the blade surface were less than 300. Since the local Reynolds number was about 10^8, the encryption of the surface mesh was reasonable for the calculation requirements of the SST k-ω model. The total number of mesh nodes was 2.6 million, and the number of mesh nodes per component is shown in Table 2. Based on the grid independence verification performed in the authors' previous study [7], this grid density is sufficient to resolve the internal flow patterns within the runner and draft tube.

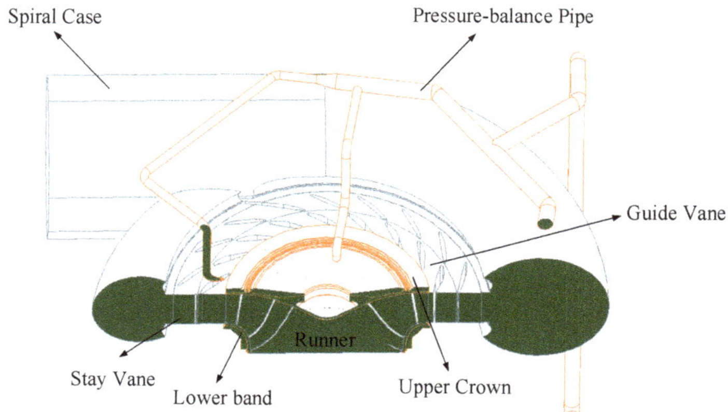

Figure 1. Schematic of the computational domain.

Table 2. Number of mesh nodes in partial fluid domain.

Fluid Domain	Spiral Case	Stay Vane	Guide Vane	Runner	Draft Tube
Number of Nodes ($\times 10^4$)	13.2	36.2	26.5	156	26.6

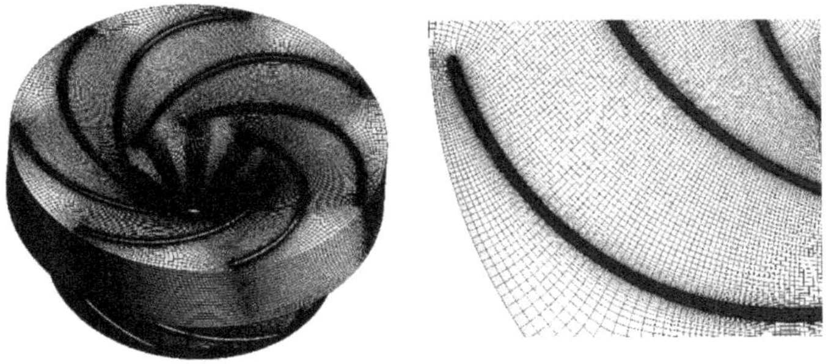

(**a**) (**b**)

Figure 2. Mesh division of the runner area. (**a**) Global mesh. (**b**) Details near the blades.

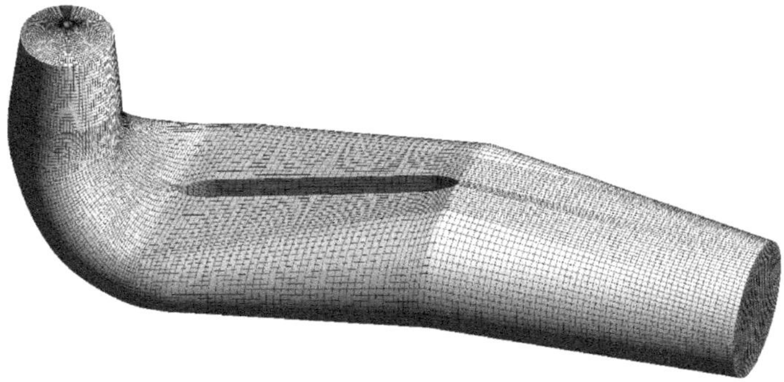

Figure 3. Mesh division of the draft tube area.

Numerical simulations were carried out in ANSYS CFX using the Reynolds averaged Navier–Stokes (RANS) equations [31] for the internal flow characteristics of the pump turbine.

$$\frac{\partial \rho}{\partial t} + \frac{\partial}{\partial x_j}(\rho u_j) = 0 \tag{2}$$

$$\frac{\partial}{\partial t}(\rho u_i) + \frac{\partial}{\partial x_j}(\rho u_i u_j) = -\frac{\partial p}{\partial x_i} + \frac{\partial}{\partial x_j}\left(\mu \frac{\partial u_i}{\partial t} - \rho \overline{u_i' u_j'}\right) + S_M \tag{3}$$

where ρ is water density, p is the pressure, and u is the flow velocity. The range of indicators for i and j is (1, 2, 3), μ is the dynamic viscosity, S_M is the external momentum source term, and $\tau_{ij} = \rho\overline{u_i' u_j'}$ are the Reynolds stresses. The SST k-ω model was used to solve the Reynolds stresses.

Boundary conditions: A pressure inlet and a pressure outlet were employed, since there are large reservoirs upstream and downstream of the pump turbine, specifying pressures is closer to the real case than giving flow rates. In the present case, the inlet pressure was 1.058 MPa and the outlet pressure was atmospheric. No-slip boundary conditions were used on all stationary walls. The rotating and stationary regions were modeled with a frozen rotor boundary and transient rotor–stator boundary for steady and unsteady simulations, respectively. For the unsteady simulations, the results from the steady simulations were used as the initial conditions. Within each runner rotation cycle, 360 time steps were calculated, giving a time step size of 8.33×10^{-4} s, and 70 rotation cycles were computed in total. Reference [32] presents extensive sensitivity tests on the grid and time step size for a pump turbine similar to that in the present study. It demonstrates that a time step equivalent to 3° of runner rotation per step can capture the flow features of the main working points, while 1.8° per step resolves the finer details under low-flow-rate conditions. Therefore, the selection in this work of 1° runner rotation per time step is good enough.

3. Results and Discussion

3.1. Field Test Results of Pressure Fluctuation

In previous work, we conducted field tests at the studied power plant and obtained pressure data from several measurement points. In turbine mode, we tested the pressure fluctuations at the elbow of the draft tube under various unit loads. The frequency domain results are shown in Figure 4. It can be observed that, at around 50% load, distinct low-frequency components emerged at the draft tube elbow, which were absent at lower or higher loads. We separately examined the operating points at 50% and 100% loads, whose signals and spectrum are shown in Figure 5. The pressure coefficient, C_p, here is specified as Equation (4) [33]. The $p - \overline{p}$ indicates the removal of the DC component from the original pressure signal. For the convenience of comparison, all frequency values below are expressed as the ratio of frequency to rated rotation frequency, that is, f/f_n, where f_n = 20.94 Hz.

$$C_p = \frac{p - \overline{p}}{\rho g H} \quad (4)$$

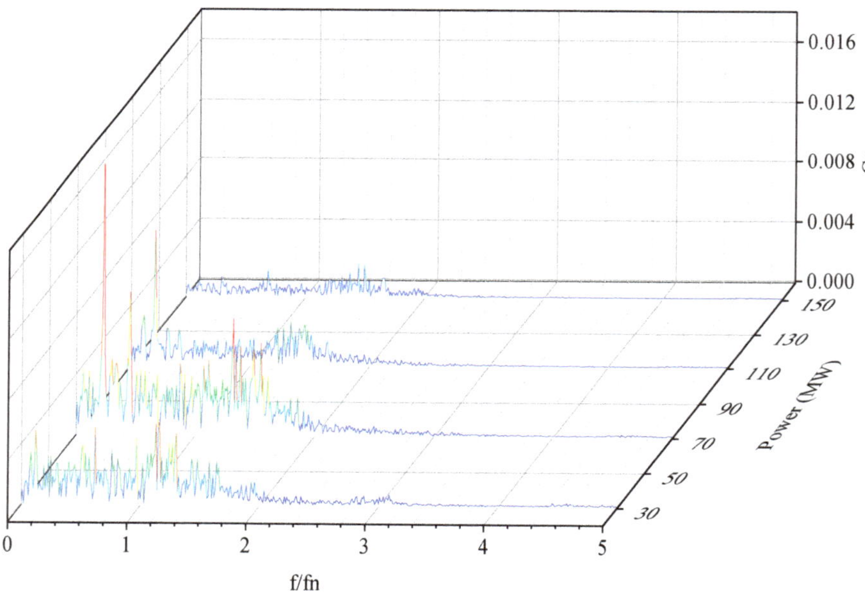

Figure 4. Pressure signal spectrum of the draft tube elbow section under different loads. Red means a higher amplitude.

At 50% load, the dominant frequency of pressure fluctuation at the draft tube elbow was $0.23f_n$, with remarkably high amplitude. Based on our engineering experience, such a dominant frequency generally originates from the vortex rope in the draft tube. However, for the power plant, most pressure sensors are used to monitor the DC component, i.e., the instantaneous mean pressure, and due to the long pressure measuring lines and low sampling rates, the high-frequency components are limited. Therefore, after summarizing the preliminary field test results, more high-frequency contents (including rotor–stator interaction, RSI) still need to be analyzed using the CFD simulation results, especially the working point at 50% load.

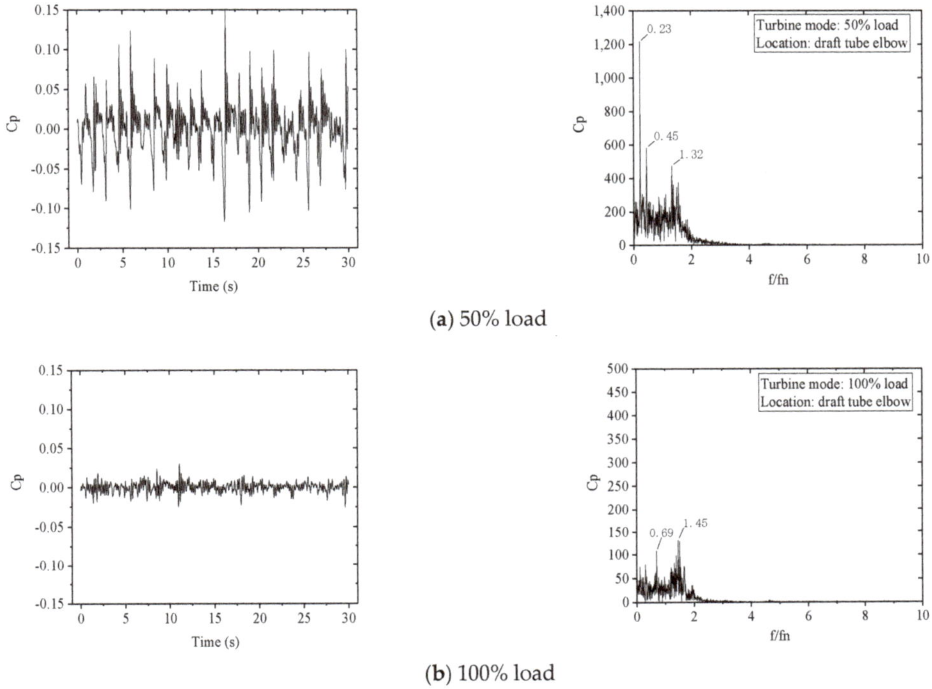

Figure 5. Pressure signal and spectrum of the draft tube elbow section under 100% and 50% loads.

3.2. Internal Flow Characteristic of Ring and Runner

The internal flow characteristics excluding the draft tube should be shown first. Figure 6 shows the pressure distribution on the cross section of spanwise = 0.5 in the guide vane region under 100% and 50% loads. It is evident that the pressure near the outer edge of the guide vane is higher than that near the inner edge, indicating certain pressure loss around the guide vanes. As the load decreases, the pressure near the outer edge increases markedly, leading to a larger pressure difference between the outer and inner edges. This implies an increase in hydraulic loss in the guide vane region at partial load.

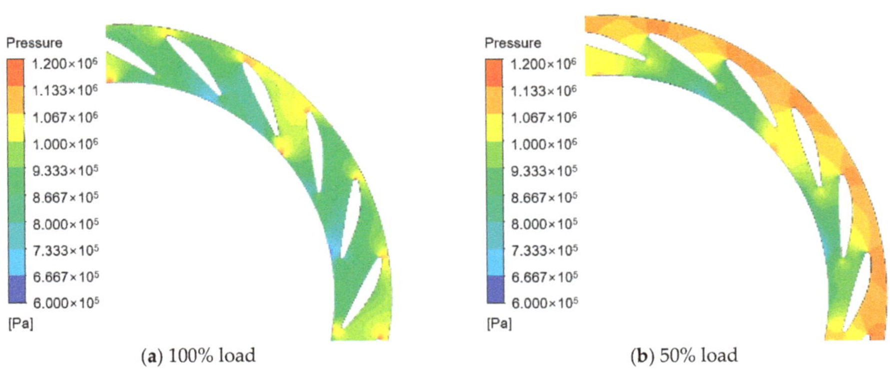

Figure 6. Pressure distribution of the guide vane. Spanwise = 0.5.

Figures 7 and 8 show the streamline distributions in the isometric view and top view of the runner, respectively. At 100% load, the streamlines align well with the runner geometry, whereas at 50% load, the flow becomes turbulent to some extent. Flow separation vortices appear on the suction side near the leading edge of three blades, and an undeveloped vortex occurs on one blade. Since there are seven blades in total, it can be said that vortices exist in half of the flow passages, and the vortices locate at the mid-span of the blade's leading edge.

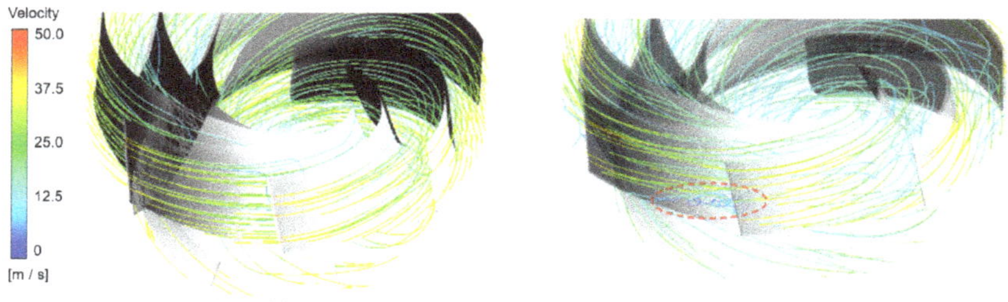

(**a**) 100% (**b**) 50%

Figure 7. The three-dimensional streamline distribution of the runner.

(**a**) 100% (**b**) 50%

Figure 8. Streamline distribution in the middle section of the runner.

3.3. Internal Flow Characteristic of Draft Tube

Since the draft tube in this study has a single pier, we define the flow passages on both sides of the pier as Passage A and Passage B, as shown in Figure 9. It is evident that Passages A and B exhibited significantly different flow characteristics, i.e., asymmetry. At 100% load, the flow in Passage A and Passage B aligned better with the geometry of the draft tube, while the flow in Passage B was somewhat turbulent. At 50% load, the flow velocity near the outer wall at the inlet of the draft tube was high, with an obvious swirling motion, resulting in an uneven flow distribution between Passages A and B, and the flow characteristics became more complex. Judging from the flow streamline density, the majority of flow was concentrated on the left side of Passage A, while a large vortex formed on the right side of Passage A near the pier. In Passage B, low-speed spiral motion dominated. Near the outlet of the draft tube where the pier disappeared, the spiral flow in Passage B from the bottom mixed with the high-speed region in Passage A.

The difference in streamline density between the two passages means a difference in the flow rate. It intuitively comes from the geometry: as the center-symmetric vortex flow turns from vertical to horizontal inside the draft tube elbow, the curvature means shorter effective lengths near the top of the draft tube elbow and longer effective lengths near the bottom. Based on the direction of rotation, the top fluid mainly flows from Passage A to Passage B, which means fluid obstructed by the prematurely encountered pier accumulates in Passage A. The bottom fluid mainly flows from Passage B to Passage A, entering the pier later and allowing more fluid transfer across passages. The combined result is visibly greater discharge in Passage A over Passage B. The abstract diagram is shown in Figure 10.

There is obviously more fluid on the side of Passage A (dashed blue line) than on the side of Passage B (solid blue line).

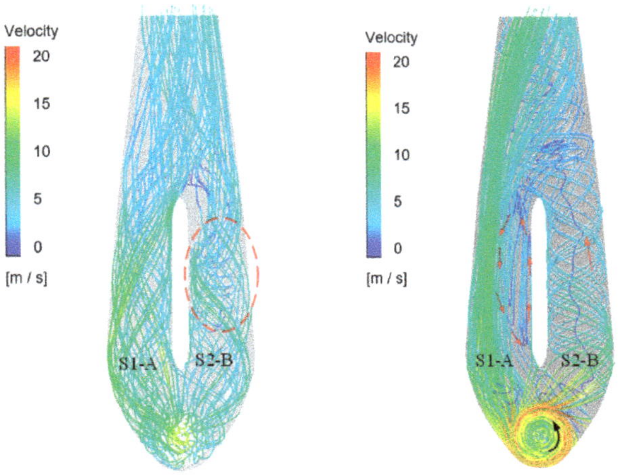

(**a**) 100% (**b**) 50%

Figure 9. The three-dimensional streamline distribution of the draft tube.

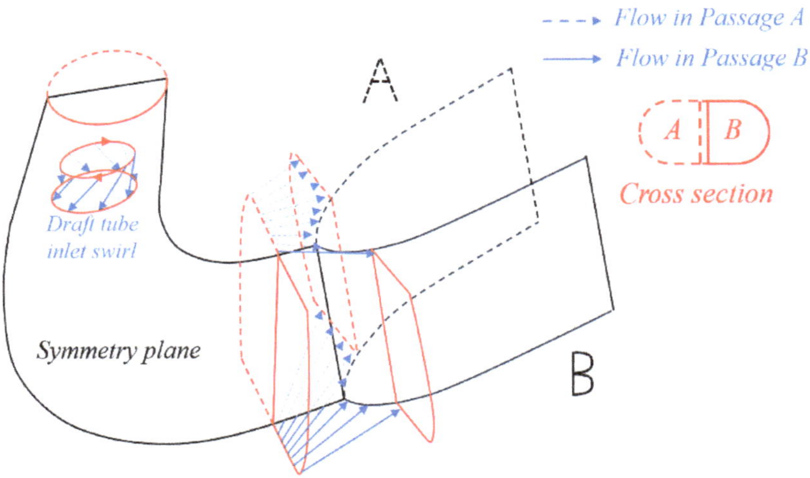

Figure 10. Schematic diagram of flow asymmetry caused by draft tube elbow bending.

In Figure 11, two cross sections S1-A and S2-B are selected along the flow direction to observe the flow characteristics inside the vertical plane. The surface streamlines are shown in Figure 12. At 100% load, only a small back-flow area appeared on the outer side of the elbow in the draft tube, with a relatively low vortex intensity. Even though spiral motion existed in Passage B to some extent, the back-flow area was small. At 50% load, the vortex area at the elbow increased, distributing closer to the center of the flow cross section. Passages A and B exhibited significant differences: a large vortex formed near the pier in Passage A, while a large vortex appeared near the outlet of the draft tube in Passage B. Judging from the streamline distribution, the presence of the pier has such an effect: splitting the high-speed swirling flow at the inlet of the draft tube into two parts, one recirculating and dissipating in Passage A, while most of the angular momentum is stored in the slower outflow in Passage B, then mixing and dissipating after the pier ends.

To some extent, this design reduces the likelihood of intense pressure fluctuation at specific frequencies, but the resulting changes in pressure distribudtion characteristics need further examination.

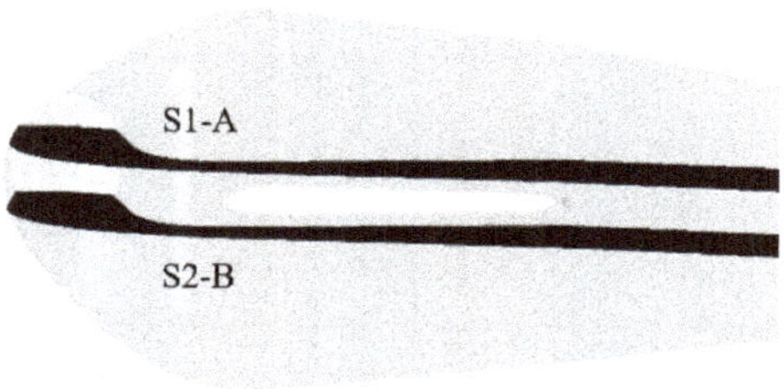

Figure 11. Two cross sections S1-A and S2-B.

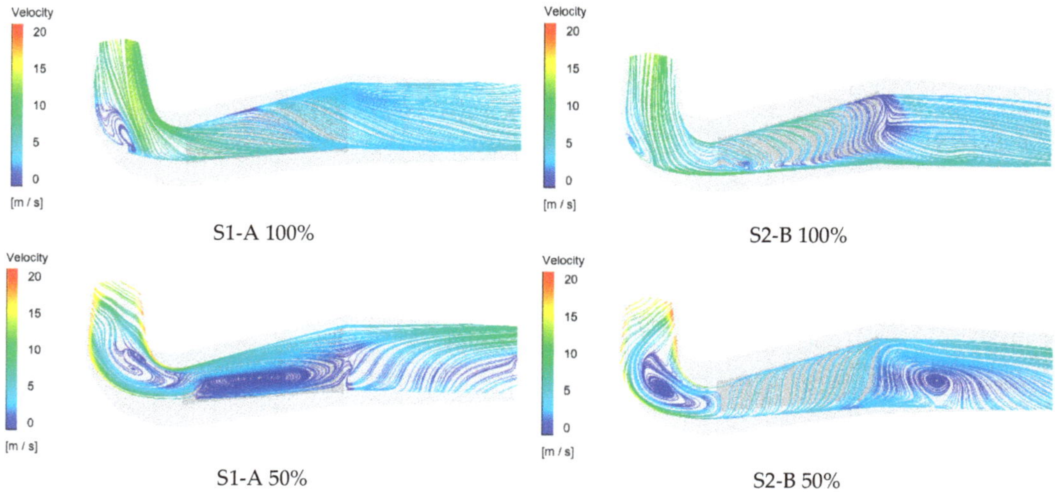

Figure 12. The streamline distribution of the draft tube at S1-A and S2-B.

Intuitively, the flow asymmetry would lead to differences in the pressure distributions between the two passages, with the load likely also impacting these differences. However, unexpectedly, the results for the average pressure distributions at S1-A and S2-B shown in Figure 13 did not exhibit quite significant discrepancies, except for the larger high-pressure zone on the outer side of the draft tube elbow at 50% load. This may be an advantage conferred by the presence of the pier—despite the apparent differences in flow characteristics between the two passages, strong pressure differentials are not induced, which is important for improving the hydraulic efficiency and fatigue life of structural components.

To observe this further, we chose several sections along the mainstream direction to check the process of pressure differential change. Their pressure contour maps are displayed in Figure 14, which allows us to further see how the pier played a role in this process. At 100% load, the two passages had relatively symmetric pressure distributions. Although there was a high-pressure zone near Passage B at section S2, both passages maintained a similar concentric circular pressure distribution when entering sections S3 and S4. With the generation of the vortex rope at 50% load, the pressure distribution at section S2 showed an evident low-pressure zone at the inlet of Passage A and a high-pressure zone at the inlet of Passage B. However, this difference was mitigated suddenly when entering section S3, at the cost of Passage B losing the concentric circular pressure distribution. Judging from the results, Passage B reduced the global pressure differential with Passage A through localized pressure adjustments.

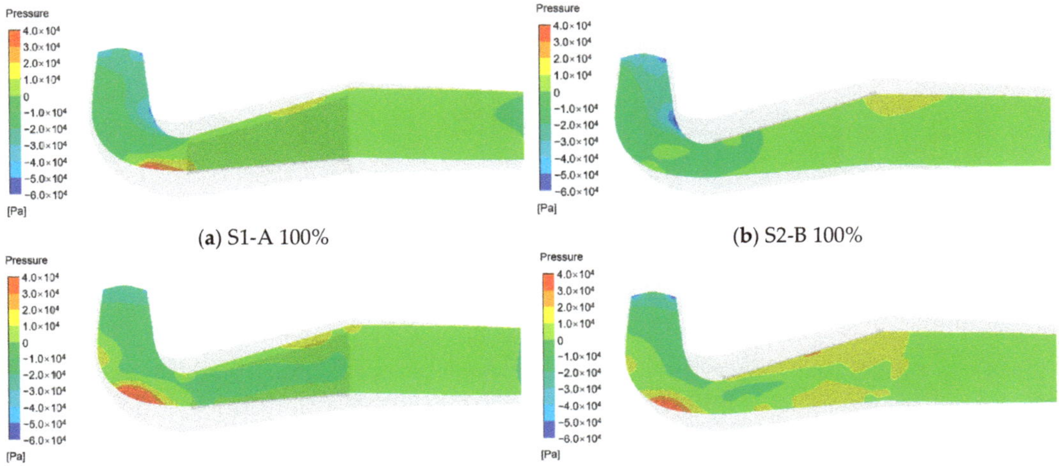

(**a**) S1-A 100% (**b**) S2-B 100%

(**c**) S1-A 50% (**d**) S2-B 50%

Figure 13. Pressure distribution of the draft tube at S1-A and S2-B.

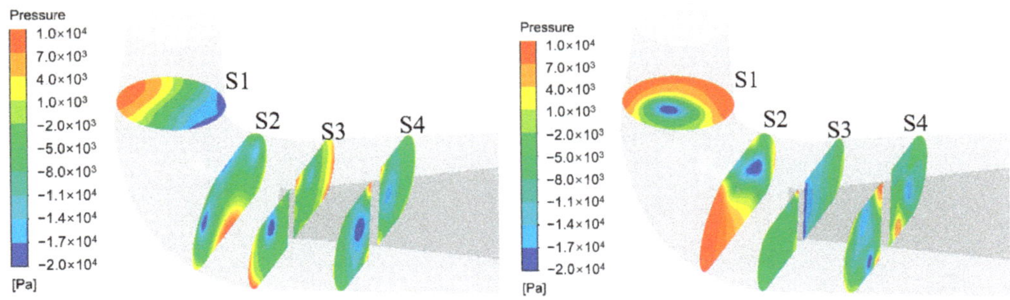

(**a**) 100% (**b**) 50%

Figure 14. Pressure distribution of the draft tube at different main flow cross sections.

From the distribution of turbulent kinetic energy in Figure 15, it can be seen that under 50% load, the flow inside the draft tube was unstable, the velocity fluctuation increased, and the turbulent kinetic energy increased and presented an uneven distribution. Overall, the turbulent kinetic energy decreased significantly after entering the passage on both sides of the pier. Additionally, at the S4 section in the middle of the pier, the turbulent kinetic energy in channel A was significantly higher than that in channel B. The occurrence of these results is understandable in light of the preceding analysis. At 50% load, a vortex rope developed in the draft tube. Regions approaching red in the color contours denote areas significantly influenced by the vortex rope. Such high-turbulence fluids divide—one part entering Passage A, another entering Passage B. However, lower mainstream velocities in Passage B afford ample time for turbulence dissipation. Thus, by cross section S4, these fluids demonstrated drastically lower turbulence than in Passage A. Fundamentally, dissipation constitutes energy loss, eventually manifesting pressure decreases. This further validates the rapid pressure differential reduction between Passages A and B—Passage B experienced immense dissipation around cross section S3.

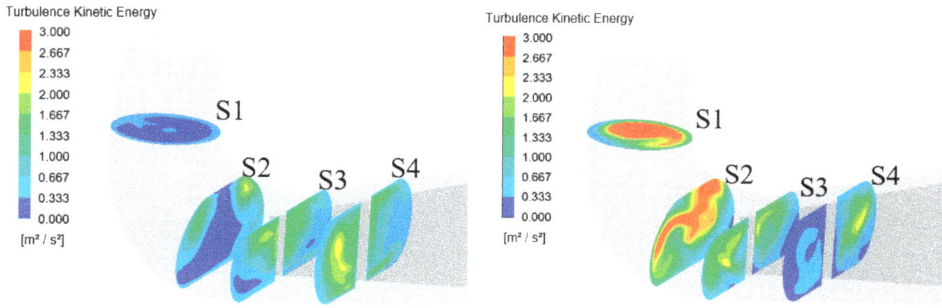

(**a**) 100% (**b**) 50%

Figure 15. Turbulent kinetic energy of the draft tube at main flow cross section.

3.4. Spectral Characteristics of Draft Tube Pressure Fluctuation

Figure 16 shows the distribution of monitoring points in the draft tube. Points dt1 to dt3 were at different heights in the draft tube; dt5, dt7, and dt9 were in Passage A of the draft tube; and dt4, dt6, and dt8 were in Passage B of the draft tube.

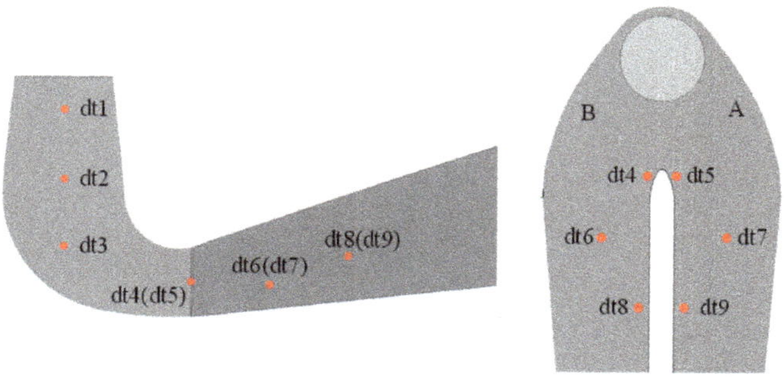

(**a**) Front view (**b**) Top view

Figure 16. The distribution of monitoring points in the draft tube.

Figure 17 shows the pressure fluctuation spectra of monitoring points in the draft tube at 100% load. The results indicate that at 100% load, no draft tube vortex rope was generated. The monitoring points were mainly affected by rotor–stator interaction, exhibiting frequencies like $7f_n$, $14f_n$, and $21f_n$. The amplitude of rotor–stator interaction frequency decreased as the distance from the runner increased. However, near the pier in Passage B, dt8 showed a low-frequency vortex frequency of $0.1f_n$, and near the pier in Passage A, dt9 also exhibited $0.2f_n$. Looking back at Figure 12a,b again, the reason for the discrepancy in the dominant frequencies of the two low-frequency components can be identified. The streamlines in Passage A are relatively smoother with less backflow, while Passage B has an adverse pressure gradient, and the streamlines bend twice. In the front section of the pier, the same pressure fluctuation may occur twice, resulting in a higher dominant frequency compared to Passage A.

Figure 18 shows the pressure fluctuation spectra of monitoring points in the draft tube at 50% load. At all monitoring points, $21f_n$ lot its dominance, $7f_n$ and $14f_n$ almost disappeared. Instead, low-frequency peaks emerged. At most points, the low-frequency component centered at $0.2f_n$ had a relatively prominent peak, but some monitoring points still exhibited slight differences. The dt5 point had more low-frequency components with a wider frequency band, corresponding to the large back-flow area on the pier side of Passage A and the confluence point where the mainstream and recirculation flow merged. It is imaginable that complex vortex characteristics are generated there. Point dt8 was exceptionally different, with a very low amplitude at $0.2f_n$, while $0.13f_n$ and $0.4f_n$ appeared as compensation, though the peaks of these two frequencies remained insignificant. From Figure 14b, two local minimums can be observed in the pressure distribution on the cross section S4 of the passage, indicating that the $0.2f_n$ vortex in Passage B is split into $0.13f_n$ and $0.4f_n$ vortices at the middle of the pier, and the amplitudes are low because the energy is shared. This phenomenon is highly likely due to kinematic progression caused by flow evolution—it is also essentially due to the position of the pier, which is important for improving the understanding of flow inside draft tubes with a pier.

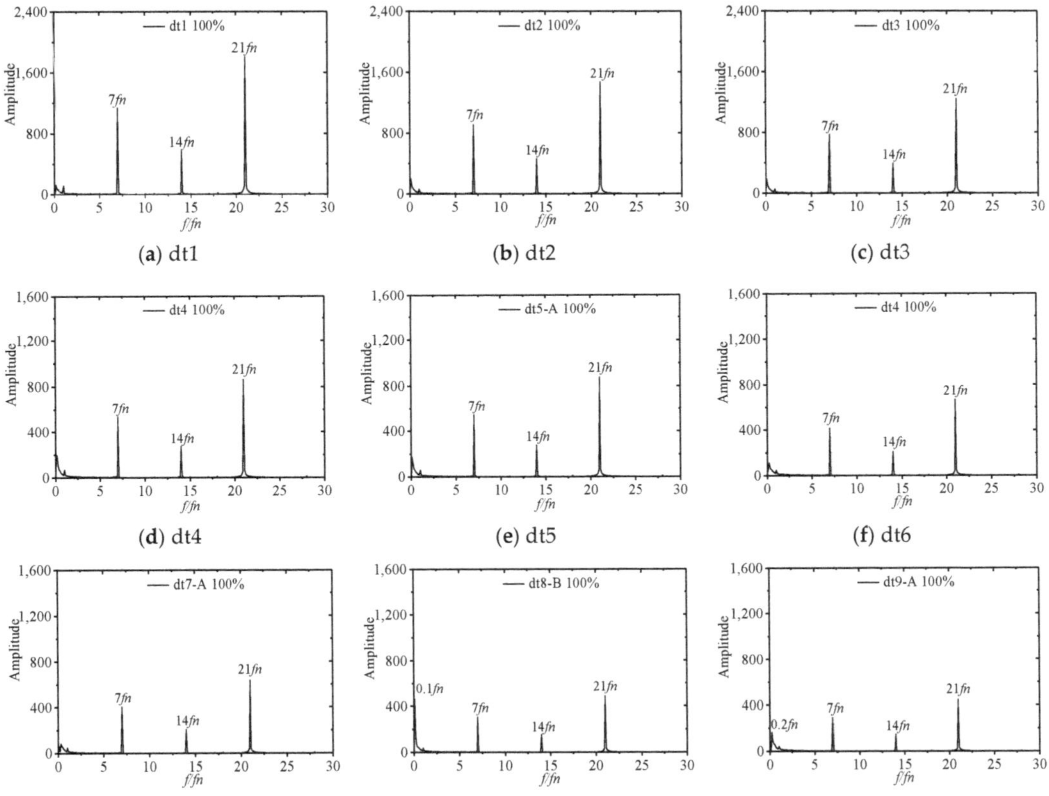

Figure 17. Pressure fluctuation spectrum of the draft tube under 100% load.

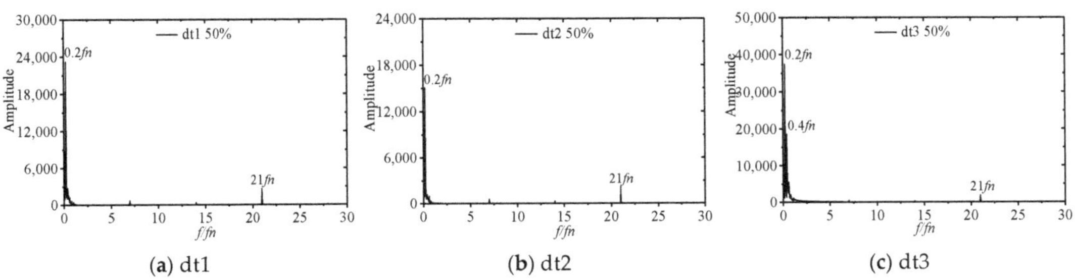

Figure 18. *Cont.*

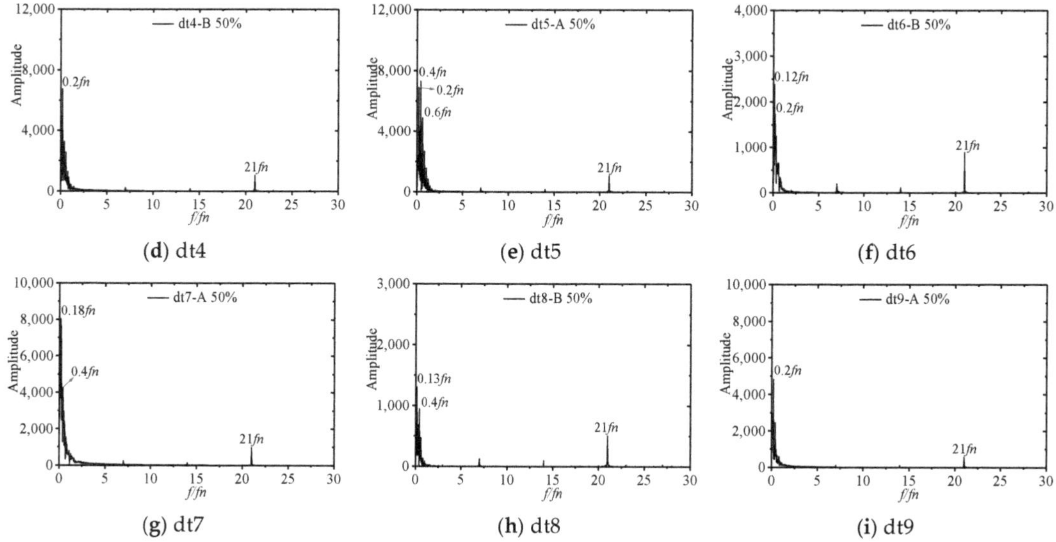

Figure 18. Pressure fluctuation spectrum of the draft tube under 50% load.

3.5. Discussion

3.5.1. Evolution Process of Vortex Rope

As a supplement, the morphology of the draft tube vortex rope can be observed to discuss the cause of $0.2f_n$ under 50% load. The vortex rope edge is represented by the iso-surface of velocity $\lambda_2 = 0$ as illustrated in Figure 19. The rotation period of the runner $T = 1/f_n = 0.3$s, and it can be seen that the draft tube vortex rope exhibited a periodic evolution pattern over five rotation periods of the runner. At 0T, vortex ropes existed at both the inlet and elbow of the draft tube. Afterward, the vortex rope volume at the inlet decreased, while the one at the elbow increased. Until 2/3T, the inlet vortex rope disappeared. Then, the elbow vortex rope split into two parts, one moving toward the inlet, the other moving toward the pier. The part moving toward the pier gradually dissipated when approaching the pier, and completely disappeared at 5/3T. The part moving toward the inlet became shorter and thicker during 5/3T to 13/3T, with reducing volume, until a new vortex rope was generated at the elbow again at 5T, similar to the structure at 0T. This fully demonstrates that the $0.2f_n$ frequency component originated from the periodic generation, movement, and decay of the draft tube vortex rope.

3.5.2. Complementary Shape of Vortex Rope and Recirculation Zone

Figure 20 shows the recirculation zone corresponding to the draft tube vortex ropes and the flow lines inside the runner at different moments, which can be used to observe the influence of the vortex rope evolution. It is evident that the vortex rope and recirculation zone exhibited a complementary morphology in their distributions: At 0T and 2/3T, the large vortex rope volume at the elbow caused a flow blockage there, resulting in extensive recirculation in the center region of the draft tube's straight cone section. Meanwhile, vortex motion also occurred in the center region inside the runner. As the main vortex rope body moved toward the inlet, the recirculation zone decreased, and the flow inside the runner also gradually improved, especially at 10/3T when the recirculation zone almost disappeared. Afterward, the recirculation zone emerged again at the elbow and started the next cycle.

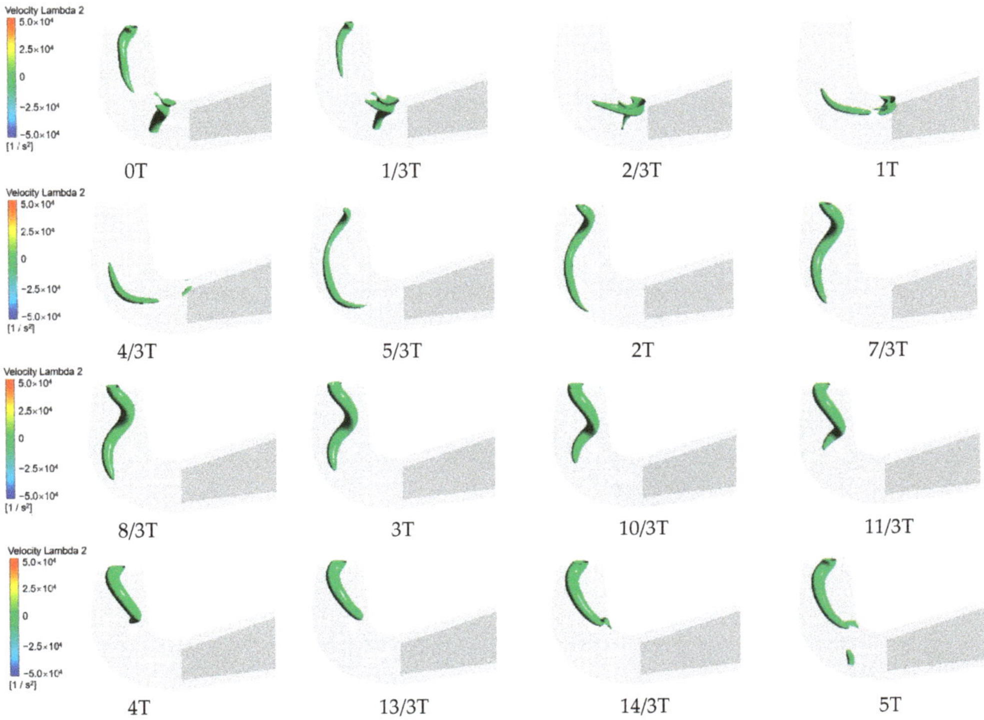

Figure 19. The vortex rope changes in the draft tube. T = $1/f_n$ = 0.3 s.

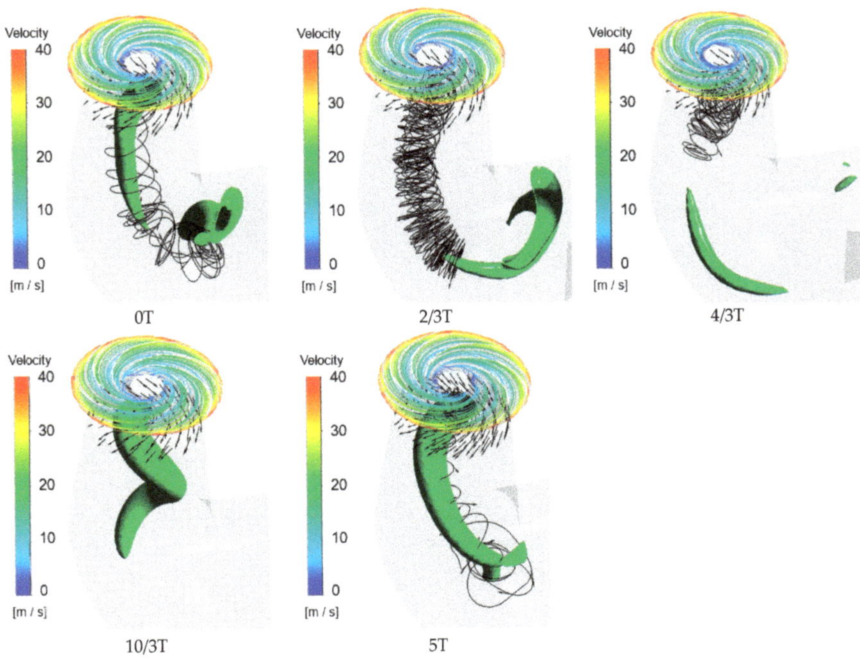

Figure 20. The recirculation zone of the draft tube vortex rope. T = $1/f_n$ = 0.3s.

3.5.3. Effect on the Runner

We also evaluated the influence of vortex rope evolution on the upstream runner through spectral analysis. In the rotating coordinate system, two typical pressure fluctuation monitoring points were selected inside the runner, as shown in Figure 21: rn1 located near the inlet of the runner, and rn3 located near the outlet of the runner. From the pressure fluctuation spectra in Figure 22, it can be seen that the runner inlet was dominated by the rotor–stator interaction with a main frequency of $20f_n$, while the runner outlet was dominated by the draft tube vortex rope frequency of 0.8fn, which presents a considerably large amplitude. This indicates that, within one evolution cycle of the draft tube vortex rope, the interior of the runner experienced four pressure pulsations induced by the draft tube vortex rope. It is intricately linked to the alternate impingement on the runner outlet by the vortex rope and the recirculation zone in the cone section.

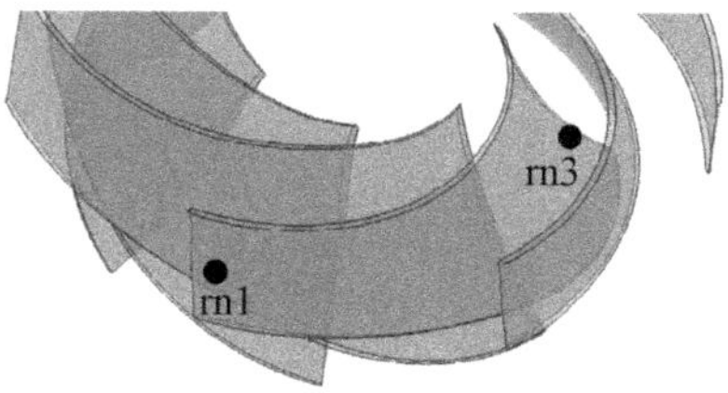

Figure 21. Monitoring point distributions of the runner.

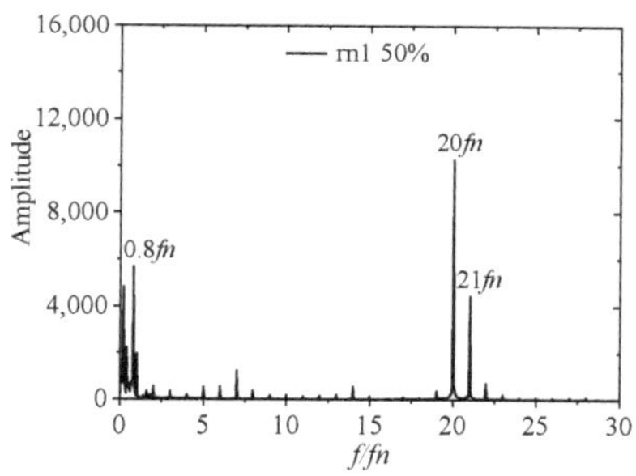

(**a**) rn1

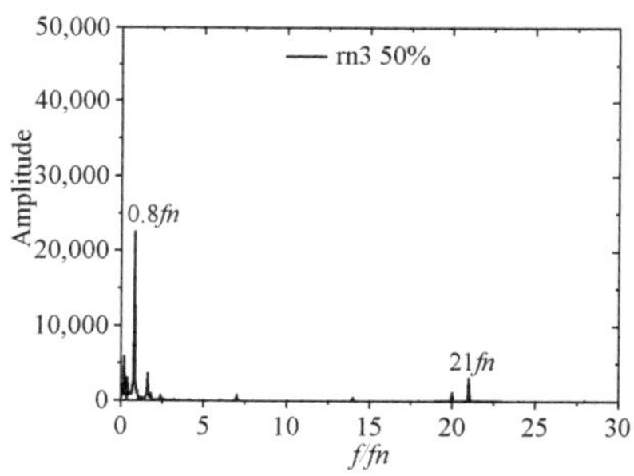

(**b**) rn3

Figure 22. Pressure fluctuation spectrum of the runner under 50% load.

3.5.4. Preliminary Analysis of Fluid Kinematic Progression

When discussing Figure 18 previously, we mentioned the potential existence of kinematic progression. Here, a preliminary explanation for this progression phenomenon is provided based on vortex theory. Influenced by geometric potential, the vortex tube experiences twisting, causing the vortex tube center to deviate from its initial rotation center and revolve around it. The definitions of various parameters are illustrated in Figure 23. Inside the vortex tube ($r \leq r_0$), the flow is dominated by a forced vortex, while outside ($r > r_0$), it is dominated by a free vortex. Here, for ease of analysis, we neglected the transition region in between. According to the energy equations of vortex theory [34] (Equation (5)), the circumferential velocity distributions inside and outside the vortex tube, namely u_{forced} and u_{free}, follow Equations (6) and (7):

$$\frac{dH}{dr} = \frac{u}{g}\left(\frac{du}{dr} + \frac{u}{r}\right) \quad (5)$$

$$u_{forced} = \omega r \quad (6)$$

$$u_{free} = \frac{c}{r} \quad (7)$$

where c is a constant. Then, for a vortex tube with a revolution angular velocity of ω_0, the relative velocity v_r and the convected velocity v_c are described in Equations (8) and (9), respectively. Additionally, the distance L between any fluid particle and the origin O is given by Equation (10):

$$v_r = \begin{cases} (\omega r\sin\theta, \omega r\cos\theta) & r \leq r_0 \\ (\frac{c}{r}\sin\theta, \frac{c}{r}\cos\theta) & r > r_0 \end{cases} \quad (8)$$

$$v_c = (\omega_0 R\sin\alpha, \omega_0 R\cos\alpha) \quad (9)$$

$$L = (R\cos\alpha + r\sin\theta, R\cos\alpha + r\cos\theta) \quad (10)$$

here $c = \omega r_0^2$. According to Helmholtz's theorems, vortex tubes demonstrate conservativity. In the present analysis, this manifests as the integral conservation of velocity moments inside the control volume, as given in Equation (11):

$$\int_{r=0}^{r\prime}\int_{\theta=0}^{2\pi} (v_r + v_c) \cdot L d\theta dr = const. \quad (11)$$

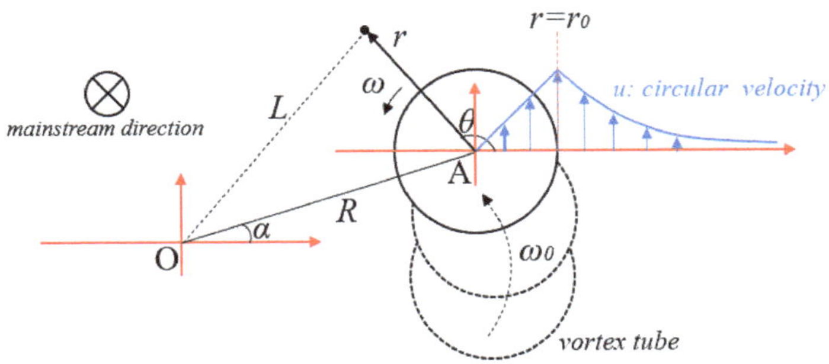

Figure 23. Theoretical flow field analysis diagram in the vicinity of the vortex tube.

After simplification, we obtain Equation (12) as follows:

$$r'^3 \cdot \omega + 6R^2 \cdot \omega_0 = const. \quad (12)$$

where $r\prime$ is the upper limit of the control volume's radius. In theoretical analysis, this value is infinite, but in actual fluid flows, we can still select an appropriate upper limit based on the situation, representing the maximum dominant range of the vortex tube in complex flow fields. Meanwhile, the mainstream velocity, namely the flow velocity along the vortex tube's extending direction, will also affect $r\prime$ through the continuity equation. This may hold significance for more in-depth analyses, but Equation (12) alone can already offer us a clear explanation for the occurrence of splitting at $0.2f_n$ in Passage B—the vortex tube experiences twisting under the influence of both mainstream and geometric potential, with its twisting frequency and fluid vortex frequency demonstrating opposite variation trends.

3.5.5. Comparability and Limitation of Experimental and Numerical Simulations

From researchers' expectations, greater alignment of numerical results with experiments is preferred. This consistency is more likely to occur in model tests predicting pump turbines' overall energy characteristics, while for more delicate, local traits like pressure fluctuation in a prototype test, the situation is much worse. Fully comprehensive prototype data comparable to numerical simulations are rarely attainable. The authors provide two examples here. First, the power plant involved used ~10m long measuring pipes, severely attenuating high-frequency components in pressure fluctuation signals. In Figure 5, absence of rotor–stator interaction (RSI) at the draft tube elbow seems fallacious when considering common knowledge or contrasting with numerical results (Figure 17). This reflects one limitation of prototype testing—the majority of power plants were constructed sans considerations for intricate measurement requisites.

As a second example, the said plant (and numerous others) adopted annular pipelines or aggregated chambers for pressure extraction, necessitating multiple boreholes per cross section. For instance, tapping at 0°, 90°, 180°, and 270° orientations of a draft tube cross-section to get four boreholes, then connecting them and entering the same sensor for an "average" value. Aside from hampering pulsation measurement accuracy, such schemes inherently disturb the internal flow. Thus, consistently evaluating the relative accuracy between testing and simulation proves difficult—perhaps the CAD modeling, turbulence model, and boundary conditions of numerical simulation will have an impact on the accuracy, but there are also many unexpected factors in the test. Nonetheless, clearly capturing fluid phenomena bears significance for the current work, at least underscoring the necessity of future model tests with pier inclusion.

The only stronger consistency manifests in the draft tube vortex rope frequency. The test showed $0.23 f_n$ at 50% load versus $0.2 f_n$ in simulations. This minor numeric deviation matters less as the author's past study [35] demonstrates test variability in this frequency. The extremes of spectrum amplitude signify the absolute dominance of the vortex rope near a 50% load. The field tests relayed data more akin to Figure 18c's simulation at dt3, where the vortex rope's second harmonic appeared in both.

Thus, the simulations essentially expand upon the field test data presently. Long-term prototype pump turbine research requires optimization from the design stage—improving test accuracy for enhanced comparability against simulations.

4. Conclusions

This paper uses numerical simulation methods to study the internal flow characteristics of a pump turbine with a single pier in the draft tube. The results are as follows:

(1) The asymmetry of the flow is preordained. At 100% load, the flow in Passage A and Passage B on both sides of the pier was relatively consistent with the geometric contour of the draft tube. At 50% load, Passage A had a larger flow rate and generated a large area of vortex near the pier. Passage B mainly exhibited a low-speed spiral motion, mixed with the high-speed area of Passage A, and formed a large-scale vortex at the outlet of the draft tube. The existence of the pier made Passage B reduce the global pressure difference with Passage A through localized pressure adjustments and also reduced the turbulent kinetic energy of the fluid, especially in the front section of Passage B.

(2) Through the analysis of pressure fluctuation, it was found that, at 50% load, the rotor–stator interaction component that dominated at 100% load basically disappeared, and the more pronounced component was the low frequency ($0.2f_n$), which is the evolution period of the process of the generation, movement, and decay of the vortex rope. At the monitoring point in the middle of the pier, the main frequency split into $0.13f_n$ and $0.4f_n$, which may be due to the kinematic precession originated from the evolution of vortex rope.

(3) The distribution of the vortex rope in the draft tube and the recirculation zone showed a complementary shape. The frequency of the vortex rope had a significant impact on the pressure fluctuation of the runner, and the point near trailing edge of the runner blade showed a high-amplitude vortex rope frequency of $0.8f_n$.

Prospection: The pier exerts conspicuous benefits on turbine mode—vortex rope fragmentation, draft tube diffuser pressure balance, etc. This underscores the meaningfulness of reconsidering pier design in subsequent pump-turbine model tests, encompassing the pier effects on pump mode.

Author Contributions: Conceptualization, H.H. and Z.W.; data curation, H.H. and M.X.; formal analysis, H.H. and M.X.; investigation, H.H., X.S. and M.Q.; methodology, H.H. and M.X.; resources, M.Q.; software, X.S. and M.Q.; supervision, Z.W.; validation, M.X., X.S. and Z.W.; visualization, M.X. and X.S.; writing—original draft, H.H. and M.X.; writing—review and editing, X.S., Z.W. and M.Q. All authors have read and agreed to the published version of the manuscript.

Funding: This research was funded by the project "Research Project on Strong Vibrations in the Head Cover of Baishan Pumped Storage Power Station (No. 525784270002)" of State Grid Xinyuan Co., Ltd.

Data Availability Statement: The data presented in this study are available on request from the corresponding author. The data are not publicly available due to the owner's security requirements for engineering data.

Acknowledgments: The authors thank the cooperation of the leadership and field staff of the Baishan Power Station in the testing work. In addition, the author thanks Hunan Shuhua Technology for the support of testing equipment and debugging technology.

Conflicts of Interest: Author Haiping Hu was employed by the Baishan Storage Power Station, State Grid Xinyuan Co., Ltd., and author Ming Xia was employed by the Department of Energy and Power Engineering, Tsinghua University. The remaining authors declare that the research was conducted in the absence of any commercial or financial relationships that could be construed as a potential conflict of interest.

References

1. Zhao, J.F.; Oh, U.J.; Park, J.C.; Park, E.S.; Im, H.B.; Lee, K.Y.; Choi, J.S. A Review of World-wide Advanced Pumped Storage Hydropower Technologies. *IFAC Pap.* **2022**, *55*, 170–174. [CrossRef]
2. Hu, J.Y.; Wang, Q.B.; Song, H.G.; Mao, J.H. Effect of wave disturbance on the fluid dynamics of the pump turbine in a seawater pumped storage hydroplant. *Front. Earth Sci.* **2023**, *11*, 1220842. [CrossRef]
3. Hu, J.Y.; Wang, Q.B.; Meng, Z.Z.; Song, H.G.; Chen, B.W.; Shen, H. Numerical Study of the Internal Fluid Dynamics of Draft Tube in Seawater Pumped Storage Hydropower Plant. *Sustainability* **2023**, *15*, 8327. [CrossRef]
4. Zuo, Z.G.; Liu, S.H. Flow-Induced Instabilities in Pump-Turbines in China. *Engineering* **2017**, *3*, 504–511. [CrossRef]
5. Yan, J.P.; Seidel, U.; Koutnik, J. Numerical simulation of hydrodynamics in a pump-turbine at off-design operating conditions in turbine mode. *IOP Conf. Ser. Earth Environ. Sci.* **2013**, *15*, 032041. [CrossRef]
6. Zhu, D.; Yan, W.; Guang, W.L.; Wang, Z.W.; Tao, R. Influence of Guide Vane Opening on the Runaway Stability of a Pump-Turbine Used for Hydropower and Ocean Power. *J. Mar. Sci. Eng.* **2023**, *11*, 1218. [CrossRef]
7. Hu, H.P.; Xia, M.; Qiao, M.; Wang, W.; Wang, J.Z.; Wang, Z.W. A Simulation Study of Hydraulic Vibration caused by Clearance Flow in a Pump Turbine. *IOP Conf. Ser. Earth Environ. Sci.* **2022**, *1079*, 012032. [CrossRef]
8. Asomani, S.N.; Yuan, J.P.; Wang, L.Y.; Appiah, D.; Zhang, F. Geometrical effects on performance and inner flow characteristics of a pump-as-turbine: A review. *Adv. Mech. Eng.* **2020**, *12*, 1687814020912149. [CrossRef]
9. Fu, X.L.; Li, D.Y.; Wang, H.J.; Zhang, G.H.; Li, Z.G.; Wei, X.Z. Influence of the clearance flow on the load rejection process in a pump-turbine. *Renew. Energy* **2018**, *127*, 310–321. [CrossRef]
10. Yang, H.; He, Q.; Bi, H.; Huang, X.; Yang, M.; Zou, M.; Wang, Z. Head-Cover vibration investigation of a Prototype Reversible Pump-Turbine Unit during Start-up in Pump mode. Part I: Fluid dynamic analysis. *IOP Conf. Ser. Earth Environ. Sci.* **2022**, *1037*, 012055. [CrossRef]
11. Yang, M.Q.; Zhao, W.Q.; Bi, H.L.; Yang, H.X.; He, Q.L.; Huang, X.X.; Wang, Z.W. Flow-Induced Vibration of Non-Rotating Structures of a High-Head Pump-Turbine during Start-Up in Turbine Mode. *Energies* **2022**, *15*, 8743. [CrossRef]
12. Zhang, W.W.; Chen, Z.M.; Zhu, B.S.; Zhang, F. Pressure fluctuation and flow instability in S-shaped region of a reversible pump-turbine. *Renew. Energy* **2020**, *154*, 826–840. [CrossRef]
13. Kim, S.J.; Suh, J.W.; Choi, Y.S.; Park, J.; Park, N.H.; Kim, J.H. Inter-Blade Vortex and Vortex Rope Characteristics of a Pump-Turbine in Turbine Mode under Low Flow Rate Conditions. *Water* **2019**, *11*, 2554. [CrossRef]
14. Tridon, S.; Barre, S.; Ciocan, G.D.; Tomas, L. Experimental analysis of the swirling flow in a Francis turbine draft tube: Focus on radial velocity component determination. *Eur. J. Mech.-B-Fluids* **2010**, *29*, 321–335. [CrossRef]
15. Lin, T.; Li, J.; Xie, B.F.; Zhang, J.R.; Zhu, Z.C.; Yang, H.; Wen, X.M. Vortex-Pressure Fluctuation Interaction in the Outlet Duct of Centrifugal Pump as Turbines (PATs). *Sustainability* **2022**, *14*, 15250. [CrossRef]
16. Pang, S.; Zhu, B.; Shen, Y.; Chen, Z. Study on cavitating vortex rope characteristics of reversible pump-turbine under part load turbine condition. *Phys. Fluids* **2023**, *35*, 085131. [CrossRef]
17. Vu, T.C.; Devals, C.; Zhang, Y.; Nennemann, B.; Guibault, F. Steady and unsteady flow computation in an elbow dr aft tube with experimental validation. *IOP Conf. Ser. Earth Environ. Sci.* **2010**, *12*, 012045. [CrossRef]

18. Ciocan, G.D.; Iliescu, M.S.; Vu, T.C.; Nennemann, B.; Avellan, F. Experimental Study and Numerical Simulation of the FLINDT Draft Tube Rotating Vortex. *J. Fluids Eng.* **2006**, *129*, 146–158. [CrossRef]
19. Foroutan, H.; Yavuzkurt, S. Unsteady Numerical Simulation of Flow in Draft Tube of a Hydroturbine Operating under Various Conditions Using a Partially Averaged Navier–Stokes Model. *J. Fluids Eng.* **2015**, *137*, 061101. [CrossRef]
20. Gagnon, J.-M.; Aeschlimann, V.; Houde, S.; Flemming, F.; Coulson, S.; Deschenes, C. Experimental Investigation of Draft Tube Inlet Velocity Field of a Propeller Turbine. *J. Fluids Eng.* **2012**, *134*, 101102. [CrossRef]
21. Gouin, P.; Deschênes, C.; Iliescu, M.; Ciocan, G.D. Experimental investigation of draft tube flow of an axial turbine by laser doppler velocimetry. In Proceedings of the 3rd IAHR International Meeting of the Workgroup on Cavitation and Dynamic Problems in Hydraulic Machinery and Systems, Brno, Czech Republic, 14–16 October 2009; pp. 87–97.
22. Lemay, S.; Fraser, R.; Ciocan, G.D.; Aeschlimann, V.; Deschênes, C. Flow field study in a bulb turbine runner using LDV and endoscopic S-PIV measurements. *IOP Conf. Ser. Earth Environ. Sci.* **2014**, *22*, 022015. [CrossRef]
23. Vuillemard, J.; Aeschlimann, V.; Fraser, R.; Lemay, S.; Deschênes, C. Experimental investigation of the draft tube inlet flow of a bulb turbine. *IOP Conf. Ser. Earth Environ. Sci.* **2014**, *22*, 032010. [CrossRef]
24. Tridon, S.; Barre, S.; Ciocan, G.D.; Leroy, P.; Ségoufin, C. Experimental investigation of draft tube flow instability. *IOP Conf. Ser. Earth Environ. Sci.* **2010**, *12*, 012044. [CrossRef]
25. Tridon, S.; Barre, S.; Ciocan, G.; Ségoufin, C.; Leroy, P. Discharge Imbalance Mitigation in Francis Turbine Draft-Tube Bays. *J. Fluids Eng.* **2012**, *134*, 041102. [CrossRef]
26. Cedric, D.; Balarac, G.; Metais, O.; Sylvain, T.; Stephane, B.; Ciocan, G.; Thomas, L.; Laurent, T. Large-Eddy Simulations of Draft Tube Flow and Validation from Experimental Results. In Proceedings of the 3rd IAHR International Meeting of the Workgroup on Cavitation and Dynamic Problems in Hydraulic Machinery and Systems, Brno, Czech Republic, 14–16 October 2009.
27. Li, H.; Xu, L.; Zhao, F.; Zhang, F.; Deng, Y.; Li, H.; Cheng, Y.; Zheng, Y.; Kan, K. Investigation of Flow Characteristics of Flow Passage Components in Francis Turbine under No-load and 50% Load operating conditions. *J. Phys. Conf. Ser.* **2022**, *2285*, 012040. [CrossRef]
28. Wang, L.; Cui, J.; Shu, L.; Jiang, D.; Xiang, C.; Li, L.; Zhou, P. Research on the Vortex Rope Control Techniques in Draft Tube of Francis Turbines. *Energies* **2022**, *15*, 9280. [CrossRef]
29. Paik, J.; Sotiropoulos, F.; Sale Michael, J. Numerical Simulation of Swirling Flow in Complex Hydroturbine Draft Tube Using Unsteady Statistical Turbulence Models. *J. Hydraul. Eng.* **2005**, *131*, 441–456. [CrossRef]
30. Pasche, S.; Gallaire, F.; Avellan, F. Origin of the synchronous pressure fluctuations in the draft tube of Francis turbines operating at part load conditions. *J. Fluid Struct.* **2019**, *86*, 13–33. [CrossRef]
31. Temam, R. *Navier-Stokes Equations: Theory and Numerical Analysis*; American Mathematical Society: Providence, RI, USA, 2001; Volume 343.
32. Hu, J.; Luo, X.; Zhao, Y.; Liang, Q.; Liu, D.; He, Q.; Wang, Z.; Guan, Z. Evolution mechanism of unsteady internal flow of an ultra-high head pump-turbine in pump mode. *J. Energy Storage* **2023**, *74*, 109309. [CrossRef]
33. Zhang, Y.L.; Li, J.F.; Zhu, Z.C. The acceleration effect of pump as turbine system during starting period. *Sci. Rep.* **2023**, *13*, 4913. [CrossRef]
34. Chen, W.; Zydek, N.; Parma, F. Evaluation of hydrocyclone models for practical applications. *Chem. Eng. J.* **2000**, *80*, 295–303. [CrossRef]
35. Hu, H.P.; Xia, M.; Song, X.H.; Zhao, W.Q.; Wang, W.; Wang, Z.W. A Field Investigation of Stability Characteristics of Pressure Fluctuation and Vibration in Prototype Pump Turbine at Multiple Working Points. *Water* **2023**, *15*, 3378. [CrossRef]

water

MDPI

Article

A Field Investigation of Stability Characteristics of Pressure Fluctuation and Vibration in Prototype Pump Turbine at Multiple Working Points

Haiping Hu [1], Ming Xia [2], Xianghui Song [1], Weiqiang Zhao [2], Wei Wang [3] and Zhengwei Wang [2,*]

1 Baishan Storage Power Station, State Grid Xinyuan Co., Ltd., Huadian, Jilin 132400, China; sea3219@163.com (H.H.); sxh17@163.com (X.S.)
2 Department of Energy and Power Engineering, Tsinghua University, Beijing 100084, China; xiam17@mails.tsinghua.edu.cn (M.X.); zhaoweiqiang@mail.tsinghua.edu.cn (W.Z.)
3 School of Water Conservancy and Hydropower Engineering, North China Electric Power University, Beijing 100084, China; wweihd@ncepu.edu.cn
* Correspondence: wzw@mail.tsinghua.edu.cn

Abstract: In practical operation, pump turbines typically operate far from their designed working points, which has a significant impact on the stability of the unit's operation. In this paper, we conducted a field test to investigate the stability characteristics of prototype pump turbines at different working points. By adjusting the given power of the generator in a stepwise manner to control its working point, we obtained the statistical and spectral characteristics of pressure signals and acceleration signals. In turbine mode, the result shows that, at low, medium, and high power, the variation in pressure fluctuation characteristics is influenced by three different factors, while vibration generally reaches its maximum value at approximately 50 MW. In pump mode, variations in pressure were observed among different measurement points in the low-frequency range, and the characteristics of vibration acceleration were influenced by both the rotor–stator interaction (RSI) and the structural modal frequencies. We emphasized that the high-frequency bands have influences on the unit comparable in magnitude to those of the rotor–stator interaction, which has rarely been mentioned in previous studies. Through detailed testing and analysis of the unit's actual operation, we can gain a better understanding of its behavior and performance in the turbine and pump modes, and these results hold significant importance for ensuring the stability and reliability of the unit.

Keywords: pump turbine; field test; pressure fluctuation; acceleration; stability characteristics; rotor–stator interaction

Citation: Hu, H.; Xia, M.; Song, X.; Zhao, W.; Wang, W.; Wang, Z. A Field Investigation of Stability Characteristics of Pressure Fluctuation and Vibration in Prototype Pump Turbine at Multiple Working Points. *Water* **2023**, *15*, 3378. https://doi.org/10.3390/w15193378

Academic Editor: Aonghus McNabola

Received: 4 September 2023
Revised: 17 September 2023
Accepted: 18 September 2023
Published: 27 September 2023

1. Introduction

With the gradual depletion of fossil fuel reserves and increasing environmental concerns associated with their use, the development of renewable energy has been rapidly advancing. Renewable energy accounted for 47.3% of China's total installed power capacity in 2022, among which wind and solar energy have seen continuous growth, accounting for 8.8% and 4.9% of total installed power capacity, respectively [1]. However, the inherent drawback of wind and solar energy lies in their short-term output instability, which can present challenges to grid stability [2]. Studies have shown that, when the combined share of wind and solar energy (also called Renewable Energy Penetration Level) exceeds 10% on the grid, without additional control measures, their short-term fluctuations can significantly impact grid security [3–5]. To address the issue of unstable output from renewable energy sources, pumped-storage hydropower plants play a crucial role as a relatively mature energy storage solution. The pump turbine is the key component of pumped-storage hydropower, and its operation principle is simple. During periods of excess electricity generation, water is pumped from a lower reservoir to an upper reservoir. During periods of electricity shortage, the stored water in the upper reservoir is released through turbines to generate electricity. This meets the usage needs of both the energy storage and power generation directions.

Due to the remote control center's overall dispatch of the power grid, pumped-storage power stations often play a role in peak shaving and valley filling, leading pump turbines with frequent start–stop operations and wide operating ranges to encounter greater challenges in terms of stability compared to large-scale Francis units that maintain steady operation [6]. The frequent changes in working points can lead to severe hydraulic instability, such as pressure fluctuation and increasing vibrations [7,8], cavitation in the draft tube [9], and vortex-induced vibration due to multiphase flow [10,11], especially when the unit operates at low flow rates in turbine mode [12], referred to as an S-shaped characteristic. Many studies have been dedicated to analyzing the internal flow pattern and pressure fluctuation associated with the S-shaped characteristic [13,14]. In one study, model tests and numerical simulations were conducted on a model pump turbine [15], and some low-frequency components were found in the vaneless space, which may not originate from specific vortex structures [16], but rather from the combined effects of the rotating stall, circumferential propagation of fluid blockage, and vortex rope [17]. In terms of improving the stability of pump turbines, researchers have attempted jet injection in the vaneless space [18], but a more reliable approach is the use of misaligned guide vanes (MGVs), which can change the characteristics of the turbine brake and reverse pump regions in the four-quadrant performance curves of the pump turbine [19]. In recent years, research on splitter blades has become increasingly popular. This design alternates long and short blades within the runner, resulting in significant differences in pressure fluctuation characteristics compared to previously used pump turbines [20–22], and greatly improving the stability of the unit. In pump mode, the risk of extending the operating range will rise due to the hump characteristic [23,24], so the working points of most pump turbines in pump mode are completely fixed.

Most of the studies mentioned above have focused on comparing experimental data from model tests with numerical simulations. However, the flow characteristics of prototype units in engineering practice are often different from those of model units, especially in low-flow conditions [25]. Unlike laboratory experiments that employ visualization techniques such as PIV, field tests of prototypes typically involve the installation of high-stability sensors within the unit, such as pressure sensors, vibration sensors, acoustic sensors, strain gauges, etc. [26,27]. Through indirect analysis, these sensors enable the investigation of potential flow phenomena occurring inside the turbine unit. Currently, there is a limited amount of detailed field research on pump turbines [28–30]. Zhang et al. identified the cause of strong vibrations in a pumped-storage hydropower plant as rotor–stator interaction (RSI) by analyzing the pressure fluctuation at critical locations within the unit [31]. They proposed increasing the stay vaneless space as an improvement and verified the effectiveness of this measure through a model test. Hu et al. conducted a field test [30] and obtained pressure fluctuations at three key measurement points for subsequent frequency domain analysis. They identified high-frequency noise, possibly caused by the RSI, as well as low-frequency components in the draft tube, but these frequencies are not limited to typical blade-passing frequencies or draft tube vortex rope frequencies, and occur with greater randomness. Therefore, these phenomena are often difficult to capture, even with sufficiently refined grids, simulation duration, and short time steps in numerical simulation works. In fact, numerical simulations excel in capturing prominently periodic vortex structures [32] and may not be able discern whether there are any components in the signal-related generator electromagnetics.

In this study, we conducted a field test on a prototype pump turbine in Huadian, Jilin, China. Multiple pressure sensors were installed at the end of the pressure measurement pipeline, while accelerometers were installed at key points in structural components. This allowed for the assessment of differences in signal statistical indicators and spectral characteristics of the unit at different operating conditions, thus providing insights into the operating performance and stability of the pump turbine.

2. Experimental Methods and Data Acquisition

2.1. Experimental Design of Working Points

Typical parameters of the pump turbine are shown in Table 1. We conducted field tests on the pumped-storage hydropower unit under various working points to obtain pressure and vibration acceleration data. The turbine operating conditions are all power points, but at intervals of 10 MW, including typical power ranges representing low-load operating conditions (10–40 MW), medium-load operating conditions (70–90 MW), and near-full-load operating conditions (100–130 MW). The pump operating conditions also include zero-flow operating conditions and normal pumping operating conditions. All working points are listed in Table 2.

Table 1. Typical pump turbine parameters.

Geometric parameters	Volute inlet diameter	~4200 mm
	Runner inlet diameter (D_1)	~5200 mm
	Runner outlet diameter (D_2)	~4100 mm
Turbine mode	Specific speed (n_s)	228
	Rated head	105.8 m
	Rated output power	150 MW
	Rated speed	200 rpm
	Rated discharge	148.7 m^3/s
Pump mode	Design head	126.7 m
	Maximum discharge	138 m^3/s
	Maximum input power	157.95 MW

Table 2. Tested working points.

Turbine mode	low load (10–40 MW)
	medium load (70–90 MW)
	full load (100–130 MW)
Pump mode	zero-flow
	normal pumping

2.2. Experimental Equipment and Installation

To measure pressure, and potential pressure fluctuation, high-performance pressure sensors were installed at the end of the existing pressure measurement pipelines on the unit; the correspondence between the physical meaning of the signal and the serial number is shown in Figure 1. Meanwhile, we installed IEPE accelerometers near the upper, lower, and water guide bearings of the unit. The sensors were as close to the bearings as possible, but were inevitably placed on the bracket and head cover, as shown in Figure 2. All of these sensors were connected to their respective data acquisition units, and all of the acquisition units were time-aligned.

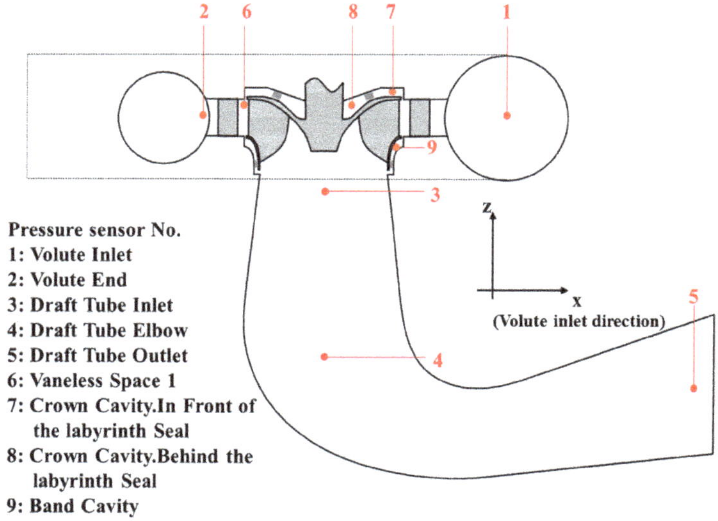

Figure 1. Installation locations of pressure sensors.

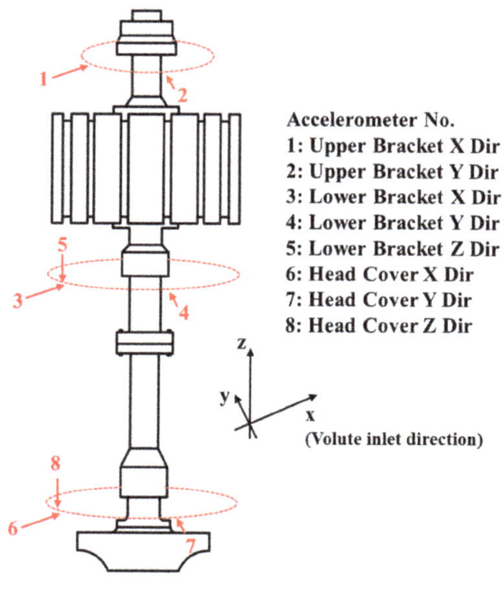

Figure 2. Installation locations of accelerometers.

2.3. Data Acquisition Configuration

A continuous sampling method was utilized to acquire the vibration and pressure pulsation signals from the sensors installed on the pump turbine unit. The data was acquired with a high-speed A/D conversion system and stored directly on the computer for subsequent analysis. A sampling frequency of 4096 Hz was selected to ensure the frequency range analyzed (below 1600 Hz) was much higher than the frequencies of potential hydraulic instability phenomena, thus avoiding significant spectral leakage.

After 1 min of steady operation in each working point, we continuously collected data for at least 30 s. Since the unit rated speed is 3.33 Hz, we can achieve a frequency resolution of 0.01 times the rotational frequency when analyzing within the frequency domain. Since all signals were continuously collected over a long duration, there is no comparison of results from multiple measurements. However, prior to finalizing the signal samples to be used for spectrum analysis, we examined the coherence between different signal segments under the same working point to ensure that the unit is truly in a steady state of operation. Poor coherence results can also occur, especially in pump mode, which will be explained later in the text.

3. Results and Discussion

3.1. Turbine Mode

3.1.1. Pressure Fluctuation in Turbine Mode

In turbine mode, the amplitude of pressure fluctuation tends to decrease, and stability improves as the unit operates closer to its designed working point. The variation in pressure signal with output power is shown in Figure 3, and the signal spectra of all measured points at four representative working points are shown in Figure 4. The characteristics of pressure fluctuation show three stages of gradual rise in output power.

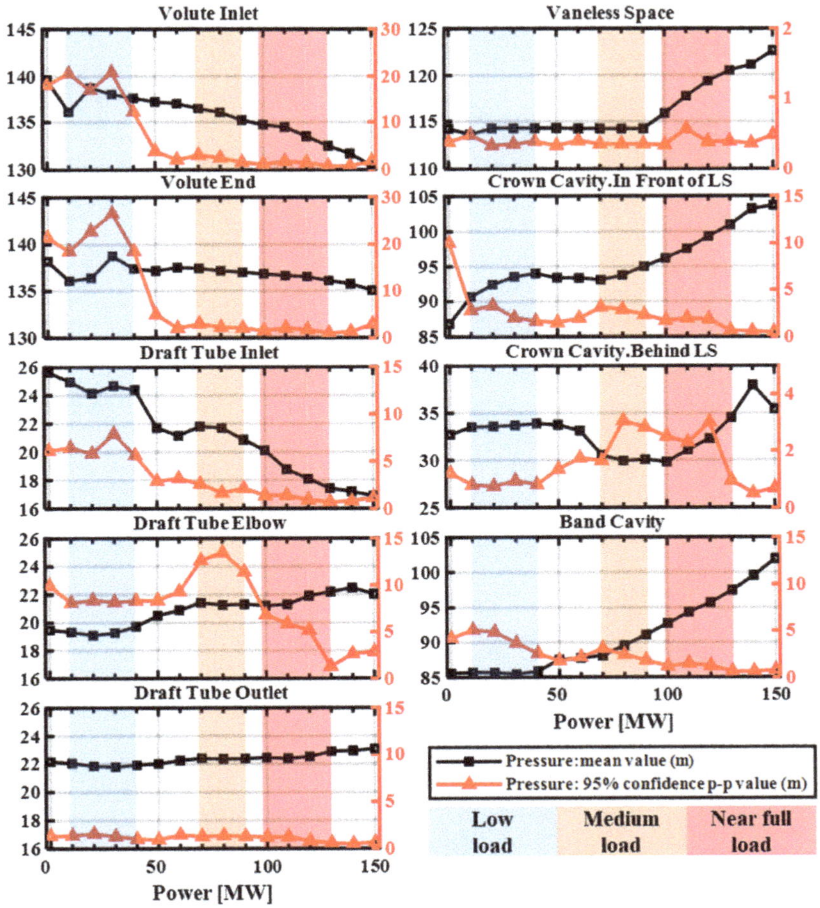

Figure 3. Pressure signal statistical indices at different power conditions.

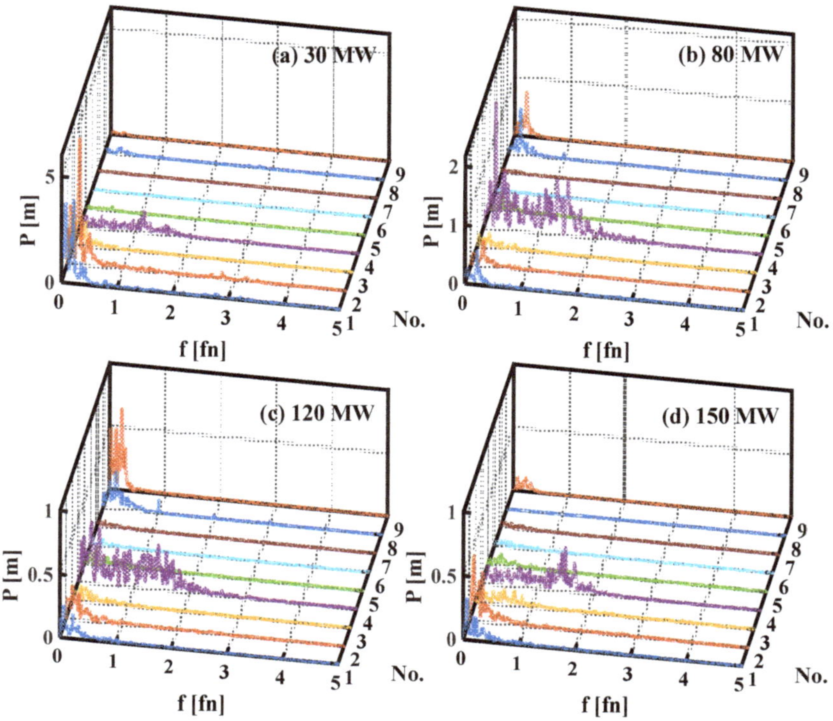

Figure 4. Pressure signal spectra under typical power conditions: (**a**) low load; (**b**) medium load; (**c**) near-full load; (**d**) full load.

(1) Under low-load (10 MW–40 MW) conditions, significant pressure fluctuation was observed in the volute, with a 95% confidence peak-to-peak value of approximately 20 m, accounting for 20% of the unit's head. Frequency spectrum analysis revealed that the intense pressure fluctuation was primarily dominated by low-frequency components (<0.5 fn) with unstable peak values and pronounced nonlinearity. The data from the slave computer for the governor also identified the phenomena of unstable output and difficulty in maintaining output power stability, as shown in Figure 5. Figure 5 also shows that the governor data changes throughout the test, where the power rises in steps. These observations may be attributed to the presence of flow blockages near the guide vanes, due to the rotating stall when the unit operates at lower flow rates and smaller guide vane openings [33].

(2) Under medium-load (70 MW–90 MW) conditions, pronounced pressure fluctuation was observed in the elbow section of the draft tube, with a 95% confidence peak-to-peak value exceeding 10 m, accounting for 10% of the unit's head. It should be noted that the pressure fluctuation conditions in the draft tube were unfavorable across all operating conditions, with a 0–2 fn frequency band indicating the presence of intense broadband noise within the draft tube. In the process of checking the signal, we found that the coherence of this low-frequency band is poor; that is, when different signal fragments are selected, the relative relationship between the amplitudes of each frequency point in the low-frequency band is different. This may be related to the strong randomness of flow at the elbow of the draft tube. However, in the medium-load range, a distinct main peak emerged in this frequency band, around 0.23 fn, which slightly decreased as the power increased, exhibiting an increasing-then-decreasing trend in amplitude, as shown in Figure 6. This phenomenon can be attributed to the vortex rope in the draft tube. As the flow rate increases, the vorticity of the fluid near the outer wall of the draft tube gradually decreases, resulting in reduced

rotating speed of the vortex rope inside the draft tube. Nevertheless, this component was not prominent in working points below 60 MW or above 110 MW, due to insufficient flow conditions for vortex rope generation and the presence of stronger pressure fluctuation from other sources [17], thereby demonstrating that the contribution of vortex-rope-induced pressure fluctuation is secondary.

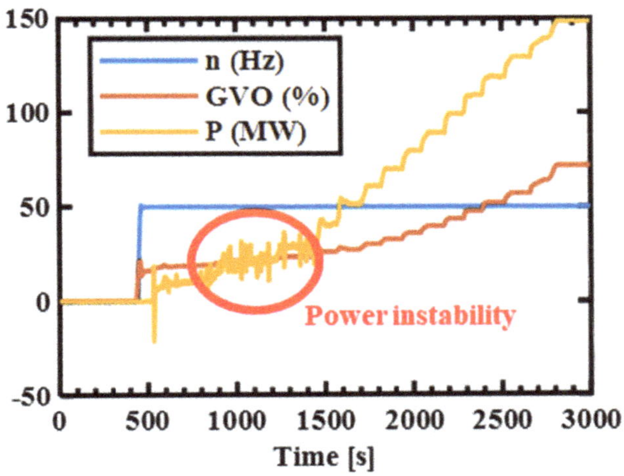

Figure 5. Governor slave computer monitoring data of whole test process.

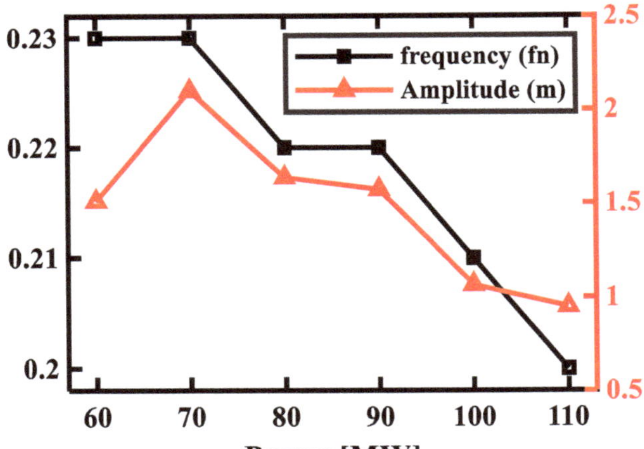

Figure 6. Pressure amplitude and frequency of vortex rope.

(3) Under near-full-load (100 MW–130 MW) conditions, an enhanced pressure fluctuation was observed behind the upper crown cavity sealing. The dominant frequency primarily occurred in the low-frequency range (<1 Hz) and exhibited unstable behavior, with significant nonlinearity. That is, even if we choose different time signal pieces at the same working point, the low-frequency band is not the same. This can be attributed to the interaction between the newly generated frequencies caused by the flow passing through the labyrinth seal and the frequency of the vortex rope, which corresponds to the results of a simple study the authors previously conducted [34]. This phenomenon briefly emerged under near-full-load conditions, disappearing after 150 MW. Concurrently, the frequency

distribution of draft tube pressure fluctuation became more concentrated as full load was attained, exhibiting a frequency band centered at 1.3 fn.

Overall, the characteristics of pressure fluctuation in turbine mode showed some similarities with those reported in reference [35]. However, under near-full-load conditions, the reference identified pressure modes corresponding to rotor–stator interaction (RSI), which were not distinctly observed in this study. One possible reason for this discrepancy is the attenuating effect of the pressure measurement pipelines. Nevertheless, information from the vibration signals can serve as a supplement at higher frequencies.

3.1.2. Vibration Acceleration in Turbine Mode

On the one hand, in terms of measurement location, the closer a measurement point is to the runner, the higher the vibration intensity, whatever the direction of the vibration. That is, the vibration in the upper bracket is the smallest, followed by the lower bracket, and the vibration in the head cover is the strongest. On the other hand, in terms of working points, lower unit output power corresponds to higher vibration levels in general. The vibration in the upper bracket is the highest at the no-load point, decreases as the power increases, and then increases again under a near-full load, but with a slight absolute value change. Vibrations in the lower bracket and head cover are generally higher at low power, with the maximum peak-to-peak value observed around 50 MW, reaching up to 50 m/s^2 at head cover. Subsequently, the vibration intensity rapidly decreases and reaches a minimum value at 100 MW, with a slight increase during the load ramp-up process, from 100 MW to 150 MW. All of these characteristics can be found summarized in Figure 7. We selected two key measurement points for detailed discussion.

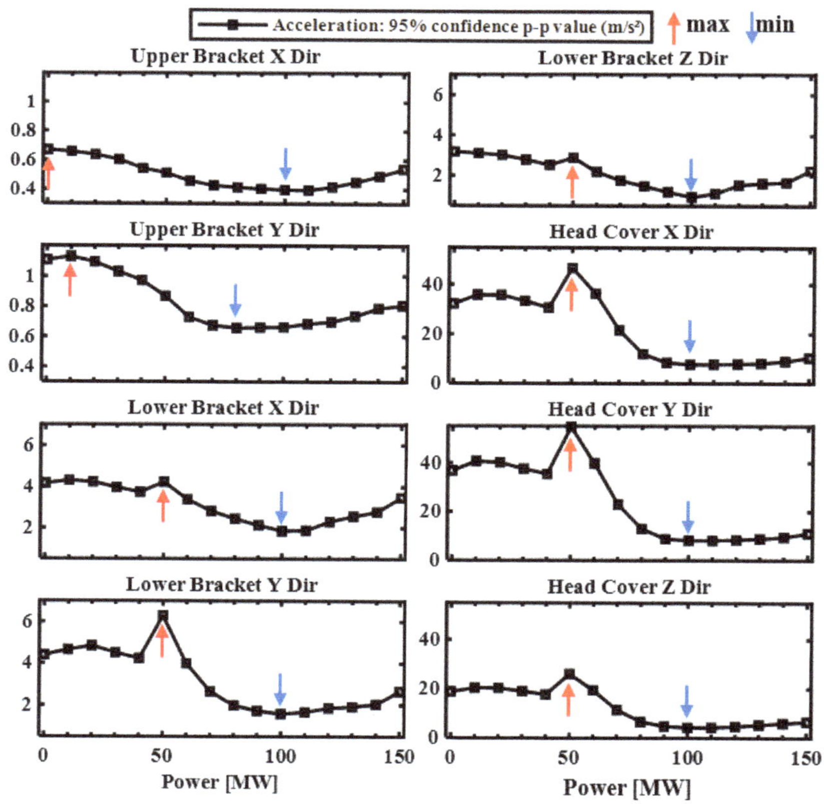

Figure 7. Acceleration signal statistical indices at different power levels.

Low bracket Y direction (LBY): Owing to the pronounced amplitude fluctuations exhibited by the vibration sensor installed in the low bracket in the Y direction, this measurement location was selected as a representative case with which to perform an in-depth analysis on the spectral composition (Figure 8):

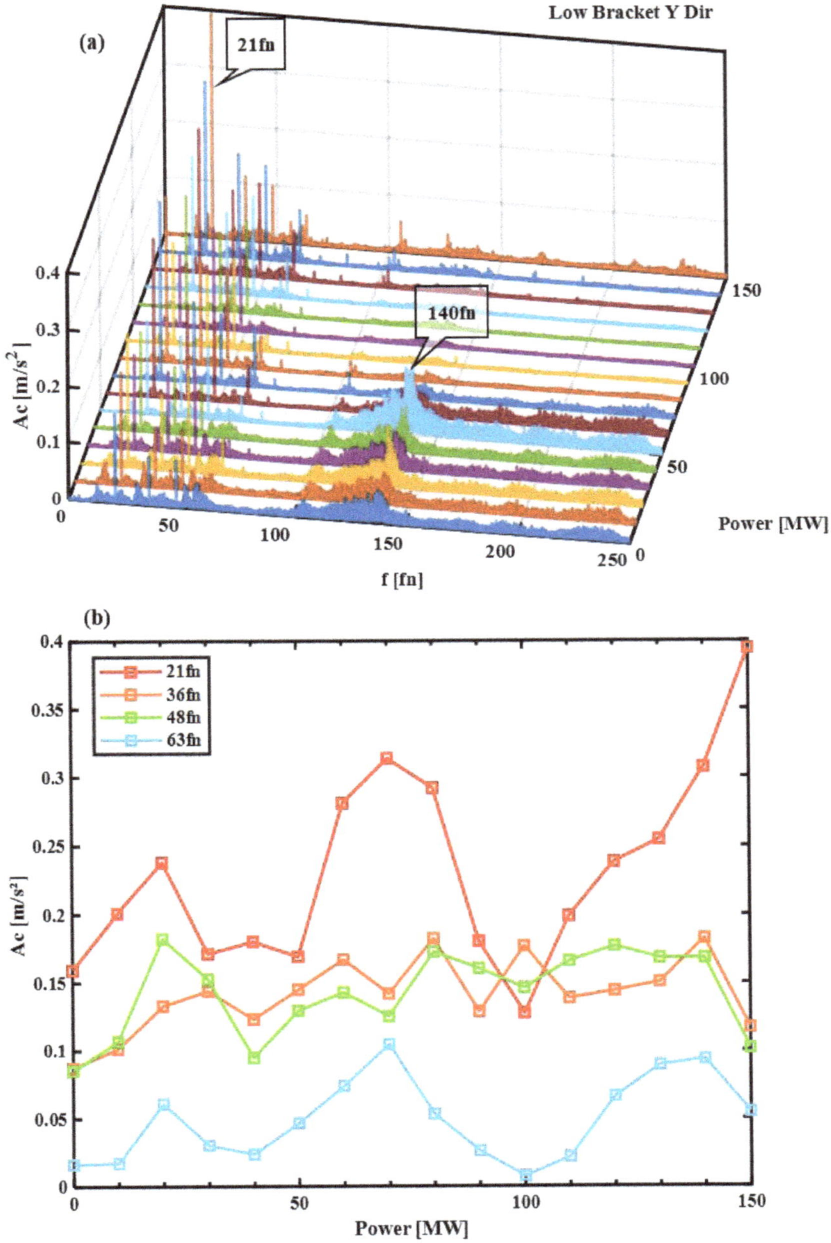

Figure 8. Acceleration signal spectra of different power levels in the LBY: (**a**) all frequencies and (**b**) comparison of typical frequency amplitudes.

(1) Compared to other output power levels, the spectrum at 50 MW shows a prominent peak, located at the 140 fn frequency band, which is the main reason for the higher vibration near the lower guide bearing. Considering that the unit has 7 blades and 20 movable guide vanes, this frequency band is likely caused by pressure waves resulting from the interaction between the rotating and stationary components. With 7 × 20 = 140 being the simplest calculation, each blade excites the stationary component 20 times within one cycle, and the excitations of the seven blades are completely out of asynchronous. The phase of causes for these excitations is completely different, so resonance does not occur, leading to a relatively smaller amplitude and less stable frequency, eventually forming a frequency band near 140 fn.

(2) In most working points, there are distinct components at 21 fn, 36 fn, 48 fn, and 63 fn. Among these components, 21 fn and 63 fn are pressure modal rotation frequencies caused by the rotor–stator interaction (RSI). According to the formula from [36]:

$$\nu = kZ_b - mZ_s \quad (1)$$

For a 7/20 unit, the first, second, and third modal frequencies in the vaneless space are 21 fn, 42 fn, and 63 fn, respectively, as shown in Table 3. However, the sources of the 36 fn and 48 fn components remain unclear, and these two frequencies are also widely present in other measurement points.

Table 3. Theoretical values of pressure modal rotation.

k	m	v	Mode Shape
3	1	1	1 ND
6	2	2	2 ND
9	3	3	3 ND

(3) By comparing the variations in the 140 fn frequency band and those in the 21 fn band (and its sub-harmonics) under different loads, a shift in their dominance can be observed. The 140 fn band reached its maximum at 50 MW and rapidly decayed as power increased further. In contrast, the 21 fn peaked at 70 MW, bottomed out at 100 MW, and rebounded thereafter. Despite the 21 fn band having the highest amplitude, the 140 fn band contributed more to the signal power, due to its broad frequency distribution. Hence, the peak vibration at 50 MW was dominated by the 140 fn band, and the rapid decline from 50 to 100 MW also resulted from its attenuation. Above 100 MW, with the near-disappearance of the 140 fn band, vibrations were governed by the 21 fn band, exhibiting a mildly rising trend. In summary, contrary to previous studies, we emphasized the significance of high-frequency bands, which posed a greater threat to the unit compared to rotor–stator interaction under low-load conditions.

Head cover Z direction (HCZ): In practical operation, the plant staff perceived significant vibrations when working near the head cover. As evident from the vertical axis of Figure 7, compared to other locations, the peak-to-peak value of vibration near the head cover was extremely pronounced, with the average close to 40 m/s^2. Therefore, the Z-direction vibration in the head cover was selected for additional analysis (Figure 9):

(1) Aside from the 140 fn frequency band, the main frequency component of the vibration is concentrated at 21 fn, with a small presence of 42 fn and 14 fn components. The 36 fn and 48 fn components are not obvious.

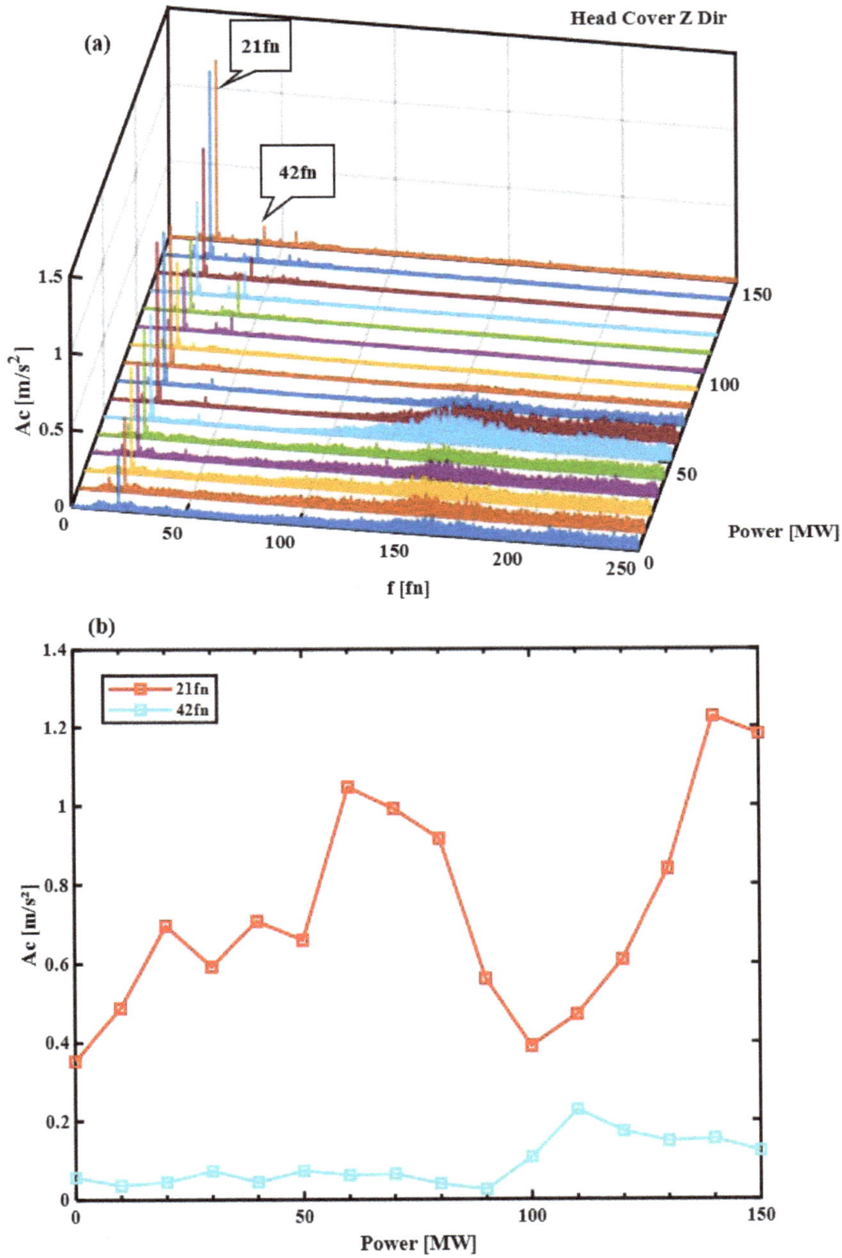

Figure 9. Acceleration signal spectra of different power in the HCZ: (**a**) all frequencies and (**b**) comparison of typical frequencies amplitude.

(2) At 50 MW, the 140 fn frequency band also exhibits higher vibration intensity, representing the working point with the strongest vibration. As the power continues to increase, both the 140 fn and 21 fn bands decrease, resulting in reduced vibration and reaching a minimum vibration level at 100 MW. At a near-full load, the 21 fn band increases again, but due to the near-disappearance of the 140 fn band, the peak-to-peak value of the vibration does not increase significantly. Due to the simpler frequency components of the head cover vibration, we can more easily draw a conclusion: in turbine mode, vibration is influenced by the 140 fn frequency band as simple excitation, and by the 21 fn band as pressure modal rotation, with a dominance in the 140 fn band over the 21 fn one.

(3) Compared to the LBY, the 21 fn (and its sub-harmonic) component at HCZ was significantly increased, being three times that of the LBY. This substantial growth in the 21 fn band also accounted for the higher vibration intensity observed in the HCZ. Meanwhile, the high-frequency band peak corresponding to 140 fn became less pronounced, overall resembling high-frequency random noise. However, the dominant relationships between the frequency components did not change.

3.2. Pump Mode

3.2.1. Pressure Fluctuation in Pump Mode

A comparison was made between zero-flow pumping and normal pumping. In the low-frequency range, the stability of pressure fluctuation is relatively worse after opening the guide vanes. After the guide vanes are opened, the 1fn component becomes more prominent, while higher amplitudes occur at 0.19 fn in the volute inlet, 0.09 fn in the volute end, and 0.37 fn in the draft tube inlet, as well as a frequency band near 1.3 fn in the elbow section of the draft tube and a noticeable 0.5 fn in the draft tube outlet. Inside the area behind the labyrinth seal at the upper crown cavity, a frequency component of 0.3 fn is observed, while the pressure fluctuation spectrum does not show significant changes in vaneless space, although this is somewhat against common sense, as shown in Figure 10. Due to the excessive length of the pressure measurement pipeline, the high-frequency components of the pressure fluctuation were not captured by the pressure sensors, but we can analyze vibration signals as a supplement to high-frequency information.

3.2.2. Vibration Acceleration in Pump Mode

In the normal pumping stage after the guide vanes were opened, the vibration in the unit significantly increased, and the vibration amplification near the water guide bearing was the most obvious. The peak-to-peak value of Y direction vibration at the head cover increased from 1.3 m/s^2 to 14.36 m/s^2, representing a 1004% increase, as shown in Figure 11. From the frequency spectrum analysis, the vibration at the upper bracket is mainly attributed to the 90 fn frequency component. Considering that the generator has 15 pole pairs, and this frequency does not appear at other locations, we attribute it to the generator. In addition, at all measurement points, the dominant frequency is 21 fn, which is the modal rotation frequency caused by the RSI.

At the lower bracket, the frequencies of 36 fn, 48 fn, 60 fn, and 72 fn are also noticeable, but there are no clues regarding the causes of these frequencies. During normal pumping operation, the amplitudes of these frequencies weaken, with a slight increase at 20 fn, which is the blade-passing frequency (BPF). For the vibration at the head cover, the 21 fn component during zero-flow pumping is significantly higher than that during normal pumping, but the high-frequency band components during normal pumping are higher compared to during the zero-flow pumping condition, as shown in Figure 12. Ultimately, the peak-to-peak value of vibration during normal pumping is stronger, indicating that the main factors enhancing the vibration near the water guide bearing are the high-frequency band components, rather than the 21 fn component. These high-frequency band components exhibit a continuous frequency band with several peaks, presenting a typical frequency response function (FRF) shape, especially in the HCX and HCY as two different directions to show a completely different shape, suggesting that the head cover is subjected to broad-frequency excitations that partially excite the natural frequencies of the structural components. Through the above analysis, it can be observed that the studied pump turbine experiences intense vibrations in the head cover, influenced by both the pressure mode generated by the RSI and the structural mode derived from the natural frequencies of the structural components. Each of these factors dominates under different working points of the unit.

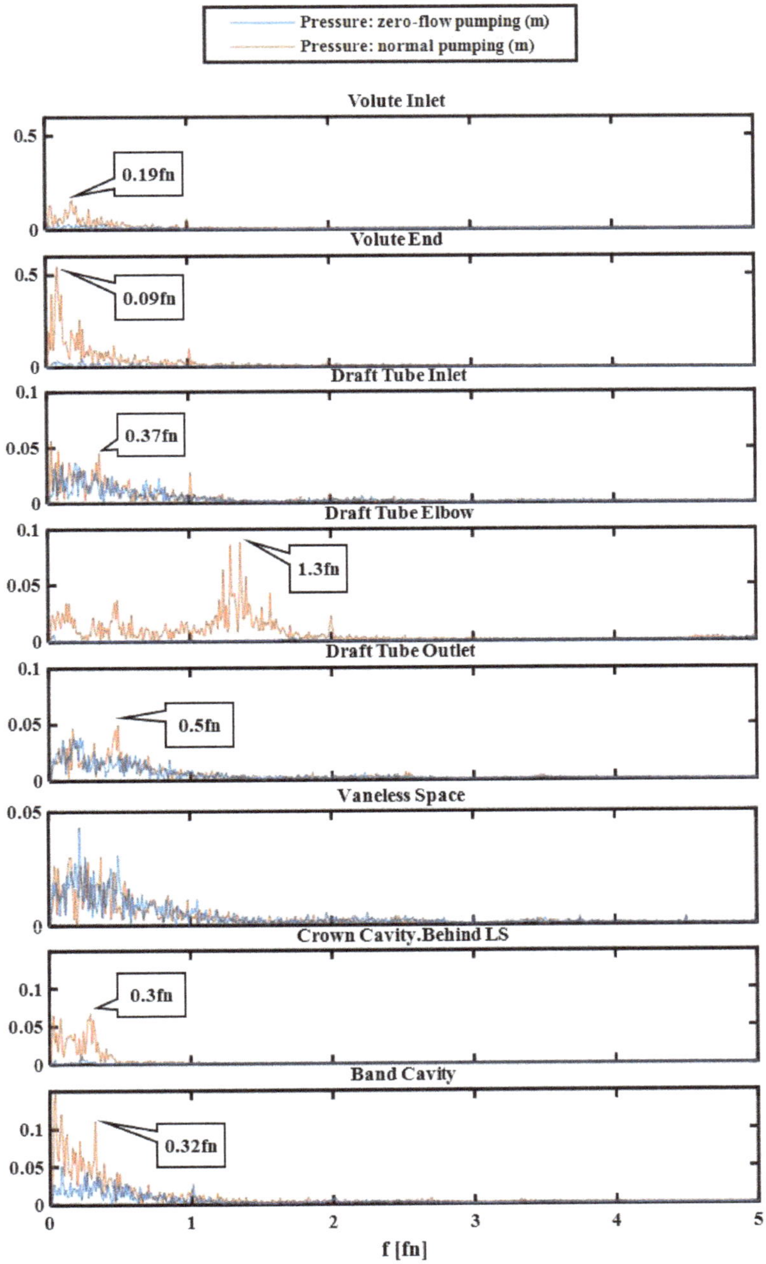

Figure 10. Comparison of pressure signal spectra between zero-flow pumping and normal pumping.

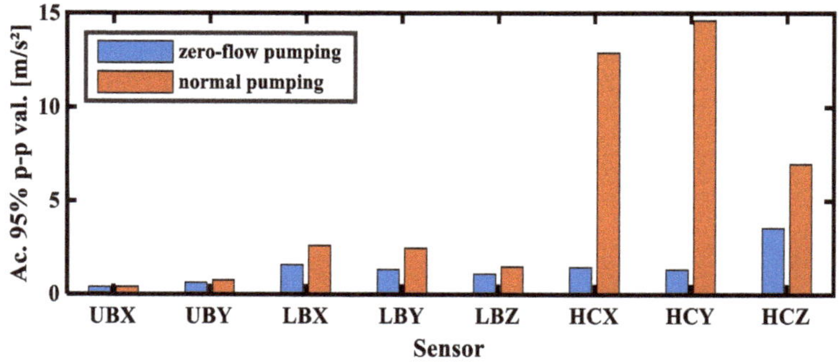

Figure 11. Comparison of peak-to-peak value of acceleration between zero-flow pumping and normal pumping.

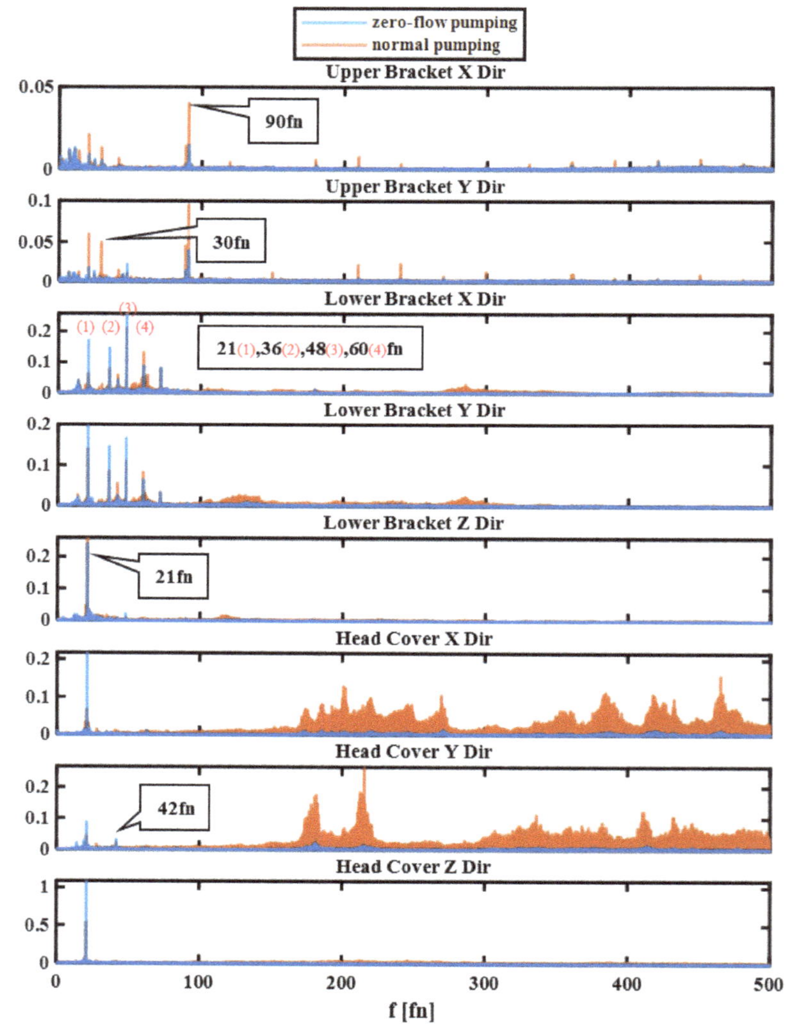

Figure 12. Comparison of acceleration signal spectra between zero-flow pumping and normal pumping.

4. Conclusions

In this study, a field measurement was conducted to investigate the stability characteristics of a pump turbine at multiple working points. Pressure sensors and accelerometers were installed to capture information in both the low-frequency and high-frequency ranges. The main research findings are summarized as follows.

In turbine mode, the pressure fluctuation stability of the unit presents three stages as the output power increases from 0 to 150 MW. At low, medium, and high loads, the main components of the pressure fluctuation come from the flow in the volute, the vortex rope of the draft tube, and the clearance flow in the crown cavity. The vibration in the unit is primarily affected by the RSI, leading to frequencies of pressure modal rotation (21 fn) and a simple pressure wave (140 fn). The 140 fn component induces a frequency band with higher power, and dominates at low loads. Most measurement points exhibit the highest vibration intensity at 50 MW and the lowest at 100 MW. This indicates that the design point (150 MW) is not the optimal solution in terms of operational stability, although it may have higher hydraulic efficiency.

In pump mode, the stability of normal pumping is worse than that of zero-flow pumping. The pressure fluctuation in the pressure measurement points exhibits distinct characteristics in the low-frequency range. The frequency spectrum shows the coexistence of the 21 fn band and continuous frequency bands with several peaks, which means that vibration characteristics are mainly affected by the pressure modal rotation caused by the RSI and the structural modal frequencies.

These experimental results will be further compared with the flow field results of CFD simulation and structural modal tests in future work to gain a better understanding of the specific causes and processes behind these frequencies.

Author Contributions: H.H. and X.S.: conceptualization, supervision, and field test command and propulsion; M.X.: data curation, formal analysis, and writing—original draft. W.Z.: investigation, methodology, and data acquisition. W.W.: writing—review and editing. Z.W.: methodology, validation, and funding acquisition. All authors have read and agreed to the published version of the manuscript.

Funding: This study was supported by the project "Research on Transient Intense Vibration in Head Cover During Transient Processes in Baishan Hydropower Station" (project no. 462132-9003001-F101).

Data Availability Statement: For the sake of information security, the original data used in this paper will not be disclosed.

Acknowledgments: The authors of this paper would like to express gratitude to the leadership and staff of Baishan Hydropower Station for their support in arranging the testing and installation of equipment. Special thanks are also extended to Shuhua Technology of Hunan Province, China for providing the acquisition modality and technical support for this research.

Conflicts of Interest: The authors declare no conflict of interest.

References

1. China Renewable Energy Engineering Institute. China Renewable Energy Development Report 2022. 2023. Available online: http://www.creei.cn/web/content.html?id=5429 (accessed on 17 September 2023).
2. Zhao, X.X.; Ren, Y.; Sha, Y.B.; Zhang, L.L.; Hou, S.C.; Xiao, F.M.; Chen, F.M.; Chen, S.D.; He, K.D.; Luo, L.J.; et al. Research on the Impact of Hydro-PV Complementary System Operation on Power Grid Based on New Energy Consumption. *Math. Probl. Eng.* **2022**, *2022*, 8459948. [CrossRef]
3. Guerra, K.; Haro, P.; Gutierrez, R.E.; Gomez-Barea, A.A. Facing the high share of variable renewable energy in the power system: Flexibility and stability requirements. *Appl. Energy* **2022**, *310*, 118561. [CrossRef]
4. Morabito, A.; Hendrick, P. Pump as turbine applied to micro energy storage and smart water grids: A case study. *Appl. Energy* **2019**, *241*, 567–579. [CrossRef]
5. Heard, B.P.; Brook, B.W.; Wigley, T.M.L.; Bradshaw, C.J.A. Burden of proof: A comprehensive review of the feasibility of 100% renewable-electricity systems. *Renew. Sustain. Energy Rev.* **2017**, *76*, 1122–1133. [CrossRef]
6. Zuo, Z.G.; Liu, S.H. Flow-Induced Instabilities in Pump-Turbines in China. *Engineering* **2017**, *3*, 504–511. [CrossRef]

7. Asomani, S.N.; Yuan, J.P.; Wang, L.Y.; Appiah, D.; Zhang, F. Geometrical effects on performance and inner flow characteristics of a pump-as-turbine: A review. *Adv. Mech. Eng.* **2020**, *12*, 1687814020912149. [CrossRef]
8. Egusquiza, E.; Valero, C.; Huang, X.X.; Jou, E.; Guardo, A.; Rodriguez, C. Failure investigation of a large pump-turbine runner. *Eng. Fail Anal.* **2012**, *23*, 27–34. [CrossRef]
9. Yu, A.; Wang, Y.S.; Tang, Q.H.; Lv, R.R.; Yang, Z.P. Investigation of the vortex evolution and hydraulic excitation in a pump-turbine operating at different conditions. *Renew. Energy* **2021**, *171*, 462–478. [CrossRef]
10. Li, L.; Tan, D.; Yin, Z.; Wang, T.; Fan, X.; Wang, R. Investigation on the multiphase vortex and its fluid-solid vibration characters for sustainability production. *Renew. Energy* **2021**, *175*, 887–909. [CrossRef]
11. Li, L.; Xu, W.X.; Tan, Y.F.; Yang, Y.S.; Yang, J.G.; Tan, D.P. Fluid-induced vibration evolution mechanism of multiphase free sink vortex and the multi-source vibration sensing method. *Mech. Syst. Signal Proces.* **2023**, *189*, 110058. [CrossRef]
12. Kim, S.J.; Suh, J.W.; Yang, H.M.; Park, J.; Kim, J.H. Internal flow phenomena of a Pump-Turbine model in turbine mode with different Thoma numbers. *Renew. Energy* **2022**, *184*, 510–525. [CrossRef]
13. Lin, T.; Li, J.; Xie, B.F.; Zhang, J.R.; Zhu, Z.C.; Yang, H.; Wen, X.M. Vortex-Pressure Fluctuation Interaction in the Outlet Duct of Centrifugal Pump as Turbines (PATs). *Sustainability* **2022**, *14*, 15250. [CrossRef]
14. Xia, L.S.; Cheng, Y.G.; Yang, J.D.; Cai, F. Evolution of flow structures and pressure fluctuations in the S-shaped region of a pump-turbine. *J. Hydraul. Res.* **2019**, *57*, 107–121. [CrossRef]
15. Pang, S.J.; Zhu, B.S.; Shen, Y.D.; Chen, Z.M. S-shaped characteristics of pump turbine with large guide vane opening by experimental and numerical analysis. *AIP Adv.* **2023**, *13*, 015201. [CrossRef]
16. Gentner, C.; Sallaberger, M.; Widmer, C.; Bobach, B.J.; Jaberg, H.; Schiffer, J.; Senn, F.; Guggenberger, M. Comprehensive experimental and numerical analysis of instability phenomena in pump turbines. In Proceedings of the 27th Iahr Symposium on Hydraulic Machinery and Systems (Iahr 2014), Montreal, QC, Canada, 22–26 September 2014; Volume 22. [CrossRef]
17. Zhang, W.W.; Chen, Z.M.; Zhu, B.S.; Zhang, F. Pressure fluctuation and flow instability in S-shaped region of a reversible pump-turbine. *Renew. Energy* **2020**, *154*, 826–840. [CrossRef]
18. Deniz, S.; Asaro, F. Improvements of Flow Control with Fluid Injection for the Suppression of Flow Instabilities in Pump-Turbines. In Proceedings of the Asme 2021 Fluids Engineering Division Summer Meeting (Fedsm 2021), Virtual, 10–12 August 2021; Volume 2.
19. Liu, J.T.; Wu, Y.L.; Wang, L.Q. Instability Analysis of a Model Pump-Turbine with MGV Based on Nonlinear Partially Averaged Navier-Stokes Methods. *Adv. Mech. Eng.* **2013**, *5*, 710769. [CrossRef]
20. Chen, H.; Lu, Y.; Liu, K.; Zhang, Z.; Li, H.; Huang, X.; Zhao, W.; Wang, Z. Study on the Internal Flow Characteristics of Long and Short Blade Runners of a 1000 MW Francis Turbine under Different Opening Conditions. *Processes* **2023**, *11*, 1796. [CrossRef]
21. Hu, J.H.; Yang, J.B.; He, X.H.; Zeng, W.; Zhao, Z.G.; Yang, J.D. Transition of amplitude-frequency characteristic in rotor-stator interaction of a pump-turbine with splitter blades. *Renew. Energy* **2023**, *205*, 663–677. [CrossRef]
22. Song, H.Q.; Zhang, J.F.; Huang, P.; Cai, H.K.; Cao, P.Y.; Hu, B. Analysis of Rotor-Stator Interaction of a Pump-Turbine with Splitter Blades in a Pump Mode. *Mathematics* **2020**, *8*, 1465. [CrossRef]
23. Qin, Y.L.; Li, D.Y.; Zhu, Y.T.; Wang, H.J.; Wei, X.Z. Influence of geometric factors at runner outlet on the hump characteristics of a pump-turbine. *Sustain. Energy Technol. Assess.* **2022**, *51*, 101890. [CrossRef]
24. Yang, G.; Shen, X.; Shi, L.; Meng, J.; Luo, W.; Zhang, D.; van Esch, B.P.M. Unsteady numerical investigations of the effect of guide vane openings on the hydrodynamic characteristics under stall conditions in a pump-turbine pump mode. *Energy Convers. Manag.* **2023**, *293*, 117499. [CrossRef]
25. Barrio, R.; Fernandez, J.; Blanco, E.; Parrondo, J.; Marcos, A. Performance characteristics and internal flow patterns in a reverse-running pump-turbine. *Proc. Inst. Mech. Eng. Part C J. Mech. Eng. Sci.* **2012**, *226*, 695–708. [CrossRef]
26. Valentin, D.; Presas, A.; Bossio, M.; Egusquiza, M.; Egusquiza, E.; Valero, C. Feasibility of Detecting Natural Frequencies of Hydraulic Turbines While in Operation, Using Strain Gauges. *Sensors* **2018**, *18*, 174. [CrossRef] [PubMed]
27. Valentin, D.; Presas, A.; Valero, C.; Egusquiza, M.; Egusquiza, E. Detection of Hydraulic Phenomena in Francis Turbines with Different Sensors. *Sensors* **2019**, *19*, 4053. [CrossRef]
28. Hu, J.H.; Yang, J.B.; Zeng, W.; Zhao, Z.G.; Yang, J.D. Hydraulic interaction of two parallel pump-turbines in constant-speed oscillation: Measurement, simulation, and sensitivity analysis. *Renew. Energy* **2021**, *176*, 269–279. [CrossRef]
29. Zhang, Y.N.; Chen, T.; Li, J.W.; Yu, J.X. Experimental Study of Load Variations on Pressure Fluctuations in a Prototype Reversible Pump Turbine in Generating Mode. *J. Fluid Eng.* **2017**, *139*, 074501. [CrossRef]
30. Hu, J.H.; Yang, J.D.; Zeng, W.; Yang, J.B. Transient Pressure Analysis of a Prototype Pump Turbine: Field Tests and Simulation. *J. Fluid Eng.* **2018**, *140*, 071102. [CrossRef]
31. Zhang, F.; Lowys, P.Y.; Houdeline, J.B.; Guo, X.D.; Hong, P.; Laurant, Y. Pump-turbine Rotor-Stator Interaction Induced Vibration: Problem Resolution and Experience. In Proceedings of the 30th Iahr Symposium on Hydraulic Machinery and Systems (Iahr 2020), Lausanne, Switzerland, 21–26 March 2021; Volume 774. [CrossRef]
32. Kim, S.J.; Suh, J.W.; Choi, Y.S.; Park, J.; Park, N.H.; Kim, J.H. Inter-Blade Vortex and Vortex Rope Characteristics of a Pump-Turbine in Turbine Mode under Low Flow Rate Conditions. *Water* **2019**, *11*, 2554. [CrossRef]
33. Zhang, Y.N.; Zhang, Y.N.; Wu, Y.L. A review of rotating stall in reversible pump turbine. *Proc. Inst. Mech. Eng. Part C J. Mech. Eng. Sci.* **2017**, *231*, 1181–1204. [CrossRef]

34. Hu, H.P.; Xia, M.; Qiao, M.; Wang, W.; Wang, J.Z.; Wang, Z.W. A Simulation Study of Hydraulic Vibration caused by Clearance Flow in a Pump Turbine. *IOP Conf. Ser. Earth Environ. Sci.* **2022**, *1079*, 012032. [CrossRef]
35. Zheng, X.; Zhang, S.; Zhang, Y.; Li, J.; Zhang, Y. Dynamic characteristic analysis of pressure pulsations of a pump turbine in turbine mode utilizing variational mode decomposition combined with Hilbert transform. *Energy* **2023**, *280*, 128148. [CrossRef]
36. Doerfler, P. On the role of phase resonance in vibrations caused by blade passage in radial hydraulic turbomachines. IAHRSection Hydraulic Machinery, Equipment, and Cavitation. In Proceedings of the 12th IAHR Symposium on Hydraulic Machineryand System, Stirling, UK, 27–30 August 1984.

water

MDPI

Article

A Study on the Transient Characteristics of the Power-Off Transition Process of a Double-Volute Centrifugal Pump

Lifeng Lu [1], Ziwei Ren [2], Zhongzan Wang [3], Wenjie Zhou [1], Siwei Li [3], Jin Dai [1], Chunxia Yang [2,*] and Mengfan Dang [3]

1 Hunan Provincial Water Resources Development and Investment Co., Ltd., Changsha 410007, China; 18163697368@163.com (L.L.); zwj6900@163.com (W.Z.); chc_dj@163.com (J.D.)
2 School of Electrical and Power Engineering, Hohai University, Nanjing 211100, China; 221606040014@hhu.edu.cn
3 Hunan Water Resources and Hydropower Survey, Design, Planning and Research, Co., Ltd., Changsha 410007, China; wangzhongzan@126.com (Z.W.); siwei0055@126.com (S.L.); 18229835971@163.com (M.D.)
* Correspondence: yangchunxia@hhu.edu.cn

Abstract: A double-volute centrifugal pump is a very important pump type; the internal flow field of a centrifugal pump will change drastically during the transition process of power failure, which will affect the safety and stability of the pump's operation. In this paper, the CFD numerical simulation method is used, and the UDF procedure is developed to realize the continuous update of the impeller speed at each time step. The working parameters, such as the torque and flow rate at the instantaneous moment, are obtained through the sequential iteration of each small step, and a numerical simulation of the power-off transient is carried out on a double-volute centrifugal pump; additionally, the changes in the external characteristic parameters and the internal flow field of the centrifugal pump are analyzed in detail. The results show that the double-volute centrifugal pump experienced four different modes after power failure, namely pump mode, braking mode, turbine mode, and runaway mode, and the absolute values of the runaway speed and runaway flow rate are 1.465 times and 1.21 times the initial values, respectively. Through the analysis of the flow field in different regions, the change processes of the generation, development, and disappearance of the vortex at each position of the centrifugal pump are obtained, and the change and development processes of the internal velocity gradient of the centrifugal pump are obtained. In addition, it is found that the high-speed area located in the second volute runner is larger than that of the first volute runner because the second volute runner is shorter and narrower than the first volute runner.

Keywords: double-volute centrifugal pump; power-off transition process; internal flow field characteristics; external characteristic parameters

Citation: Lu, L.; Ren, Z.; Wang, Z.; Zhou, W.; Li, S.; Dai, J.; Yang, C.; Dang, M. A Study on the Transient Characteristics of the Power-Off Transition Process of a Double-Volute Centrifugal Pump. *Water* **2024**, *16*, 1707. https://doi.org/10.3390/w16121707

Academic Editor: Giuseppe Oliveto

Received: 16 May 2024
Revised: 9 June 2024
Accepted: 13 June 2024
Published: 15 June 2024

1. Introduction

A double-volute centrifugal pump is a very important pump type, which uses a splitter to divide the traditional single volute runner into two volute runners, so it has better hydraulic efficiency and head, it can better balance the radial force on the impeller, and it can effectively suppress pressure pulsation [1,2]. Li Q et al. [3] investigated the entropy generation of single and double-volute molten salt pumps, and they discovered that the double-volute molten salt pump experiences a less radial hydraulic force. In general, the double-volute scheme reduces energy loss and ensures better structural stability. Shim H S et al. [4,5] numerically calculated the hydraulic efficiency and radial force of double-volute centrifugal pumps with different volute geometries and found that a reasonable double-volute structure can effectively reduce the radial force on the impeller. Khalifa A E et al. [6] found that there is a relationship between pressure fluctuation and vibration in a double-volute centrifugal pump and found an effective way to reduce the vibration in the double-volute pump. Xiao R et al. [7] found that adding a reasonable splitter structure in a single-volute double-suction pump can make the double-volute double-suction pump maintain its original hydraulic performance and reduce the radial force of the impeller. Yang J et al. [8] calculated and analyzed the transient flow during the start-up process of the double-volute centrifugal pump and found that the pressure fluctuation on the outside of the splitter was relatively stable relative to the inside of the splitter. Yang M et al. [9] numerically simulated the pressure pulsation characteristics and radial force of the impeller of the double-volute centrifugal pump and found that there was obvious pressure pulsation in the double-volute pump.

In the process of power failure and during the shutdown and start-up of the pump, the internal flow field will change drastically, which will affect the safety and stability of the pump's operation. At present, some progress has been made in the internal transient flow of centrifugal pumps under transient operating conditions. In the 1980s, Tsukamoto et al. [10,11] systematically studied the transient transition process of a small volute centrifugal pump during the rapid start-up and shutdown process, and they found that because of the existence of pulsating pressure, the dimensionless head coefficient was much higher than the calculated value under a quasi-steady state. Chalghoum I et al. [12] theorized transient flow during centrifugal pump start-up. The effects of the start-up time, impeller diameter, number of blades, and blade height on pressurization were analyzed. The numerical results show that the pressure increment is inversely proportional to the start-up time. Tanaka T et al. [13] conducted experiments and CFD studies on the transient characteristics of centrifugal pumps during rapid start-up, and the results showed that the transient characteristics in the experiment were greater than the quasi-steady state change at the beginning of the transient period and then gradually approached the quasi-steady state change. Zhou D et al. [14] revealed variations in parameters such as the speed, flow, torque, and pressure at the measuring point with time through the numerical simulation of the axial flow pump device model during runaway caused from power failure. Wang W et al. [15] analyzed the unsteady internal flow characteristics and time-frequency characteristics of pressure fluctuation of a pump turbine in turbine mode (PAT) during power failure, and the results showed that the flow field between the blades was extremely unstable under braking conditions, and the main frequency of volute pressure fluctuation was related to the rotation speed, which was mainly caused by dynamic and static interaction. Wang W [16] et al. studied the pressure pulsation characteristics of a mixed-flow pump under turbine and runaway modes on an open test bench and analyzed the frequency domain characteristics of pressure pulsation based on FFT (Fourier Frequency Transform), and they found that in order to avoid pressure pulsation under runaway mode, the mixed-flow pump is best operated under small flow conditions such as those in a hydraulic turbine. Feng J et al. [17] simulated the power-off process of the centrifugal pump, and the results showed that the characteristic curves and runaway parameters of the pump predicted by the numerical simulation were in good agreement with the experimental results. The transient process after power failure mainly goes through four working modes: pump, brake, turbine, and runaway. Dong W et al. [18] studied the internal flow characteristics of centrifugal pumps under two linear start-up schemes, 0.1 s and 0.3 s, and found that the vorticity distribution of the 0.3 s start-up scheme was more regular, and the slender high-vorticity region was only distributed near the blade side. Zhang Y L et al. [19] established a closed-loop pipeline system including a centrifugal pump to achieve self-coupling solving, which revealed the transient characteristics of the centrifugal pump during shutdown and provided a good reference for the transient behavior of other impeller machinery.

When the pump unit is shut down due to sudden power failure or maloperation, it will eventually enter runaway mode, and when the pump unit is running in runaway mode, some parts of the pump unit may be damaged due to the mass imbalance of the rotating body and the hydraulic instability inside the unit [20]. More and more scholars have begun to conduct experiments and numerical simulation studies on the runaway mode of the pump unit [21,22]. Kan K et al. [23] analyzed in detail the flow characteristics and energy loss of the internal flow field during the runaway process of an axial flow pump unit. Wang G [24] et al. conducted an experimental study on the hydraulic characteristics of the inertial tank after the failure of the pump power and analyzed the influence of structural parameters, including CPD and the transition section, on the hydraulic characteristics of the inertial tank. Yang F et al. [25] conducted runaway characteristic tests on multiple groups of pump systems, and the measurement results showed that the runaway speed of the unit was different under different blade angles and increased with the increase in blade angles.

The accidental shutdown of a pumping station unit caused by power failure is a huge threat to the safe operation of the pumping station. Due to safety, financial, or technical constraints, some pump transient processes cannot be evaluated experimentally. In order to ensure the safe and stable operation of a pump unit in daily life, it is necessary to study the transient flow of the centrifugal pump after power failure. In this paper, the power-off transition process of the double-volute centrifugal pump is simulated by introducing the UDF procedure, and the transient change process of the external characteristics and internal flow field of the double-volute centrifugal pump are analyzed and revealed. This study provides a certain theoretical basis for the safe operation of a pumping station.

2. Numerical Methodology

2.1. Three-Dimensional Modeling of Centrifugal Pump

The centrifugal pump model is mainly composed of the intake pipe, impeller, double volute, and outlet diffusion section, as shown in Figure 1. The design parameters of the double-volute centrifugal pump are shown in Table 1. The specific speed used in this study is defined as follows:

$$n_s = \frac{3.65 N_d \sqrt{Q_d}}{H_d^{0.75}} \quad (1)$$

where n_s, N_d, Q_d, and H_d denote the specific speed, design rotational speed, design flow rate, and design head, respectively.

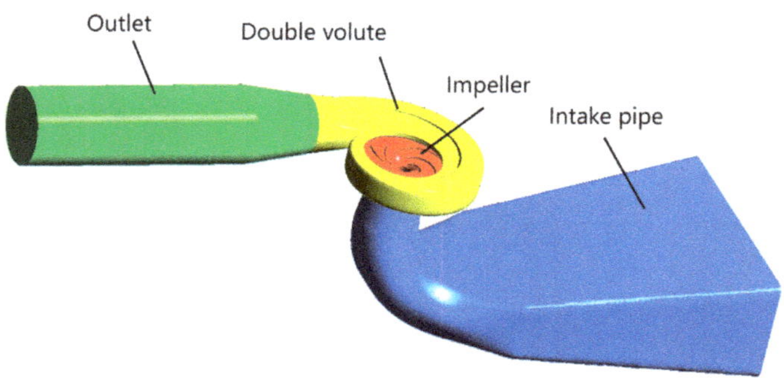

Figure 1. Three-dimensional model of double-volute centrifugal pump.

Table 1. Main parameters of centrifugal pump.

Parameter Name	Numerical Value
Design flow rate, Q_d (m^3/s)	0.321
Design head, H_d (m)	40.3
Design rotational speed, N_d (r/min)	1480
Specific speed, n_s	191.3
Impeller inlet diameter, D_1 (mm)	283.7
Impeller outlet diameter, D2 (mm)	399.8
Number of impeller blades, Z_1	6

2.2. Mesh Generation

In order to obtain more accurate numerical simulation results, this paper performs hexahedral structured meshing of each part of the double-volute centrifugal pump and encrypts the local mesh of key parts such as the double-volute tongue and the watershed around the runner blades. A schematic view of the grid of the individual components is shown in Figure 2, and the grid around an impeller blade is shown in Figure 3. Grid independence verification based on head and efficiency parameters is shown in Figure 4. When the number of grids is greater than 8.2 million, the efficiency value and head value tend to be stable. Therefore, the total number of calculated grid cells in the fluid area of the entire centrifugal pump is 8202648, and the number of grid cells for each component is listed in Table 2.

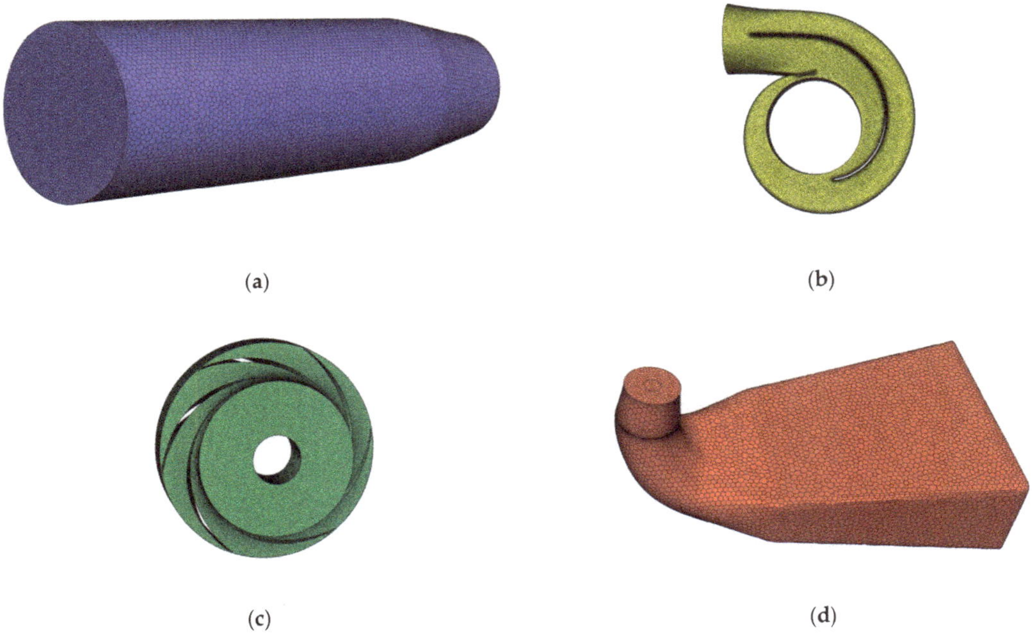

Figure 2. Grid view of each component: (**a**) outlet, (**b**) double volute, (**c**) impeller, and (**d**) intake pipe.

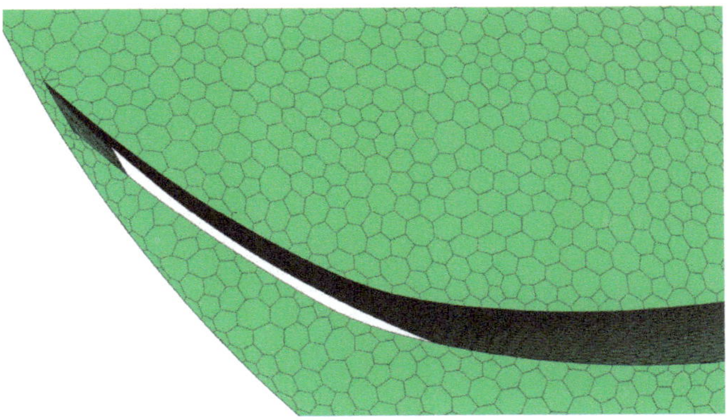

Figure 3. The grid around an impeller blade.

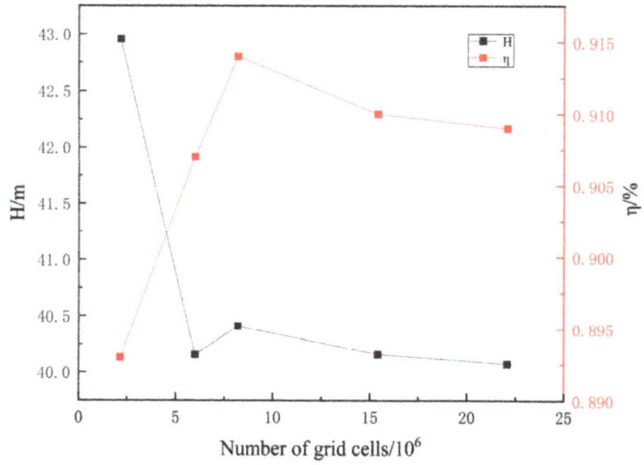

Figure 4. Grid independence verification diagram.

Table 2. Number of grid cells for each component.

Component Name	Intake Pipe	Impeller	Double Volute	Outlet Pipe
Number of grid cells	921,713	3,571,714	3,563,816	145,405

2.3. Setting of Boundary Conditions

In this paper, the transient calculation and analysis of the power-off transition process of a centrifugal pump are carried out based on CFD numerical simulation, and the SST k-ω turbulence model is selected. Compared to the k-ε model, the SST k-ω model can better deal with the transport of turbulent shear stress in the reverse pressure gradient and separation boundary layer. The centrifugal pump unit has a strong reverse pressure gradient, flow separation, reflux, and other abnormal flow phenomena during the shutdown process, so the SST k-ω model is used to calculate the subsequent transition process [26,27].

The inlet is adopted as the pressure inlet boundary condition, and the outlet is adopted as the pressure outlet boundary condition. The impeller is defined as a dynamic grid area. SIMPLEC is used for the pressure and velocity coupling algorithm; the pressure term in the calculation process adopts the second-order format, and the turbulent kinetic energy term and convection term are both in the second-order upwind style. The time step of the transient simulation is set to 4.5×10^{-4} s, which represents the time it takes for the impeller to rotate by 4 degrees. The number of iterations during each time step is set to 20, considering the influence of gravity, and the user-defined function (UDF) is used to calculate and control the angular speed of the impeller area during the entire shutdown process.

During power failure, the impeller speed and torque are determined by the angular momentum balance equation [28] represented by Equation (2):

$$M_t - M_g = J\frac{d\omega}{dt} \tag{2}$$

where M_t is the resultant torque acting on the impeller, N·m; M_g is the drag torque, N·m; J is the moment of inertia of the impeller of the centrifugal pump, kg·m^2; and ω is the angular velocity, rad/s.

When the centrifugal pump is suddenly powered off, the drag torque is quickly reduced to zero, $M_g = 0$, that is, the rotation angular velocity of the impeller at any time is shown in Equation (3):

$$\omega_{i+1} = \omega_i + \frac{M_t^i}{J} \times \Delta t \tag{3}$$

where Δt is the time step of the simulation calculation, and the angular velocity ω_i and torque M_t^i at the ith moment can be obtained to iteratively calculate the speed of the next moment so as to realize the continuous update of the impeller speed at each time step. The sequential iteration of each small step is carried out to obtain working parameters such as the torque and flow rate at the instantaneous moment so as to realize the whole process calculation of the power-off transient.

2.4. Validation of Numerical Methods

In order to verify the accuracy of the numerical simulations, tests were carried out on a high-precision test bench, and the characteristic parameters of the pumps were obtained; the test bench is shown in Figure 5. The test process and methods strictly followed the international standard IEC 60193 [29]. By changing the speed of the pump to adjust the flow rate, the test data under multiple sets of pump conditions with different flow rates were measured, and Figure 6 shows the comparison of the simulation and experimental characteristic parameters, including the head (H), efficiency (η), and power (P). From the graph, it can be observed that the calculated value deviates from the experimental value. Under the condition of a small flow rate, the calculated values of the head, power, and efficiency are lower than the experimental values. Under the condition of a large flow rate, the calculated values of the head and power are higher than the experimental values, and the efficiency is lower than the experimental value. The errors are all within 5%, and they could have been caused by subtle geometric differences between the simulated model and the physical test model, including gaps that were overlooked during the simulation. In general, the results of the numerical calculations are basically consistent with the testing data, so the method of numerically simulating the transient power-off process of the centrifugal pump is reliable.

Figure 5. Test bench.

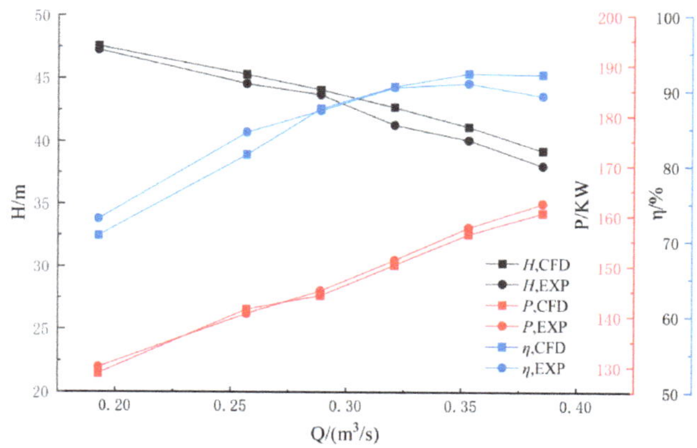

Figure 6. Comparison of simulation and experimental characteristic parameters.

3. Results

3.1. Variation of External Characteristics

In order to more intuitively and clearly observe the variation in each external characteristic parameter of a centrifugal pump with time, the three values of torque, flow rate, and speed are dimensionless. Three relative values are introduced, which are the relative value of torque (M_{rel}), the relative value of the flow rate (Q_{rel}), and the relative value of speed (N_{rel}), where the subscripts t and 0 represent the instantaneous value and the initial value, respectively, as shown in Equation (4):

$$\begin{aligned} M_{rel} &= M_t / M_0 \\ Q_{rel} &= Q_t / Q_0 \\ N_{rel} &= N_t / N_0 \end{aligned} \tag{4}$$

As shown in Figure 7, the centrifugal pump has experienced four working modes, namely pump, brake, turbine, and runaway, during the power failure process. Under the pump mode (t = 0.212 s), the centrifugal pump is in the steady-state operation state under the design working condition during the 0–0.1 s period. The accident power failure occurs at 0.1 s, the parameters decrease rapidly, the relative value of the speed decreases by about 45%, the torque decreases to about 23.6% of the initial value, and the flow rate drops to 0 at t = 0.212 s. Then, it enters the braking mode, and the braking time is about 0.148 s (t = 0.212–0.36 s), the reverse flow rate begins to appear in the centrifugal pump and increases, the torque reaches the trough at t = 0.237 s, and then it begins to rise gradually. Under the action of reverse flow, the speed decreases faster than that of the pump under the action of reverse flow, and the speed drops to 0 at t = 0.36 s. As the direction of the rotation of the impeller changes, the centrifugal pump enters the turbine mode (t = 0.36–0.886 s), and the torque reaches its peak at t = 0.42 s, at which time M_{rel} = 1.02. Due to the influence of its own inertia, the flow rate decreases with the increase in speed after t = 0.531 s, and the speed, N_{rel}, gradually increases from 0 and then decreases slightly, finally reaching runaway speed. At t = 0.732 s, the reverse speed reaches the maximum value, and then because the centrifugal pump has experienced a small fluctuation from the turbine working condition to the hydraulic turbine dynamic condition and then to the hydraulic turbine working mode, the torque has a small fluctuation in a short time near 0. When the torque is stable at about 0 fluctuations, it enters a stable runaway mode (t > 0.886 s). At this time, the torque, flow rate, and rotational speed tend to be stable values. The absolute value of the speed is 1.465 times of the design rotational speed, and the absolute value of the flow rate is 1.21 times of the design flow rate, that is, the runaway speed is −2168.2 r/min, and the runaway flow rate is −0.38841 m^3/s.

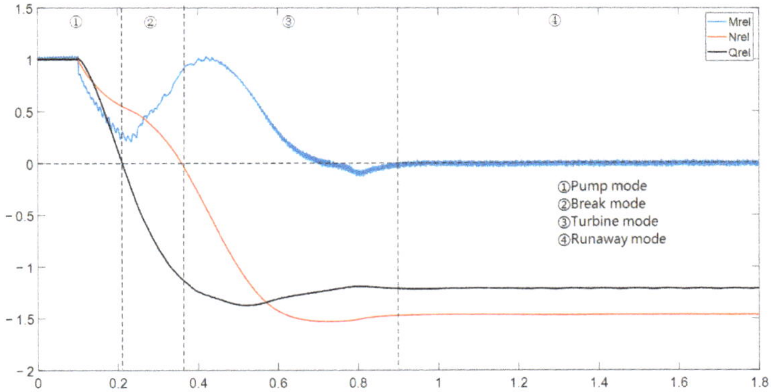

Figure 7. Pump characteristic parameters during power failure.

3.2. Evolution of Internal Flow

3.2.1. Flow State of Intake Pipe

In this section, the flow field inside the inlet pipe is studied by analyzing the streamline and velocity distribution of the intake pipe, as shown in Figure 8.

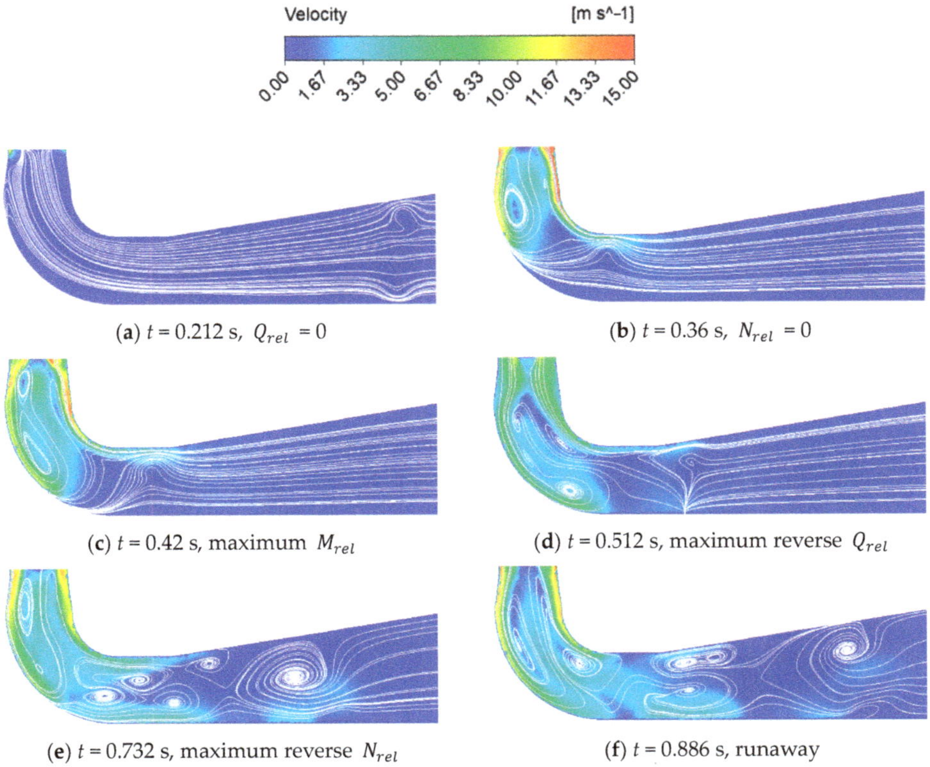

(**a**) $t = 0.212$ s, $Q_{rel} = 0$ (**b**) $t = 0.36$ s, $N_{rel} = 0$ (**c**) $t = 0.42$ s, maximum M_{rel} (**d**) $t = 0.512$ s, maximum reverse Q_{rel} (**e**) $t = 0.732$ s, maximum reverse N_{rel} (**f**) $t = 0.886$ s, runaway

Figure 8. Streamline and velocity distributions in intake pipe at various times.

It is known from Figure 8a that when the flow rate of the centrifugal pump is 0, the streamline in the inlet pipe is relatively stable as a whole. But the streamline in the intake pipe and the inlet area of the impeller begins to rotate, indicating that the flow state in the inlet pipe is about to be unstable. When the speed of the impeller drops to 0, combined with Figure 8b, the straight cone pipe section close to the impeller begins to appear as an unstable vortex. The high flow velocity area is mainly concentrated in the wall surface of the straight cone pipe section. When the impeller reverses and enters the turbine mode, the swirl movement inside the intake pipe gradually increases. As shown in Figure 8c, the vortex of the straight cone section moves towards the elbow section and creates a new vortex in the straight cone section, and the high flow velocity area also moves towards the wall of the elbow section. Then, the maximum reverse flow is reached. As shown in Figure 8d, the original large vortex developed into several small vortices and began to move towards the diffusion pipe section, and the flow velocity of the pipe wall decreased. As shown in Figure 8e, with the continuous increase in the reversal speed, the previous small vortices gradually moved outward and, at the same time, produced new vortices in the straight cone pipe section. The high-speed area at the straight cone pipe wall reappeared. As shown in Figure 8f, in the stable runaway mode, there was a large range of vortices on the inside of the straight cone pipe section and the elbow pipe section, and some large vortices moved to the inlet of the intake pipe.

3.2.2. Flow State of Double Volute and Outlet

In order to better understand the content of this section, Figure 9 marks the locations of the tongue, the splitter, the first volute runner, and the second volute runner.

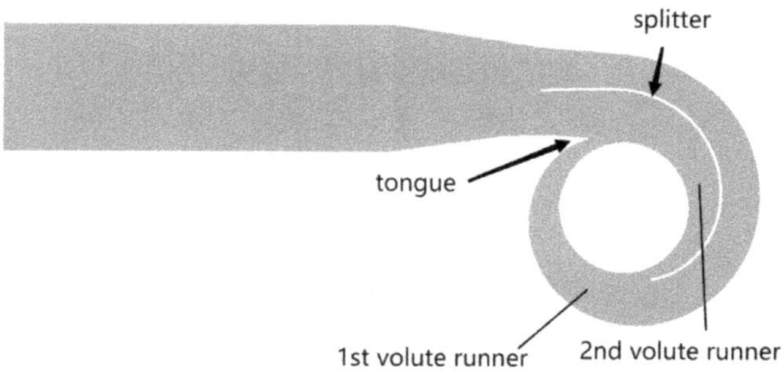

Figure 9. Locations of the tongue, the splitter, the first volute runner, and the second volute runner.

Figure 10 illustrates the velocity and streamline distribution in the double-volute and outlet diffusion section at various times. As can be seen in Figure 10a, when the flow rate of the centrifugal pump is 0, the centrifugal pump enters the braking mode. The flow regime of the double-volute area and the outlet diffusion section is very complex and disordered, a large number of vortices are located in the outlet diffusion section, and the vortex in the double-volute area is mainly located at the tongue and the tail end of the splitter. Due to the influence of dynamic and static interference, the high-speed area is mainly located at the beginning of the tongue and the splitter, and the speed at the tongue is higher. When the impeller speed drops to 0, it begins to enter the turbine mode; as seen in Figure 10b,c, the flow regime of the double volute area and the outlet diffusion section becomes relatively stable, the vortex disappears, and the high-speed area appears in a small area at the tail of the impeller blade. Because the second volute runner is shorter and narrower than the first volute runner, the area of the high-speed area located in the second volute runner is 30–50% larger than the first volute runner. This is when the maximum value of the reverse flow rate is reached. As shown in Figure 10d, the high-speed area develops from a block-like distribution to a continuous band-like distribution. As shown in Figure 10e,f, the high-speed area develops into a discontinuous band-like distribution divided by blades as the reversal speed increases until it reaches the runaway speed.

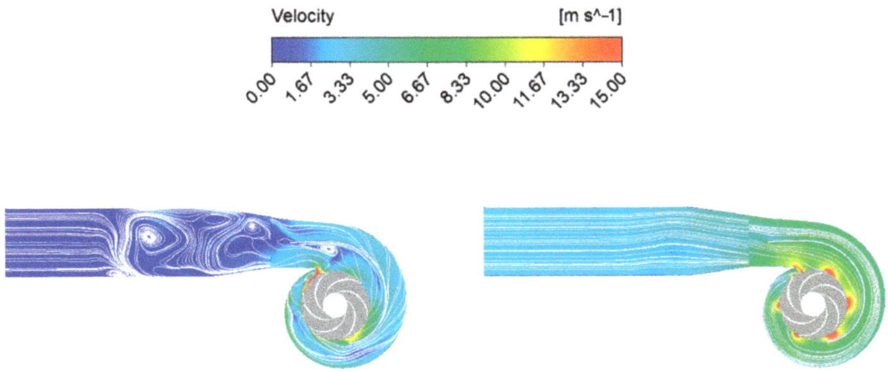

(**a**) $t = 0.212$ s, $Q_{rel} = 0$ (**b**) $t = 0.36$ s, $N_{rel} = 0$

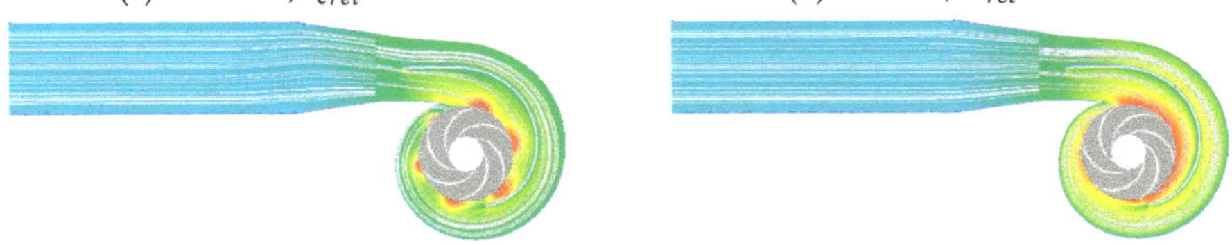

(**c**) $t = 0.42$ s, maximum M_{rel} (**d**) $t = 0.512$ s, maximum reverse Q_{rel}

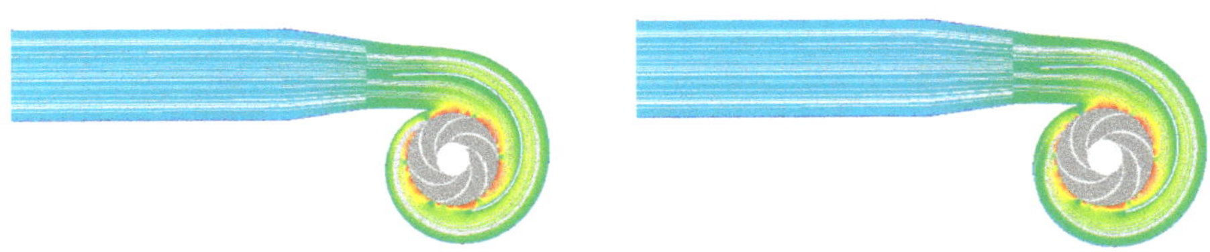

(**e**) $t = 0.732$ s, maximum reverse N_{rel} (**f**) $t = 0.886$ s, runaway

Figure 10. Streamline and velocity distributions in double volute and outlet at various times.

3.2.3. Flow State of Impeller

The impeller is the most important component of the centrifugal pump, and the direction of the impeller will change during the transient process of stopping the pump. In order to facilitate the description of the suction surface and pressure surface of the impeller and the inlet and outlet of the impeller during the transition process, the initial pump working conditions are taken as the reference. That is, the interface between the impeller and the intake pipe is defined as the impeller inlet, the interface between the impeller and the volute is defined as the impeller outlet, the convex surface of the blade is the pressure surface, and the concave surface of the blade is the suction surface. Figure 11 shows the velocity and streamline distribution of the central section of the impeller at various times.

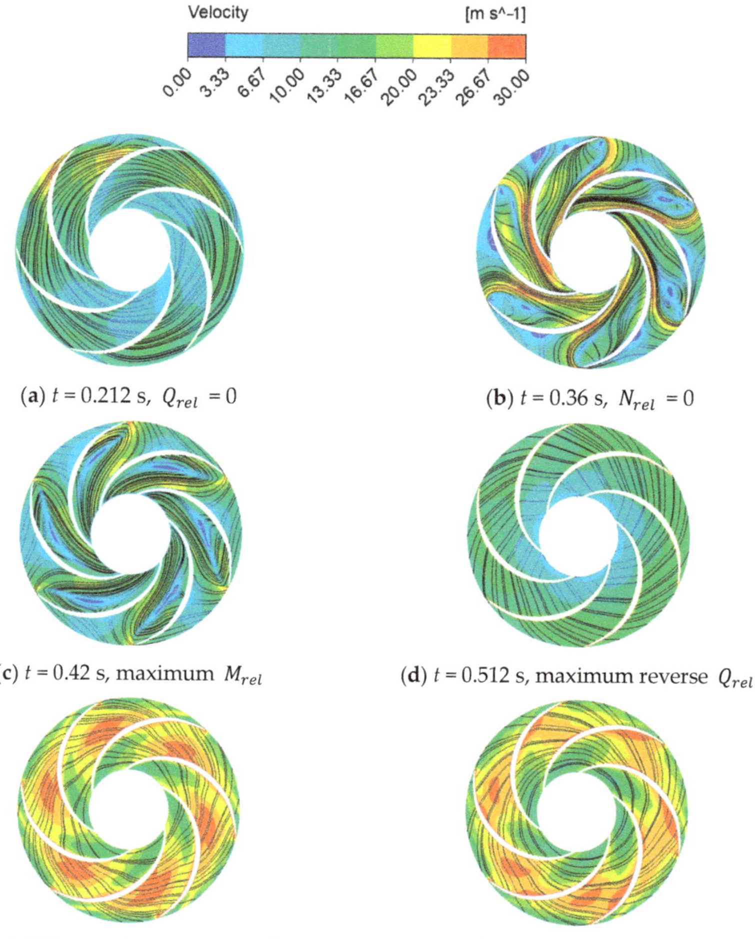

(**a**) $t = 0.212$ s, $Q_{rel} = 0$ (**b**) $t = 0.36$ s, $N_{rel} = 0$

(**c**) $t = 0.42$ s, maximum M_{rel} (**d**) $t = 0.512$ s, maximum reverse Q_{rel}

(**e**) $t = 0.732$ s, maximum reverse N_{rel} (**f**) $t = 0.886$ s, runaway

Figure 11. Streamline and velocity distributions in impeller at various times.

As can be seen in Figure 11a, the speed and flow rate of the centrifugal pump continue to decrease after power failure. When the flow rate of the centrifugal pump is 0, the streamline in the impeller area is relatively stable, and the speed gradient is not obvious. But there is a small high-speed area near the tongue due to the influence of dynamic and static interference, and there is a small low-speed area at the entrance of the impeller, which may be the place where the vortex of the inlet pipe is generated. When the speed of the impeller drops to 0, as seen in Figure 11b, the flow regime in the impeller area changes drastically. It can clearly be seen that there is a large number of vortices at the tail of the impeller suction surface, and the velocity gradient in the impeller area has a drastic difference. There are both high-speed and low-speed areas; the high-speed area is mainly concentrated in the front end of the impeller pressure surface, and the low-speed area is located in the small area of the rear end of the impeller pressure surface and the center of the vortex at the tail of the impeller suction surface. When the impeller begins to reverse into the turbine mode. As shown in Figure 11c, the vortex begins to move towards the front end of the impeller suction surface, the velocity gradient at the front end of the impeller pressure surface decreases significantly, and the low-speed area also moves with the vortex towards the impeller inlet. When the maximum value of reverse flow is reached. As shown in Figure 11d, the vortex in the runner area gradually disappears due to the gradual decrease in the impact of the water flow on the blades. The internal flow regime gradually stabilizes, the velocity gradient is no longer obvious, and only a small low-speed area is located in the runner inlet area. As shown in Figure 11e, with the continuous increase in the reverse speed, the streamline is no longer smooth, and there is a tendency to rotate but no new vortex is formed. The high-speed area reappears, which is located in a leafless zone near the middle of the impeller pressure surface. As shown in Figure 11f, the internal flow regime does not change significantly in the stable runaway mode, and the high-speed area begins to move towards the tail of the impeller suction surface.

4. Conclusions

In this paper, the numerical simulation of the power-off transient of the centrifugal pump is carried out by using the CFD numerical simulation and the UDF procedure that introduces the torque balance equation to control the speed of the centrifugal pump. The changes in the external characteristic parameters and internal flow field of the centrifugal pump are analyzed in detail, and the following conclusions are drawn:

(1) After the double-volute centrifugal pump is powered off, it experienced four different modes, namely pump mode, braking mode, turbine mode, and runaway mode. The impeller torque, flow rate, and speed of the centrifugal pump underwent drastic changes after power failure and finally became stable under runaway mode. The absolute values of the runaway speed and the runaway flow rate are 1.465 times and 1.21 times the initial values, respectively.

(2) Due to the huge change in the direction of speed and flow rate in the process of power failure, the flow regime inside the whole centrifugal pump became very complicated, and a large number of unstable flows of secondary flow, reflux, and flow separation were produced in the flow runner. Through the analysis of the flow field in the intake pipe area, the impeller area, the double-volute area, and the outlet diffusion section, we can see the change process of the generation, development, and disappearance of the vortex at each position of the centrifugal pump. We can also see the change and development process of the internal speed gradient of the centrifugal pump.

(3) For the double-volute centrifugal pump, not only the dynamic and static interference of the tongue, but also the dynamic and static interference at the splitter will cause certain changes in the internal flow regime of the centrifugal pump. Because the second volute runner is shorter and narrower than the first volute runner, the high-speed area located in the second volute runner will be larger than that of the first volute runner.

Author Contributions: L.L.: Conceptualization, Formal Analysis, and Methodology. Z.R.: Software, Data Curation, and Writing—Original Draft Preparation. Z.W.: Investigation, Validation, and Methodology. W.Z.: Resources, Visualization, and Supervision. S.L.: Project Administration, Software, and Visualization. J.D.: Validation and Writing—Review and Editing. C.Y.: Funding Acquisition and Writing—Review and Editing. M.D.: Investigation and Software. All authors have read and agreed to the published version of the manuscript.

Funding: This project was supported by the Key Program of the National Natural Science Foundation of China (No. 52339006) and the National Natural Science Foundation of China (No. 52271275).

Data Availability Statement: The data presented in this study are available on request from the corresponding author.

Conflicts of Interest: Author Lifeng Lu, Wenjie Zhou, and Jin Dai were employed by the company Hunan Provincial Water Resources Development and Investment Co., Ltd.; and Zhongzan Wang, Siwei Li, and Mengfan Dang were employed by the company Hunan Water Resources and Hydropower Survey, Design, Planning and Research, Co., Ltd. The remaining authors declare that the research was conducted in the absence of any commercial or financial relationships that could be construed as a potential con-flict of interest.

References

1. Zhang, X.; Yuan, S.; Zhang, J. Effects of spiral volute type on performance of high specific speed centrifugal pump. *J. Drain. Irrig. Mach. Eng.* **2017**, *35*, 25–31.
2. Zhang, Z.; Wang, F.; Yao, Z.; Xiao, R. Effect of Double-Volute Casing on Impeller Radial Force for a Large Double-Suction Centrifugal Pump. In *Fluids Engineering Division Summer Meeting*; American Society of Mechanical Engineers: New York, NY, USA, 2015; Volume 57212, p. V001T33A016.
3. Li, Q.; Zhang, R.; Xu, H. Effects of volute structure on energy performance and rotor operational stability of molten salt pumps. *J. Appl. Fluid Mech.* **2023**, *16*, 1615–1626.
4. Shim, H.S.; Kim, K.Y. Numerical investigation on hydrodynamic characteristics of a centrifugal pump with a double volute at off-design conditions. *Int. J. Fluid Mach. Syst.* **2017**, *10*, 218–226. [CrossRef]
5. Shim, H.S.; Afzal, A.; Kim, K.Y.; Jeong, H.S. Three-objective optimization of a centrifugal pump with double volute to minimize radial thrust at off-design conditions. *Proc. Inst. Mech. Eng. Part A J. Power Energy* **2016**, *230*, 598–615. [CrossRef]
6. Khalifa, A.E.; Al-Qutub, A.M.; Ben-Mansour, R. Study of pressure fluctuations and induced vibration at blade-passing frequencies of a double volute pump. *Arab. J. Sci. Eng.* **2011**, *36*, 1333–1345. [CrossRef]
7. Xiao, R.; Lu, T.; Wang, F. Influence of rib structure in double-volute centrifugal pumps on the impeller radial force. *Nongye Jixie Xuebao (Trans. Chin. Soc. Agric. Mach.)* **2011**, *42*, 85–88.
8. Yang, J.; Zhou, R.; Chen, H.; Yu, T. Transient flow characteristics and pressure pulsation characteristics at splitter of double-volute centrifugal pump during startup. *J. Jiangsu Univ. (Nat. Sci. Ed.)* **2021**, *42*, 278–283.
9. Yang, M.; Min, S.M.; Wang, F.J. Numerical simulation of pressure fluctuation and radial force in a double volute pump. *Trans. Chin. Soc. Agric. Mach.* **2009**, *40*, 83–88.
10. Tsukamoto, H.; Ohashi, H. Transient Characteristics of a Centrifugal Pump During Starting Period. *J. Fluids Eng.* **1982**, *104*, 6–13. [CrossRef]
11. Tsukamoto, H.; Matsunaga, S.; Yoneda, H.; Hata, S. Transient Characteristics of a Centrifugal Pump During Stopping Period. *J. Fluids Eng.* **1986**, *108*, 392–399. [CrossRef]
12. Chalghoum, I.; Elaoud, S.; Akrout, M.; Taieb, E.H. Transient behavior of a centrifugal pump during starting period. *Appl. Acoust.* **2016**, *109*, 82–89. [CrossRef]
13. Tanaka, T.; Takatsu, N. *Transient Characteristics of a Centrifugal Pump at Rapid Startup*; IOP Publishing Ltd.: Bristol, UK, 2019; Volume 240, p. 052016.
14. Zhou, D.; Zhong, L.; Zheng, Y.; Mao, Y. Numerical simulation of transient flow in axial-flow pump unit model during runaway caused from power failure. *J. Drain. Irrig. Mach. Eng./Paiguan Jixie Gongcheng Xuebao* **2012**, *30*, 401–406.
15. Wang, W.; Guo, H.; Zhang, C.; Shen, J.; Pei, J.; Yuan, S. Transient characteristics of PAT in micro pumped hydro energy storage during abnormal shutdown process. *Renew. Energy* **2023**, *209*, 401–412. [CrossRef]
16. Wang, W.; Guo, H.; Zhang, C.; Shen, J.; Pei, J.; Yuan, S. Experimental investigation on pressure fluctuation characteristics of a mixed-flow pump as turbine at turbine and runaway conditions. *J. Energy Storage* **2022**, *55*, 105562. [CrossRef]
17. Feng, J.; Ge, Z.; Zhang, Y.; Zhu, G.; Wu, G.; Lu, J.; Luo, X. Numerical investigation on characteristics of transient process in centrifugal pumps during power failure. *Renew. Energy* **2021**, *170*, 267–276. [CrossRef]
18. Dong, W.; Fan, X.; Dong, Y.; He, F. Transient Internal Flow Characteristics of Centrifugal Pump During Rapid Start-Up. *Iran. J. Sci. Technol. Trans. Mech. Eng.* **2023**. [CrossRef]
19. Zhang, Y.L.; Zhu, Z.C.; Cui, B.L.; Li, Y.; Tao, R.H. Numerical simulation of unsteady flow in centrifugal pump during stopping period. *J. Eng. Thermophys.* **2012**, *33*, 2096–2099.
20. Kan, K.; Zheng, Y.; Chen, H.; Zhou, D.; Dai, J.; Binama, M.; Yu, A. Numerical simulation of transient flow in a shaft extension tubular pump unit during runaway process caused by power failure. *Renew. Energy* **2020**, *154*, 1153–1164. [CrossRef]
21. Walseth, E.C.; Nielsen, T.K.; Svingen, B. Measuring the dynamic characteristics of a low specific speed pump—Turbine model. *Energies* **2016**, *9*, 199. [CrossRef]
22. Ye, C.; Tang, Y.; An, D.; Wang, F.; Zheng, Y.; van Esch, B.P.M. Investigation on stall characteristics of marine centrifugal pump considering transition effect. *Ocean Eng.* **2023**, *280*, 114823. [CrossRef]
23. Kan, K.; Xu, Z.; Chen, H.; Xu, H.; Zheng, Y.; Zhou, D.; Muhirwa, A.; Maxime, B. Energy loss mechanisms of transition from pump mode to turbine mode of an axial-flow pump under bidirectional conditions. *Energy* **2022**, *257*, 124630. [CrossRef]
24. Wang, G.; Wang, B.; Wen, J.; Tian, R.; Niu, Z.; Liu, X. Experimental study on the hydraulic characteristics of inertia tank after the failure of pump power. *Ann. Nucl. Energy* **2021**, *151*, 107885. [CrossRef]
25. Yang, F.; Liu, C.; Tang, F.P.; Zhou, J.R. *Experimental Study on Runaway Characteristics of Pump System*; IOP Conference Series: Materials Science and Engineering; IOP Publishing: Bristol, UK, 2013; Volume 52, p. 022021.

26. Menter, F.R. Two-equation eddy-viscosity turbulence models for engineering applications. *AIAA J.* **1994**, *32*, 1598–1605. [CrossRef]
27. Mao, J.Y.; Yuan, S.Q.; Pei, J.; Zhang, J.F.; Wang, W.J. *Applications of Different Turbulence Models in Simulations of a Large Annular Volute-Type Pump with the Diffuser*; IOP Conference Series: Earth and Environmental Science; IOP Publishing: Bristol, UK, 2014; Volume 22, p. 022019.
28. Fu, S.; Zheng, Y.; Kan, K.; Chen, H.; Han, X.; Liang, X.; Liu, H.; Tian, X. Numerical simulation and experimental study of transient characteristics in an axial flow pump during start-up. *Renew. Energy* **2020**, *146*, 1879–1887. [CrossRef]
29. *IEC 60193*; Hydraulic Turbines, Storage Pumps and Pump-Turbines-Model Acceptance Tests. IEC: Geneva, Switzerland, 2019.

MDPI AG
Grosspeteranlage 5
4052 Basel
Switzerland
Tel.: +41 61 683 77 34

Water Editorial Office
E-mail: water@mdpi.com
www.mdpi.com/journal/water

www.ingramcontent.com/pod-product-compliance
Lightning Source LLC
Chambersburg PA
CBHW061612220926
51850CB00005B/492
9783725816507